Lecture Notes in Artificial Intelligence   11940

Subseries of Lecture Notes in Computer Science

Series Editors

Randy Goebel
  University of Alberta, Edmonton, Canada
Yuzuru Tanaka
  Hokkaido University, Sapporo, Japan
Wolfgang Wahlster
  DFKI and Saarland University, Saarbrücken, Germany

Founding Editor

Jörg Siekmann
  DFKI and Saarland University, Saarbrücken, Germany

More information about this series at http://www.springer.com/series/1244

Nahla Ben Amor · Benjamin Quost ·
Martin Theobald (Eds.)

Scalable Uncertainty Management

13th International Conference, SUM 2019
Compiègne, France, December 16–18, 2019
Proceedings

Springer

Editors
Nahla Ben Amor
Institut Supérieur de Gestion de Tunis
Bouchoucha, Tunisia

Benjamin Quost
University of Technology of Compiègne
Compiègne, France

Martin Theobald
University of Luxembourg
Esch-Sur-Alzette, Luxembourg

ISSN 0302-9743   ISSN 1611-3349   (electronic)
Lecture Notes in Artificial Intelligence
ISBN 978-3-030-35513-5   ISBN 978-3-030-35514-2   (eBook)
https://doi.org/10.1007/978-3-030-35514-2

LNCS Sublibrary: SL7 – Artificial Intelligence

This Springer imprint is published by the registered company Springer Nature Switzerland AG
The registered company address is: Gewerbestrasse 11, 6330 Cham, Switzerland

Preface

These are the proceedings of the 13th International Conference on Scalable Uncertainty Management (SUM 2019) held during December 16–18, 2019, in Compiègne, France. The SUM conferences are annual events which gather researchers interested in the management of imperfect information from a wide range of fields, such as artificial intelligence, databases, information retrieval, machine learning, and risk analysis, and with the aim of fostering the collaboration and cross-fertilization of ideas from different communities.

The first SUM conference was held in Washington DC in 2007. Since then, the SUM conferences have successively taken place in Napoli in 2008, Washington DC in 2009, Toulouse in 2010, Dayton in 2011, Marburg in 2012, Washington DC in 2013, Oxford in 2014, Québec in 2015, Nice in 2016, Granada in 2017, and Milano in 2018.

The 25 full, 4 short, 4 tutorial, 2 invited keynote papers gathered in this volume were selected from an overall amount of 44 submissions (5 of which were desk-rejected or withdrawn by the authors), after a rigorous peer-review process by at least 3 Program Committee members. In addition to the regular presentations, the technical program of SUM 2019 also included invited lectures by three outstanding researchers: Cassio P. de Campos (Eindhoven University of Technology, The Netherlands) on "Scalable Reliable Machine Learning Using Sum-Product Networks," Jérôme Lang (CNRS, Paris, France) on "Computational Social Choice," and Wolfgang Gatterbauer (Northeastern University, Boston, USA) on "Algebraic approximations of the Probability of Boolean Functions."

An originality of the SUM conferences is the care for dedicating a large space of their programs to invited tutorials about a wide range of topics related to uncertainty management, to further embrace the aim of facilitating interdisciplinary collaboration and cross-fertilization of ideas. This edition includes five tutorials, for which we thank Christophe Gonzales, Thierry Denœux, Marie-Jeanne Lesot, Maximilian Schleich, and the Kay R. Amel working group for preparing and presenting these tutorials (four of these tutorials have a companion paper included in this volume).

We would like to thank all of the authors, invited speakers, and tutorial speakers for their valuable contributions. We in particular also express our gratitude to the members of the Program Committee as well as to the external reviewers for their constructive comments on the submissions. We would like to extend our appreciation to all participants of SUM 2019 for their great contribution and the success of the conference. We are grateful to the Steering Committee for their suggestions and support, and to the Organization Committee for their support in the organization for the great work accomplished. We are also very grateful to the Université de Technologie de Compiègne (UTC) for hosting the conference, to the Heudiasyc laboratory and the

MS2T laboratory of excellence for their financial and technical support, and to Springer for sponsoring the Best Paper Award as well as for the ongoing support of its staff in publishing this volume.

December 2019   Nahla Ben Amor
                                                                     Benjamin Quost
                                                                     Martin Theobald

Organization

General Chair

Benjamin Quost   Université de Technologie de Compiègne, France

Program Committee Chairs

Nahla Ben Amor   LARODEC - Institut Supérieur de Gestion Tunis, Tunisia

Martin Theobald   University of Luxembourg, Luxembourg

Steering Committee

Didier Dubois   IRIT-CNRS, France
Lluis Godo   IIIA-CSIC, Spain
Eyke Hüllermeier   Universität Paderborn, Germany
Anthony Hunter   University College London, UK
Henri Prade   IRIT-CNRS, France
Steven Schockaert   Cardiff University, UK
V. S. Subrahmanian   University of Maryland, USA

Program Committee

Nahla Ben Amor (PC Chair)   Institut Supérieur de Gestion de Tunis and LARODEC, Tunisia

Martin Theobald (PC Chair)   University of Luxembourg, Luxembourg
Sébastien Destercke   CNRS, Heudiasyc, France
Henri Prade   CNRS-IRIT, France
John Grant   Towson University, USA
Leila Amgoud   CNRS-IRIT, France
Benjamin Quost   Université de Technologie de Compiègne, Heudiasyc, France

Thomas Lukasiewicz   University of Oxford, UK
Pierre Senellart   DI, École Normale Supérieure, Université PSL, France
Francesco Parisi   DIMES, University of Calabria, Italy
Davide Ciucci   Università di Milano-Bicocca, Italy
Fernando Bobillo   University of Zaragoza, Spain
Salem Benferhat   UMR CNRS 8188, Université d'Artois, France
Silviu Maniu   Université Paris-Sud, France

Rafael Peñaloza	University of Milano-Bicocca, Italy
Fabio Cozman	University of São Paulo, Brazil
Umberto Straccia	ISTI-CNR, Italy
Lluis Godo	Artificial Intelligence Research Institute, IIIA-CSIC, Spain
Philippe Leray	LINA/DUKe, Université de Nantes, France
Zied Elouedi	Institut Supérieur de Gestion de Tunis, Tunisia
Olivier Pivert	IRISA Laboratory, ENSSAT, France
Didier Dubois	CNRS-IRIT, France
Olivier Colot	Université Lille I, France
Leopoldo Bertossi	Relational AI Inc. and Carleton University, Canada
Manuel Gómez-Olmedo	University of Granada, Spain
Andrea Pugliese	University of Calabria, Italy
Alessandro Antonucci	IDSIA, Switzerland
Maurice van Keulen	University of Twente, The Netherlands
Thierry Denœux	Université de Technologie de Compiègne, France
Sebastian Link	University of Auckland, New Zealand
Christoph Beierle	FernUniversität Hagen, Germany
Cassio De Campos	Utrecht University, The Netherlands
Andrea Tettamanzi	Université de Nice-Sophia-Antipolis, France
Rainer Gemulla	Universität Mannheim, Germany
Daniel Deutch	Tel Aviv University, Israel
Raouia Ayachi	LARODEC, Institut Supérieur de Gestion de Tunis, Tunisia
Imen Boukhris	LARODEC, Institut Supérieur de Gestion de Tunis, Tunisia

Organization Committee

Yonatan Carlos Carranza Alarcon	Université de Technologie de Compiègne, France
Sébastien Destercke	CNRS, Université de Technologie de Compiègne, France
Marie-Hélène Masson	Université de Picardie Jules Verne, France
Benjamin Quost (General Chair)	Université de Technologie de Compiègne, France
David Savourey	Université de Technologie de Compiègne, France

Contents

An Experimental Study on the Behaviour of Inconsistency Measures. 1
  Matthias Thimm

Inconsistency Measurement 9
  Matthias Thimm

Using Graph Convolutional Networks for Approximate Reasoning
with Abstract Argumentation Frameworks: A Feasibility Study. 24
  Isabelle Kuhlmann and Matthias Thimm

The Hidden Elegance of Causal Interaction Models. 38
  Silja Renooij and Linda C. van der Gaag

Computational Models for Cumulative Prospect Theory: Application
to the Knapsack Problem Under Risk 52
  Hugo Martin and Patrice Perny

On a New Evidential C-Means Algorithm with Instance-Level Constraints. . . 66
  Jiarui Xie and Violaine Antoine

Hybrid Reasoning on a Bipolar Argumentation Framework 79
  Tatsuki Kawasaki, Sosuke Moriguchi, and Kazuko Takahashi

Active Preference Elicitation by Bayesian Updating
on Optimality Polyhedra 93
  Nadjet Bourdache, Patrice Perny, and Olivier Spanjaard

Selecting Relevant Association Rules From Imperfect Data 107
  *Cécile L'Héritier, Sébastien Harispe, Abdelhak Imoussaten,
  Gilles Dusserre, and Benoît Roig*

Evidential Classification of Incomplete Data via Imprecise Relabelling:
Application to Plastic Sorting 122
  *Lucie Jacquin, Abdelhak Imoussaten, François Trousset,
  Jacky Montmain, and Didier Perrin*

An Analogical Interpolation Method for Enlarging a Training Dataset 136
  Myriam Bounhas and Henri Prade

Towards a Reconciliation Between Reasoning and Learning -
A Position Paper. 153
  Didier Dubois and Henri Prade

CP-Nets, π-pref Nets, and Pareto Dominance   169
  Nic Wilson, Didier Dubois, and Henri Prade

Measuring Inconsistency Through Subformula Forgetting.   184
  Yakoub Salhi

Explaining Hierarchical Multi-linear Models.   192
  Christophe Labreuche

Assertional Removed Sets Merging of DL-Lite Knowledge Bases.   207
  Salem Benferhat, Zied Bouraoui, Odile Papini, and Eric Würbel

An Interactive Polyhedral Approach for Multi-objective Combinatorial
Optimization with Incomplete Preference Information   221
  Nawal Benabbou and Thibaut Lust

Open-Mindedness of Gradual Argumentation Semantics.   236
  Nico Potyka

Approximate Querying on Property Graphs   250
  Stefania Dumbrava, Angela Bonifati, Amaia Nazabal Ruiz Diaz,
  and Romain Vuillemot

Learning from Imprecise Data: Adjustments of Optimistic
and Pessimistic Variants.   266
  Eyke Hüllermeier, Sébastien Destercke, and Ines Couso

On Cautiousness and Expressiveness in Interval-Valued Logic   280
  Sébastien Destercke and Sylvain Lagrue

Preference Elicitation with Uncertainty: Extending Regret Based
Methods with Belief Functions   289
  Pierre-Louis Guillot and Sebastien Destercke

Evidence Propagation and Consensus Formation in Noisy Environments   310
  Michael Crosscombe, Jonathan Lawry, and Palina Bartashevich

Order-Independent Structure Learning of Multivariate Regression
Chain Graphs.   324
  Mohammad Ali Javidian, Marco Valtorta, and Pooyan Jamshidi

Comparison of Analogy-Based Methods for Predicting Preferences   339
  Myriam Bounhas, Marc Pirlot, Henri Prade, and Olivier Sobrie

Using Convolutional Neural Network in Cross-Domain Argumentation
Mining Framework   355
  Rihab Bouslama, Raouia Ayachi, and Nahla Ben Amor

ConvNet and Dempster-Shafer Theory for Object Recognition   368
  Zheng Tong, Philippe Xu, and Thierry Denœux

On Learning Evidential Contextual Corrections from Soft Labels Using
a Measure of Discrepancy Between Contour Functions   382
  Siti Mutmainah, Samir Hachour, Frédéric Pichon, and David Mercier

Efficient Möbius Transformations and Their Applications to D-S Theory   390
  Maxime Chaveroche, Franck Davoine, and Véronique Cherfaoui

Dealing with Continuous Variables in Graphical Models   404
  Christophe Gonzales

Towards Scalable and Robust Sum-Product Networks   409
  Alvaro H. C. Correia and Cassio P. de Campos

Learning Models over Relational Data: A Brief Tutorial   423
  *Maximilian Schleich, Dan Olteanu, Mahmoud Abo-Khamis,
  Hung Q. Ngo, and XuanLong Nguyen*

Subspace Clustering and Some Soft Variants   433
  Marie-Jeanne Lesot

Invited Keynotes

From Shallow to Deep Interactions Between Knowledge Representation,
Reasoning and Machine Learning   447
  Kay R. Amel

Algebraic Approximations for Weighted Model Counting   449
  Wolfgang Gatterbauer

Author Index   451

An Experimental Study on the Behaviour of Inconsistency Measures

Matthias Thimm[✉]

University of Koblenz-Landau, Koblenz, Germany
thimm@uni-koblenz.de

Abstract. We apply a selection of 19 inconsistency measures from the literature on artificially generated knowledge bases and study the distribution of their values and their pairwise correlation. This study augments previous analytical evaluations on the expressivity and the pairwise incompatibility of these measures and our findings show that (1) many measures assign only few distinct values to many different knowledge bases, and (2) many measures, although founded on different theoretical concepts, correlate significantly.

1 Introduction

An inconsistency measure $\mathcal{I}$ is a function that assigns to a knowledge base $\mathcal{K}$ (usually assumed to be formalised in propositional logic) a non-negative real value $\mathcal{I}(\mathcal{K})$ such that $\mathcal{I}(\mathcal{K}) = 0$ iff $\mathcal{K}$ is consistent and larger values of $\mathcal{I}(\mathcal{K})$ indicate "larger" inconsistency in $\mathcal{K}$ [3,5,12]. Thus, each inconsistency measure $\mathcal{I}$ formalises a notion of a degree of inconsistency and a lot of different concrete approaches have been proposed so far, see [11–13] for some surveys. The quest for the "right" way to measure inconsistency is still ongoing and many (usually controversial) rationality postulates to describe the desirable behaviour of an inconsistency measure have been proposed so far [2,12].

Our study aims at providing a new perspective on the analysis of existing approaches to inconsistency measurement by *experimentally* analysing the behaviour of inconsistency measures. More precisely, our study provides a quantitative analysis of two aspects of inconsistency measures:

A1 the distribution of inconsistency values on actual knowledge bases, and
A2 the correlation of different inconsistency measures.

Regarding the first item, [11] investigated the theoretical expressivity of inconsistency measures, i. e., the number of different inconsistency values a measure attains when some dimension of the knowledge base is bounded (such as the number of formulas or the size of the signature). One result in [11] is that e. g. the measure $\mathcal{I}_{\mathrm{dalal}}^{\Sigma}$ (see Sect. 3) has maximal expressivity and the number of different inconsistency values is not bounded if only one of these two dimensions is bounded. However, [11] does not investigate the *distribution* of inconsistency values. It may be the case that, although a measure can attain many different

© Springer Nature Switzerland AG 2019
N. Ben Amor et al. (Eds.): SUM 2019, LNAI 11940, pp. 1–8, 2019.
https://doi.org/10.1007/978-3-030-35514-2_1

values, most inconsistent knowledge bases are clustered on very few inconsistency values. Regarding the second item, previous works have shown—see [12] for an overview—that all inconsistency measures developed so far are "essentially" different. More precisely, for each pair of measures one can find a property that is satisfied by one measure but not by the other. Moreover, for each pair of inconsistency measures one can find knowledge bases that are ordered different wrt. their inconsistency. However, until now it has not been investigated how "significant" the difference between measures actually is. It may be the case that two measures order all but just a very few knowledge bases differently (or the other way around). In order to analyse these two aspects we applied 19 different inconsistency measures from the literature on artificially generated knowledge bases and performed a statistical analysis on the results. After a brief review of necessary preliminaries in Sect. 2 and the considered inconsistency measures in Sect. 3, we provide some details on our experiments and our findings in Sect. 4 and conclude in Sect. 5.

2   Preliminaries

Let At be some fixed propositional signature, i.e., a (possibly infinite) set of propositions, and let $\mathcal{L}(\mathsf{At})$ be the corresponding propositional language constructed using the usual connectives $\wedge$ (*and*), $\vee$ (*or*), and $\neg$ (*negation*).

Definition 1. *A knowledge base $\mathcal{K}$ is a finite set of formulas $\mathcal{K} \subseteq \mathcal{L}(\mathsf{At})$. Let $\mathbb{K}$ be the set of all knowledge bases.*

If X is a formula or a set of formulas we write $\mathsf{At}(X)$ to denote the set of propositions appearing in X. Semantics to a propositional language is given by *interpretations* and an *interpretation* ω on At is a function $\omega : \mathsf{At} \to \{\mathsf{true}, \mathsf{false}\}$. Let $\Omega(\mathsf{At})$ denote the set of all interpretations for At. An interpretation ω *satisfies* (or is a *model* of) an atom $a \in \mathsf{At}$, denoted by $\omega \models a$, if and only if $\omega(a) = \mathsf{true}$. The satisfaction relation $\models$ is extended to formulas in the usual way.

For $\Phi \subseteq \mathcal{L}(\mathsf{At})$ we also define $\omega \models \Phi$ if and only if $\omega \models \phi$ for every $\phi \in \Phi$. Define furthermore the set of models $\mathsf{Mod}(X) = \{\omega \in \Omega(\mathsf{At}) \mid \omega \models X\}$ for every formula or set of formulas X. By abusing notation, a formula or set of formulas X_1 *entails* another formula or set of formulas X_2, denoted by $X_1 \models X_2$, if $\mathsf{Mod}(X_1) \subseteq \mathsf{Mod}(X_2)$. Two formulas or sets of formulas X_1, X_2 are *equivalent*, denoted by $X_1 \equiv X_2$, if $\mathsf{Mod}(X_1) = \mathsf{Mod}(X_2)$. If $\mathsf{Mod}(X) = \emptyset$ we also write $X \models \perp$ and say that X is *inconsistent*.

3   Inconsistency Measures

Let $\mathbb{R}_{\geq 0}^{\infty}$ be the set of non-negative real values including ∞. Inconsistency measures are functions $\mathcal{I} : \mathbb{K} \to \mathbb{R}_{\geq 0}^{\infty}$ that aim at assessing the severity of the inconsistency in a knowledge base $\mathcal{K}$. The basic idea is that the larger the inconsistency in $\mathcal{K}$ the larger the value $\mathcal{I}(\mathcal{K})$. We refer to [11–13] for surveys.

$$\mathcal{I}_d(\mathcal{K}) = \begin{cases} 1 \text{ if } \mathcal{K} \models \perp \\ 0 \text{ otherwise} \end{cases}$$

$$\mathcal{I}_{\mathsf{MI}}(\mathcal{K}) = |\mathsf{MI}(\mathcal{K})|$$

$$\mathcal{I}_{\mathsf{MI}c}(\mathcal{K}) = \sum_{M \in \mathsf{MI}(\mathcal{K})} \frac{1}{|M|}$$

$$\mathcal{I}_\eta(\mathcal{K}) = 1 - \max\{\xi \mid \exists P \in \mathcal{P}(\mathsf{At}) : \forall \alpha \in \mathcal{K} : P(\alpha) \geq \xi\}$$

$$\mathcal{I}_c(\mathcal{K}) = \min\{|v^{-1}(B)| \mid v \models^3 \mathcal{K}\}$$

$$\mathcal{I}_{mc}(\mathcal{K}) = |\mathsf{MC}(\mathcal{K})| + |\mathsf{SC}(\mathcal{K})| - 1$$

$$\mathcal{I}_p(\mathcal{K}) = |\bigcup_{M \in \mathsf{MI}(\mathcal{K})} M|$$

$$\mathcal{I}_{hs}(\mathcal{K}) = \min\{|H| \mid H \subseteq \Omega(\mathsf{At}), \forall \phi \in \mathcal{K} \exists \omega \in H : \omega \models \phi\} - 1$$

$$\mathcal{I}_{\mathrm{dalal}}^{\Sigma}(\mathcal{K}) = \min\{\sum_{\alpha \in \mathcal{K}} d_d(\mathsf{Mod}(\alpha), \omega) \mid \omega \in \Omega(\mathsf{At})\}$$

$$\mathcal{I}_{\mathrm{dalal}}^{\max}(\mathcal{K}) = \min\{\max_{\alpha \in \mathcal{K}} d_d(\mathsf{Mod}(\alpha), \omega) \mid \omega \in \Omega(\mathsf{At})\}$$

$$\mathcal{I}_{\mathrm{dalal}}^{\mathrm{hit}}(\mathcal{K}) = \min\{|\{\alpha \in \mathcal{K} \mid d_d(\mathsf{Mod}(\alpha), \omega) > 0\}| \mid \omega \in \Omega(\mathsf{At})\}$$

$$\mathcal{I}_{D_f}(\mathcal{K}) = 1 - \Pi_{i=1}^{|\mathcal{K}|}(1 - R_i(\mathcal{K})/i)$$

$$\mathcal{I}_{mv}(\mathcal{K}) = \frac{|\bigcup_{M \in \mathsf{MI}(\mathcal{K})} \mathsf{At}(M)|}{|\mathsf{At}(\mathcal{K})|}$$

$$\mathcal{I}_{nc}(\mathcal{K}) = |\mathcal{K}| - \max\{n \mid \forall \mathcal{K}' \subseteq \mathcal{K} : |\mathcal{K}'| = n \Rightarrow \mathcal{K}' \not\models \perp\}$$

$$\mathcal{I}_{mcsc}(\mathcal{K}) = |\mathcal{K}| - \lambda(\mathcal{C})$$

$$\mathcal{I}_{\mathsf{CSP}}(\mathcal{K}) = \max\{\mathcal{W}(\mathcal{P}) \mid \mathcal{P} \in \mathcal{P}_{\mathsf{MI}(\mathcal{K})}\}$$

$$\mathcal{I}_{\mathrm{forget}}(\mathcal{K}) = \min\{k \mid (\bigwedge \mathcal{K})[a_1, i_1 \to \phi_1; \ldots; a_k, i_k \to \phi_k] \not\models \perp,$$
$$\phi_j \in \{\perp, \top\}\}$$

$$\mathcal{I}_{\mathsf{CC}}(\mathcal{K}) = \max\{n \mid \{\mathcal{K}_1, \ldots, \mathcal{K}_n\} \text{ is a CI partition of } \mathcal{K}\}$$

$$\mathcal{I}_{\mathrm{is}}(\mathcal{K}) = \log |\{M \subseteq \mathsf{MI}(\mathcal{K}) \mid M \text{ is pairwise disjoint}\}|$$

Fig. 1. Definitions of the considered measures

The formal definitions of the considered inconsistency measures can be found in Fig. 1 while the necessary notation for understanding these measures follows below. Please see the above-mentioned surveys and the original papers referenced therein for explanations and examples.

A set $M \subseteq \mathcal{K}$ is called *minimal inconsistent subset* (MI) of $\mathcal{K}$ if $M \models \perp$ and there is no $M' \subset M$ with $M' \models \perp$. Let $\mathsf{MI}(\mathcal{K})$ be the set of all MIs of $\mathcal{K}$. Let furthermore $\mathsf{MC}(\mathcal{K})$ be the set of maximal consistent subsets of $\mathcal{K}$, i. e., $\mathsf{MC}(\mathcal{K}) = \{\mathcal{K}' \subseteq \mathcal{K} \mid \mathcal{K}' \not\models \perp \wedge \forall \mathcal{K}'' \supsetneq \mathcal{K}' : \mathcal{K}'' \models \perp\}$, and let $\mathsf{SC}(\mathcal{K})$ be the set of self-contradictory formulas of $\mathcal{K}$, i. e., $\mathsf{SC}(\mathcal{K}) = \{\phi \in \mathcal{K} \mid \phi \models \perp\}$.

A probability function P is of the form $P : \Omega(\mathsf{At}) \to [0,1]$ with $\sum_{\omega \in \Omega(\mathsf{At})} P(\omega) = 1$. Let $\mathcal{P}(\mathsf{At})$ be the set of all those probability functions and for a given probability function $P \in \mathcal{P}(\mathsf{At})$ define the probability of an arbitrary formula ϕ via $P(\phi) = \sum_{\omega \models \phi} P(\omega)$.

A three-valued interpretation v on At is a function $v : \text{At} \to \{T, F, B\}$ where the values T and F correspond to the classical true and false, respectively. The additional truth value B stands for *both* and is meant to represent a conflicting truth value for a proposition. Taking into account the *truth order* $\prec$ defined via $T \prec B \prec F$, an interpretation v is extended to arbitrary formulas via $v(\phi_1 \land \phi_2) = \min_\prec(v(\phi_1), v(\phi_2))$, $v(\phi_1 \lor \phi_2) = \max_\prec(v(\phi_1), v(\phi_2))$, and $v(\neg T) = F$, $v(\neg F) = T$, $v(\neg B) = B$. An interpretation v satisfies a formula α, denoted by $v \models^3 \alpha$ if either $v(\alpha) = T$ or $v(\alpha) = B$.

The *Dalal distance* d_d is a distance function for interpretations in $\Omega(\text{At})$ and is defined as $d(\omega, \omega') = |\{a \in \text{At} \mid \omega(a) \neq \omega'(a)\}|$ for all $\omega, \omega' \in \Omega(\text{At})$. If $X \subseteq \Omega(\text{At})$ is a set of interpretations we define $d_\text{d}(X, \omega) = \min_{\omega' \in X} d_\text{d}(\omega', \omega)$ (if $X = \emptyset$ we define $d_\text{d}(X, \omega) = \infty$). We consider the inconsistency measures $\mathcal{I}^\Sigma_\text{dalal}$, $\mathcal{I}^\text{max}_\text{dalal}$, and $\mathcal{I}^\text{hit}_\text{dalal}$ from [4] but only for the Dalal distance. Note that in [4] these measures were considered for arbitrary distances and that we use a slightly different but equivalent definition of these measures.

For every knowledge base $\mathcal{K}$, $i = 1, \ldots, |\mathcal{K}|$ define $\text{MI}^{(i)}(\mathcal{K}) = \{M \in \text{MI}(\mathcal{K}) \mid |M| = i\}$ and $\text{CN}^{(i)}(\mathcal{K}) = \{C \subseteq \mathcal{K} \mid |C| = i \land C \not\models \bot\}$. Furthermore define $R_i(\mathcal{K}) = 0$ if $|\text{MI}^{(i)}(\mathcal{K})| + |\text{CN}^{(i)}(\mathcal{K})| = 0$ and otherwise $R_i(\mathcal{K}) = |\text{MI}^{(i)}(\mathcal{K})|/(|\text{MI}^{(i)}(\mathcal{K})| + |\text{CN}^{(i)}(\mathcal{K})|)$. Note that the definition of $\mathcal{I}_{D_f}$ in Table 1 is only one instance of the family studied in [9], other variants can be obtained by different ways of aggregating the values $R_i(\mathcal{K})$.

A set of maximal consistent subsets $\mathcal{C} \subseteq \text{MC}(\mathcal{K})$ is called an MC-*cover* [1] if $\bigcup_{C \in \mathcal{C}} C = K$. An MC-cover $\mathcal{C}$ is *normal* if no proper subset of $\mathcal{C}$ is an MC-cover. A normal MC-cover is maximal if $\lambda(\mathcal{C}) = |\bigcap_{C \in \mathcal{C}} C|$ is maximal for all normal MC-covers.

For a formula ϕ let $\phi[a_1, i_1 \to \psi_1; \ldots; a_k, i_k \to \psi_k]$ denote the formula ϕ where the i_jth occurrence of the proposition a_j is replaced by the formula ψ_j, for all $j = 1, \ldots, k$.

A set $\{K_1, \ldots, K_n\}$ of pairwise disjoint subsets of $\mathcal{K}$ is called a *conditional independent MUS (CI) partition* of $\mathcal{K}$ [6], iff each K_i is inconsistent and $\text{MI}(K_1 \cup \ldots \cup K_n)$ is the disjoint union of all $\text{MI}(K_i)$.

An ordered set $\mathcal{P} = \{P_1, \ldots, P_n\}$ with $P_i \subseteq \text{MI}(\mathcal{K})$ for $i = 1, \ldots, n$ is called an *ordered CSP-partition* [7] of $\text{MI}(\mathcal{K})$ if 1.) $\text{MI}(\mathcal{K})$ is the disjoint union of all P_i for $i = 1, \ldots, n$, 2.) each P_i is a conditional independent MUS partition of $\mathcal{K}$ for $i = 1, \ldots, n$, and 3.) $|P_i| \geq |P_{i+1}|$ for $i = 1, \ldots, n - 1$. For such $\mathcal{P}$ define furthermore $\mathcal{W}(\mathcal{P}) = \sum_{i=1}^n |P_i|/i$.

4   Experiments

In the following, we give some details on our experiments, the evaluation methodology, and our findings.

4.1   Knowledge Base Generation

Due to the lack of a dataset of real-world knowledge bases with a significantly rich profile of inconsistencies, we used artificially generated knowledge bases. In order to avoid biasing our study on random instances of a specific probabilistic model for knowledge base generation, we developed an algorithm that enumerates all syntactically different knowledge bases with increasing size and considered the first 188900 bases generated this way. For example, the first five knowledge bases generated this way are $\emptyset, \{x_1\}, \{\neg x_1\}, \{\neg\neg x_1\}, \{x_1, x_2\}$ and, e. g., number 72793 is $\{x_1, x_2, \neg x_2, \neg(\neg x_2 \wedge \neg\neg x_2)\}$. From the 188900 generated knowledge bases, 127814 are consistent and 61086 are inconsistent. For the remainder of this paper, let $\hat{K}$ denote the set of all 188900 knowledge bases and let $\hat{K}^{\perp} \subseteq \hat{K}$ be only the inconsistent ones.

The implementation[1] for this algorithm is available in the Tweety project[2] [10]. The generated knowledge bases and their inconsistency values wrt. each of considered inconsistency measures are available online[3].

4.2 Evaluation Measures

In order to evaluate A1, we apply the *entropy* on the distribution of inconsistency values of each measure. For $K \subseteq \mathbb{K}$ let $\mathcal{I}(K) = \{\mathcal{I}(\mathcal{K}) \mid \mathcal{K} \in K\}$ denote the image of K wrt. $\mathcal{I}$.

Definition 2. *Let K be a set of knowledge bases and $\mathcal{I}$ be an inconsistency measure. The* entropy $H_K(\mathcal{I})$ *of $\mathcal{I}$ wrt. K is defined via*

$$H_K(\mathcal{I}) = -\sum_{x \in \mathcal{I}(K)} \frac{|\mathcal{I}^{-1}(x)|}{|K|} \ln \frac{|\mathcal{I}^{-1}(x)|}{|K|}$$

where $\ln x$ denotes the natural logarithm with $0 \ln 0 = 0$.

For example, if a measure $\mathcal{I}^*$ assigns to a set K^* of 10 knowledge bases 5 times the value X, 3 times the value Y, and 2 times the value Z, we have

$$H_{K^*}(\mathcal{I}^*) = -\frac{5}{10} \ln \frac{5}{10} - \frac{3}{10} \ln \frac{3}{10} - \frac{2}{10} \ln \frac{2}{10} \approx 1.03$$

The interpretation behind the entropy here is that a larger value $H_K(\mathcal{I})$ indicates a more uniform distribution of the inconsistency values on elements of K, a value $H_K(\mathcal{I}) = 0$ indicates that all elements are assigned the same inconsistency value. Thus, the larger $H_K(\mathcal{I})$ the "more use" the measure makes of its available inconsistency values.

In order to evaluate A2, we use a specific notion of a *correlation coefficient*. For two measures $\mathcal{I}_1$ and $\mathcal{I}_2$ and two knowledge bases $\mathcal{K}_1$ and $\mathcal{K}_2$ we say that $\mathcal{I}_1$ and $\mathcal{I}_2$ are *order-compatible* wrt. $\mathcal{K}_1$ and $\mathcal{K}_2$, denoted by $\mathcal{I}_1 \sim_{\mathcal{K}_1, \mathcal{K}_2} \mathcal{I}_2$ iff

$$\mathcal{I}_1(\mathcal{K}_1) > \mathcal{I}_1(\mathcal{K}_2) \wedge \mathcal{I}_2(\mathcal{K}_1) > \mathcal{I}_2(\mathcal{K}_2)$$
$$\text{or} \quad \mathcal{I}_1(\mathcal{K}_1) < \mathcal{I}_1(\mathcal{K}_2) \wedge \mathcal{I}_2(\mathcal{K}_1) < \mathcal{I}_2(\mathcal{K}_2)$$
$$\text{or} \quad \mathcal{I}_1(\mathcal{K}_1) = \mathcal{I}_1(\mathcal{K}_2) \wedge \mathcal{I}_2(\mathcal{K}_1) = \mathcal{I}_2(\mathcal{K}_2)$$

---

[1] http://mthimm.de/r/?r=tweety-ckb.
[2] http://tweetyproject.org.
[3] http://mthimm.de/misc/exim_mt.zip.

Table 1. Entropy values of the investigated measures wrt. $\hat{K}^\perp$ (rounded to two decimals and sorted by increasing entropy).

	$\mathcal{I}_d$	$\mathcal{I}_{CC}$	$\mathcal{I}_{dalal}^{hit}$	$\mathcal{I}_c$	$\mathcal{I}_{mc}$	$\mathcal{I}_{forget}$	$\mathcal{I}_{MI}$	$\mathcal{I}_{is}$	$\mathcal{I}_{dalal}^{max}$	$\mathcal{I}_{CSP}$
$H_{\hat{K}^\perp}(\mathcal{I})$	0	0.08	0.09	0.12	0.13	0.18	0.24	0.28	0.29	0.29
	$\mathcal{I}_{hs}$	$\mathcal{I}_\eta$	$\mathcal{I}_{dalal}^\Sigma$	$\mathcal{I}_{MIC}$	$\mathcal{I}_{mv}$	$\mathcal{I}_{mcsc}$	$\mathcal{I}_p$	$\mathcal{I}_{nc}$	$\mathcal{I}_{D_f}$	
$H_{\hat{K}^\perp}(\mathcal{I})$	0.29	0.33	0.36	0.37	0.45	0.48	0.51	0.52	0.78	

Table 2. Correlation coefficients $C_{\hat{K}^\perp}(\cdot,\cdot)$ of the investigated measures wrt. $\hat{K}^\perp$ (rounded to two decimals).

	$\mathcal{I}_d$	$\mathcal{I}_{MI}$	$\mathcal{I}_{MIC}$	$\mathcal{I}_\eta$	$\mathcal{I}_c$	$\mathcal{I}_{mc}$	$\mathcal{I}_p$	$\mathcal{I}_{hs}$	$\mathcal{I}_{dalal}^\Sigma$	$\mathcal{I}_{dalal}^{max}$	$\mathcal{I}_{dalal}^{hit}$	$\mathcal{I}_{D_f}$	$\mathcal{I}_{mv}$	$\mathcal{I}_{nc}$	$\mathcal{I}_{mcsc}$	$\mathcal{I}_{CSP}$	$\mathcal{I}_{forget}$	$\mathcal{I}_{CC}$	$\mathcal{I}_{is}$
$\mathcal{I}_d$	1	0.69	0.44	0.5	0.86	0.87	0.35	0.52	0.47	0.52	0.9	0.22	0.48	0.33	0.37	0.68	0.76	0.92	0.67
$\mathcal{I}_{MI}$		1	0.54	0.37	0.72	0.74	0.65	0.38	0.41	0.38	0.76	0.28	0.41	0.47	0.52	0.99	0.7	0.75	0.99
$\mathcal{I}_{MIC}$			1	0.72	0.47	0.51	0.53	0.7	0.73	0.7	0.52	0.49	0.41	0.43	0.84	0.55	0.51	0.5	0.55
$\mathcal{I}_\eta$				1	0.47	0.48	0.36	0.98	0.93	0.98	0.49	0.53	0.39	0.33	0.84	0.37	0.48	0.5	0.37
$\mathcal{I}_c$					1	0.85	0.4	0.49	0.53	0.49	0.88	0.25	0.45	0.38	0.42	0.72	0.88	0.87	0.72
$\mathcal{I}_{mc}$						1	0.45	0.48	0.48	0.48	0.95	0.26	0.45	0.39	0.39	0.75	0.8	0.94	0.75
$\mathcal{I}_p$							1	0.36	0.39	0.36	0.43	0.25	0.32	0.43	0.5	0.64	0.42	0.41	0.64
$\mathcal{I}_{hs}$								1	0.95	0.99	0.51	0.52	0.4	0.32	0.85	0.38	0.5	0.52	0.38
$\mathcal{I}_{dalal}^\Sigma$									1	0.95	0.51	0.53	0.4	0.34	0.89	0.42	0.54	0.5	0.42
$\mathcal{I}_{dalal}^{max}$										1	0.5	0.52	0.4	0.32	0.85	0.38	0.5	0.52	0.38
$\mathcal{I}_{dalal}^{hit}$											1	0.26	0.46	0.4	0.41	0.77	0.85	0.98	0.77
$\mathcal{I}_{D_f}$												1	0.53	0.19	0.56	0.29	0.29	0.26	0.29
$\mathcal{I}_{mv}$													1	0.25	0.39	0.41	0.43	0.46	0.41
$\mathcal{I}_{nc}$														1	0.39	0.47	0.4	0.39	0.47
$\mathcal{I}_{mcsc}$															1	0.53	0.44	0.4	0.53
$\mathcal{I}_{CSP}$																1	0.71	0.76	0.99
$\mathcal{I}_{forget}$																	1	0.82	0.71
$\mathcal{I}_{CC}$																		1	0.76
$\mathcal{I}_{is}$																			1

Let $\|A\|$ be the indicator function, which is defined as $\|A\| = 1$ iff A is true and $\|A\| = 0$ otherwise.

Definition 3. *Let K be a set of knowledge bases and $\mathcal{I}_1, \mathcal{I}_2$ be two inconsistency measures. The correlation coefficient $C_K(\mathcal{I}_1, \mathcal{I}_2)$ of $\mathcal{I}_1$ and $\mathcal{I}_2$ wrt. K is defined via*

$$C_K(\mathcal{I}_1, \mathcal{I}_2) = \frac{\sum_{\mathcal{K}, \mathcal{K}' \in K, \mathcal{K} \neq \mathcal{K}'} \|\mathcal{I}_1 \sim_{\mathcal{K}, \mathcal{K}'} \mathcal{I}_2\|}{|K|(|K| - 1)}$$

In other words, $C_K(\mathcal{I}_1, \mathcal{I}_2)$ gives the ratio of how much $\mathcal{I}_1$ and $\mathcal{I}_2$ agree on the inconsistency order of any pair of knowledge bases from K.[4] Observe that $C_K(\mathcal{I}_1, \mathcal{I}_2) = C_K(\mathcal{I}_2, \mathcal{I}_1)$.

---

[4] Note that C_K is equivalent to the Kendall's tau coefficient [8] but scaled onto $[0, 1]$.

4.3   Results

Tables 1 and 2 show the results of analysing the considered measures on $\hat{K}^{\perp}$ wrt. the two evaluation measures from before[5].

Regarding A1, it can be seen that $\mathcal{I}_d$ has minimal entropy (by definition). However, also measures $\mathcal{I}_{\mathrm{dalal}}^{\mathrm{hit}}$ and $\mathcal{I}_{\mathrm{CC}}$ and to some extent most of the other measures are quite indifferent in assigning their values. For example, out of 61086 inconsistent knowledge bases, $\mathcal{I}_{\mathrm{CC}}$ assigns to 58523 of them the same value 1. On the other hand, measure $\mathcal{I}_{D_f}$ has maximal entropy among the considered measures.

Regarding A2, we can observe some surprising correlations between measures, even those which are based on different concepts. For example, we have $C_{\hat{K}^{\perp}}(\mathcal{I}_{\mathrm{dalal}}^{\mathrm{max}}, \mathcal{I}_{hs}) \approx 0.99$ indicating a high correlation between $\mathcal{I}_{\mathrm{dalal}}^{\mathrm{max}}$ and $\mathcal{I}_{hs}$ although $\mathcal{I}_{\mathrm{dalal}}^{\mathrm{max}}$ is defined using distances and $\mathcal{I}_{hs}$ is defined using hitting sets. Equally high correlations can be observed between the three measures $\mathcal{I}_{\mathrm{MI}}$, $\mathcal{I}_{\mathrm{CSP}}$, and $\mathcal{I}_{\mathrm{is}}$. Further high correlations (e. g. above 0.8) can be observed between many other measures. On the other hand, the measure $\mathcal{I}_{D_f}$ has (on average) the smallest correlation to all other measures, backing up the observation from before.

5   Conclusion

Our experimental analysis showed that many existing measures have low entropy on the distribution of inconsistency values and correlate significantly in their ranking of inconsistent knowledge bases. A web application for trying out all the discussed inconsistency measures can be found on the website of TWEETY-PROJECT[6], cf. [10]. Most of these measures have been implemented using naive algorithms and research on the algorithmic issues of inconsistency measure is still desirable future work, see also [13].

Acknowledgements. The research reported here was partially supported by the Deutsche Forschungsgemeinschaft (grant DE 1983/9-1).

References

1. Ammoura, M., Raddaoui, B., Salhi, Y., Oukacha, B.: On measuring inconsistency using maximal consistent sets. In: Destercke, S., Denoeux, T. (eds.) ECSQARU 2015. LNCS (LNAI), vol. 9161, pp. 267–276. Springer, Cham (2015). https://doi.org/10.1007/978-3-319-20807-7_24
2. Besnard, P.: Revisiting postulates for inconsistency measures. In: Fermé, E., Leite, J. (eds.) JELIA 2014. LNCS (LNAI), vol. 8761, pp. 383–396. Springer, Cham (2014). https://doi.org/10.1007/978-3-319-11558-0_27
3. Grant, J., Hunter, A.: Measuring inconsistency in Knowledge bases. J. Intell. Inf. Syst. **27**, 159–184 (2006)

---

[5] We only considered the inconsistent knowledge bases from $\hat{K}$ as all measures assign degree 0 to the consistent ones anyway.

[6] http://tweetyproject.org/w/incmes/.

4. Grant, J., Hunter, A.: Distance-based measures of inconsistency. In: van der Gaag, L.C. (ed.) ECSQARU 2013. LNCS (LNAI), vol. 7958, pp. 230–241. Springer, Heidelberg (2013). https://doi.org/10.1007/978-3-642-39091-3_20

5. Hunter, A., Konieczny, S.: Approaches to measuring inconsistent information. In: Bertossi, L., Hunter, A., Schaub, T. (eds.) Inconsistency Tolerance. LNCS, vol. 3300, pp. 191–236. Springer, Heidelberg (2005). https://doi.org/10.1007/978-3-540-30597-2_7

6. Jabbour, S., Ma, Y., Raddaoui, B.: Inconsistency measurement thanks to mus decomposition. In: Scerri, L., Huhns, B. (eds.) Proceedings of the 13th International Conference on Autonomous Agents and Multiagent Systems (AAMAS 2014), pp. 877–884 (2014)

7. Jabbour, S., Ma, Y., Raddaoui, B., Sais, L., Salhi, Y.: On structure-based inconsistency measures and their computations via closed set packing. In: Proceedings of the 14th International Conference on Autonomous Agents and Multiagent Systems (AAMAS 2015), pp. 1749–1750 (2015)

8. Kendall, M.: A new measure of rank correlation. Biometrika **30**(1–2), 81–89 (1938)

9. Mu, K., Liu, W., Jin, Z., Bell, D.: A syntax-based approach to measuring the degree of inconsistency for belief bases. Int. J. Approximate Reasoning **52**(7), 978–999 (2011)

10. Thimm, M.: Tweety - a comprehensive collection of Java Libraries for logical aspects of artificial intelligence and knowledge representation. In: Proceedings of the 14th International Conference on Principles of Knowledge Representation and Reasoning (KR 2014), pp. 528–537, July 2014

11. Thimm, M.: On the expressivity of inconsistency measures. Artif. Intell. **234**, 120–151 (2016)

12. Thimm, M.: On the compliance of rationality postulates for inconsistency measures: a more or less complete picture. Künstliche Intell. **31**(1), 31–39 (2017)

13. Thimm, M., Wallner, J.P.: Some complexity results on inconsistency measurement. In: Proceedings of the 15th International Conference on Principles of Knowledge Representation and Reasoning (KR 2016), pp. 114–123, April 2016

Inconsistency Measurement

Matthias Thimm$^{(\boxtimes)}$

University of Koblenz-Landau, Koblenz, Germany
`thimm@uni-koblenz.de`

Abstract. The field of *Inconsistency Measurement* is concerned with the development of principles and approaches to quantitatively assess the severity of inconsistency in knowledge bases. In this survey, we give a broad overview on this field by outlining its basic motivation and discussing some of these core principles and approaches. We focus on the work that has been done for classical propositional logic but also give some pointers to applications on other logical formalisms.

1 Introduction

Inconsistency is a ubiquitous phenomenon whenever knowledge[1] is compiled in some formal language. The notion of *inconsistency* refers (usually) to multiple pieces of information and represents a conflict between those, i.e., they cannot hold at the same time. The two statements "It is sunny outside" and "It is not sunny outside" represent inconsistent information and in order to draw meaningful conclusions from a knowledge base containing these statements, this conflict has to be resolved somehow. In applications such as decision-support systems, a knowledge base is usually compiled by merging the formalised knowledge of many different experts. It is unavoidable that different experts contradict each other and that the merged knowledge base becomes inconsistent. The field of *Knowledge Representation and Reasoning* (KR) [7] is the subfield of *Artificial Intelligence* (AI) that deals with the issues of logical formalisations of information and the modelling of rational reasoning behaviour, in particular in light of inconsistent or uncertain information. One paradigm to deal with inconsistent information is to abandon classical inference and define new ways of reasoning. Some examples of such formalisms are, e.g., paraconsistent logics [6], default logic [34], answer set programming [15], and, more recently, computational models of argumentation [1]. Moreover, the fields of belief revision [21] and belief merging [10,28] deal with the particular case of inconsistencies in dynamic settings.

The field of *Inconsistency Measurement*—see the seminal work [20] and the recent book [19]—provides an *analytical* perspective on the issue of inconsistency. Its aim is to quantitatively assess the *severity* of inconsistency in order

---

[1] We use the term *knowledge* to refer to *subjective knowledge* or *beliefs*, i.e., pieces of information that may not necessary be true in the real world but are only assumed to be true for the agent(s) under consideration.

© Springer Nature Switzerland AG 2019
N. Ben Amor et al. (Eds.): SUM 2019, LNAI 11940, pp. 9–23, 2019.
https://doi.org/10.1007/978-3-030-35514-2_2

to both guide automatic reasoning mechanisms and to help human modellers in identifying issues and compare different alternative formalisations. Consider the following two knowledge bases $\mathcal{K}_1$ and $\mathcal{K}_2$ formalised in classical propositional logic (see Sect. 2 for the formal background) modelling some information about the weather:

$$\mathcal{K}_1 = \{\text{sunny}, \neg\text{sunny}, \text{hot}, \neg\text{hot}\}$$
$$\mathcal{K}_2 = \{\neg\text{hot}, \text{sunny}, \text{sunny} \rightarrow \text{hot}, \text{humid}\}$$

Both $\mathcal{K}_1$ and $\mathcal{K}_2$ are classically inconsistent, i.e., there is no interpretation satisfying any of them. But looking closer into the structure of the knowledge bases one can identify differences in the severity of the inconsistency. In $\mathcal{K}_1$ there are two "obvious" contradictions, i.e., $\{\text{sunny}, \neg\text{sunny}\}$ and $\{\text{hot}, \neg\text{hot}\}$ are directly conflicting formulas. In $\mathcal{K}_2$, the conflict is a bit more hidden. Here, three formulas are necessary to produce a contradiction ($\{\neg\text{hot}, \text{sunny}, \text{sunny} \rightarrow \text{hot}\}$). Moreover, there is one formula in $\mathcal{K}_2$ (humid), which is not participating in any conflict and one could still infer meaningful information from this by relying on e.g. paraconsistent reasoning techniques [6]. In conclusion, one should regard $\mathcal{K}_1$ as *more inconsistent* than $\mathcal{K}_2$. So a decision-maker should prefer using $\mathcal{K}_2$ instead of $\mathcal{K}_1$.

The analysis of the severity of inconsistency in the knowledge bases $\mathcal{K}_1$ and $\mathcal{K}_2$ above was informal. Formal accounts to the problem of assessing the severity of inconsistency are given by *inconsistency measures* and there have been a lot of proposals of those in recent years. Up to today, the concept of *severity of inconsistency* has not been axiomatised in a satisfactory manner and the series of different inconsistency measures approach this challenge from different points of view and focus on different aspects on what constitutes *severity*. Consider the next two knowledge bases (with abstract propositions a and b)

$$\mathcal{K}_3 = \{a, \neg a, b\} \qquad\qquad \mathcal{K}_4 = \{a \vee b, \neg a \vee b, a \vee \neg b, \neg a \vee \neg b\}$$

Again both $\mathcal{K}_3$ and $\mathcal{K}_4$ are inconsistent, but which one is more inconsistent than the other? Our reasoning from above cannot be applied here in the same fashion. The knowledge base $\mathcal{K}_3$ contains an apparent contradiction ($\{a, \neg a\}$) but also a formula not participating in the inconsistency ($\{b\}$). The knowledge base $\mathcal{K}_4$ contains a "hidden" conflict as four formulas are necessary to produce a contradiction, but all formulas of $\mathcal{K}_4$ are participating in this. In this case, it is not clear how to assess the inconsistency of these knowledge bases and different measures may order these knowledge bases differently. More generally speaking, it is not universally agreed upon which so-called *rationality postulates* should be satisfied by a reasonable account of inconsistency measurement, see [3,5,41] for a discussion. Besides concrete approaches to inconsistency measurement the community has also proposed a series of those rationality postulates in order to describe general desirable behaviour and the classification of inconsistency measures by the postulates they satisfy is still one the most important ways to evaluate the quality of a measure, even if the set of desirable postulates is not universally accepted. For example, one of the most popular rationality postulates is *monotony* which states that for any $\mathcal{K} \subseteq \mathcal{K}'$, the knowledge base $\mathcal{K}$ cannot be regarded as more inconsistent as $\mathcal{K}'$. The justification for this demand is that inconsistency cannot be resolved when adding new information but only increased[2]. While this is usually regarded as a reasonable demand there are also situations where *monotony* may be seen as counterintuitive, even in monotonic logics. Consider the next two knowledge bases

$$\mathcal{K}_5 = \{a, \neg a\} \qquad\qquad \mathcal{K}_6 = \{a, \neg a, b_1, \ldots, b_{998}\}$$

We have $\mathcal{K}_5 \subseteq \mathcal{K}_6$ and following *monotony*, $\mathcal{K}_6$ should be regarded as least as inconsistent as $\mathcal{K}_5$. However, when judging the content of the knowledge bases "relatively", $\mathcal{K}_5$ may seem more inconsistent: $\mathcal{K}_5$ contains no useful information and all formulas of $\mathcal{K}_5$ are in conflict with another formula. In $\mathcal{K}_6$, however, only 2 out of 1000 formulas are participating in the contradiction. So it may also be a reasonable point of view to judge $\mathcal{K}_5$ more inconsistent than $\mathcal{K}_6$.

In this survey paper, we give a brief overview on formal accounts to inconsistency measurement. We focus on approaches building on classical propositional logic but also briefly discuss approaches for other formalisms. A more technical survey of inconsistency measures can be found in [41] and the book [19] captures the recent state-of-the-art as a whole. An older survey can also be found in [22].

The remainder of this paper is organised as follows. In Sect. 2 we give some necessary technical preliminaries. Section 3 introduces the concept of inconsistency measures formally and discusses rationality postulates. In Sect. 4 we discuss some of the most important concrete approaches to inconsistency measurement for classical propositional logic and in Sect. 5 we give an overview on approaches for other formalisms. Section 6 concludes.

2 Preliminaries

Let At be some fixed set of propositions and let $\mathcal{L}(\mathsf{At})$ be the corresponding propositional language constructed using the usual connectives $\wedge$ (*conjunction*), $\vee$ (*disjunction*), $\rightarrow$ (*implication*), and $\neg$ (*negation*).

Definition 1. *A knowledge base $\mathcal{K}$ is a finite set of formulas $\mathcal{K} \subseteq \mathcal{L}(\mathsf{At})$. Let $\mathbb{K}$ be the set of all knowledge bases.*

If X is a formula or a set of formulas we write $\mathsf{At}(X)$ to denote the set of propositions appearing in X.

Semantics for a propositional language is given by *interpretations* where an *interpretation* ω on At is a function $\omega : \mathsf{At} \rightarrow \{\mathsf{true}, \mathsf{false}\}$. Let $\Omega(\mathsf{At})$ denote the set of all interpretations for At. An interpretation ω *satisfies* (or is a *model* of) a proposition $a \in \mathsf{At}$, denoted by $\omega \models a$, if and only if $\omega(a) = \mathsf{true}$. The satisfaction relation $\models$ is extended to formulas in the usual way.

---

[2] At least in monotonic logics; for a discussion about inconsistency measurement in non-monotonic logics see [9,43] and Sect. 5.3.

For $\Phi \subseteq \mathcal{L}(\mathsf{At})$ we also define $\omega \models \Phi$ if and only if $\omega \models \phi$ for every $\phi \in \Phi$. A formula or set of formulas X_1 *entails* another formula or set of formulas X_2, denoted by $X_1 \models X_2$, if and only if $\omega \models X_1$ implies $\omega \models X_2$. If there is no ω with $\omega \models X$ we also write $X \models \perp$ and say that X is *inconsistent*.

3 Measuring Inconsistency

Let $\mathbb{R}_{\geq 0}^{\infty}$ be the set of non-negative real values including infinity. The most general form of an inconsistency measure is as follows.

Definition 2. *An inconsistency measure $\mathcal{I}$ is any function $\mathcal{I} : \mathbb{K} \rightarrow \mathbb{R}_{\geq 0}^{\infty}$.*

The above definition is, of course, under-constrained for the purpose of providing a quantitative means to measure inconsistency. The intuition we intend to be behind any concrete approach to inconsistency measure $\mathcal{I}$ is that a larger value $\mathcal{I}(\mathcal{K})$ for a knowledge base $\mathcal{K}$ indicates more severe inconsistency in $\mathcal{K}$ than lower values. Moreover, we wish to reserve the minimal value (0) to indicate the complete absence of inconsistency. This is captured by the following postulate [23]:

Consistency $\mathcal{I}(\mathcal{K}) = 0$ iff $\mathcal{K}$ is consistent.

Satisfaction of the *consistency* postulate is a basic demand for any reasonable inconsistency measure and is satisfied by all known concrete approaches [39, 41]. Beyond the *consistency* postulates a series of further postulates has been proposed in the literature [41]. We only recall the basic ones initially proposed in [23]. In order to state these postulates we need two further definitions.

Definition 3. *A set $M \subseteq \mathcal{K}$ is a* minimal inconsistent subset *of $\mathcal{K}$ iff $M \models \perp$ and there is no $M' \subsetneq M$ with $M' \models \perp$. Let $\mathsf{MI}(\mathcal{K})$ be the set of all minimal inconsistent subsets of $\mathcal{K}$.*

Definition 4. *A formula $\alpha \in \mathcal{K}$ is called* free formula *if $\alpha \notin \bigcup \mathsf{MI}(\mathcal{K})$. Let $\mathsf{Free}(\mathcal{K})$ be the set of all free formulas of $\mathcal{K}$.*

In other words, a minimal inconsistent subset characterises a minimal conflict in a knowledge base and a free formula is a formula that is not directly participating in any derivation of a contradiction. Let $\mathcal{I}$ be any function $\mathcal{I} : \mathbb{K} \rightarrow \mathbb{R}_{\geq 0}^{\infty}$, $\mathcal{K}, \mathcal{K}' \in \mathbb{K}$, and $\alpha, \beta \in \mathcal{L}(\mathsf{At})$. The remaining rationality postulates from [23] are:

Normalisation $0 \leq \mathcal{I}(\mathcal{K}) \leq 1$.
Monotony If $\mathcal{K} \subseteq \mathcal{K}'$ then $\mathcal{I}(\mathcal{K}) \leq \mathcal{I}(\mathcal{K}')$.
Free-formula independence If $\alpha \in \mathsf{Free}(\mathcal{K})$ then
    $\mathcal{I}(\mathcal{K}) = \mathcal{I}(\mathcal{K} \setminus \{\alpha\})$.
Dominance If $\alpha \not\models \perp$ and $\alpha \models \beta$ then $\mathcal{I}(\mathcal{K} \cup \{\alpha\}) \geq \mathcal{I}(\mathcal{K} \cup \{\beta\})$.

The postulate *normalisation* states that the inconsistency value is always in the unit interval, thus allowing inconsistency values to be comparable even if knowledge bases are of different sizes. *Monotony* requires that adding formulas to the knowledge base cannot decrease the inconsistency value. *Free-formula independence* states that removing free formulas from the knowledge base should not change the inconsistency value. The motivation here is that free formulas do not participate in inconsistencies and should not contribute to having a certain inconsistency value. *Dominance* says that substituting a consistent formula α by a weaker formula β should not increase the inconsistency value. Here, as β carries less information than α there should be less opportunities for inconsistencies to occur.

The five postulates from above are independent (no single postulates entails another one) and compatible (as e. g. the drastic measure $\mathcal{I}_d$, see below, satisfies all of them). However, they do not characterise a single concrete approach but leave ample room for various different approaches. Moreover, for all rationality postulates (except *consistency*) there is at least one inconsistency measure in the literature that does not satisfy it [41] and there is no general agreement on whether these postulates are indeed *desirable* at all [3,5,41]. We already gave an example why *monotony* may not be desirable in the introduction. Here is another example for *free-formula independence* taken from [3].

Example 1. Consider the knowledge base $\mathcal{K}_7$ defined via

$$\mathcal{K}_7 = \{a \wedge c, b \wedge \neg c, \neg a \vee \neg b\}$$

Notice that $\mathcal{K}_7$ has a single minimal inconsistent subset $\{a \wedge c, b \wedge \neg c\}$ and $\neg a \vee \neg b$ is a free formula. If $\mathcal{I}$ satisfies *free-formula independence* we have $\mathcal{I}(\mathcal{K}_7) = \mathcal{I}(\mathcal{K}_7 \setminus \{\neg a \vee \neg b\})$. However, $\neg a \vee \neg b$ adds another "conflict" about the truth of propositions a and b.

We will continue the discussion on rationality postulates later in Sect. 6. But first we will have a look at some concrete approaches.

4 Approaches

There is a wide variety of inconsistency measures in the literature, the work [41] alone lists 22 measures in 2018 and more have been proposed since then[3]. In this paper we consider only a few to illustrate the main concepts.

The measure $\mathcal{I}_d$ is usually referred to as a baseline for inconsistency measures as it only distinguishes between consistent and inconsistent knowledge bases.

---

[3] Implementations of most of these measures can also be found in the *Tweety Libraries for Artificial Intelligence* [40] and an online interface is available at http:// tweetyproject.org/w/incmes.

Definition 5 ([24]). *The* drastic inconsistency measure $\mathcal{I}_d : \mathbb{K} \rightarrow \mathbb{R}^\infty_{\geq 0}$ *is defined as*

$$\mathcal{I}_d(\mathcal{K}) = \begin{cases} 1 & \text{if } \mathcal{K} \models \perp \\ 0 & \text{otherwise} \end{cases}$$

for $\mathcal{K} \in \mathbb{K}$.

While not being particularly useful for the purpose of actually differentiating between inconsistent knowledge bases, the measure $\mathcal{I}_d$ already satisfies the basic five postulates from above [24].

In [22] several dimensions for measuring inconsistency have been discussed. A particular observation from this discussion is that inconsistency measures can be roughly divided into two categories: *syntactic* and *semantic* approaches. While this distinction is not clearly defined[4] it has been used in following works to classify many inconsistency measures. Using this categorisation, *syntactic* approaches refer to inconsistency measures that make use of syntactic objects such as minimal inconsistent sets (or maximal consistent sets). On the other hand, *semantic* approaches refer to measures employing non-classical semantics for that purpose. However, there are further measures which fall into neither (or both) categories. In the following, we will look at some measures from each of these categories.

4.1   Measures Based on Minimal Inconsistent Sets

A minimal inconsistent subset M of a knowledge base $\mathcal{K}$ represents the "essence" of a single conflict in $\mathcal{K}$. Naturally, a simple approach to measure inconsistency is to take the number of minimal inconsistent subsets as a measure.

Definition 6 ([24]). *The* MI-inconsistency measure $\mathcal{I}_{MI} : \mathbb{K} \rightarrow \mathbb{R}^\infty_{\geq 0}$ *is defined as $\mathcal{I}_{MI}(\mathcal{K}) = |\mathsf{MI}(\mathcal{K})|$ for $\mathcal{K} \in \mathbb{K}$.*

The above measure complies with the postulates of *consistency*, *monotony*, and *free-formula independence* but fails to satisfy *dominance* and *normalisation* (although a normalised variant that suffers from other shortcomings can easily be defined). Table 2 below gives an overview on the compliance of the measures formally considered in this paper with the basic postulates from above, see [41] for proofs or references to proofs. The idea behind the MI-inconsistency measure can be refined in several ways, taking e.g. the sizes of the individual minimal inconsistent sets and how they overlap into account [13,25,26]. One example being the following measure.

Definition 7 ([24]). *The* MI^c-inconsistency measure $\mathcal{I}_{MI^c} : \mathbb{K} \rightarrow \mathbb{R}^\infty_{\geq 0}$ *is defined as*

$$\mathcal{I}_{MI^c}(\mathcal{K}) = \sum_{M \in \mathsf{MI}(\mathcal{K})} \frac{1}{|M|}$$

for $\mathcal{K} \in \mathbb{K}$.

---

[4] And in this author's opinion also a bit mislabelled.

The MI^c-*inconsistency measure* takes also the sizes of the individual minimal inconsistent subsets into account. The intuition here is that larger minimal inconsistent subsets represent less inconsistency (as the conflict is more "hidden") and small minimal inconsistent subsets represent more inconsistency (as it is more "apparent").

Example 2. Consider again knowledge bases $\mathcal{K}_1$ and $\mathcal{K}_2$ from before defined via

$$\mathcal{K}_1 = \{\text{sunny}, \neg\text{sunny}, \text{hot}, \neg\text{hot}\}$$
$$\mathcal{K}_2 = \{\neg\text{hot}, \text{sunny}, \text{sunny} \rightarrow \text{hot}, \text{humid}\}$$

Here we have

$$\mathcal{I}_{\mathsf{MI}}(\mathcal{K}_1) = 2 \qquad\qquad \mathcal{I}_{\mathsf{MI}}(\mathcal{K}_2) = 1$$
$$\mathcal{I}_{\mathsf{MI}^c}(\mathcal{K}_1) = 1 \qquad\qquad \mathcal{I}_{\mathsf{MI}^c}(\mathcal{K}_2) = 1/3$$

Observe that, while $\mathcal{I}_{\mathsf{MI}}$ and $\mathcal{I}_{\mathsf{MI}^c}$ disagree on the exact values of the inconsistency in $\mathcal{K}_1$ and $\mathcal{K}_2$ they do agree on their order ($\mathcal{K}_1$ is more inconsistent than $\mathcal{K}_2$). This is not generally true, consider

$$\mathcal{K}_8 = \{a, \neg a\}$$
$$\mathcal{K}_9 = \{a_1, \neg a_1 \vee b_1, \neg b_1 \vee c_1, \neg \vee d_1, \neg d_1 \vee \neg a_1,$$
$$a_2, \neg a_2 \vee b_2, \neg b_2 \vee c_2, \neg \vee d_2, \neg d_2 \vee \neg a_2\}$$

$$\mathcal{I}_{\mathsf{MI}}(\mathcal{K}_8) = 1 \qquad\qquad \mathcal{I}_{\mathsf{MI}}(\mathcal{K}_9) = 2$$
$$\mathcal{I}_{\mathsf{MI}^c}(\mathcal{K}_8) = 1/2 \qquad\qquad \mathcal{I}_{\mathsf{MI}^c}(\mathcal{K}_9) = 2/5$$

where $\mathcal{K}_8$ is less inconsistent than $\mathcal{K}_9$ according to $\mathcal{I}_{\mathsf{MI}}$ and the other way around for $\mathcal{I}_{\mathsf{MI}^c}$.

4.2   Measures Based on Non-classical Semantics

Measures based on minimal inconsistent subsets provide a *formula-centric* view on the matter of inconsistency [22]. If a formula (as a whole) is part of a conflict, it is taken into account for measuring inconsistency. Another possibility is to focus on propositions rather than formulas. Consider again the knowledge base $\mathcal{K}_7 = \{a \wedge c, b \wedge \neg c, \neg a \vee \neg b\}$ from Example 1 which possesses one minimal inconsistent subset $\{a \wedge c, b \wedge \neg c\}$. However, it is clear that there is also a conflict involving the propositions a and b, which is not "detected" by measures based on minimal inconsistent subsets. Thus, another angle for measuring inconsistency consists in counting how many propositions participate in the inconsistency. A possible means for doing this is by relying on non-classical semantics. The *contension* measure [17] makes use of Priest's logic of paradox, which has a paraconsistent semantics that we briefly recall now. A *three-valued interpretation* v on At is a function $v : \mathsf{At} \rightarrow \{T, F, B\}$ where the values T and F correspond to the classical true and false, respectively. The additional truth value B stands for *both* and is meant to represent a conflicting truth value for a proposition. The function v is extended to arbitrary formulas as shown in Table 1. An interpretation v satisfies a formula α, denoted by $v \models^3 \alpha$ if either $v(\alpha) = T$ or $v(\alpha) = B$. Define $v \models^3 \mathcal{K}$ for a knowledge base $\mathcal{K}$ accordingly. Now inconsistency can be measured by seeking an interpretation v that assigns B to a minimal number of propositions.

Definition 8 ([17]). *The* contension inconsistency measure $\mathcal{I}_c : \mathbb{K} \to \mathbb{R}^\infty_{\geq 0}$ *is defined as*

$$\mathcal{I}_c(\mathcal{K}) = \min\{|v^{-1}(B) \cap \mathsf{At}| \mid v \models^3 \mathcal{K}\}$$

for $\mathcal{K} \in \mathbb{K}$.

Note that $\mathcal{I}_c$ is well-defined as for every knowledge $\mathcal{K}$ there is always at least one interpretation v satisfying it, e. g., the interpretation that assigns B to all propositions.

Table 1. Truth tables for propositional three-valued logic.

α	β	$v(\alpha \wedge \beta)$	$v(\alpha \vee \beta)$	α	$v(\neg\alpha)$
T	T	T	T	T	F
T	B	B	T	B	B
T	F	F	T	F	T
B	T	B	T		
B	B	B	B		
B	F	F	B		
F	T	F	T		
F	B	F	B		
F	F	F	F		

A further approach—that is in contrast to $\mathcal{I}_c$ still formula-centric—is to make use of probability logic to define an inconsistency measure [27]. A *probability function* P on $\mathcal{L}(\mathsf{At})$ is a function $P : \Omega(\mathsf{At}) \to [0, 1]$ with $\sum_{\omega \in \Omega(\mathsf{At})} P(\omega) = 1$. We extend P to assign a probability to any formula $\phi \in \mathcal{L}(\mathsf{At})$ by defining

$$P(\phi) = \sum_{\omega \models \phi} P(\omega)$$

Let $\mathcal{P}(\mathsf{At})$ be the set of all those probability functions.

Definition 9 ([27]). *The η-inconsistency measure $\mathcal{I}_\eta : \mathbb{K} \to \mathbb{R}^\infty_{\geq 0}$ is defined as*

$$\mathcal{I}_\eta(\mathcal{K}) = 1 - \max\{\xi \mid \exists P \in \mathcal{P}(\mathsf{At}) : \forall \alpha \in \mathcal{K} : P(\alpha) \geq \xi\}$$

for $\mathcal{K} \in \mathbb{K}$.

The measure $\mathcal{I}_\eta$ looks for a probability function P that maximises the minimum probability of all formulas in $\mathcal{K}$. The larger this probability the less inconsistent $\mathcal{K}$ is assessed (if there is a probability function assigning 1 to all formulas then $\mathcal{K}$ is obviously consistent).

Example 3. Consider again knowledge bases $\mathcal{K}_1$ and $\mathcal{K}_2$ from before defined via

$$\mathcal{K}_1 = \{\text{sunny}, \neg\text{sunny}, \text{hot}, \neg\text{hot}\}$$
$$\mathcal{K}_2 = \{\neg\text{hot}, \text{sunny}, \text{sunny} \rightarrow \text{hot}, \text{humid}\}$$

Here we have

$$\mathcal{I}_c(\mathcal{K}_1) = 2 \qquad\qquad \mathcal{I}_c(\mathcal{K}_2) = 1$$
$$\mathcal{I}_\eta(\mathcal{K}_1) = 0.5 \qquad\qquad \mathcal{I}_\eta(\mathcal{K}_2) = 1/3$$

where, in particular, $\mathcal{I}_c$ also agrees with $\mathcal{I}_{\mathsf{MI}}$ (see Example 2). Consider now

$$\mathcal{K}_{10} = \{a, \neg a\} \qquad\qquad \mathcal{K}_{11} = \{a \wedge b \wedge c, \neg a \wedge \neg b \wedge \neg c\}$$

where

$$\mathcal{I}_c(\mathcal{K}_1) = 1 \qquad\qquad \mathcal{I}_c(\mathcal{K}_2) = 3$$
$$\mathcal{I}_\eta(\mathcal{K}_1) = 0.5 \qquad\qquad \mathcal{I}_\eta(\mathcal{K}_2) = 0.5$$
$$\mathcal{I}_{\mathsf{MI}}(\mathcal{K}_1) = 1 \qquad\qquad \mathcal{I}_c(\mathcal{K}_2) = 1$$

So $\mathcal{I}_c$ looks inside formulas to determine the severity of inconsistency.

While $\mathcal{I}_c$ makes use of paraconsistent logic and $\mathcal{I}_\eta$ of probability logic other logics can be used for that purpose as well. In [38] a general framework is established that allows to plugin any many-valued logic (such as fuzzy logic) to define inconsistency measures.

4.3   Further Measures

There are further ways to define inconsistency measures that do not fall strictly in one of the two paradigms above. We have a look at some now.

A simple approach to obtain a more proposition-centric measure (as $\mathcal{I}_c$) while still relying on minimal inconsistent sets is the following measure.

Definition 10 ([44]). *The* mv *inconsistency measure* $\mathcal{I}_{mv} : \mathbb{K} \rightarrow \mathbb{R}_{\geq 0}^\infty$ *is defined as*

$$\mathcal{I}_{mv}(\mathcal{K}) = \frac{|\bigcup_{M \in \mathsf{MI}(\mathcal{K})} \mathsf{At}(M)|}{|\mathsf{At}(\mathcal{K})|}$$

for $\mathcal{K} \in \mathbb{K}$.

In other words, $\mathcal{I}_{mv}(\mathcal{K})$ is the ratio of the number of propositions that appear in at least one minimal inconsistent set and the number of all propositions.

Another approach that makes no use of either minimal inconsistent sets or non-classical semantics is the following one. A subset $H \subseteq \Omega(\mathsf{At})$ is called a *hitting set* of $\mathcal{K}$ if for every $\phi \in \mathcal{K}$ there is $\omega \in H$ with $\omega \models \phi$.

Definition 11 ([37]). *The* hitting-set inconsistency measure $\mathcal{I}_{hs} : \mathbb{K} \to \mathbb{R}_{\geq 0}^{\infty}$ *is defined as*

$$\mathcal{I}_{hs}(\mathcal{K}) = \min\{|H| \mid H \text{ is a hitting set of } \mathcal{K}\} - 1$$

for $\mathcal{K} \in \mathbb{K}$ *with* $\min \emptyset = \infty$.

So $\mathcal{I}_{hs}$ seeks a minimal number of (classical) interpretations such that for each formula there is at least one model in this set.

Example 4. Consider again knowledge bases $\mathcal{K}_1$ and $\mathcal{K}_2$ from before defined via

$$\mathcal{K}_1 = \{\text{sunny}, \neg\text{sunny}, \text{hot}, \neg\text{hot}\}$$
$$\mathcal{K}_2 = \{\neg\text{hot}, \text{sunny}, \text{sunny} \to \text{hot}, \text{humid}\}$$

Here we have

$$\mathcal{I}_{mv}(\mathcal{K}_1) = 1 \qquad\qquad \mathcal{I}_{mv}(\mathcal{K}_2) = 2/3$$
$$\mathcal{I}_{hs}(\mathcal{K}_1) = 1 \qquad\qquad \mathcal{I}_{hs}(\mathcal{K}_2) = 1$$

Moreover, Grant and Hunter [18] define new families of inconsistency measures based on *distances* of classical interpretations to being models of a knowledge base. Besnard [4] counts how many propositions have to be *forgotten*—i.e. removed from the underlying signature of the knowledge base—to turn an inconsistent knowledge base into a consistent one.

Table 2. Compliance of inconsistency measures with rationality postulates *consistency* (CO), *normalisation* (NO), *monotony* (MO), *free-formula independence* (IN), and *dominance* (DO)

$\mathcal{I}$	CO	NO	MO	IN	DO
$\mathcal{I}_d$	✓	✓	✓	✓	✓
$\mathcal{I}_{\mathsf{MI}}$	✓	✗	✓	✓	✗
$\mathcal{I}_{\mathsf{MI}^c}$	✓	✗	✓	✓	✗
$\mathcal{I}_c$	✓	✗	✓	✓	✓
$\mathcal{I}_\eta$	✓	✓	✓	✓	✓
$\mathcal{I}_{mv}$	✓	✓	✗	✗	✗
$\mathcal{I}_{hs}$	✓	✗	✓	✓	✓

5   Beyond Propositional Logic

While most work in the field of inconsistency measurement is concerned with using propositional logic as the knowledge representation formalism, there are some few works, which consider measuring inconsistency in other logics. We will have a brief overview on some of these works now, see [19] for some others.

5.1   First-Order and Description Logic

In [16], first-order logic is considered as the base logic. Allowing for objects and quantification brings new challenges to measuring inconsistency as one should distinguish in a more fine-grained manner how *much* certain formulas contribute to inconsistency. For example, a formula $\forall X : bird(X) \rightarrow flies(X)$—which models that all birds fly—is probably the culprit of some inconsistency in any knowledge base. However, depending on how many objects actually satisfy/violate the implication, the severity of the inconsistency of the overall knowledge base may differ (compare having a knowledge base with 10 flying birds and 1 non-flying bird to a knowledge base with 1000 flying birds and 1 non-flying bird). [16] address this challenge by proposing some new inconsistency measures for first-order logic.

There are also several works—see e. g. [29,45]—that deal with measuring inconsistency in ontologies formalised in certain description logics.

5.2   Probabilistic Logic

In probabilistic logic, classical propositional formulas are augmented by probabilities yielding statements such as (sunny ∧ humid)[0.7] meaning "it will be sunny and humid with probability 0.7". Semantics are given to such a logic by means of probability distributions over sets of propositions. Inconsistencies in modelling with such a logic can appear, in particular, when "the numbers do not add up". In addition to the previous formula consider (humid)[0.5] which states that "it will be humid with probability 0.5". Both formulas together are inconsistent as it cannot be the case the probability of being humid is at least 0.7 (which is implied by the first formula) and 0.5 at the same time. Measures for probabilistic logic, see the recent survey [12], focus on measuring distances of the probabilities of the formulas to a consistent state or propose weaker notions of satisfying probability distributions and measure distances between those and classical probability distributions.

5.3   Non-monotonic Logics

In non-monotonic logics, inconsistency in a knowledge base may be resolved by adding formulas. Consider e. g. the following rules in answer set programming [8]: $\{b \leftarrow, \neg b \leftarrow \text{not } a\}$. Informally, these rules state that b is the case and that if a is not the case, $\neg b$ is the case. The negation "not" is a negation-as-failure and the whole program is inconsistent as both b and $\neg b$ can be derived. However, adding the rule $a \leftarrow$ stating that a is the case, makes the program consistent again as the second rule is not applicable any more. An implication of this observation is that consistent programs may have inconsistent subsets, which make the application of classical measures based on minimal inconsistent sets useless. In [9] a stronger notion for minimal inconsistent sets for non-monotonic logics is proposed that is used for inconsistency measurement in [43], and, in particular, for answer set programming in [42].

6   Summary and Discussion

In this paper we gave a brief overview on the field of inconsistency measurement. We motivated the field, discussed several rationality postulates for concrete measures, and surveyed some of its basic approaches. We also gave a short overview on approaches that use formalisms other than propositional logic as the base knowledge representation formalism.

Inconsistency measures can be used to compare different formalisations of knowledge, to help debug flawed knowledge bases, and guide automatic repair methods. For example, inconsistency measures have been used to estimate reliability of agents in multi-agent systems [11], to allow for inconsistency-tolerant reasoning in probabilistic logic [33], or to monitor and maintain quality in database settings [14].

Inconsistency measurement is a problem that is not easily defined in a formal manner. Many approaches have been proposed, in particular in recent years, each taking a different perspective on this issue. We discussed rationality postulates as a means to prescribe general desirable behaviour of an inconsistency measure and there have also been a lot of proposals in the recent past, [41] lists an additional 13 compared to the five postulates we discussed here. Many of them are mutually exclusive, describe orthogonal requirements, and are not generally accepted in the community. Besides rationality postulates, other dimensions for comparing inconsistency measures are their *expressivity* and their *computational complexity*. Expressivity [36,41] refers to the capability of an inconsistency to differentiate between *many* inconsistent knowledge base. For example, the drastic inconsistency measure—which assigns 1 to every inconsistent knowledge base—has minimal expressivity as it can only differentiate between consistency and inconsistency. On the other hand, the contension measure $\mathcal{I}_c$ can differentiate up to $n + 1$ different states of inconsistency, where n is the number of propositions appearing in the signature. As for computational complexity, it is clear that all problems related to inconsistency measurement are coNP-hard, as the identification of unsatisfiability is always part of the definition. In fact, the decision problem of deciding whether a certain value is a lower bound for the actual inconsistency value of a given inconsistency measure, is coNP-complete for many measures such as $\mathcal{I}_c$ [35,41]. However, the problem is harder for other measures, e. g., the same problem for $\mathcal{I}_{mv}$ is already Σ_2^p-complete [44].

This paper points to a series of open research questions that may be interesting to pursue. For example, the discussion on the "right" set of postulates is not over. What is needed is a characterising definition of an inconsistency measure using few postulates, as the *entropy* is characterised by few simple properties as an information measure. However, we are currently far away from a complete understanding of what an inconsistency measure constitutes. Moreover, the *algorithmic study* of inconsistency measurement has (almost) not been investigated at all. Although straightforward prototype implementations of most measures are available[5], those implementations do not necessarily optimise runtime performance. Only a few papers [2,30–32,37] have addressed this challenge previously, mainly by developing approximation algorithms. Besides more work on approximation algorithms, another venue for future work is also to develop algorithms that work effectively on certain language fragments—such as certain description logics—and thus may work well in practical applications.

Acknowledgements. The research reported here was partially supported by the Deutsche Forschungsgemeinschaft (grant DE 1983/9-1).

References

1. Baroni, P., Gabbay, D., Giacomin, M., van der Torre, L. (eds.): Handbook of Formal Argumentation. College Publications, London (2018)
2. Bertossi, L.: Repair-based degrees of database inconsistency. In: Balduccini, M., Lierler, Y., Woltran, S. (eds.) LPNMR 2019. LNCS, vol. 11481, pp. 195–209. Springer, Cham (2019). https://doi.org/10.1007/978-3-030-20528-7_15
3. Besnard, P.: Revisiting postulates for inconsistency measures. In: Fermé, E., Leite, J. (eds.) JELIA 2014. LNCS (LNAI), vol. 8761, pp. 383–396. Springer, Cham (2014). https://doi.org/10.1007/978-3-319-11558-0_27
4. Besnard, P.: Forgetting-based inconsistency measure. In: Schockaert, S., Senellart, P. (eds.) SUM 2016. LNCS (LNAI), vol. 9858, pp. 331–337. Springer, Cham (2016). https://doi.org/10.1007/978-3-319-45856-4_23
5. Besnard, P.: Basic postulates for inconsistency measures. In: Hameurlain, A., Küng, J., Wagner, R., Decker, H. (eds.) Transactions on Large-Scale Data- and Knowledge-Centered Systems XXXIV. LNCS, vol. 10620, pp. 1–12. Springer, Heidelberg (2017). https://doi.org/10.1007/978-3-662-55947-5_1
6. Béziau, J.-Y., Carnielli, W., Gabbay, D. (eds.): Handbook of Paraconsistency. College Publications, London (2007)
7. Brachman, R.J., Levesque, H.J.: Knowledge Representation and Reasoning. Morgan Kaufmann Publishers, Massachusetts (2004)
8. Brewka, G., Eiter, T., Truszczynski, M.: Answer set programming at a glance. Commun. ACM **54**(12), 92–103 (2011)
9. Brewka, G., Thimm, M., Ulbricht, M.: Strong inconsistency. Artif. Intell. **267**, 78–117 (2019)
10. Cholvy, L., Hunter, A.: Information fusion in logic: a brief overview. In: Gabbay, D.M., Kruse, R., Nonnengart, A., Ohlbach, H.J. (eds.) ECSQARU/FAPR-1997. LNCS, vol. 1244, pp. 86–95. Springer, Heidelberg (1997). https://doi.org/10.1007/BFb0035614
11. Cholvy, L., Perrussel, L., Thevenin, J.M.: Using inconsistency measures for estimating reliability. Int. J. Approximate Reasoning **89**, 41–57 (2017)

---

[5] see http://tweetyproject.org/w/incmes.

12. De Bona, G., Finger, M., Potyka, N., Thimm, M.: Inconsistency measurement in probabilistic logic. In: Measuring Inconsistency in Information, College Publications (2018)
13. De Bona, G., Grant, J., Hunter, A., Konieczny, S.: Towards a unified framework for syntactic inconsistency measures. In: Proceedings of AAAI 2018 (2018)
14. Decker, H., Misra, S.: Database inconsistency measures and their applications. In: Damaševičius, R., Mikašytė, V. (eds.) ICIST 2017. CCIS, vol. 756, pp. 254–265. Springer, Cham (2017). https://doi.org/10.1007/978-3-319-67642-5_21
15. Gelfond, M., Leone, N.: Logic programming and knowledge representation - the a-prolog perspective. Artif. Intell. **138**(1–2), 3–38 (2002)
16. Grant, J., Hunter, A.: Analysing inconsistent first-order knowledgebases. Artif. Intell. **172**(8–9), 1064–1093 (2008)
17. Grant, J., Hunter, A.: Measuring consistency gain and information loss in stepwise inconsistency resolution. In: Liu, W. (ed.) ECSQARU 2011. LNCS (LNAI), vol. 6717, pp. 362–373. Springer, Heidelberg (2011). https://doi.org/10.1007/978-3-642-22152-1_31
18. Grant, J., Hunter, A.: Analysing inconsistent information using distance-based measures. Int. J. Approximate Reasoning **89**, 3–26 (2017)
19. Grant, J., Martinez, M.V. (eds.): Measuring Inconsistency in Information. College Publications, London (2018)
20. Grant, J.: Classifications for inconsistent theories. Notre Dame J. Form. Log. **19**(3), 435–444 (1978)
21. Hansson, S.O.: A Textbook of Belief Dynamics. Kluwer Academic Publishers, Dordrecht (2001)
22. Hunter, A., Konieczny, S.: Approaches to measuring inconsistent information. In: Bertossi, L., Hunter, A., Schaub, T. (eds.) Inconsistency Tolerance. LNCS, vol. 3300, pp. 191–236. Springer, Heidelberg (2005). https://doi.org/10.1007/978-3-540-30597-2_7
23. Hunter, A., Konieczny, S.: Shapley inconsistency values. In: Proceedings of KR 2006, pp. 249–259 (2006)
24. Hunter, A., Konieczny, S.: Measuring inconsistency through minimal inconsistent sets. In: Proceedings of KR 2008, pp. 358–366 (2008)
25. Jabbour, S., Ma, Y., Raddaoui, B.: Inconsistency measurement thanks to MUS decomposition. In: Proceedings of AAMAS 2014, pp. 877–884 (2014)
26. Jabbour, S.: On inconsistency measuring and resolving. In: Proceedings of ECAI 2016, pp. 1676–1677 (2016)
27. Knight, K.M.: Measuring inconsistency. J. Philos. Log. **31**, 77–98 (2001)
28. Konieczny, S., Pino Pérez, R.: On the logic of merging. In: Proceedings of KR 1998 (1998)
29. Ma, Y., Hitzler, P. : Distance-based measures of inconsistency and incoherency for description logics. In: Proceedings of DL 2010 (2010)
30. Ma, Y., Qi, G., Xiao, G., Hitzler, P., Lin, Z.: An anytime algorithm for computing inconsistency measurement. In: Karagiannis, D., Jin, Z. (eds.) KSEM 2009. LNCS (LNAI), vol. 5914, pp. 29–40. Springer, Heidelberg (2009). https://doi.org/10.1007/978-3-642-10488-6_7
31. Ma, Y., Qi, G., Xiao, G., Hitzler, P., Lin, Z.: Computational complexity and anytime algorithm for inconsistency measurement. Int. J. Softw. Inform. **4**(1), 3–21 (2010)
32. McAreavey, K., Liu, W., Miller, P.: Computational approaches to finding and measuring inconsistency in arbitrary knowledge bases. Int. J. Approximate Reasoning **55**, 1659–1693 (2014)

33. Potyka, N., Thimm, M.: Inconsistency-tolerant reasoning over linear probabilistic knowledge bases. Int. J. Approximate Reasoning **88**, 209–236 (2017)
34. Reiter, R.: A logic for default reasoning. Artif. Intell. **13**, 81–132 (1980)
35. Thimm, M., Wallner, J. P.: Some complexity results on inconsistency measurement. In: Proceedings of KR 2016, pp. 114–123 (2016)
36. Thimm, M.: On the expressivity of inconsistency measures. Artif. Intell. **234**, 120–151 (2016)
37. Thimm, M.: Stream-based inconsistency measurement. Int. J. Approximate Reasoning **68**, 68–87 (2016)
38. Thimm, M.: Measuring inconsistency with many-valued logics. Int. J. Approximate Reasoning **86**, 1–23 (2017)
39. Thimm, M.: On the compliance of rationality postulates for inconsistency measures: a more or less complete picture. Künstliche Intell. **31**(1), 31–39 (2017)
40. Thimm, M.: The tweety library collection for logical aspects of artificial intelligence and knowledge representation. Künstliche Intell. **31**(1), 93–97 (2017)
41. Thimm, M.: On the evaluation of inconsistency measures. In: Measuring Inconsistency in Information. College Publications (2018)
42. Ulbricht, M., Thimm, M., Brewka, G.: Inconsistency measures for disjunctive logic programs under answer set semantics. In: Measuring Inconsistency in Information. College Publications (2018)
43. Ulbricht, M., Thimm, M., Brewka, G.: Measuring strong inconsistency. In: Proceedings of AAAI 2018, pp. 1989–1996 (2018)
44. Xiao, G., Ma, Y.: Inconsistency measurement based on variables in minimal unsatisfiable subsets. In: Proceedings of ECAI 2012 (2012)
45. Zhou, L., Huang, H., Qi, G., Ma, Y., Huang, Z., Qu, Y.: Measuring inconsistency in DL-lite ontologies. In: Proceedings of WI-IAT 2009, pp. 349–356 (2009)

Using Graph Convolutional Networks for Approximate Reasoning with Abstract Argumentation Frameworks: A Feasibility Study

Isabelle Kuhlmann and Matthias Thimm[✉]

University of Koblenz-Landau, Koblenz, Germany
thimm@uni-koblenz.de

Abstract. We employ graph convolutional networks for the purpose of determining the set of acceptable arguments under preferred semantics in abstract argumentation problems. While the latter problem is complexity-wise one of the hardest problems in reasoning with abstract argumentation problems, approximate methods are needed here in order to obtain a practically relevant runtime performance. This first study shows that deep neural network models such as graph convolutional networks significantly improve the runtime while keeping the accuracy of reasoning at about 80% or even more.

Keywords: Neural network · Reasoning · Abstract argumentation

1 Introduction

Computational models of argumentation [3] are approaches for non-monotonic reasoning that focus on the interplay between arguments and counterarguments in order to reach conclusions. These approaches can be divided into either *abstract* or *structured* approaches. The former encompass the classical abstract argumentation frameworks following Dung [9] that model argumentation scenarios by directed graphs, where vertices represent arguments and directed links represent attacks between arguments. In these graphs one is usually interested in identifying *extensions*, i.e., sets of arguments that are mutually acceptable and thus provide a coherent perspective on an outcome of the argumentation. On the other hand, structured argumentation approaches consider arguments to be collections of formulas and/or rules which entail some conclusion. The most prominent structured approaches are ASPIC+ [21], ABA [26], DeLP [13], and *deductive argumentation* [4]. These approaches consider a knowledge base of formulas and/or rules as a starting point.

In this paper, we are interested in approximate methods to reasoning with abstract argumentation approaches. Previous works on reasoning with abstract argumentation focus mostly on *sound* and *complete* methods, see e.g. [5] for

© Springer Nature Switzerland AG 2019
N. Ben Amor et al. (Eds.): SUM 2019, LNAI 11940, pp. 24–37, 2019.
https://doi.org/10.1007/978-3-030-35514-2_3

a recent survey and the International Competition on Computational Models of Argumentation[1] (ICCMA) [12,25] for actual implementations. To the best of our knowledge, the only incomplete algorithms for abstract argumentation are [22,24] that use stochastic local search. Here, we use deep neural networks to model the problem of deciding (credulous) acceptability of arguments wrt. preferred semantics as a classification problem. We train a graph convolutional neural network [17]—a special form of a convolutional neural network that is tailored towards processing of graphs—with data obtained by random generation of abstract argumentation frameworks and annotated by a sound and complete solver (in our case *CoQuiAAS* [19]). After training, the obtained classifier can be used to solve the acceptability problem in *constant* time. However, the obtained classifier provides only an approximation to the actual answer. Our experiments showed that approximation quality is about 80% in general, while it can be up to 99% in certain cases.

The remainder of this paper is structured as follows. In Sect. 2, the basic concepts of abstract argumentation and artificial neural networks are recalled. Section 3 explains the approach of representation the acceptability problems as a classification problem. Section 4 describes our experimental evaluation and discusses its results. We conclude in Sect. 5 with a discussion and summary.

2   Preliminaries

In the following, we recall basic definitions of abstract argumentation and artificial neural networks.

2.1   Abstract Argumentation

An abstract argumentation framework [9] AF is a tuple AF $= (\mathsf{Arg}, \rightarrow)$ where Arg is a set of arguments and $\rightarrow \subseteq \mathsf{Arg} \times \mathsf{Arg}$ is the attack relation.

Semantics are given to abstract argumentation frameworks by means of *extensions*. A set of arguments $E \subseteq \mathsf{Arg}$ is called an extension if it fulfils certain conditions. There are various types of extensions, however this paper will be focused on the four classical types proposed by Dung [9]. Namely, these are *complete, grounded, preferred,* and *stable* semantics. All of these types of extensions must be *conflict-free*. A set of arguments $E \subseteq \mathsf{Arg}$ in an argumentation framework AF $= (\mathsf{Arg}, \rightarrow)$ is conflict-free, iff there are no arguments $\mathcal{A}, \mathcal{B} \in E$ with $\mathcal{A} \rightarrow \mathcal{B}$.

Moreover, an argument $\mathcal{A}$ is called *acceptable* with respect to a set of arguments $E \subseteq \mathsf{Arg}$ iff for every $\mathcal{B} \in \mathsf{Arg}$ with $\mathcal{B} \rightarrow \mathcal{A}$ there is an argument $\mathcal{A}' \in E$ with $\mathcal{A}' \rightarrow \mathcal{B}$. Based on these definitions, the four different types of extensions are defined for an argumentation framework AF $= (\mathsf{Arg}, \rightarrow)$ as follows:

1. **Complete extension:** A set of arguments $E \subseteq \mathsf{Arg}$ is called a complete extension iff it is conflict-free, all arguments $\mathcal{A} \in E$ are acceptable with

---

[1] http://argumentationcompetition.org.

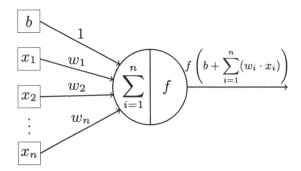

Fig. 1. Artificial neuron, adapted from https://inspirehep.net/record/1300728/plots

respect to E and there is no argument $\mathcal{B} \in$ Arg $\setminus E$ that is acceptable with respect to E.

2. **Grounded extension:** A set of arguments $E \subseteq$ Arg is called a grounded extension iff it is complete and E is minimal with respect to set inclusion.
3. **Preferred extension:** A set of arguments $E \subseteq$ Arg is called a preferred extension iff it is complete and E is maximal with respect to set inclusion.
4. **Stable extension:** A set of arguments $E \subseteq$ Arg is called a stable extension iff it is complete and $\forall \mathcal{A} \in$ Arg$\setminus E : \exists \mathcal{B} \in E$ with $\mathcal{B} \to \mathcal{A}$.

2.2   Artificial Neural Networks and Graph Convolutional Networks

An *artificial neural network* (henceforth also referred to as *neural network* or simply *network*) generally consists of multiple artificial neurons that are connected with each other. In biology, a neuron is a nerve cell that occurs, for example, in the brain or in the spinal cord. Neurons are specialised on conducting and transferring stimuli [23]. In computer science, (artificial) neurons denote a data structure that was developed to work similarly to their biological example. It is to be noted that there exist different models of artificial neurons and neural networks. Due to its contextual relevance in this paper, solely the structure and functionality of the *multilayer perceptron* model [14] will be described.

An artificial neuron can have multiple inputs $x_i \in \mathbb{R}$ with $i \in \{1, \ldots, n\}$ that form the input vector $\boldsymbol{x} = (x_1, \ldots, x_n)^\top$. Each of the n inputs is multiplied by a weight w_i. In addition to the regular inputs, there are so-called *bias* inputs b. They serve the purpose of stabilising the computation. As visualised in Fig. 1, an activation function $f(\cdot)$ is applied to the sum of all weighted inputs. The result of the function is the neuron's output [8,16].

Analogously to the biological prototype, artificial neurons are connected to networks. Such networks are usually arranged in layers that consist of at least one neuron. There is one input layer, one or more so-called *hidden layers*, and one output layer. It is to be noted that the input layer is considered a layer only for convenience, because it only passes the input values to the next layer without further processing [16,20]. Neural networks can be understood as graphs, with neurons as nodes and their connections as edges. For training neural networks, the *back-propagation* algorithm is used in most cases. Back-propagation is a supervised learning method, meaning that at all times during training, the output corresponding to the current input must be known. The goal is to find the most exact mapping of the input vectors to their output vectors. This is realised by adjusting the weights on the edges of the graph, see [16] for details.

In the context of graph theory, Kipf et al. [17] introduce graph convolutional networks that are able to directly use graphs as input instead of a vector of reals. More precisely, they introduce a layer-wise propagation rule for neural networks that operates directly on graphs. It is formulated as follows:

$$H^{(l+1)} = \sigma \left(\tilde{D}^{-\frac{1}{2}} \tilde{A} \tilde{D}^{-\frac{1}{2}} H^{(l)} W^{(l)} \right) \tag{1}$$

$H^{(l)} \in \mathbb{R}^{N \times D}$ denotes the matrix of activations in the lth layer. $\sigma(\cdot)$ is an activation function, such as ReLU (*Rectified Linear Units*) [18]. Moreover, $D_{ii} = \sum_j \tilde{A}_{ij}$ and $\tilde{A} = A + I_N$, where A is the adjacency matrix of the graph and I_N is the identity matrix. $W^{(l)}$ denotes a layer-specific trainable weight matrix. Spectral convolutions on graphs are defined as

$$g_\theta * x = U g_\theta U^\top x. \tag{2}$$

A signal $x \in \mathbb{R}^N$ (a scalar for every node) is multiplied by a filter $g_\theta = \text{diag}(\theta)$, which is parameterised by θ in the Fourier domain. U is the matrix of Eigenvectors of the normalised graph Laplacian $L = I_N - D^{-\frac{1}{2}} A D^{-\frac{1}{2}} = U \Lambda U^\top$, where Λ is a diagonal matrix of the Laplacian's Eigenvalues. $U^\top x$ is the graph Fourier transform of x [17].

For a number of reasons, evaluating Eq. (2) is computationally expensive. For example, computing the Eigendecomposition of L might become rather expensive for large graphs. Hammond et al. [15] suggest that $g_\theta(\Lambda)$ can be approximated by a truncated expansion in terms of Chebyshev polynomials in order to avoid this problem:

$$g_{\theta'}(\Lambda) \approx \sum_{k=0}^{K} \theta'_k T_k(\tilde{\Lambda}) \tag{3}$$

$T_k(x)$ denotes the Chebyshev polynomials up to K^{th} order. The matrix Λ is rescaled to $\tilde{\Lambda} = \frac{2}{\lambda_{\max}} \Lambda - I_N$, where $\lambda_{\max}$ describes the largest Eigenvalue of L. Besides, $\theta' \in \mathbb{R}^K$ is now a vector of Chebyshev coefficients. Integrating this approximation into the definition of a convolution of a signal x with a filter $g_{\theta'}$ yields

$$g_{\theta'} * x \approx \sum_{k=0}^{K} \theta'_k T_k(\tilde{L}) x, \tag{4}$$

with $\tilde{L} = \frac{2}{\lambda_{\max}} L - I_N$ [17]. Because this convolution is a Kth-order polynomial in the Laplacian, it is K-localized. This means, it depends only on a certain neighbourhood—more specifically: it only depends on nodes which are at maximum K steps away from the central node.

Stacking multiple convolutional layers in the form of Eq. (4) (each layer followed by a point-wise non-linearity) leads to a neural network model that can directly process graphs.

3   Casting the Acceptability Problem as a Classification Problem

In abstract argumentation there are several interesting decision problems with varying complexity [10]. For example, the problem CRED_σ with σ being either complete, grounded, preferred, or stable semantics, asks for a given $\text{AF} = (\text{Arg}, \rightarrow)$ and an argument $\mathcal{A} \in \text{Arg}$, whether $\mathcal{A}$ is contained in at least one σ-extension of AF. For preferred semantics this is an NP-complete problem [10]. For our first feasibility study here, we will focus on this problem, i.e., CRED_{PR}.

In order to represent CRED_{PR} as a classification problem, we assume that for any given input argumentation framework $\text{AF} = (\text{Arg}, \rightarrow)$ we have an arbitrary but fixed order of the arguments, i.e., $\text{Arg} = \{\mathcal{A}_1, \ldots, \mathcal{A}_n\}$. Moreover, let $\mathfrak{A}$ denote the set of all abstract argumentation frameworks and V the set of all vectors with values in [0,1] of arbitrary dimension. Conceptually, our classifier C then will be a function of the type $C : \mathfrak{A} \rightarrow V$ with $|C(\text{Arg}, \rightarrow)| = |\text{Arg}|$, i.e., on an input argumentation framework with n arguments we get an n-dimensional real vector as the result.[2] The interpretation of this output then is that the i-th entry of $C(\text{Arg}, \rightarrow)$ denotes the likelihood of argument $\mathcal{A}_i$ being credulously accepted wrt. preferred semantics. Of course, a sound and complete classifier C should output 1 whenever this is true and 0 otherwise. However, as we will only approximate the true solution, all values in the interval [0,1] are possible.

The function C, in our case represented by a graph convolutional network, will be trained on benchmark graphs where the gold standard, i.e. the *true* solutions, is available, e.g., by means of asking a complete oracle solver. Given enough and diverse benchmark graphs for training, our main hypothesis is that C approximates the intended behaviour.

4   Experimental Evaluation

The framework for graph convolutional networks (GCNs) offered by Kipf et al. [17], which is realised with the aid of Google's *TensorFlow* [1], is designed to find labels for certain nodes of a given graph and is thus a reasonable starting point for examining if it is possible to decide whether an argument is credulously accepted wrt. preferred semantics by the use of neural networks.

---

[2] Note that implementation-wise this is not completely true as the size of the output vector has to be fixed.

4.1   Datasets

An essential part of any machine learning task is collecting sufficient training and test data. The probo[3] [7] benchmark suite can be used to generate graphs with different properties. A solver such as CoQuiAAS [19] can then be used to compute the corresponding extensions. The suite offers three different graph generators that each yield graphs with different properties. The first one, the *GroundedGenerator*, produces graphs that have a large grounded extension. The *SccGenerator* produces graphs that are likely to have many strongly connected components. Lastly, the *StableGenerator* generates graphs that are likely to have many stable, preferred, and complete extensions. To provide even more diversity in the data, we use *AFBenchGen*[4] [6] as a second graph generator. It generates random scale-free graphs by using the *Barabási-Albert model* [2], as well as graphs using the *Watts-Strogatz model* [27], and the *Erdős-Rényi model* [11].

In order to examine the impact of the training set size on the classification results, a number of different-sized datasets is generated. It is to be noted that each dataset contains the next smaller dataset in addition to some new data. This strategy is supposed to keep changes in the character of the dataset minimal. The test set is, of course, an exception from this rule. Moreover, each dataset (including the test set) is composed of equal shares of all six previously described types of graphs, and all graphs have between 100 and 400 nodes. Table 1 gives an overview.

In addition to the specifically generated test set, a fraction of the benchmark dataset used in the International Competition on Computational Models of Argumentation (ICCMA) 2017 [12] is used in order to examine how a trained model performs on external data. Said fraction consists of 45 graphs of group B (the only one designated for solvers of CRED_{PR}) that were chosen from all five difficulty categories.

Table 1. Dataset overview.

ID	Number of graphs	Total number of nodes
5-of-each	30	5,461
10-of-each	60	12,056
25-of-each	150	32,026
50-of-each	300	73,717
75-of-each	450	108,050
100-of-each	600	149,130
Test	120	30,603

---

[3] https://sourceforge.net/projects/probo/.
[4] https://sourceforge.net/p/afbenchgen/wiki/Home/.

4.2   Experimental Setup

The GCN framework [17] was designed to perform node-wise classification on a single large graph in a semi-supervised fashion. In order to use the GCN framework in its intended way, three different matrices need to be provided: an $N \times N$ adjacency matrix (N: number of nodes), an $N \times D$ feature matrix (D: number of features per node), and an $N \times F$ binary label matrix (F: number of classes).

For this work, the training process should be supervised rather than semi-supervised. However, the set of unlabeled nodes can be left empty. Because all nodes consequently have a known label, the training process becomes supervised instead of semi-supervised. Besides, instead of one single graph with some nodes to be classified, entire sets of graphs are supposed to provide the training and test sets. To realise this, the graphs in both training and test set are considered one big graph. This yields an adjacency matrix that essentially contains the adjacency matrices of all graphs. The graphs belonging to the test set make up the set of nodes that are to be classified.

The feature matrix can be used to provide additional information on the content of the nodes that could be used to improve classification. However, defining an appropriate feature matrix is a rather difficult matter in our application scenario, because the nodes do not contain any information, in contrast to, for example, social networks or citation networks. In Sect. 4.3, two different solutions are explored. The first one is a simple $N \times 1$ matrix that contains the same constant for every node (which means that no additional features are provided for the nodes). For the second option, the number of incoming and outgoing attacks per argument are used as features, resulting in an $N \times 2$ matrix (one column for each incoming and outgoing attacks).

4.3   Results

When dealing with artificial neural networks, quite a few parameters can influence the outcome of the training process. The following section describes various experimental results in which the impact of different factors on the quality of the classification process is examined. Those factors include, for instance, the size and nature of the training set, the learning rate, and the number of epochs being used to train the neural network model. Finally, we report on some runtime comparison with a sound and complete solver.

Feature Matrix. As explained in Sect. 4.2, there are two different types of feature matrix that may be used in the training process. While training with the feature matrix that does not contain any features (henceforth referred to as FM1) always results in an accuracy of 77.0%, training with the matrix that encodes incoming and outgoing attacks as features (henceforth referred to as FM2) offers slightly better results (up to 80.3%). *Accuracy* is measured by dividing the number of correct predictions by the total number of predictions. The

Table 2. Accuracy per class for both feature matrix types.

FM1		FM2	
Accuracy YES	Accuracy NO	Accuracy YES	Accuracy NO
0.0000	1.0000	0.1499	0.9846
0.0000	1.0000	0.2025	0.9810
0.0000	1.0000	0.2083	0.9803

Table 3. Training results for individual graph types and parameter settings for training. Additional parameters were set as follows: number of epochs: 500, learning rate: 0.001, dropout: 0.05.

	Barabási-Albert	Erdős-Rényi	Grounded	Scc	Stable	Watts-Strogatz
Accuracy YES	1.0000	0.0000	0.0771	0.0000	0.0000	0.0000
Accuracy No	0.0000	1.0000	0.9950	1.0000	1.0000	1.0000
Accuracy total	0.8421	0.8152	0.7109	0.9886	0.8421	0.9988
F1 Score	0.0000	0.0000	0.1417	0.0000	0.0000	0.0000

accuracy value for class YES can also be viewed as the *recall* value, which is calculated by dividing the number of true positives by the sum of true positives and false negatives. Moreover, by calculating the *precision* (true positives divided by the sum of true positives and false positives), the *F1 score* can be obtained as follows:

$$F_1 = 2 \cdot \frac{\text{Precision} \cdot \text{Recall}}{\text{Precision} + \text{Recall}} \tag{5}$$

Moreover, because it seems unusual that multiple different training setups all return the same value, it is important to also look into the class-specific accuracies. Table 2 reveals that the network only learned to classify all nodes as No when trained with FM1. Incorporating FM2 into the training process leads to an accuracy of class YES of up to 20.8%. Whereas this result still needs optimisation, it shows that using FM2 is the more promising approach. In all following experiments, FM2 is used.

Graph Types. In order to further investigate the background of the prior results, the different graph types are examined. Six additional datasets that consist of one graph type each, are created. Each one contains 100 graphs for training and 20 graphs for testing. Essentially, the 100-of-each training set and the test set are split into six subsets consisting of only one graph type per set.

In Table 3, the training results, alongside the settings that were used to retrieve these values, are presented. Several observations can be made from the results. Firstly, a set of parameter settings does not work equally well on all graph types. While four out of six graph types only learn to decide on one class for all instances, Grounded and Stable graphs show first signs of a deeper learning

Table 4. Classification results after training with different-sized training sets. Parameter settings: epochs: 500, learning: 0.01, dropout rate: 0.05. However, a difference in training set size might require different settings. For example, a larger dataset might need more epochs to converge than a smaller ones.

Dataset	Accuracy YES	Accuracy NO	Accuracy total	F1 score
5-of-each	0.0000	1.0000	0.7701	0.0000
10-of-each	0.1869	0.9795	0.7972	0.2976
25-of-each	0.2025	0.9810	0.8020	0.3199
50-of-each	0.2170	0.9797	0.8043	0.3377
75-of-each	0.2174	0.9793	0.8041	0.3380
100-of-each	0.2210	0.9786	0.8044	0.3419

process. Increasing the number of epochs to 1000 yields exactly the same accuracies for Barabási-Albert, Erdős-Rényi, Scc, and Watts-Strogatz graphs, but improves the values for Grounded and Stable. This leads to the assumption that the graph types are of different difficulty for the network to learn. The fact that 98.86% (Scc) or even 99.89% (Watts-Strogatz) of the graphs' nodes belong to one class supports this assumption. Classifying such unevenly distributed classes is quite a difficult task for a neural network.

Another observation is that the set of Barabási-Albert graphs is the only one where the majority of instances is in the class YES. This might help creating a dataset with more evenly distributed classes. Generally, it is certainly helpful to have some graphs with more YES instances in a dataset in order to generate more diversity. Having a diverse dataset is a vital aspect when training neural networks. Otherwise, the network might overfit to irrelevant features or might not work for some application scenarios.

Dataset Size. Besides the influence of a dataset's diversity, the amount of data also has an impact on the training process. Table 4 shows some classification results for the different datasets described in Sect. 4.1. As expected, it indicates that bigger training sets have a greater potential to improve classification results. Nonetheless, utilizing more training data does not automatically mean better results. As displayed in Table 4, adding more than 50 graphs of each type does not yield a significant increase in accuracy. The values for overall accuracy and accuracy for class NO do not change much at all (both less than 3.5%) when adding more training data. It is, however, crucial to look into the accuracy of class YES as well as the F1 scores, because it indicates that the network actually learned some features of a preferred extension, instead of guessing NO for all instances. Training with 25 graphs per type (150 in total) already results in 20.25% accuracy of class YES—only 1.85% less than a training with a total of 600 graphs yields. Training with 50 graphs per type increases the accuracy for YES by another 1.45%, which may still be regarded as significant when considering that the difference to the next bigger training set is merely 0.04%. In summary,

Table 5. Classification results after training with a more balanced dataset in regard to instances per class.

Number of epochs	Learning rate	Dropout	Accuracy YES	Accuracy NO	Accuracy total	F1 score
500	0.1	0.05	0.2488	0.9705	0.8045	0.3693
500	0.01	0.05	0.2589	0.9669	0.8041	0.3781
500	0.001	0.05	0.2372	0.9735	0.8042	0.3578
250	0.01	0.05	0.2659	0.9644	0.8037	0.3839
750	0.01	0.05	0.2728	0.9622	0.8037	0.3899
500	0.01	0.01	0.2682	0.9637	0.8038	0.3859
500	0.01	0.1	0.2494	0.9697	0.8041	0.3693

the increase in accuracy for class YES rather quickly starts stagnating when more data is added.

Optimisation. Training a neural network is a task that demands careful adjustment of various parameters and other aspects. This section describes several approaches that may optimise the results gathered so far.

The main problem with the previous results is that the model seems to underfit. A reason for that might be that the training set is badly balanced in terms of number of instances per class. A dataset where the two classes are about equally distributed might lead to an improvement. Therefore, an additional training set is generated, which consists of 100 Barabási-Albert graphs and a total of 100 graphs of the other types (20 graphs of each). The results for training with this dataset under different parameter settings (regarding the learning rate, number of epochs, and dropout rate) are displayed in Table 5. It becomes clear that the overall accuracy does not improve significantly in comparison to the previous results. Nevertheless, the accuracy of class YES increased to values between 23.72% (500 epochs, learning rate 0.001, dropout 0.05) and 27.28% (750 epochs, learning rate 0.01, dropout 0.05). So, these results might be considered a slight improvement, because they are more evenly distributed than the former ones. Another observation is that changes in number of epochs, learning rate, or dropout rate do not lead to any significant improvements in total accuracy. In fact, most alterations in parameter settings yield slightly worse results.

Looking into the actual numbers of instances of YES and NO reveals that instances of the latter class are still the majority (54.4%). To further equalize the number of instances per class, the training set is augmented by 27 more Barabási-Albert graphs (7300 arguments). The distribution of ground truth labels is now 50.6% YES and 49.4% NO, respectively. Training the neural network with this dataset (parameters are set to 500 epochs, a learning rate of 0.01, and a dropout rate of 0.05) results in a total accuracy of 80.0%. However, the accuracy of class YES increased to 29.7%, while the corresponding value for class NO marginally decreased to 95.0%. This demonstrates that using a more balanced training set

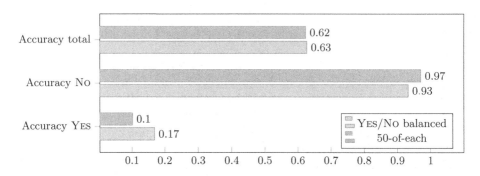

Fig. 2. Results for testing with benchmark data.

(in respect of instances per class) also leads to more balanced results. Since the test set consists of 77.0% instances of class No, the total accuracy does not increase, though.

Competition Data. In order to get a sense of how the training results transfer to other data, two differently trained models are tested on the competition data (see Sect. 4.1). The first model is trained with the 50-of-each dataset. The learning rate is set to 0.01, dropout to 0.05, and number of epochs to 500. The second model uses the same settings, but is trained with the more balanced dataset containing 127 Barabási-Albert graphs and 100 others as illustrated above. Figure 2 displays a comparison of the results. The overall accuracy is very similar for both training sets: about 17% lower than for the regular test set, and the class-specific accuracy values are lower, too. This might be due to the benchmark dataset containing graphs that are smaller or larger than the ones in the training set. Also, additional types of graphs are included in the benchmark dataset.

Runtime Performance. Aside from the quality of the classification results, another aspect that needs to be considered is the time efficiency. In order to put GCN's efficiency into perspective, it is compared to CoQuiAAS, the SAT-based argumentation solver used to provide ground truth labels for the training and test sets.

For the GCN approach, only the time for evaluating the test set is measured, since a neural network can, once it is trained, classify as many arguments as one wishes. Both methods are evaluated on classifying the entire test set (see Sect. 4.1) using the same hardware. The difference is enormous: While the GCN classifies the entire test set within <0.5 s, CoQuiAAS needs about an hour (60.98 min). It is to be noted that the value for testing using a trained GCN varies a bit depending on the training conditions. For example, a measurement taken after training with the biggest training set (600 graphs) is 0.22 s. Training with half the data lead to 0.13 s.

Table 6 reveals the big fluctuations in the amounts of time CoQuiAAS needs to decide for a single argument whether it is included in a preferred extension

Table 6. Time measurements in comparison.

Method	Property	Time in seconds
CoQuiAAS	Maximum	19.274452
CoQuiAAS	Mean	0.119561
CoQuiAAS	Minimum	0.002222
GCN	Mean	0.000007

or not. While the lowest value is at 0.002 s, the highest one is at 19.27 s—which is about 8674 times as much. It is also worth noting that, if evaluating the whole test set takes the GCN 0.22 s, it takes an average of $7 \cdot 10^{-6} = 0.000007$ s. That means, the minimal amount of time CoQuiAAS needed to evaluate an argument is still 317 times as much as the average amount of time the GCN takes. We only report on the mean runtime for the GCN approach as classification is independent of the instance, it is only polynomial in the size of the trained network. It follows that the GCN approach has constant runtime wrt. the size of the instance.

Of course, one needs to consider that a neural network also needs time for training and possibly for preprocessing. Using the GCN framework, the training process took approximately between 20 min and two hours—depending on the dataset size and the parameter settings such as number of epochs or learning rate. For other network models and frameworks, training might take a lot longer. Nonetheless, once sufficient data is provided and the network is trained, it can be used for any test set and it is extremely fast.

5 Conclusion

All in all, the attempt of training a graph-convolutional network on abstract argumentation frameworks in order to decide whether an argument is included in a preferred extension or not was rather moderate. The overall accuracy did under no circumstances exceed 80.5%. When testing with benchmark data, it was even lower (63%). However, extending the diversity of the training set, for instance, by adding different-sized graphs or by adding new types of graphs, might improve this result.

Furthermore, training a neural network model involves adjusting a great number of parameters. Also, some of these parameters depend on each other. Considering that training a neural network requires careful adaption of the training data, the parameter settings, and the network architecture itself, and that some aspects also affect others, examining all reasonable possibilities exceeds the extent of this work.

The training results are moderate: On the one hand, the overall classification accuracy does not exceed 80.5%, which is not good enough for practical applications, but on the other hand, it proves that the network learned at least some rudimental features of a preferred extension. The fact that instances from both classes can be classified correctly reinforces this statement. The accuracy for class YES is far lower (<30%) than the accuracy for class NO (>90%) in all training procedures. A reason for this effect may be that the majority of the training data is not included in an extension and thus labelled as NO. Using a training set where the distribution of instances per class is more balanced, counteracts this effect to some degree. Using benchmark data for testing leads to an overall accuracy of about 63%. The decrease in accuracy in comparison to the specifically generated test set might be due to graph sizes and types that are unknown to the network model, as they were not included in the training data.

Moreover, a GCN's classification process is very time efficient: the entire test set (30,603 arguments) is classified in <0.5 s. For comparison: the SAT solver CoQuiAAS takes about an hour for the same dataset.

Generally, neural networks seem to be suited to perform the task of classifying arguments as "included in a preferred extension" or "not included in a preferred extension". After all, it did work to a certain degree. Nevertheless, the chosen network architecture seems to be inadequate for the task of abstract argumentation. It is quite possible that a different network architecture leads to better results. For example, an increased number of layers in a network or more neurons per layer may increase the network's ability to learn more complex features. The results gathered in this paper show signs of underfitting, so a deeper network would be a plausible strategy. Besides, GCNs were originally constructed to process undirected graphs, yet argumentation frameworks are represented as directed graphs. If a better suited neural network is found, the next step could be to expand the classification problem to a regression problem by training the network to predict entire extensions, or even all possible extensions of an argumentation framework.

Acknowledgements. The research reported here was partially supported by the Deutsche Forschungsgemeinschaft (grant KE 1686/3-1).

References

1. Abadi, M., et al.: Tensorflow: a system for large-scale machine learning. OSDI **16**, 265–283 (2016)
2. Albert, R., Barabási, A.L.: Statistical mechanics of complex networks. Rev. Mod. Phys. **74**(1), 47 (2002)
3. Atkinson, K., et al.: Toward artificial argumentation. AI Mag. **38**(3), 25–36 (2017)
4. Besnard, P., Hunter, A.: Constructing argument graphs with deductive arguments: a tutorial. Argum. Comput. **5**(1), 5–30 (2014)
5. Cerutti, F., Gaggl, S.A., Thimm, M., Wallner, J.P.: Foundations of implementations for formal argumentation. In: Baroni, P., Gabbay, D., Giacomin, M., van der Torre, L. (eds.) Handbook of Formal Argumentation, chap. 15. College Publications, London (2018)
6. Cerutti, F., Giacomin, M., Vallati, M.: Generating challenging benchmark AFs. In: COMMA, vol. 14, pp. 457–458 (2014)
7. Cerutti, F., Oren, N., Strass, H., Thimm, M., Vallati, M.: A benchmark framework for a computational argumentation competition. In: COMMA, pp. 459–460 (2014)

8. Ding, B.N.K.L.: Neural Network Fundamentals with Graphs, Algorithms and Applications. Mac Graw-Hill, New York (1996)
9. Dung, P.M.: On the acceptability of arguments and its fundamental role in nonmonotonic reasoning, logic programming and n-person games. Artif. Intell. **77**(2), 321–357 (1995)
10. Dvořák, W., Dunne, P.E.: Computational problems in formal argumentation and their complexity. In: Baroni, P., Gabbay, D., Giacomin, M., van der Torre, L. (eds.) Handbook of Formal Argumentation, chap. 14. College Publications, London (2018)
11. Erdos, P., Rényi, A.: On the evolution of random graphs. Publ. Math. Inst. Hung. Acad. Sci. **5**(1), 17–60 (1960)
12. Gaggl, S.A., Linsbichler, T., Maratea, M., Woltran, S.: Summary report of the second international competition on computational models of argumentation. AI Mag. **39**(4), 77–79 (2018)
13. García, A.J., Simari, G.R.: Defeasible logic programming: DeLP-servers, contextual queries, and explanations for answers. Argum. Comput. **5**(1), 63–88 (2014)
14. Gardner, M.W., Dorling, S.: Artificial neural networks (the multilayer perceptron) – a review of applications in the atmospheric sciences. Atmos. Environ. **32**(14), 2627–2636 (1998)
15. Hammond, D.K., Vandergheynst, P., Gribonval, R.: Wavelets on graphs via spectral graph theory. Appl. Comput. Harmon. Anal. **30**(2), 129–150 (2011)
16. Jain, A.K., Mao, J., Mohiuddin, K.M.: Artificial neural networks: a tutorial. Computer **29**(3), 31–44 (1996)
17. Kipf, T.N., Welling, M.: Semi-supervised classification with graph convolutional networks. arXiv preprint arXiv:1609.02907 (2016)
18. Krizhevsky, A., Sutskever, I., Hinton, G.E.: ImageNet classification with deep convolutional neural networks. In: Advances in Neural Information Processing Systems, pp. 1097–1105 (2012)
19. Lagniez, J.M., Lonca, E., Mailly, J.G.: CoQuiAAS: a constraint-based quick abstract argumentation solver. In: 2015 IEEE 27th International Conference on Tools with Artificial Intelligence (ICTAI), pp. 928–935. IEEE (2015)
20. Michie, D., Spiegelhalter, D.J., Taylor, C.C.: Machine learning, neural and statistical classification. Citeseer (1994)
21. Modgil, S., Prakken, H.: The ASPIC+ framework for structured argumentation: a tutorial. Argum. Comput. **5**, 31–62 (2014)
22. Niu, D., Liu, L., Lü, S.: New stochastic local search approaches for computing preferred extensions of abstract argumentation. AI Commun. **31**(4), 369–382 (2018)
23. Schmidt, R.F., Lang, F., Heckmann, M.: Physiologie des menschen: mit pathophysiologie. Springer, Heidelberg (2011). https://doi.org/10.1007/978-3-540-32910-7
24. Thimm, M.: Stochastic local search algorithms for abstract argumentation under stable semantics. In: Modgil, S., Budzynska, K., Lawrence, J. (eds.) Proceedings of the Seventh International Conference on Computational Models of Argumentation (COMMA 2018). Frontiers in Artificial Intelligence and Applications, Warsaw, Poland, vol. 305, pp. 169–180, September 2018
25. Thimm, M., Villata, S.: The first international competition on computational models of argumentation: results and analysis. Artif. Intell. **252**, 267–294 (2017)
26. Toni, F.: A tutorial on assumption-based argumentation. Argum. Comput. **5**(1), 89–117 (2014)
27. Watts, D.J., Strogatz, S.H.: Collective dynamics of 'small-world' networks. Nature **393**(6684), 440 (1998)

The Hidden Elegance of Causal Interaction Models

Silja Renooij[1(✉)] and Linda C. van der Gaag[1,2]

[1] Department of Information and Computing Sciences, Utrecht University, Utrecht, The Netherlands
s.renooij@uu.nl
[2] Dalle Molle Institute for Artificial Intelligence, Lugano, Switzerland
linda.vandergaag@idsia.ch

Abstract. Causal interaction models such as the noisy-OR model, are used in Bayesian networks to simplify probability acquisition for variables with large numbers of modelled causes. These models essentially prescribe how to complete an exponentially large probability table from a linear number of parameters. Yet, typically the full probability tables are required for inference with Bayesian networks in which such interaction models are used, although inference algorithms tailored to specific types of network exist that can directly exploit the decomposition properties of the interaction models. In this paper we revisit these decomposition properties in view of general inference algorithms and demonstrate that they allow an alternative representation of causal interaction models that is quite concise, even with large numbers of causes involved. In addition to forestalling the need of tailored algorithms, our alternative representation brings engineering benefits beyond those widely recognised.

Keywords: Bayesian networks · Causal interaction models · Maintenance robustness

1 Introduction

The use of *causal interaction models* has become popular as a technique for simplifying probability acquisition upon building Bayesian networks for real-world applications. These interaction models essentially impose specific patterns of interaction among the causal influences on an effect variable, by means of a *parameterised conditional probability table* for the latter variable. The number of parameters involved in this table typically is linear in the number of causes involved, where the full table itself is exponentially large in this number. Various different causal interaction models have been designed for use in Bayesian networks, the best known among which are the (leaky) noisy-OR model and its generalisations (see for example [4,11,17]).

While a causal interaction model describes a conditional probability table for the effect variable in a causal mechanism by a linear number of parameters, most software packages for inference with the embedding Bayesian network

© Springer Nature Switzerland AG 2019
N. Ben Amor et al. (Eds.): SUM 2019, LNAI 11940, pp. 38–51, 2019.
https://doi.org/10.1007/978-3-030-35514-2_4

require the fully specified table. This full probability table is then generated from the parameters and the definition of the interaction model used, prior to the inference. Using fully expanded probability tables is associated with two serious disadvantages, however. Firstly, the size of the full table is exponential in the number of cause variables involved in a causal mechanism, which induces both the specification size of the network and the runtime complexity of inference to increase substantially. Secondly, using full tables has the engineering disadvantage that the modelling decision to impose a specific pattern of causal interaction is no longer explicit in the representation, as a consequence of which the intricate dependencies between the cells of the table are effectively hidden.

For richly-connected Bayesian networks with large numbers of cause variables per effect variable, as found for example from probabilistic relational models [7], inference scales poorly and quickly becomes infeasible. Over the last decades therefore, researchers have addressed ways to ameliorate the representational and inferential complexity of using fully expanded probability tables with causal interaction models. One such approach has focused on the design of tailored inference algorithms for noisy-OR Bayesian networks, which trade off general applicability and runtime efficiency; these algorithms in essence exploit the structured specification of the noisy-OR model for all variables upon inference (see for example [5,6,8,12,15]). While experimental results underline their scalability for noisy-OR networks, these tailored algorithms are not easily integrated with current algorithms for probabilistic inference in general. Another approach to tackling the representational and inferential complexity of using fully expanded probability tables for causal interaction models, has focused on the design of more concise representations of causal mechanisms; these alternative representations in essence are distilled automatically from the interaction models at hand and allow use of general inference algorithms (see for example [9,10,16,18,19]).

In this paper we reconsider and integrate some of the early work in which causal mechanisms with interaction models are represented by alternative graphical structures and probability tables. We demonstrate that interaction models with specific decomposition properties can be represented efficiently by an alternative structure with associated small tables that have an intuitively appealing semantics. This alternative structure can be readily embedded in a general Bayesian network and thereby allows for inference without the necessity of pre-processing tables or using tailored algorithms. We further argue that this alternative representation induces elegant properties from an engineering perspective which allow more ready maintenance and safer fine-tuning of parameters than the use of fully expanded probability tables in causal mechanisms.

The paper is organised as follows. In Sect. 2, we briefly review causal interaction models, and the (leaky) noisy-OR model more specifically. In Sect. 3, we reconsider the partition of causal interaction models into a deterministic function and associated independent noise variables, and demonstrate when and how the underlying deterministic function can be decomposed. Based on these insights, we derive our alternative cascading representation and study its properties in Sect. 4. We conclude the paper in Sect. 5.

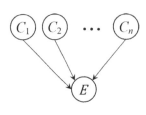

$\mathbf{c}$	$\Pr(e \mid \mathbf{c})$
$\bar{c}_1, \bar{c}_2, \bar{c}_3$	0
$c_1, \bar{c}_2, \bar{c}_3$	p_1
$\bar{c}_1, c_2, \bar{c}_3$	p_2
$\bar{c}_1, \bar{c}_2, c_3$	p_3
$c_1, c_2, \bar{c}_3$	$1 - (1 - p_1) \cdot (1 - p_2)$
$c_1, \bar{c}_2, c_3$	$1 - (1 - p_1) \cdot (1 - p_3)$
$\bar{c}_1, c_2, c_3$	$1 - (1 - p_2) \cdot (1 - p_3)$
c_1, c_2, c_3	$1 - (1 - p_1) \cdot (1 - p_2) \cdot (1 - p_3)$

Fig. 1. A causal mechanism $\mathcal{M}(n)$ with n cause variables C_i and the effect variable E (*left*); a conditional probability table imposed by the noisy-OR model, for $n = 3$ (*right*).

2   Preliminaries

We briefly review causal interaction models for Bayesian networks and thereby introduce our notational conventions. In this paper, we focus on binary random variables, which are denoted by (possibly indexed) capital letters X. The values of such a variable X are denoted by small letters; more specifically, we write $\bar{x}$ and x to denote absence and presence, respectively, of the concept modelled by X. (Sub)sets of variables are denoted by bold-face capital letters $\mathbf{X}$ and their joint value combinations by bold-face small letters $\mathbf{x}$; $\Omega(\mathbf{X})$ is used to denote the domain of all value combinations of $\mathbf{X}$. We further consider joint probability distributions Pr over sets of variables, represented by a Bayesian network.

Within Bayesian networks, we consider *causal*[1] *mechanisms* $\mathcal{M}(n)$ composed of a single effect variable E and one or more cause variables C_i, $i = 1, \ldots, n$, with arcs pointing to E; Fig. 1 (*left*) illustrates the basic idea of such a mechanism. For the effect variable E of a causal mechanism, a conditional probability table is specified, with distributions $\Pr(E \mid \mathbf{C})$ over E for each joint value combination $\mathbf{c}$ for its set $\mathbf{C}$ of cause variables; this table thus specifies a number of distributions that is exponential in the number of cause variables involved.

A *causal interaction model* for a causal mechanism $\mathcal{M}(n)$ takes the form of a parameterised probability table for the effect variable involved. The noisy-OR model [17], which is the best known among these interaction models, defines the conditional probability table for the effect variable E of $\mathcal{M}(n)$ through

- the conditional probability $\Pr(e \mid \bar{c}_1, \ldots, \bar{c}_n) = 0$;
- the *parameters* $p_i = \Pr(e \mid \bar{c}_1, \ldots, \bar{c}_{i-1}, c_i, \bar{c}_{i+1}, \ldots, \bar{c}_n)$, for all $i = 1, \ldots, n$;
- the *definitional rule* $\Pr(e \mid \mathbf{c}) = 1 - \prod_{i \in I_\mathbf{c}} (1 - p_i)$ for the probabilities given the remaining value combinations $\mathbf{c}$ involving the presence of two or more causes, where $I_\mathbf{c}$ is the set of indices of the present causes c_i in $\mathbf{c}$.

Figure 1 (*right*) illustrates the parameterised table of the noisy-OR model for a mechanism with three cause variables. For a causal mechanism $\mathcal{M}(n)$, the model

---

[1] Although we do not make any claim with respect to causal interpretation, we adopt the terminology commonly used.

defines a full probability table over $n + 1$ variables, specifying a total of $2 \cdot 2^n$ probabilities; half of these are derived from $\Pr(e \mid \mathbf{c}) + \Pr(\overline{e} \mid \mathbf{c}) = 1$ and, hence, are redundant. Of the 2^n non-redundant probabilities, the noisy-OR model allows the values of only the n parameter probabilities p_i to be chosen freely. The model further forces the distribution $\Pr(E \mid \overline{c}_1, \ldots, \overline{c}_n)$ to be degenerate.

Since its introduction, the noisy-OR model has given rise to several variants and generalisations (see [4] for an overview). Of these, we briefly review here the *leaky noisy-OR model*. This model differs from the noisy-OR model in that it includes an additional *leak parameter* $p_L = \Pr(e \mid \overline{c}_1, \ldots, \overline{c}_n)$ that captures the probability of the effect e occurring in the absence of all modelled causes. Different interpretations of the noisy-OR parameters in view of this leak probability have given rise to different definitional rules for the remaining probabilities [4,11]. Without loss of generality, we adopt in this paper the interpretation proposed by Díez [4], and use the rule $\Pr(e \mid \mathbf{c}) = 1 - (1 - p_L) \cdot \prod_{i \in I_{\mathbf{c}}} (1 - p_i)$ for the probabilities given arbitrary joint value combinations $\mathbf{c}$ with multiple present causes, where $I_{\mathbf{c}}$ again is the set of indices of the causes present in $\mathbf{c}$.

3   Decomposition of Causal Interaction Models

Causal interaction models are often viewed as combining a deterministic function f with independent noise variables Z_i per cause variable (see for example [10, 14, 17]); Fig. 2 (*left*) illustrates this view for the (leaky) noisy-OR model. The noise variables Z_i are associated with the probabilities $\Pr(z_i \mid c_i) = p_i$, $\Pr(z_i \mid \overline{c}_i) = 0$, where the p_i are the model's parameters; in the leaky variant of the noisy-OR model, the prior probability $\Pr(z_L) = p_L$ for the designated noise variable Z_L is the leak parameter. The deterministic function f equals the logical OR and is encoded in the probability table $\Pr(E \mid \mathbf{Z})$ for the effect variable E through degenerate distributions. The variable E thereby is a deterministic variable and, by convention, is indicated by a double border in our figure. Slightly abusing notation, we will further write $E = f(\mathbf{Z})$.

The representation in Fig. 2 (*left*) was introduced originally to indicate how a causal interaction model could ease the task of knowledge acquisition for causal mechanisms involving large numbers of variables [9]: by making independence of the causal influences explicit, the partition into a deterministic part and a probabilistic noise part underlines the requirement of actually just a limited number of parameters. While indeed easing the task of knowledge acquisition for practical applications, the partition of a causal interaction model does not reduce the actual size of its representation for use with general inference algorithms. In fact, embedding the partition of a causal mechanism $\mathcal{M}(n)$ in a Bayesian network will increase the total number of variables involved by n and still require the specification of exponentially many probabilities for the effect variable E.

Specific types of causal interaction model however, actually do allow a reduced representation [10]. More formally, it are specific decomposability properties of the deterministic function f that provide for a reduction of the size of the conditional probability table(s) for the effect variable(s) in a causal mechanism.

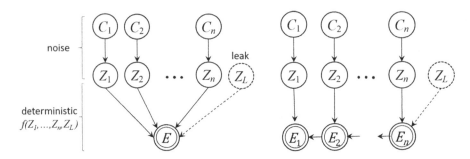

Fig. 2. Partition of a causal interaction model into a probabilistic noise part and a deterministic functional part (*left*); a chain decomposition for a commutative and associative deterministic function (*right*).

Such decomposability properties of functions are widely used in mathematics and computing science to simplify functions by their hidden structure: a function $f(\cdot)$ on a set of entities is called *self-decomposable* if, for any two disjoint subsets $\mathbf{X}, \mathbf{Y}$, the property $f(\mathbf{X} \cup \mathbf{Y}) = f(\mathbf{X}) \diamond f(\mathbf{Y})$ holds, for some commutative and associative merge operator $\diamond$ (cf. [13]). Commutative and associative logical operators, such as AND and OR, are self-decomposable Boolean functions. Now, if the deterministic function f modelled for the effect variable E in the partition in Fig. 2 (*left*) is self-decomposable, it can be split into a sequence of function applications, each to a subset of E's cause variables. Each such application can then be described by an auxiliary effect variable E_i with fewer parents than E. The set of auxiliary variables resulting from such a functional decomposition can be organised in various different graphical structures. In this paper the chained organisation from Fig. 2 (*right*) will be used and referred to as a *chain decomposition*. We would like to note that the idea of introducing additional variables to reduce the number of parents for a variable is a general modelling technique for Bayesian networks, known as *parent divorcing* [16].

We consider again the partition of a causal interaction model into a probabilistic part with noise variables Z_i, $i = 1, \ldots, n$, and a deterministic part $E = f(Z_1, \ldots, Z_n)$ for some self-decomposable deterministic function f. The chain decomposition of the model replaces the effect variable E of this partition by n auxiliary variables E_i, $i = 1, \ldots, n$, such that

- E_n has the noise variable Z_n for its single parent and encodes the function application $E_n = f(Z_n, I)$, where the variable I captures identity under f;
- for all $i = 1, \ldots, n - 1$, the variable E_i has Z_i and E_{i+1} for its parents and encodes $E_i = f(Z_i, E_{i+1})$.

If the interaction model includes a leak variable Z_L the identity variable I in the function application $f(Z_n, I)$ is replaced by Z_L, to give $E_n = f(Z_n, Z_L)$. We note that the number of variables in the chain decomposition has increased, from $2 \cdot n + 1$ in the original partition, to $3 \cdot n$. The total number of non-redundant probabilities required for the probability tables for the variables E_i in the chain

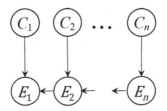

Fig. 3. The cascading representation of a causal interaction model, which results from marginalising out the noise variables Z_i from its chain decomposition.

equals $4 \cdot n - 2$ however, instead of the 2^n probabilities required for the effect variable E in the original partition. For an interaction model with a leak variable, the number of required probabilities for the effect variable(s) is reduced from 2^{n+1} to $4 \cdot n$. We will return to these observations in further detail in Sect. 4.

While the original motivation for partitioning causal interaction models was to underline their induced ease of knowledge acquisition, Heckerman noted that the introduction of the hidden noise variables Z_i in fact made probability elicitation harder rather than easier, as "assessments are easier to elicit (and presumably more reliable) when a person makes them in terms of observable variables" [9]. Following this insight, he proposed a *temporal interpretation* of independence of causal influences for causal interaction models in which a cause C_i is assumed to occur (or not) at time i and has associated its own effect variable E_i indicating the effect after the presence or absence of the first i causes have been observed. With this temporal interpretation, the hidden noise variables are no longer required and the effect variables E_i have in fact become observable variables with a clear semantics supporting probability elicitation. As noted already by Heckerman himself, this temporal interpretation for causal interaction models has reduced applicability for its main drawback [9,10].

4 Properties of a Cascading Representation

We propose a representation of causal interaction models that is quite similar to Heckerman's temporal representation, yet without the temporal interpretation. We will argue that our representation has a clear semantics and in addition allows for easy maintenance in the event of changes in the parameters of the represented interaction model. Before demonstrating the latter in Sect. 4.2, we now first detail our *cascading representation* of causal interaction models.

4.1 The Cascading Representation and Its Equivalence Property

We focus on causal mechanisms with an underlying self-decomposable deterministic function f as reviewed in the previous section, and consider their chain decomposition as illustrated in Fig. 2 (*right*). Instead of building on a temporal interpretation as suggested by Heckerman, we propose to sum out the noise variables Z_i by marginalisation. We note that, by doing so, the effect variables E_i, $i = 1, \ldots, n$, become stochastic rather than deterministic. The resulting representation, called the *cascading representation* of a causal interaction model, is illustrated for a mechanism $\mathcal{M}(n)$ in Fig. 3, where

- the variable E_n, with the cause variable C_n for its single parent, has the probability table derived from the chain decomposition as

$$\Pr(e_n \mid C_n) = \sum_{z_n' \in \Omega(Z_n)} \Pr(e_n \mid z_n') \cdot \Pr(z_n' \mid C_n) \tag{1}$$

or, in the presence of a leak probability, as

$$\begin{aligned} \Pr(e_n \mid C_n) \quad &= \quad p_L \cdot \sum_{z_n' \in \Omega(Z_n)} \Pr(e_n \mid z_n', z_L) \cdot \Pr(z_n' \mid C_n) \\ &\quad + (1 - p_L) \cdot \sum_{z_n' \in \Omega(Z_n)} \Pr(e_n \mid z_n', \overline{z}_L) \cdot \Pr(z_n' \mid C_n) \end{aligned} \tag{2}$$

- the variables E_i, $i = 1, \ldots, n - 1$, with the parents C_i and E_{i+1}, have the probability table derived as

$$\Pr(e_i \mid C_i, E_{i+1}) = \sum_{z_i' \in \Omega(Z_i)} \Pr(e_i \mid z_i', E_{i+1}) \cdot \Pr(z_i' \mid C_i) \tag{3}$$

We note that all probabilities conditioned on a value of a noise variable originate from the degenerate distributions modelling the deterministic function f of the interaction model. We further note that the inclusion of a leak probability affects only the cells of the probability table for the variable E_n, whereas it affects, through the definitional rule of the interaction model at hand, *all* cells in the fully expanded table for the variable E in the causal mechanism.

To ensure that our cascading representation of an interaction model is equivalent to its original representation in a causal mechanism, the variable E_1 in our representation should represent the exact same information as the effect variable E in a mechanism $\mathcal{M}(n)$. Any probability $\Pr(\overline{e} \mid \mathbf{c}) = 1 - \Pr(e \mid \mathbf{c})$ specified in the full probability table for E should therefore be the same as the probability $\Pr(\overline{e}_1 \mid \mathbf{c})$ that is computed from the cascading representation as

$$\Pr(\overline{e}_1 \mid \mathbf{c}) = \sum_{\mathbf{e}^- \in \Omega(\mathbf{E}^-)} \Pr(\overline{e}_1 \mid c_1', e_2') \cdot \prod_{k=2}^{n-1} \Pr(e_k' \mid c_k', e_{k+1}') \cdot \Pr(e_n' \mid c_n') \tag{4}$$

where $\Omega(\mathbf{E}^-)$ is the domain of the variable set $\mathbf{E}^- = \{E_2, \ldots, E_n\}$, and where $e_k' \in \Omega(E_k)$, $k = 2, \ldots, n$, is consistent with $\mathbf{e}^-$ and $c_k' \in \Omega(C_k)$, $k = 1, \ldots, n$, is consistent with $\mathbf{c}$. We emphasize that we focus on the value $\overline{e}_1$ of the variable E_1 rather than on the value e_1, to simplify our arguments in the sequel.

We now illustrate the derivation of the probability tables for the cascading representations of the noisy-OR and leaky noisy-OR models, and demonstrate their equivalence to the standard causal-mechanism representation.

The cascading noisy-OR. We begin with constructing the conditional probability tables to be specified for the noisy-OR model in its cascading representation. For the variables E_i, $i = 1, \ldots, n - 1$, we find from Eq. 3 that

$$\Pr(e_i \mid \overline{c}_i, \overline{e}_{i+1}) = 1 \cdot 0 + 0 \cdot 1 = 0$$
$$\Pr(e_i \mid c_i, \overline{e}_{i+1}) = 1 \cdot p_i + 0 \cdot (1 - p_i) = p_i$$
$$\Pr(e_i \mid \overline{c}_i, e_{i+1}) = 1 \cdot 0 + 1 \cdot 1 = 1$$
$$\Pr(e_i \mid c_i, e_{i+1}) = 1 \cdot p_i + 1 \cdot (1 - p_i) = 1$$

where $p_i = \Pr(e \mid \overline{c}_1, \ldots, \overline{c}_{i-1}, c_i, \overline{c}_{i+1}, \ldots, \overline{c}_n)$ coincides with a regular noisy-OR parameter. For the variable E_n we similarly find from Eq. 2 that

$$\Pr(e_n \mid \overline{c}_n) = 1 \cdot 0 + 0 \cdot 1 = 0$$
$$\Pr(e_n \mid c_n) = 1 \cdot p_n + 0 \cdot (1 - p_n) = p_n$$

where p_n is again a regular noisy-OR parameter. We observe that each parameter p_i, $i = 1, \ldots, n$, occurs in the specification of exactly *one* table, which is the table for the variable E_i. In addition to this single associated noisy-OR parameter, the probability table for the variable E_i further specifies just zeroes and ones.

We now show that the cascading representation, with the probability specification above, correctly captures the noisy-OR model. To this end, we observe that for a summand of Eq. 4 to actually contribute to the computation of $\Pr(\overline{e}_1 \mid \mathbf{c})$, it should be a product composed of just non-zero terms. Such non-zero terms are found only with the following probabilities:

- $\Pr(\overline{e}_n \mid \overline{c}_n)$ or $\Pr(e'_n \mid c_n)$, for the variable E_n;
- $\Pr(e_i \mid c'_i, e_{i+1})$, $\Pr(\overline{e}_i \mid c'_i, \overline{e}_{i+1})$, and $\Pr(e_i \mid c_i, \overline{e}_{i+1})$, for the variable E_i, $i = 1, \ldots, n - 1$;

with $e'_i \in \Omega(E_j)$ and $c'_i \in \Omega(C_i)$, $i = 1, \ldots, n$. Close examination of these non-zero probabilities shows that for the value $\overline{e}_1$ of E_1 under consideration, only value combinations $\mathbf{e}^-$ for $\mathbf{E}^- = \{E_2, \ldots, E_n\}$ consistent with $\overline{e}_2$ can possibly contribute a non-zero term to a summand of Eq. 4. By iteratively applying this argument to the variables $E_3, \ldots, E_n$, we conclude that only the value combination $\mathbf{e}^- = \overline{e}_2, \ldots, \overline{e}_n$ contributes a non-zero summand to the probability $\Pr(\overline{e}_1 \mid \mathbf{c})$. For the cascading representation of the noisy-OR model therefore, Eq. 4 reduces to:

$$\Pr(\overline{e}_1 \mid \mathbf{c}) = \prod_{i=1}^{n-1} \Pr(\overline{e}_i \mid c'_i, \overline{e}_{i+1}) \cdot \Pr(\overline{e}_n \mid c'_n) \tag{5}$$

To show that the cascading representation correctly captures the noisy-OR model, we now consider the three different cases distinguished by this model:

- Where the noisy-OR model has $\Pr(e \mid \mathbf{c}) = 0$ for $\mathbf{c} = \overline{c}_1, \ldots, \overline{c}_n$, we find in the cascading representation from $\Pr(\overline{e}_j \mid \overline{c}_j, \overline{e}_{j+1}) = 1$ for $j = 1, \ldots, n - 1$ and $\Pr(\overline{e}_n \mid \overline{c}_n) = 1$, that $\Pr(\overline{e}_1 \mid \mathbf{c}) = 1$ and, hence, $\Pr(e_1 \mid \mathbf{c}) = 0$.

Table 1. For the two representations of the noisy-OR model for a causal mechanism $\mathcal{M}(n)$: the number of variables (#*variables*), the number of non-redundant probabilities for the effect variable(s) (#*probabilities*), and of those, the number of free parameters to be acquired (#*free*) and the number of zeroes and ones (#0/1).

Representation	#variables	#probabilities	#free	#0/1
Full table	$n+1$	2^n	n	1
Cascade	$2 \cdot n$	$4 \cdot n - 2$	n	$3 \cdot n - 2$

- Where the noisy-OR model has $\Pr(e \mid \mathbf{c}) = p_i$ for $\mathbf{c}$ including the *single* present cause c_i, we have in the cascading representation that the product term contributed for the variable E_i has the probability $\Pr(\bar{e}_i \mid c_i, \bar{e}_{i+1}) = 1 - p_i$ or, in case $i = n$, $\Pr(\bar{e}_n \mid c_n) = 1 - p_n$. As all other terms in the product of Eq. 5 equal 1, we find that $\Pr(\bar{e}_1 \mid \mathbf{c}) = 1 - p_i$ and, hence, $\Pr(e_1 \mid \mathbf{c}) = p_i$.

- For any value combination $\mathbf{c}$ including multiple present causes, with their indices in $I_{\mathbf{c}}$, the noisy-OR model has $\Pr(e \mid \mathbf{c}) = 1 - \prod_{i \in I_{\mathbf{c}}}(1 - p_i)$. In the cascading representation, the product term contributed by any E_j with $j \notin I_{\mathbf{c}}$ equals 1 and the term by any E_i with $i \in I_{\mathbf{c}}$ is $1 - p_i$. We thus find that $\Pr(\bar{e}_1 \mid \mathbf{c}) = \prod_{i \in I_{\mathbf{c}}}(1 - p_i)$ and, hence, $\Pr(e_1 \mid \mathbf{c}) = 1 - \prod_{i \in I_{\mathbf{c}}}(1 - p_i)$.

From the three cases above, we conclude that the cascading representation indeed correctly captures the noisy-OR model and, hence, that the cascading representation is equivalent with the fully expanded probability table for the effect variable E in a causal mechanism with a noisy-OR model.

The cascading representation of the noisy-OR model is a more efficient representation than a causal mechanism $\mathcal{M}(n)$ with a full probability table for the effect variable E, despite the increase in number of variables to $2 \cdot n$ compared to the $n + 1$ variables in the standard representation. More specifically, the cascading representation requires $4 \cdot (n - 1) + 2$ conditional probability distributions in total for the variables E_i, of which $3 \cdot (n - 1) + 1$ are degenerate. For ease of reference, Table 1 summarises a comparison of the size of the cascading representation with that of the standard representation. We note that the cascading representation is more concise when a causal mechanism would include $n \geq 4$ cause variables for the effect variable of interest.

The cascading leaky noisy-OR. We now briefly address the cascading representation of the noisy-OR model in the presence of a leak probability, which differs from that of the standard noisy-OR model only in the specification of the probability table for the variable E_n, which is derived from Eq. 2 as

$$\Pr(e_n \mid \bar{c}_n) = p_L$$
$$\Pr(e_n \mid c_n) = p_L + p_n \cdot (1 - p_L) = 1 - (1 - p_L) \cdot (1 - p_n)$$

where p_n is again a regular noisy-OR parameter and p_L is the leak probability. To show that the cascading representation with this specification correctly captures the leaky noisy-OR model, we use Eq. 5 again, now for the different cases distinguished by the leaky noisy-OR model. We observe that, while with the noisy-OR model, the variable E_n would contribute to the product either $\Pr(\bar{e}_n \mid \bar{c}_n) = 1$ or $\Pr(\bar{e}_n \mid c_n) = 1 - p_n$, it contributes either $\Pr(\bar{e}_n \mid \bar{c}_n) = 1 - p_L$ or $\Pr(\bar{e}_n \mid c_n) = (1 - p_L) \cdot (1 - p_n)$ in the cascading representation of the leaky noisy-OR model. As a consequence

- With the leaky model having $\Pr(e \mid \mathbf{c}) = p_L$ for $\mathbf{c} = \bar{c}_1, \ldots, \bar{c}_n$, we find $\Pr(\bar{e}_1 \mid \mathbf{c}) = 1 - p_L$ and, hence, $\Pr(e_1 \mid \mathbf{c}) = p_L$ from the cascading representation.
- For any value combination $\mathbf{c}$ with an arbitrary number of present causes with indices in $I_\mathbf{c}$, the leaky model has $\Pr(e \mid \mathbf{c}) = 1 - (1 - p_L) \cdot \prod_{i \in I_\mathbf{c}} (1 - p_i)$. Using the observation above, we find in the cascading representation that $\Pr(\bar{e}_1 \mid \mathbf{c}) = (1 - p_L) \cdot \prod_{j \in I_\mathbf{c}} (1 - p_j)$ and, hence, $\Pr(e_1 \mid \mathbf{c}) = 1 - (1 - p_L) \cdot \prod_{j \in I_\mathbf{c}} (1 - p_j)$.

We conclude that the probabilities computed from the cascading representation indeed coincide with the probabilities in the full probability table in a causal mechanism with the leaky noisy-OR model. We thus can construct an efficient representation for a causal mechanism $\mathcal{M}(n)$ with the leaky noisy-OR model. Of the $4 \cdot (n - 1) + 2$ conditional distributions required in total by the cascading representation, now $3 \cdot (n - 1)$ are degenerate. We note that the difference of one compared with the cascading representation of the noisy-OR model originates from the inclusion of the leak probability as a parameter.

4.2   Additional Engineering Benefits

Causal mechanisms are typically modelled straightforwardly in Bayesian networks, as in Fig. 1 (*left*). The different partitions and decompositions of causal interaction models proposed, are mostly seen as alternative representations to support probability elicitation and are hardly ever used in a network directly. Table 1 clearly illustrates the reduction in specification size that would be achieved by choosing a cascading representation for causal mechanisms with large numbers of cause variables; as this representation limits the number of parents per effect variable, it also has the potential to reduce the runtime complexity of probabilistic inference, dependent of the graphical structure of the embedding Bayesian network [10, 14]. In this section, we now argue that the cascading representation further has clear engineering benefits beyond those widely recognised.

Clear semantics. Alternative representations of causal interaction models typically rely on the introduction of additional variables. Although introducing such additional variables is commonly used for reducing the number of parents for an effect variable, it is often quite undesirable from a knowledge engineering perspective. While the additional variables have a clear meaning from a mathematics point of view, they often are quite meaningless from the perspective of the application domain and thereby hamper the interpretation of the model as a domain representation. The lack of a clear meaning is especially problematic if the probabilities for these additional variables need be elicited from experts. Now, in our cascading representation of a causal interaction model, the additional variables *do* have a clear intuitive meaning, as a consequence of the decomposability properties of the underlying deterministic function: in the cascading representation of a causal mechanism $\mathcal{M}(n)$, any variable E_i can be viewed as the effect variable in the causal mechanism $\mathcal{M}(n-i+1)$ involving the subset of causes $C_i, \ldots, C_n$. This claim is readily seen by replacing E_1 by E_i in Eq. 4:

$$\Pr(\bar{e}_i \mid \mathbf{c}) = \sum_{\mathbf{e}^- \in \Omega(\mathbf{E}^-)} \Pr(\bar{e}_i \mid c_1', e_{i+1}') \cdot \prod_{k=i+1}^{n-1} \Pr(e_k' \mid c_k', e_{k+1}') \cdot \Pr(e_n' \mid c_n')$$

where $\Omega(\mathbf{E}^-)$ now is the domain of $\mathbf{E}^- = \{E_{i+1}, \ldots, E_n\}$, and e_k', c_k' are defined as before. As each variable E_i in the cascading representation represents the effect variable in a (leaky) noisy-OR model with the cause variables $C_i, \ldots, C_n$, it has an intuitive meaning that allows for explicit embedding of the representation in a network without hampering interpretation and probability elicitation.

Maintenance robustness. The cascading representation of a causal interaction model brings yet another advantage from an engineering perspective. When using fully expanded probability tables for the effect variables in a Bayesian network, any modelling decision to employ a causal interaction model is no longer explicitly visible in the network's representation. More specifically, the dependency of multiple cells of the table on the parameters of the model employed is hidden. When a network is maintained and adapted to its changing context of application over a period of years therefore, inopportune changes to the specified probabilities can disrupt the modelled interaction pattern and, thereby, the original modelling decision. We illustrate this observation by means of a causal mechanism with a noisy-OR model for the effect variable, and show that the cascading representation of the interaction model used is more robust by preventing the occurrence of such unintended disruptions.

We address the engineering task of studying the effects, on a network's output probabilities, of changing a single probability from one of the network's probability tables. Such a sensitivity analysis is usually part of the encompassing task of fine-tuning the network's specification to attain a desired effect on the output (see for example [1–3]). In view of a causal mechanism $\mathcal{M}(n)$, we now consider the output probability of interest $\Pr(e \mid c_i, c_k)$, for some $1 \leq i < k < n$, and address how this probability changes with a change of the probability $x = \Pr(e \mid \bar{c}_1, \ldots, \bar{c}_{i-1}, c_i, \bar{c}_{i+1}, \ldots, \bar{c}_n)$ of the full probability table of the effect variable E; we note that this probability is one of the parameters of the noisy-OR model. The function $[\Pr(e \mid c_i, c_k)](x)$ describing the sensitivity of $\Pr(e \mid c_i, c_k)$ to changes in x would be constant if the modelling choice of imposing a noisy-OR interaction for the mechanism at hand is not taken into consideration:

$$[\Pr(e \mid c_i, c_k)] (x) = a, \quad \text{with} \quad a = \sum_{\mathbf{c}^- \in \Omega(\mathbf{C}^-)} \Pr(e \mid c_i, c_k, \mathbf{c}^-) \cdot \Pr(\mathbf{c}^- \mid c_i, c_k) \quad (6)$$

where $\Omega(\mathbf{C}^-)$ is the domain of the set $\mathbf{C}^- = \{C_1, \ldots, C_n\} \setminus \{C_i, C_k\}$ of cause variables for which no value is fixed by the probability of interest. We note that the computation of $\Pr(\mathbf{c}^- \mid c_i, c_k)$ does not involve any probabilities from the probability table of E; in contrast, the first term in the product for each summand in a corresponds directly to a cell from the full table for E. Since the summation does not involve parameter x directly, the analysis reveals that the output probability is not sensitive to variations of the parameter. This result however, does not correctly reflect the true sensitivity of the output probability to variations in the parameter under study: the parameter x is actually included in various cells of the full probability table of E by the definitional rule from the noisy-OR model, and thereby hidden in various summands of a.

We now consider the same sensitivity analysis in view of the cascading representation of the noisy-OR model, for essentially the same probability of interest and essentially the same parameter probability. Recall that in the cascading representation, any posterior probability distribution over the variable E_1 equals the posterior distribution given the same evidence over the original variable E with the full probability table; we therefore take the probability $\Pr(e_1 \mid c_i, c_k)$ for the probability of interest. The parameter $p_i = \Pr(e \mid \bar{c}_1, \ldots, \bar{c}_{i-1}, c_i, \bar{c}_{i+1}, \ldots, \bar{c}_n)$ of the noisy-OR model moreover occurs as $p_i = \Pr(e_i \mid c_i, \bar{e}_{i+1})$ in the model's cascading representation; we thus take $x = \Pr(e_i \mid c_i, \bar{e}_{i+1})$ as the probability that will be varied. The sensitivity analysis will in essence establish the same result as presented in Eq. 6, but now the probabilities $\Pr(e \mid c_i, c_k, \mathbf{c}^-)$ follow from the cascading representation using Eq. 5, and depend explicitly on x:

$$[\Pr(\bar{e}_1 \mid c_i, c_k, \mathbf{c}^-)] (x) = \left[(1 - p_i) \cdot (1 - p_k) \cdot \prod_{j \in I_{\mathbf{c}^-}} (1 - p_j) \right] (x)$$

$$= (1 - x) \cdot (1 - p_k) \cdot \prod_{j \in I_{\mathbf{c}^-}} (1 - p_j)$$

where $I_{\mathbf{c}^-}$ indexes all present causes in $\mathbf{C}^-$ and, for ease of exposition, we again focus on the value $\bar{e}_1$ for variable E_1. As a result, we find that

$$[\Pr(\bar{e}_1 \mid c_i, c_k)] (x) = \sum_{\mathbf{c}^- \in \Omega(\mathbf{C}^-)} (1 - x) \cdot (1 - p_k) \cdot \prod_{j \in I_{\mathbf{c}^-}} (1 - p_j) \cdot \Pr(\mathbf{c}^- \mid c_i, c_k)$$

and conclude that the function $[\Pr(e_1 \mid c_i, c_k)] (x)$ is in fact a linear function of the form $a \cdot x + b$ with constants a, b, where

$$a = (1 - p_k) \cdot \sum_{\mathbf{c}^- \in \Omega(\mathbf{C}^-)} \Pr(\mathbf{c}^- \mid c_i, c_k) \prod_{j \in I_{\mathbf{c}^-}} (1 - p_j)$$

$$b = 1 - a$$

The cascading representation of the noisy-OR model performs, during inference, the computation of the probabilities of the effect e given possible combinations of causes. That is, application of the definitional rule is in essence left to inference, resulting in the dependency of the output probability of interest on the noisy-OR parameter now being correctly taken into consideration. This observation further demonstrates that, when changing a single parameter of the noisy-OR model specification upon fine-tuning a Bayesian network, in the cascading representation just a single cell of the conditional probability table for the appropriate effect variable E_i needs to be adapted; in contrast, in the representation with a full conditional probability table, various cells that are specified using the model's definitional rule will need adaptation. The cascading representation is therefore easier to adapt without the risk of violating the properties of the underlying causal interaction model.

5   Conclusions and Further Research

In this paper we revisited part of the large volume of work on causal interaction models, and focused thereby on the representational complexity of such models. We built on this early work for the purpose of demonstrating that some of these models allow for a representation with various elegant properties that have not been recognised until now. More specifically, by exploiting the property of self-decomposability of the deterministic function underlying a causal interaction model, we arrived at an alternative cascading representation that has a clear intuitive semantics in terms of the causal mechanism itself, not requiring the inclusion of artificial unobservable variables. In addition to well-known complexity benefits of such alternative representations, this specific cascading representation has important knowledge engineering benefits, allowing easier maintenance and more robust fine-tuning of parameters. As the compactness of the cascading representation can be exploited directly by standard inference algorithms moreover, we conclude all in all that this representation of causal interaction models is quite suitable for explicit embedding in Bayesian networks.

While we used the (leaky) noisy-OR model for our example causal interaction model throughout the paper, the presented properties of the cascading representation apply straightforwardly to any interaction model involving binary-valued variables and having an underlying self-decomposable deterministic function, such as the (leaky) noisy-AND model. For our further research we aim at extending our results to causal interaction models involving multi-valued variables, such as the noisy-MAX model [5], and to other types of decomposable function.

References

1. Castillo, E., Gutiérrez, J.M., Hadi, A.S.: Sensitivity analysis in discrete Bayesian networks. IEEE Trans. Syst. Man Cybern. **27**, 412–423 (1997)
2. Chan, H., Darwiche, A.: Sensitivity analysis in Bayesian networks: From single to multiple parameters. In: Halpern, J., Meek, C. (eds.) Proceedings of the 20th Conference on Uncertainty in Artificial Intelligence, pp. 67–75 (2004)

3. Coupé, V.M.H., van der Gaag, L.C.: Properties of sensitivity analysis of Bayesian belief networks. Ann. Math. Artif. Intell. **36**, 323–356 (2002)
4. Díez, F.J., Druzdzel, M.J.: Canonical Probabilistic Models for Knowledge Engineering. Technical Report CISIAD-06-01 (2007)
5. Díez, F.J., Galán, S.F.: Efficient computation for the noisy max. Int. J. Intell. Syst. **18**, 165–177 (2003)
6. Frey, B.J., Patrascu, R., Jaakkola, T., Moran, J.: Sequentially fitting inclusive trees for inference in noisy- or networks. In: Leen, T.K., Dietterich, T.G., Tresp, V. (eds.) Advances in Neural Information Processing Systems 13, pp. 493–499. MIT Press, Cambridge (2001)
7. Getoor, L.: Learning Statistical Models from Relational Data. PhD Thesis. Stanford University (2001)
8. Heckerman, D.: A tractable inference algorithm for diagnosing multiple diseases. In: Henrion, M., Kanal, L., Lemmer, J., Shachter, R. (eds.) Proceedings of the 5th Conference on Uncertainty in Artificial Intelligence, pp. 163–172 (1989)
9. Heckerman, D.: Causal independence for knowledge acquisition and inference. In: Heckerman, D., Mamdani, E. (eds.) Proceedings of the 9th Conference on Uncertainty in Artificial Intelligence, pp. 122–127 (1993)
10. Heckerman, D., Breese, J.: Causal independence for probability assessment and inference using Bayesian networks. IEEE Trans. Syst. Man Cybern. **26**, 826–831 (1996)
11. Henrion, M.: Some practical issues in constructing belief networks. In: Kanal, L.N., Levitt, T.S., Lemmer, J.F. (eds.) Uncertainty in Artificial Intelligence 3, pp. 161–173. Elsevier (1989)
12. Huang, K., Henrion, M.: Efficient search-based inference for noisy- or belief networks: TopEpsilon. In: Horvitz, E., Jensen, F. (eds.) Proceedings of the 12th Conference on Uncertainty in Artificial Intelligence, pp. 325–331 (1996)
13. Jesus, P., Baquero, C., Almeida, P.S.: A survey of distributed data aggregation algorithms. IEEE Commun. Surv. Tutorials **17**, 381–404 (2011)
14. Koller, D., Friedman, N.: Probabilistic Graphical Models: Principles and Techniques. The MIT Press, Cambridge (2009)
15. Li, W., Poupart, P., van Beek, P.: Exploiting structure in weighted model counting approaches to probabilistic inference. J. Artif. Intell. Res. **40**, 729–765 (2011)
16. Olesen, K.G., et al.: A MUNIN network for the median nerve: a case study on loops. Appl. Artif. Intell. **3**, 385–403 (1989)
17. Pearl, J.: Probabilistic Reasoning in Intelligent Systems: Networks of Plausible Inference. Morgan Kaufmann, Burlington (1988)
18. del Sagrado, J., Salmerón, A.: Representing canonical models as probability trees. In: Conejo, R., Urretavizcaya, M., Pérez-de-la-Cruz, J.-L. (eds.) CAEPIA/TTIA -2003. LNCS (LNAI), vol. 3040, pp. 478–487. Springer, Heidelberg (2004). https://doi.org/10.1007/978-3-540-25945-9_47
19. Zhang, N.L., Yan, L.: Independence of causal influence and clique tree propagation. Int. J. Approximate Reasoning **19**, 335–349 (1998)

Computational Models for Cumulative Prospect Theory: Application to the Knapsack Problem Under Risk

Hugo Martin and Patrice Perny[✉]

Sorbonne Université, CNRS, LIP6, 75005 Paris, France
{hugo.martin,patrice.perny}@lip6.fr

Abstract. Cumulative Prospect Theory (CPT) is a well known model introduced by Kahneman and Tversky in the context of decision making under risk to overcome some descriptive limitations of Expected Utility. In particular CPT makes it possible to account for the framing effect (outcomes are assessed positively or negatively relatively to a reference point) and the fact that people often exhibit different risk attitudes towards gains and losses. We study here computational aspects related to the implementation of CPT for decision making in combinatorial domains. More precisely, we consider the Knapsack Problem under Risk that consists of selecting the "best" subset of alternatives (investments, projects, candidates) subject to a budget constraint. The alternatives' outcomes may be positive or negative (gains or losses) and are uncertain due to the existence of several possible scenarios of known probability. Preferences over admissible subsets are based on the CPT model and we want to determine the CPT-optimal subset for a risk-averse Decision Maker (DM). The problem requires to optimize a non-linear function over a combinatorial domain. In the paper we introduce two distinct computational models based on mixed-integer linear programming to solve the problem. These models are implemented and tested on randomly generated instances of different sizes to show the practical efficiency of the proposed approach.

Keywords: Cumulative Prospect Theory · Knapsack Problem · Risk aversion · Mixed-integer linear programming

1 Introduction

The increasing use of intelligent systems to support human decision-making or to drive the actions of autonomous artificial agents shows the importance of developing expressive and adaptable models to support decision making activities in complex environments. One of the major challenges is to improve our understanding and control over AI-based decisions, and also their relevance, fairness, and alignment with the organisation's values and risk proneness. In the field of decision under risk, the main problem to overcome is to compare alternatives the outcomes of which are known in probabilities, and to provide a control of risk in the selection of optimal actions.

© Springer Nature Switzerland AG 2019
N. Ben Amor et al. (Eds.): SUM 2019, LNAI 11940, pp. 52–65, 2019.
https://doi.org/10.1007/978-3-030-35514-2_5

Various mathematical models have been developed in Economics to account from observed human behaviors in decision making under risk, since the seminal works of von Neumann and Morgenstern [24] and Savage [19] on the foundations of Expected Utility Theory (EU). Despite the intuitive appeal of EU theory, several experiments have shown that sophisticated rational human behaviors are not always explainable by EU theory. In particular the experiments conducted by Kahneman and Tversky [7] have shown that violations of the Von Neumann and Morgenstern independence axiom or violations of Savage's Sure Thing Principle are frequently observed, making it impossible to explain or simulate the observed behaviors using EU. This has led to alternative models, relying on a deformation of cumulative probabilities allowing to account for violations of the above mentionned independence axioms. For example, Yaari [25] proposed a dual model to EU, based on a weighting function transforming probabilities rather than a utility function transforming payoffs. A second example is Rank-dependent Utility Theory (RDU) where both transformations (probabilities and payoff) co-exist, thus providing a more general model including EU and Yaari as special cases. Although these models provide more flexibility to model preferences and decisions, they are more complex to handle for optimization purposes due to their non-linearity (w.r.t probabilities and/or payoffs) and their parameters are more complex to elicit. This issue has been considered in AI, in various topics such as sequential decision making [5,6], state space search under risk [14], and incremental preference elicitation [4,15].

Another aspect that is worth considering is that, in the field of decision under risk, decision makers tend to think of outcomes relative to a certain reference point (often the status quo). They care generally more about negative outcomes (i.e. outcomes below the reference point) than positive ones (i.e. outcomes above the reference point) and may exhibit different attitudes towards gains and losses. This observation has motivated the development of Prospect Theory [7] and Cumulative Prospect Theory (CPT) [23] that provide decision models able to account for this phenomenon. CPT theory includes a sophistication where the overall utility of a risky prospect is decomposed as the difference between an aggregate of utilities of positive outcomes and an aggregate of utilities of negative outcomes. The aggregation operation used for the positive side can be different from the one used for the negative side, thus letting the possibility to describe more sophisticated behaviors. Although the theory is well established, the use of such models for optimization tasks under risk received less attention.

The aim of this paper is to contribute to fill the gap by proposing computational models based on CPT for the effective computation of CPT-optimal solutions on combinatorial domains. For the sake of illustration we will consider the problem of selecting projects under a budget constraint and under risk (knapsack problem with multiple scenarios).

The paper is organized as follows: In Sect. 2, we briefly survey some related work. Then, in Sect. 3, we recall some background on CPT and some important results on modeling strong risk-aversion in CPT. In Sect. 4 we propose a first linearization for the CPT model, relying on the notion of core of a capacity.

This leads us to propose a MIP formulation for the Knapsack problem under risk. This model is tested on families of instances of different sizes. In Sect. 5 we consider a special case where the probability weighting functions used in CPT are piecewise linear with a bounded number of pieces. Under this assumption, we propose another MIP formulation, more compact and easier to solve, for the same problem.

2 Related Work

CPT was already used in AI, e.g., for developing a risk sensitive reinforcement learning in a traffic signal control application [16]. CPT has also been used in a number of decision support applications. For example, an application of CPT for the multi-objective optimization of a bus network is proposed in [9]. However, in this case study, the set of alternatives is explicitly defined and does not require optimization techniques.

The Knapsack Problem (KP) under consideration in this paper consists in selecting a subset of items under a budget constraint. This problem has some links with the portfolio selection problem that can be seen as the continuous relaxation of KP under risk. The application of CPT to portfolio selection and insurance demand have been studied in finance (see e.g. [3]) with a computational model solvable under some specific assumptions (S-Shaped functions, risk free reference point and/or linear utility functions). Beside CPT, several LP-computational measures of dispersion are introduced to control the risk attached to portfolios: let us mention the mean absolute deviation, the Gini's mean difference (GMD) as basic LP computable risk measures, the worst realization (Minimax) and the Conditional Value-at-Risk (CVaR) as basic LP computable safety measures [10,11]. Moreover, in the latter reference, computational issues related to the solution of portfolio models with integrity constraints are investigated and a matheuristic called Kernel Search is proposed. These contributions do not consider the use of bipolar valuation scales as in CPT.

In multicriteria analysis there is also an increasing interest for modeling different attitudes in the aggregation depending on whether evaluations are on the positive or negative side. For example, the Choquet integral has been extended to the bipolar case in [2,8] but optimization aspects attached to general bipolar Choquet integral have not been investigated. Very recently, some LP-solvable models have been proposed [12] for a subclass of bipolar Choquet integrals named biOWA (for bipolar ordered weighted average). However, biOWA are symmetric functions of their argument and do not allow to account for decision under risk when scenarios have different probabilities. Finally an LP-solvable model was proposed for a weighted extension of OWA operators [13] but does not consider the case of bipolar scales. In this paper, we are going to introduce computational models solvable by mixed-integer linear programming to determine CPT-optimal solutions in implicit decision spaces.

3 CPT and Strong Risk Aversion

Let us consider a problem of decision making under risk with a finite set of states of nature $N = \{s_1, \ldots, s_n\}$. The states represent possible scenarios under consideration, impacting differently the outcomes of the alternatives. Let p_i denote the probability of state s_i. Any feasible alternative is seen as an act in the sense of Savage. It is therefore characterized by a vector $x = (x_1, \ldots, x_n)$ where $x_i \in \mathbb{R}$ denotes the outcome of x in state s_i. In this context, the Rank-Dependent Utility (RDU) model introduced in [17] is defined as follows:

Definition 1. *Let $x \in \mathbb{R}^n$ be the outcome vector of an alternative, the RDU model is defined by the following rank-dependent expected value:*

$$f_\varphi^u(x) = \sum_{i=1}^{n} \left[\varphi\left(\sum_{k=i}^{n} p_{(k)}\right) - \varphi\left(\sum_{k=i+1}^{n} p_{(k)}\right) \right] u(x_{(i)}) \tag{1}$$

$$= \sum_{i=1}^{n} \left[u(x_{(i)}) - u(x_{(i-1)}) \right] \varphi\left(\sum_{k=i}^{n} p_{(k)}\right) \tag{2}$$

where $\varphi : [0,1] \to [0,1]$ is a non-decreasing probability weighting function, $u : \mathbb{R} \to \mathbb{R}$ is a non-decreasing real-valued utility function, and $(.)$ is a permutation defined on N and such that $x_{(1)} \le x_{(2)} \le \ldots \le x_{(n)}$.

Example 1. We consider three different scenarios $s = (s_1, s_2, s_3)$ of probability $p = (\frac{1}{2}, \frac{1}{3}, \frac{1}{6})$ and we want to select the best solution in the set of alternatives composed of $x = (9, 4, 1)$, $y = (4, 4, 4)$ and $z = (1, 16, 1)$. We assume that the preferences of the DM can be represented by RDU with $\varphi(p) = p^2$ and $u(x) = \sqrt{(x)}$. We have the following RDU value for the three alternatives:

- $f_\varphi^u(x) = 1 + (u(4) - u(1)) \times \varphi(\frac{5}{6}) + (u(9) - u(4)) \times \varphi(\frac{1}{2}) = 1 + \frac{25}{36} + \frac{1}{4} = \frac{70}{36}$
- $f_\varphi^u(y) = u(4) + (u(4) - u(4)) \times \varphi(\frac{5}{6}) + (u(4) - u(4)) \times \varphi(\frac{1}{2}) = u(4) + 0 = 2$
- $f_\varphi^u(z) = u(1) + (u(1) - u(1)) \times \varphi(\frac{1}{2}) + (u(16) - u(1)) \times \varphi(\frac{1}{3}) = 1 + 3 \times \frac{1}{9} = \frac{4}{3}$

Thus, we have the following ranking of alternatives $y \succ x \succ z$ where $\succ$ is the preference relation induced by f_φ^u.

This model clearly generalizes the Expected Utility model that can be obtained for $\varphi(p) = p$ for all $p \in [0,1]$. Moreover it also includes the dual model of EU known as Yaari's model [25] as special case (when u is linear). Nonetheless, this model is not always sufficient to account for decision behaviors observed when decision makers think of outcomes relative to a certain reference point. The utility scale is treated as an interval scale and preferences are not impacted by positive affine transformations. Thus, 0 has no specific status in the valuation scale, nor any other constant. This may prevent to account for some sophisticated decision behaviors as illustrated in the following:

Example 2. We look for an optimal path from a source node to a sink node in a network represented by a directed graph. The arcs of the graph are endowed with vectors representing the algebraic payoff attached to the arc (which can represent a gain or a loss) under two possible scenarios of *equal probability*. For example, the valuation $(-2, 3)$ means that the outcome will be a loss of 2 in scenario 1 and a gain of 3 in scenario 2. Outcomes are assumed to be additive along a path and we assume that $u(z) = z$. This problem can represent several situations (e.g., a path planning problem or investment planning problem, both under uncertainty). Let us consider two different instances of this problem, characterized by two different graphs with nodes $\{s, a, b, t\}$ and $\{s', c, d, t'\}$ respectively. The graphs are presented below (Fig. 1).

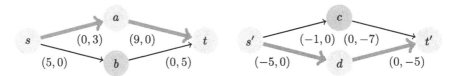

Fig. 1. Graphs considered in Example 2

On the left handside, the upper and lower s-t-paths have utilities $(9, 3)$ and $(5, 5)$ respectively. We assume here that the DM prefers the former path because she maximizes the expected outcome when all evaluations are positive. In the instance given on the right handside, the upper and lower s'-t'-paths respectively have utilities $(-1, -7)$ and $(-5, -5)$. Here the DM may exhibit a more cautious attitude towards risk due to the presence of negative outcomes. Let us assume that she prefers the latter solution due to the fact that the outcome in the worst case scenario is better. Hence, to model these preferences with RDU we must fulfill the following constraints: $f_\varphi(9, 3) > f_\varphi(5, 5)$ and $f_\varphi(-7, -1) < f_\varphi(-5, -5)$. The former inequality implies that $3 + \varphi(\frac{1}{2}) \times (9 - 3) > 5$ and therefore $\varphi(\frac{1}{2}) > \frac{1}{3}$. Moreover the latter inequality implies $-7 + \varphi(\frac{1}{2}) \times (-1 + 7) < -5$ and therefore $\varphi(\frac{1}{2}) < \frac{1}{3}$ which yields a contradiction. Hence RDU is not able to represent the observed preferences.

To overcome the descriptive limitations illustrated in the above example, we consider now the Cumulative Prospect Theory model (CPT for short), first introduced in [7].

Definition 2. *Let $x \in \mathbb{R}^n$ be the outcome vector such that $x_{(1)} \leq \cdots \leq x_{(j-1)} < 0 \leq x_{(j)} \leq \cdots \leq x_{(n)}$ with $j \in \{0, \ldots, n\}$, the Cumulative Prospect Theory is characterized by the following evaluation function:*

$$g^u_{\varphi, \psi}(x) = \sum_{i=1}^{n} w_i u(x_i) \quad \text{with} \quad w_i = \begin{cases} \varphi(\sum_{k=i}^{n} p_{(k)}) - \varphi(\sum_{k=i+1}^{n} p_{(k)}) & \text{if } (i) \geq (j) \\ \psi(\sum_{k=1}^{i} p_{(k)}) - \psi(\sum_{k=1}^{i-1} p_{(k)}) & \text{if } (i) < (j) \end{cases}$$

$$(3)$$

where φ and ψ are two real-valued increasing functions from $[0,1]$ to $[0,1]$ that assign 0 to 0 and 1 to 1, and u is a continuous and increasing real-valued utility function such that $u(0) = 0$ (hence $u(x)$ and x have the same sign).

It can easily be checked that whenever $\varphi(p) = 1 - \psi(1 - p)$ for all $p \in [0,1]$ (duality) then CPT boils down to RDU. The use of non-dual probability weighting functions φ and ψ depending on the sign of the outcomes under consideration enables to model shifts of behavior relatively to the reference point (here 0). Let us come back to Example 2 under the assumption that $u(z) = z$ for all $z \in \mathbb{R}$, we have: $g_{\varphi,\psi}(9,3) = [\varphi(1) - \varphi(\frac{1}{2})]3 + [\varphi(\frac{1}{2}) - \varphi(0)]9 = 3 + 6\varphi(\frac{1}{2})$ since $\varphi(0) = 0$ and $\varphi(1) = 1$. Similarly $g_{\varphi,\psi}(5,5) = [\varphi(1) - \varphi(\frac{1}{2})]5 + [\varphi(\frac{1}{2}) - \varphi(0)]5 = 5$. Hence $g_{\varphi,\psi}(9,3) > g_{\varphi,\psi}(5,5)$ implies $\varphi(\frac{1}{2}) > \frac{1}{3}$ (*).

On the other hand we have $g_{\varphi,\psi}(-7,-1) = [\psi(\frac{1}{2}) - \psi(0)](-7) + [\psi(1) - \psi(\frac{1}{2})](-1) = -1 - 6\psi(\frac{1}{2})$ since $\psi(0) = 0$ and $\psi(1) = 1$. Similarly $g_{\varphi,\psi}(-5,-5) = -5$. Hence $g_{\varphi,\psi}(-7,-1) < g_{\varphi,\psi}(-5,-5)$ implies $\psi(\frac{1}{2}) > \frac{2}{3}$, which does not yield any contradiction. Thus, the DM's preferences can be modeled with $g_{\varphi,\psi}$.

As CPT boils down to RDU when $\varphi(p) = 1 - \psi(1 - p)$ for all $p \in [0,1]$ it is interesting to note that under this additional constraint $\psi(\frac{1}{2}) > \frac{2}{3}$ implies $\varphi(\frac{1}{2}) < \frac{1}{3}$ which is incompatible with the constraint denoted (*) above, derived from $g_{\varphi,\psi}(9,3) > g_{\varphi,\psi}(5,5)$. This again illustrates the fact that RDU is not able to describe such preferences.

Strong Risk Aversion in CPT. In many situations decision makers are risk-averse. It is therefore useful to further specify CPT for risk-averse agents. We consider here strong risk-aversion that is standardly defined from second-order stochastic dominance. For any random variable X, let G_X be the tail distribution defined by $G_X(x) = P(X > x)$, with P a probability function. Let X, Y be two random variables, X stochastically dominates Y at the second order if and only if for all $x \in X$, $\int_{-\infty}^{x} G_X(t)dt \geq \int_{-\infty}^{x} G_Y(t)dt$. From this dominance relation, the concept of mean-preserving spread standardly used to define risk aversion can be introduced as follows. Y is said to derive from X using a mean preserving spread if and only if $E(X) = E(Y)$ and X stochastically dominates Y at the second order. We have then the following definition of strong risk aversion [18]:

Definition 3. *Let $\succsim$ be a preference relation. Strong risk aversion holds for $\succsim$ if and only if $X \succsim Y$ for all X and Y such that Y derives from X using a mean preserving spread.*

We recall now the set of conditions that CPT must fulfill to model strong risk aversion. These conditions were first established in [21].

Theorem 1. *Strong risk aversion holds in CPT if and only if φ is convex, ψ is concave, u is concave for losses and also concave for gains, and the following equation is satisfied:*

$$\left[u(x) - u(x - \frac{\delta}{q})\right]\left(\psi(q + s) - \psi(s)\right) \geq \left[u(y + \frac{\delta}{q}) - u(y)\right]\left(\varphi(p + r) - \varphi(r)\right) \quad (4)$$

for all $x \geq 0 \geq y$ and p, q, r, s such as $p + q + r + s \leq 1$, $p, q > 0$ and $r, s \geq 0$.

We remark that, when $u(z) = z$ for all z, condition (4) can be rewritten in the following simpler form: $\frac{\psi(q+s)-\psi(s)}{q} \geq \frac{\varphi(p+r)-\varphi(r)}{p}$ for all p, q, r, s such as $p + q + r + s \leq 1$, $p, q > 0$ and $r, s \geq 0$. In terms of derivative, this means that $\psi'(s) \geq \varphi'(r)$ for all $r, s \geq 0$ such that $r + s \leq 1$.

The above characterization of admissible forms of CPT for a risk-averse decision maker will be used in the next section to propose computational models for the determination of CPT-optimal solutions on implicit sets. We conclude the present section by making explicit a link between CPT and RDU model.

Linking RDU and CPT. Interestingly, CPT can be expressed as a difference of two RDU values respectively applied to the positive and negative part of the outcome vector x, using the two distinct probability weighting functions φ and ψ. This reformulation is well known in the literature on rank-dependent aggregation functions (see e.g., [2]) and reads as follows:

$$g_{\varphi,\psi}^{u}(x) = f_{\varphi}^{u^{+}}(x^{+}) - f_{\psi}^{u^{-}}(x^{-}) \tag{5}$$

where $x^{+} = \max(x,0)$, $x^{-} = \max(-x,0)$, $u^{+}(z) = u(z)$ if $z \geq 0$ and 0 otherwise, $u^{-}(-z) = -u(z)$ if $z \leq 0$ and 0 otherwise. This formulation will be useful in the next sections to propose linear reformulations of the CPT model.

The next sections are dedicated to the effective computation of CPT-optimal solutions on an implicit set of alternatives using linear programming techniques.

4 A First Linearization for CPT Optimization

We present here a first mixed-integer program to maximize function $g_{\varphi,\psi}^{u}(x)$ under linear admissibility constraints for a risk-averse agent. By Theorem 1, we know that φ must be convex and ψ must be concave to model risk aversion. These properties will be useful to establish a linearization of the CPT model. For the simplicity of presentation, we will also assume that $u(x) = x$ and notations like f_{φ}^{u} and $g_{\varphi,\psi}^{u}$ will be simplified into f_{φ} and $g_{\varphi,\psi}$. We will briefly explain later how the proposed approach can be extended to the case of a piecewise linear utility u. Let us first recall some notions linked to capacities and related concepts.

Capacities are set functions that are well known in decision theory for their ability to describe non-additive representations of beliefs or importance in decision models. Let us recall the following:

Definition 4. *A set function* $v : \mathcal{P}(N) \rightarrow [0,1]$ *is said to be a capacity if it verifies:* $v(\emptyset) = 0$ *and for all* $A, B \subseteq N, A \subseteq B \Rightarrow v(A) \leq v(B)$. *It is a normalized capacity if* $v(N) = 1$.

Among all existing capacities, some are of particular interest. In particular, a capacity v is said to be:

– convex if $v(A \cup B) + v(A \cap B) \geq v(A) + v(B)$ $\forall A, B \subseteq N$
– additive if $v(A \cup B) + v(A \cap B) = v(A) + v(B)$ $\forall A, B \subseteq N$

When v is an additive capacity it can be simply characterized by a vector $(v_1, \ldots, v_n)$ of non-negative weights such that $v(S) = \sum_{i \in S} v_i$ for all $S \subseteq N$. In the sequel we will indifferently use the same notation v for the capacity and for the weighting vector characterizing the capacity.

Let P be any probability measure on 2^N (N being the set of scenarios) and φ any probability weighting function (continuous, non-decreasing and such that $\varphi(0) = 0$ and $\varphi(1) = 1$), then the set function defined by $v(S) = (\varphi \circ P)(S) = \varphi(\sum_{i \in S} p_i)$ is a capacity. It is well known that v is convex if and only if φ is convex [1]. When v is convex, a useful property is that there exists an additive measure $\lambda(S)$ that dominates function v [22]. The set of all additive capacities dominating v is known as *the core* of v, formally defined as follows:

Definition 5. *The core of a capacity v is the set of all additive capacities dominating v, defined by $core(v) = \{\lambda : 2^N \to [0,1]$ additive $\mid \lambda(S) \geq v(S) \; \forall S \subseteq N\}$.*

Hence when φ is convex, $v = \varphi \circ P$ has a non empty core and $v(S) = \min_{\lambda \in core(v)}(\lambda(S))$. In this case, a useful result due to Schmeidler [20] that holds for general Choquet integrals used with a convex capacity implies that they can be rewritten as the minimum of a set of linear aggregation functions. When applied to $f_\varphi(x)$ (which is an instance of the Choquet integral) the result writes as follows:

Proposition 1. *If φ is convex we have* $f_\varphi(x) = \min\limits_{\lambda \in core(\varphi \circ P)} \lambda.x$

where f_φ is the Yaari's model obtained from f_φ^u when $u(z) = z$ for all z. Similarly, for a concave weighting function ψ the dual defined by $\bar{\psi}(p) = 1 - \psi(1 - p)$ for all $p \in [0, 1]$ is convex and has a non-empty core. Hence Proposition 1 can be used again to establish the following result:

Proposition 2. *If ψ is concave we have* $f_\psi(x) = \max\limits_{\lambda \in core(\bar{\psi} \circ P)} \lambda.x$

Proof. $f_\psi(x) = -f_{\bar{\psi}}(-x) = -\min\limits_{\lambda \in core(\bar{\psi} \circ P)} \lambda.(-x) = \max\limits_{\lambda \in core(\bar{\psi} \circ P)} \lambda.x.$

Using Propositions 1 and 2 and Eq. (5) we obtain a new formulation of CPT, when φ and ψ are convex and concave respectively.

Proposition 3. *Let $x \in \mathbb{R}^n$. If φ is convex and ψ is concave then we have:*

$$g_{\varphi,\psi}(x) = \min\limits_{\lambda \in core(\varphi \circ P)} \lambda \cdot x^+ - \max\limits_{\lambda \in core(\bar{\psi} \circ P)} \lambda \cdot x^-$$

Now, let us show that this new formulation can be used to optimize $g_{\varphi,\psi}(x)$ using linear programming. From Propositions 1 and 2 the values of $f_\varphi(x)$ and $f_\psi(x)$ for any outcome vector $x \in \mathbb{R}^n$ can be obtained as the solutions of the two following linear programs respectively:

$$\min \sum_{i=1}^{n} \lambda_i x_i$$
$$\varphi(P(A)) \leq \sum_{i \in A} \lambda_i \; \forall A \subseteq N$$
$$\lambda_i \geq 0, \; i = 1, .., n$$

$$\max \sum_{i=1}^{n} \lambda_i x_i$$
$$\psi(P(A)) \geq \sum_{i \in A} \lambda_i \; \forall A \subseteq N$$
$$\lambda_i \geq 0, \; i = 1, .., n$$

The left LP given above directly derives from Proposition 1. The right LP given above derives from Proposition 2 after observing that the constraints $\forall B \subseteq N, \sum_{i \in B} \lambda_i \geq \bar{\psi}(P(B))$ are equivalent to $\forall A \subseteq N, \sum_{i \in A} \lambda_i \leq \psi(P(A))$ (by setting $A = N \setminus B$). Now, if we consider x as a variable vector, we consider the dual formulations of the above LPs to get rid of the quadratic terms:

$$\max \sum_{A \subseteq N} \varphi(P(A)) \times d_A \qquad\qquad \min \sum_{A \subseteq N} \psi(P(A)) \times d_A$$

$$\sum_{A \subseteq N: i \in A} d_A \leq x_i \quad i = 1, .., n \qquad\qquad \sum_{A \subseteq N: i \in A} d_A \geq x_i \quad i = 1, .., n$$

$$d_A \geq 0 \; \forall A \subseteq N \qquad\qquad d_A \geq 0 \; \forall A \subseteq N$$

Finally, we obtain program $\mathcal{P}_1$ given below to optimize $g_{\varphi,\psi}$, with the assumptions that φ is convex, ψ is concave and that $u(x) = x$ for all $x \in \mathbb{R}^n$.

$$\max \sum_{A \subseteq N} \varphi(P(A)) \times d_A^+ - \sum_{A \subseteq N} \psi(P(A)) \times d_A^-$$

$$(\mathcal{P}_1) \begin{cases} \displaystyle\sum_{A \subseteq N: i \in A} d_A^+ \leq x_i^+ & i = 1, \ldots, n \\[2mm] \displaystyle\sum_{A \subseteq N: i \in A} d_A^- \geq x_i^- & i = 1, \ldots, n \\[2mm] x_i = x_i^+ - x_i^- & i = 1, \ldots, n \\ 0 \leq x_i^+ \leq z_i \times M & i = 1, \ldots, n \\ 0 \leq x_i^- \leq (1 - z_i) \times M & i = 1, \ldots, n \\ x \in X \end{cases}$$

$$x_i^-, x_i^+, d_A^+, d_A^- \geq 0 \; i = 1, .., n, \; \forall A \subseteq N$$
$$z_i \in \{0, 1\} \; i = 1, \ldots, n$$

The integer variables $z_i, i = 1, \ldots, n$ are used to decide whether x_i is positive or not. The M constant is used as usual to model disjunctive constraints depending on the sign of x_i. $\mathcal{P}_1$ has 2^{n+1} continuous variables, n binary variables and $5n$ constraints. It can be specialized to solve any CPT-optimization problem, by inserting the needed variables and constraints to define the set X. For example, to solve the knapsack problem under risk, we have to insert m boolean variables y_j (set to 1 iff object j is selected) subject to the constraint $\sum_{j=1}^{m} w_j y_j \leq C$, for weights $w_j, j = 1, \ldots, m$ and the knapsack capacity C. Then variables x_i are linked to variables y_j by equations of type $x_i = \sum_{j=1}^{m} u_{ij} y_j$ defining x_i as a linear utility over sets of objects for any scenario $i \in \{1, \ldots, n\}$.

We implemented the above model using the Gurobi 7.5.2 solver on a computer with 12 GB of RAM, a Intel(R) Core(TM) i7 CPU 950 @ 3.07 GHz processor. Table 2 gives the results obtained for the CPT-knapsack problem modeled as follows: m represents the number of objects, n the number of voters; utilities u_{ij} and weights w_j were randomly generated in the range $[\![-10, 10]\!]$ (resp. $[\![-100, 100]\!]$), the capacity is set to $C = (\sum_{j=1}^{m} w_j)/2$, φ and ψ are randomly drawn to satisfy the conditions of Proposition 1. Average times given in Table 2 are computed over 20 runs, with a timeout set to 1200 s. We observe that this computational model is able to solve instances with a large number of objects in a few seconds. Nonetheless, it has an exponential number of continuous variables, which may limit its applicability when the number of scenarios becomes larger. To over-

Table 1. Times (s) obtained by MIP $\mathcal{P}_1$ for the CPT-knapsack

m	$n = 3$	$n = 5$	$n = 7$
100	0.03	0.21	0.67
500	0.05	1.31	45.60
750	0.08	0.87	125.72
1000	0.13	3.28	150.48

come this limitation, we will know present a second computational model with a polynomial number of variables and constraints, which optimizes $g_{\varphi,\psi}(x)$ under some additional assumptions concerning φ and ψ (Table 1).

5 The Case of Piecewise Linear Weighting Functions

From now on, we assume that φ and ψ are piecewise-linear functions with respectively the breakpoints $0 = \alpha_0 \leq \alpha_1 \leq \alpha_2 \leq \ldots \leq \alpha_t = 1$ and $0 = \beta_0 \leq \beta_1 \leq \beta_2 \leq \ldots \leq \beta_t = 1$. This assumption is often made in different contexts of elicitation and optimization. For example, Ogryczack [13] uses a similar assumption to propose an efficient linearization of the WOWA operator. We will follow a similar idea to propose a linearization for CPT.

A piecewise-linear function has its derivative constant on each interval. Thus we define $\varphi'(u) = d_i^+$ for all $u \in [\alpha_{i-1}, \alpha_i]$ and $\psi'(u) = d_i^-$ for all $u \in [\beta_{i-1}, \beta_i]$. Moreover we assume that $d_{t+1}^+ = 0$ and $d_{t+1}^- = 0$ for convenience. For any given solution x, we define the cumulative function F_x, for all $\alpha \in [0, 1]$, by:

$$F_x(\alpha) = \sum_{i=1}^{n} p_i \delta_i(\alpha) \text{ with } \delta_i(\alpha) = \begin{cases} 1 \text{ if } x_i \leq \alpha \\ 0 \text{ otherwise} \end{cases}$$

Then we have $F_x^{(-1)}(u) = \inf\{y : F_x(y) \geq u\}$ returns the minimum performance y such that the probability of scenarios whose performance is lower than or equal to y is greater than or equal to u. Then, we define the tail function G_x, for all $\alpha \in [0, 1]$, by:

$$G_x(\alpha) = \sum_{i=1}^{n} p_i \delta_i(\alpha) \text{ with } \delta_i(\alpha) = \begin{cases} 1 \text{ if } x_i > \alpha \\ 0 \text{ otherwise} \end{cases}$$

and $G_x^{(-1)}(u) = \inf\{y : G_x(y) \leq u\}$ returns the minimum performance y such that the probability of scenarios whose performance level is greater than y is lower than or equal to u. First, we observe that the following relation holds between $G_x^{(-1)}$ and $F_x^{(-1)}$.

Proposition 4. *For all $x \in \mathbb{R}^n$ and $u \in [0, 1]$, $G_x^{(-1)}(u) = F_x^{(-1)}(1 - u)$*

Proof. According to the definition of F and G, we have $G_x(u) = 1 - F_x(u)$. We have then the following result $F_x^{(-1)}(1-u) = \inf\{y : F_x(y) \geq 1-u\} = \inf\{y : 1 - F_x(y) \leq u\} = \inf\{y : G_x(y) \leq u\} = G_x^{(-1)}(u)$ □

Then, let us show that these notions allow a new formulation of $g_{\varphi,\psi}$:

Proposition 5

$$g_{\varphi,\psi}(x) = \sum_{i=1}^{t} \left[(d_{i+1}^+ - d_i^+) \int_0^{1-\alpha_i} F_x^{(-1)}(v)dv - (d_i^- - d_{i+1}^-) \int_0^{\beta_i} G_x^{(-1)}(v)dv \right] \quad (6)$$

Proof. Let () be a permutation of scenarios such that $x_{(1)} \leq x_{(2)} \leq \cdots \leq x_{(n)}$ and $\pi_i = \sum_{k=i}^n P_{(k)}$. Let $E(x) = \int_0^1 \left(G_{x+}^{(-1)}(u)\varphi'(u) - G_{x-}^{(-1)}(u)\psi'(u) \right) du$. First, we show that $E(x) = g_{\varphi,\psi}(x)$.

$$E(x) = \int_0^1 \left(G_{x+}^{(-1)}(u)\varphi'(u) - G_{x-}^{(-1)}(u)\psi'(u) \right) du$$

$$= \sum_{i=1}^n \int_{\pi_{i+1}}^{\pi_i} G_{x+}^{(-1)}(u)\varphi'(u)du - \sum_{i=1}^n \int_{\pi_{i+1}}^{\pi_i} G_{x-}^{(-1)}(u)\psi'(u)du$$

with $\pi_{n+1} = 0$. We notice that $G_{x+}^{(-1)}(u) = x_{(i)}^+$ for all $u \in [\pi_{i+1}, \pi_i]$. We have:

$$= \sum_{i=1}^n x_{(i)}^+ \int_{\pi_{i+1}}^{\pi_i} \varphi'(u)du - \sum_{i=1}^n x_{(i)}^- \int_{\pi_{i+1}}^{\pi_i} \psi'(u)du$$

$$= \sum_{i=1}^n x_{(i)}^+ \left(\varphi(\sum_{k=i}^n P_{(k)}) - \varphi(\sum_{k=i+1}^n P_{(k)}) \right) - \sum_{i=1}^n x_{(i)}^- \left(\psi(\sum_{k=i}^n P_{(k)}) - \psi(\sum_{k=i+1}^n P_{(k)}) \right)$$

$$= g_{\varphi,\psi}(x)$$

Then, the desired result can be obtained from another formulation of $E(X)$:

$$E(x) = \int_0^1 \left(G_{x+}^{(-1)}(u)\varphi'(u) - G_{x-}^{(-1)}(u)\psi'(u) \right) du$$

$$= \sum_{i=1}^t \int_{\alpha_{i-1}}^{\alpha_i} G_{x+}^{(-1)}(u)\varphi'(u)du - \int_{\beta_{i-1}}^{\beta_i} G_{x-}^{(-1)}(u)\psi'(u)du$$

We recall that $\varphi'(u) = d_i^+$ for all $u \in [\alpha_{i-1}, \alpha_i]$ (and $d_{t+1}^+ = 0$ for convenience) and $\psi'(u) = d_i^-$ for all $u \in [\beta_{i-1}, \beta_i]$ (and $d_{t+1}^- = 0$ for convenience). We have:

$$= \sum_{i=1}^t \left[d_i^+ \int_{\alpha_{i-1}}^{\alpha_i} G_{x+}^{(-1)}(u)du - d_i^- \int_{\beta_{i-1}}^{\beta_i} G_{x-}^{(-1)}(u)du \right]$$

$$= \sum_{i=1}^t \left[d_i^+ \int_{\alpha_{i-1}}^{\alpha_i} F_{x+}^{(-1)}(1-u)du - d_i^- \int_{\beta_{i-1}}^{\beta_i} G_{x-}^{(-1)}(u)du \right] \text{(see Prop. 4)}$$

$$= \sum_{i=1}^t \left[d_i^+ \int_{1-\alpha_i}^{1-\alpha_{i-1}} F_{x+}^{(-1)}(v)dv - d_i^- \int_{\beta_{i-1}}^{\beta_i} G_{x-}^{(-1)}(u)du \right] \text{(with } v = 1 - u)$$

$$= \sum_{i=1}^{t} \left[d_i^+ \left(\int_0^{1-\alpha_{i-1}} F_{x^+}^{(-1)}(v)dv - \int_0^{1-\alpha_i} F_{x^+}^{(-1)}(v)dv \right) - d_i^- \int_{\beta_{i-1}}^{\beta_i} G_{x^-}^{(-1)}(u)du \right]$$

$$= \sum_{i=1}^{t} \left[(d_{i+1}^+ - d_i^+) \int_0^{1-\alpha_i} F_{x^+}^{(-1)}(v)dv - d_i^- \left(\int_0^{\beta_i} G_{x^-}^{(-1)}(u)du - \int_0^{\beta_{i-1}} G_{x^-}^{(-1)}(u)du \right) \right]$$

$$= \sum_{i=1}^{t} \left[(d_{i+1}^+ - d_i^+) \int_0^{1-\alpha_i} F_{x^+}^{(-1)}(v)dv - (d_i^- - d_{i+1}^-) \int_0^{\beta_i} G_{x^-}^{(-1)}(v)dv \right] \qquad \square$$

Now we introduce the two following linear programs to optimize $\int_0^{1-\alpha_k} F_x^{(-1)}(v)dv$ and $\int_0^{\alpha_k} G_x^{(-1)}(v)dv$, for a fixed x and k. The linearization of $\int_0^p F_x^{(-1)}(v)dv$ has been first proposed in [13] and is here extended to $\int_0^p G_x^{(-1)}(v)dv$:

$$\min \sum_{i=1}^{n} x_i m_i \qquad\qquad\qquad \max \sum_{i=1}^{n} x_i m_i$$
$$\begin{cases} \sum_{i=1}^{n} m_i = (1 - \alpha_k) \\ m_i \le p_i \qquad i = 1,\ldots,n \\ m_i \ge 0, \ i = 1,\ldots,n \end{cases} \qquad \begin{cases} \sum_{i=1}^{n} m_i = \alpha_k \\ m_i \le p_i \qquad i = 1,\ldots,n \\ m_i \ge 0, \ i = 1,\ldots,n \end{cases}$$

Then we consider their respective dual formulations:

$$\max(1 - \alpha_k)r - \sum_{i=1}^{n} p_i b_i \qquad\qquad \min \alpha_k r + \sum_{i=1}^{n} p_i b_i$$
$$r - b_i \le x_i \ \ i = 1,\ldots,n \qquad\qquad r + b_i \ge x_i \ \ i = 1,\ldots,n$$
$$b_i \ge 0, \ i = 1,\ldots,n \qquad\qquad\qquad b_i \ge 0, \ i = 1,\ldots,n$$

Using these formulations, we propose a mixed integer program $(\mathcal{P}_2)$ to maximize $g_{\varphi,\psi}(x)$ for any x belonging to a set X:

$$\max \sum_{k=1}^{t} d_k'^+ ((1 - \alpha_k) \times r_k^+ - \sum_{l=1}^{n} p_l^+ b_{lk}^+) - \sum_{k=1}^{t} d_k'^- (\alpha_k \times r_k^- + \sum_{l=1}^{n} p_l^- b_{lk}^-)$$

$(\mathcal{P}_2)$
$$\begin{cases} r_k^+ - b_{ik}^+ \le x_i^+ & i = 1,\ldots,n, \ k = 1,\ldots,t \\ r_k^- + b_{ik}^- \ge x_i^- & i = 1,\ldots,n, \ k = 1,\ldots,t \\ x_i = x_i^+ - x_i^- & i = 1,\ldots,n \\ 0 \le x_i^+ \le z_i \times M & i = 1,\ldots,n \\ 0 \le x_i^- \le (1 - z_i) \times M & i = 1,\ldots,n \\ x \in X \end{cases}$$
$$x_i^+, x_i^-, b_{ik} \ge 0, \ i = 1,\ldots,n, \ k = 1,\ldots,t$$
$$z_i \in \{0,1\}, \ i = 1,\ldots,n$$

with $d_k'^+ = d_{k+1}^+ - d_k^+$ and $d_k'^- = d_k^- - d_{k+1}^-$ for all $k = 1,\ldots,t$. The integer variables $z_i, i = 1,\ldots,n$ are used to decide whether x_i is positive or not. The M constant is used as usual to model disjunctive constraints depending on the sign of x_i. $\mathcal{P}_2$ contains $2nt + 3n$ constraints, n binary variables and $2nt + 2n + 2t$ continuous variables. It can be specialized to solve any CPT-optimal problem, by inserting the needed variables and constraints to define the set X, as shown for

$\mathcal{P}_1$. Table 2 gives the results obtained for the CPT-optimal knapsack problem. Functions φ and ψ are chosen piecewise linear with n breakpoints; these functions are randomly drawn to satisfy the conditions of Proposition 1. Average times given in Table 2 are computed over 20 runs, with a timeout set to 1200 s.

Table 2. Times (s) obtained by MIP $\mathcal{P}_2$ for the CPT-knapsack

m	$n = 3$	$n = 5$	$n = 7$	$n = 10$
100	0.01	0.03	0.07	0.12
500	0.04	0.13	0.19	28.22
750	0.03	0.18	2.76	107.36
1000	0.04	0.27	9.027	191.84

The linearization presented here for the case where $u(z) = z$ for all z can easily be extended to deal with piecewise linear concave utility functions u for gains and for losses (admitting a bounded number of pieces). In this case, the utility function can indeed be defined on gains as the minimum of a finite set of linear utilities which enables a linear reformulation (the same holds for losses). Note also that having a concave utility over gains and over losses is consistent with the risk-averse attitude under consideration in the paper.

6   Conclusion

CPT is a well known model in the context of decision making under risk used to overcome some descriptive limitations of both EU and RDU. In this paper, we have proposed two mixed integer programs for the search of CPT-optimal solutions on implicit sets of alternatives. We tested these computational models on randomly generated instances of the Knapsack problem involving up to 1000 objects and 10 scenarios. The second MIP formulation proposed performs significantly better due to the additional restriction to piecewise linear utility functions.

A natural extension of this work could be to address the exponential aspect of our first formulation with a Branch&Price approach. Another natural extension of this work could be to propose a similar approach for a general bipolar Choquet integral where the capacity is not necessarily defined as a weighted probability. It can easily be shown that the first linearization proposed in the paper still applies to bi-polar Choquet integrals.

References

1. Chateauneuf, A.: On the use of capacities in modeling uncertainty aversion and risk aversion. J. Math. Econ. **20**(4), 343–369 (1991)
2. Grabisch, M., Marichal, J.L., Mesiar, R., Pap, E.: Aggregation Functions, vol. 127. Cambridge University Press, Cambridge (2009)

3. He, X.D., Zhou, X.Y.: Portfolio choice under cumulative prospect theory: an analytical treatment. Manag. Sci. **57**(2), 315–331 (2011)
4. Hines, G., Larson, K.: Preference elicitation for risky prospects. In: Proceedings of the 9th International Conference on Autonomous Agents and Multiagent Systems: volume 1, vol. 1, pp. 889–896. International Foundation for Autonomous Agents and Multiagent Systems (2010)
5. Jaffray, J., Nielsen, T.: An operational approach to rational decision making based on rank dependent utility. Eur. J. Oper. Res. **169**(1), 226–246 (2006)
6. Jeantet, G., Spanjaard, O.: Computing rank dependent utility in graphical models for sequential decision problems. Artif. Intell. **175**(7–8), 1366–1389 (2011)
7. Kahneman, D., Tversky, A.: Prospect theory: an analysis of decision under risk. Econometrica **47**(2), 263–292 (1979)
8. Labreuche, C., Grabisch, M.: Generalized choquet-like aggregation functions for handling bipolar scales. Eur. J. Oper. Res. **172**(3), 931–955 (2006)
9. Li, X., Wang, W., Xu, C., Li, Z., Wang, B.: Multi-objective optimization of urban bus network using cumulative prospect theory. J. Syst. Sci. Complex. **28**(3), 661–678 (2015)
10. Mansini, R., Ogryczak, W., Speranza, M.G.: Twenty years of linear programming based portfolio optimization. Eur. J. Oper. Res. **234**(2), 518–535 (2014)
11. Mansini, R., Ogryczak, W., Speranza, M.G.: Linear and Mixed Integer Programming for Portfolio Optimization. EATOR. Springer, Cham (2015). https://doi.org/10.1007/978-3-319-18482-1
12. Martin, H., Perny, P.: Biowa for preference aggregation with bipolar scales: application to fair optimization in combinatorial domains. In: IJCAI (2019)
13. Ogryczak, W., Śliwiński, T.: On efficient wowa optimization for decision support under risk. Int. J. Approximate Reasoning **50**(6), 915–928 (2009)
14. Perny, P., Spanjaard, O., Storme, L.X.: State space search for risk-averse agents. In: IJCAI, pp. 2353–2358 (2007)
15. Perny, P., Viappiani, P., Boukhatem, A.: Incremental preference elicitation for decision making under risk with the rank-dependent utility model. In: Proceedings of Uncertainty in Artificial Intelligence (2016)
16. Prashanth, L., Jie, C., Fu, M., Marcus, S., Szepesvári, C.: Cumulative prospect theory meets reinforcement learning: prediction and control. In: International Conference on Machine Learning, pp. 1406–1415 (2016)
17. Quiggin, J.: Generalized Expected Utility Theory - The Rank-dependent Model. Kluwer Academic Publisher, Dordrecht (1993)
18. Rothschild, M., Stiglitz, J.E.: Increasing risk: I. A definition. J. Econ. Theory **2**(3), 225–243 (1970)
19. Savage, L.J.: The Foundations of Statistics. J. Wiley and Sons, New-York (1954)
20. Schmeidler, D.: Integral representation without additivity. Proc. Am. Math. Soc. **97**(2), 255–261 (1986)
21. Schmidt, U., Zank, H.: Risk aversion in cumulative prospect theory. Manag. Sci. **54**(1), 208–216 (2008)
22. Shapley, L.: Cores of convex games. Int. J. Game Theory **1**, 11–22 (1971)
23. Tversky, A., Kahneman, D.: Advances in prospect theory: cumulative representation of uncertainty. J. Risk Uncertainty **5**(4), 297–323 (1992)
24. Von Neumann, J., Morgenstern, O.: Theory of Games and Economic Behavior, 2nd edn. Princeton University Press, Princeton (1947)
25. Yaari, M.: The dual theory of choice under risk. Econometrica **55**, 95–115 (1987)

On a New Evidential C-Means Algorithm with Instance-Level Constraints

Jiarui Xie[1,2(✉)] and Violaine Antoine[1]

[1] Clermont Auvergne University, UMR 6158 CNRS, LIMOS,
63000 Clermont-Ferrand, France
jiarxie@foxmail.com, violaine.antoine@uca.fr
[2] School of Computer Science and Technology, Harbin Institute of Technology,
Harbin 150001, People's Republic of China

Abstract. Clustering is an unsupervised task whose performances can be highly improved with background knowledge. As a consequence, several semi-supervised clustering approaches have proposed to integrate prior information in the form of constraints, generally at the instance-level. Amongst them, evidential semi-supervised clustering algorithms, such as CECM or SECM algorithm, rely on the theoretical foundation of belief function which extends the probabilistic theory and allows us to express many types of uncertainty about the assignment of an object to a cluster. In this framework, no evidential clustering algorithm has ever mixed different types of instance-level constraints. We propose here to combine pairwise constraints and labeled data constraints in order to better retrieve information from the background knowledge. The new algorithm, called LPECM, shows good performances on synthetic and real data sets.

Keywords: Labeled data constraints · Pairwise constraints · Instance-level constraints · Belief function · Evidential clustering · Semi-supervised clustering

1 Introduction

Clustering is a classical data analysis method that aims at creating natural groups from a set of objects by assigning similar objects into the same cluster while separating dissimilar objects into different clusters. Clustering solutions can be expressed in the form of a partition. Amongst partitional clustering methods, some produce hard [6,18], fuzzy [10,19] and credal partitions [2–4,14]. A hard partition assigns an object to a cluster with total certainty whereas a fuzzy partition allows us to represent the class membership of an object in the form of a probabilistic distribution. The credal partition, developed in the framework of belief function theory, extends the concepts of hard and fuzzy partition. It makes possible the representation of both uncertainty and imprecision regarding the class membership of an object.

Clustering is a challenging task since various clustering solutions can be valid although distinct. In order to lead clustering methods towards a specific

© Springer Nature Switzerland AG 2019
N. Ben Amor et al. (Eds.): SUM 2019, LNAI 11940, pp. 66–78, 2019.
https://doi.org/10.1007/978-3-030-35514-2_6

and desired solution, semi-supervised clustering algorithms integrate background knowledge, generally in the form of instance-level constraints. In [2,3,19], labeled data constraints are taken into account to improve the performances of the clustering. In [4,6,10,18], two less informative constraints are introduced: the must-link constraint, which specifies that two objects have to be in the same cluster and the cannot-link constraint, which indicates that two objects should not be assigned in the same cluster.

The combination of the three types of instance-level constraints can help to retrieve as most information as possible and thus can achieve better performances. However, there exists currently very few methods able to deal with such constraints [17], more particularly, none generates a credal partition. In this paper, we propose to associate two evidential semi-supervised clustering algorithms, the first one handling pairwise constraints and the second one dealing with labeled data constraints. The goal is to create a more general algorithm that can obtain a large number of constraints from the background knowledge and that can generate a credal partition.

The rest of the paper is organized as follows. Section 2 recalls the necessary backgrounds about belief function, credal partition and evidential clustering algorithms. Section 3 introduces the new algorithm named LPECM and presents the objective function as well as the optimization steps. Several experiments are produced in Sect. 4. Finally, Sect. 5 makes a conclusion about the work.

2   Background

2.1   Belief Function and Credal Partition

Evidence theory [15] (or belief function theory) is a mathematical framework that enables to reflect the state of partial and uncertainty knowledge. Let $\mathbf{X}$ be a data set composed of n objects such that $\mathbf{x}_i \in \mathbb{R}^p$ corresponds to the ith object. Let $\Omega = \{\omega_1, \ldots, \omega_c\}$ be the set of possible clusters. The mass function $m_{ik} : 2^\Omega \to [0,1]$ applied on the instance $\mathbf{x}_i$ measures the degree of belief that the real class of $\mathbf{x}_i$ belongs to a subset $A_k \subseteq \Omega$. It satisfies:

$$\sum_{A_k \subseteq \Omega} m_{ik} = 1. \tag{1}$$

The collection $\mathbf{M} = [\mathbf{m}_1, \ldots, \mathbf{m}_n]$ such that $\mathbf{m}_i = (m_{ik})$ forms a credal partition that is a generalization of a fuzzy partition. Indeed, any subset A_k such that $m_{ik} > 0$ is named a focal set of $\mathbf{m}_i$. When all focal elements are singletons, the mass function is equivalent to a probability distribution. If such situation occurs for all objects, the credal partition $\mathbf{M}$ can be seen as a fuzzy partition.

Several transformations of a mass function $\mathbf{m}_i$ are possible in order to extract particular information. The plausibility function $pl(A) : 2^\Omega \to [0,1]$ defined in Eq. (2) corresponds to the maximal degree of belief that could be given to subset A:

$$pl(A) = \sum_{A_k \cap A \neq \emptyset} m(A_k), \quad \forall A \subseteq \Omega. \tag{2}$$

To make a decision, a mass function can also be transformed into a pignistic probability distribution [16]. Finally, a hard credal partition can be obtained by assigning each object to the subset of cluster with the highest mass. This allows us to easily detect objects located in an ambiguous region.

2.2 Evidential C-Means Algorithm

Evidential C-Means (ECM) [14] is the credibilistic version of Fuzzy C-Means algorithm (FCM) [5]. In the FCM algorithm, each cluster is represented by a point called centroid or prototype. The ECM algorithm, which generates a credal partition, generalizes the cluster representation by considering a centroid $\mathbf{v}_k$ in $\mathbb{R}^p$ for each subset $A_k \subseteq \Omega$. The objective function is:

$$J_{\text{ECM}}(\mathbf{M}, \mathbf{V}) = \sum_{i=1}^{n} \sum_{A_k \neq \emptyset} |A_k|^\alpha m_{ik}^\beta d_{ik}^2 + \sum_{i=1}^{n} \rho^2 m_{i\emptyset}^\beta, \tag{3}$$

subject to

$$\sum_{A_k \subseteq \Omega, A_k \neq \emptyset} m_{ik} + m_{i\emptyset} = 1 \quad \text{and} \quad m_{ik} \geq 0 \quad \forall i \in \{1, \dots, n\}. \tag{4}$$

where $|A_k|$ corresponds to the cardinality of the subset A_k, $\mathbf{V}$ is the set of prototypes and d_{ik}^2 represents the squared Euclidean distance between $\mathbf{x}_i$ and the centroid $\mathbf{v}_k$. Outliers are handled with masses $m_{i\emptyset}, \forall i \in 1, \dots, n$, allocated to the empty set and with the $\rho^2 > 0$ parameter. The two parameters $\alpha \geq 0$ and $\beta > 1$ are introduced to penalize the degree of belief assigned to subsets with high cardinality and to control the fuzziness of the partition.

An extension of the ECM algorithm has been proposed in order to deal with a Mahalanobis distance [4]. Such metric is adaptive and handles various ellipsoidal shapes of clusters, giving more flexibility for the algorithm to better find the inherent structure of the data. Mahalanobis distance d_{ik}^2 between a point $\mathbf{x}_i$ and a subset A_k is defined as follows:

$$d_{ik}^2 = \|\mathbf{x}_i - \mathbf{v}_k\|_{\mathbf{S}_k}^2 = (\mathbf{x}_i - \mathbf{v}_k)^T \mathbf{S}_k (\mathbf{x}_i - \mathbf{v}_k), \tag{5}$$

where $\mathbf{S}_k$ represent the evidential covariance matrix associated to subset A_k and is calculated as the average of the covariance matrices of the singletons included in subset A_k. Finally, objective function (3) has to be minimized with the respect to the credal partition matrix $\mathbf{M}$, the centroids matrix $\mathbf{V}$ and the covariance matrix $\mathbf{S} = \{\mathbf{S}_1, \dots, \mathbf{S}_c\}$ the set composed of covariance matrices dedicated to clusters.

2.3 Evidential Constrained C-Means Algorithm

Several evidential C-Means based algorithms have already been proposed [1–4,8,13] to deal with background knowledge. For each of them, constraints are expressed in the framework of a belief function and a term penalizing the constraints violation is incorporated in the objective function of the ECM algorithm.

In [2,3], labeled data constraints are introduced in the algorithms, i.e. the expert can express the uncertainty about the label of an object by assigning it to a subset. Objective functions of the algorithms are written in such a way that any mass function which partially or fully respects a constraint on a specific subset has a high weighted plausibility given to a singleton included in the subset.

$$T_{ij} = T_i(A_j) = \sum_{A_j \cap A_l \neq \emptyset} \frac{|A_j \cap A_l|^{\frac{r}{2}}}{|A_l|^r} m_{il}, \quad \forall i \in \{1 \ldots n\}, \ A_l \subseteq \Omega, \quad (6)$$

where $r \geq 0$ is a fixed parameter. Notice that if $r = 0$, then $\frac{|A_j \cap A_l|^{\frac{r}{2}}}{|A_l|^r} = 1$, which implies that T_{ij} is identical to the plausibility pl_{ij}.

In [4], authors assumed that pairwise constraints (i.e. must-link and cannot-link constraints) are available. A plausibility to belong or not to the same class is then defined. This plausibility allows us to add a penalty term having high values when there exists a high plausibility that two objects are (respectively are not) in the same cluster although they have a must-link constraint (respectively a cannot-link constraint).

$$pl_{l \times j}(\theta) = \sum_{\{A_l \times A_j \subseteq \Omega^2 | (A_l \times A_j) \cap \theta \neq \emptyset\}} m_{l \times j}(A_l \times A_j)$$
$$= \sum_{A_l \cap A_j \neq \emptyset} m_l(A_l) m_j(A_j), \quad (7)$$

$$pl_{l \times j}(\bar{\theta}) = 1 - m_{l \times j}(\emptyset) - bel_{l \times j}(\theta)$$
$$= 1 - m_{l \times j}(\emptyset) - \sum_{k=1}^{c} m_l(A_k) m_j(A_k), \quad (8)$$

where, θ denotes the event that objects x_i and x_j belong to the same class corresponds to the subset $\{(\omega_1, \omega_1), (\omega_2, \omega_2), \ldots, (\omega_k, \omega_k)\}$ within Ω^2, whereas $\bar{\theta}$ denotes the event that objects x_i and x_j do not belong to the same class corresponds to its complement.

3 The LPECM Algorithm with Instance-Level Constraints

3.1 Objective Function

We propose a new algorithm called Labeled and Pairwise constraints Evidential C-Means (LPECM), which is based on the ECM algorithm [14], handles Mahalanobis distance and combines the advantages of pairwise constraints and labeled data constraints by adding three penalty terms:

$$J_{LPECM}(\mathbf{M}, \mathbf{V}, \mathbf{S}) = \xi J_{ECM}(\mathbf{M}, \mathbf{V}, \mathbf{S}) + \gamma J_{\mathscr{M}}(\mathbf{M}) + \eta J_{\mathscr{C}}(\mathbf{M}) + \delta J_{\mathscr{L}}(\mathbf{M}), \quad (9)$$

with respect to constraints (4). Formulation of J_{ECM} corresponds to equation (3) and (5), $J_{\mathcal{M}}$ is a penalty term used for must-link constraints, $J_{\mathcal{C}}$ is dedicated to cannot-link constraints and $J_{\mathcal{L}}$ handles labeled data constraints. Coefficients ξ, γ, η and δ allow us to give more importance to the structure of the data, the pairwise constraints or the labeled data constraints, respectively.

Penalty terms for pairwise constraints and labeled data constraints are defined similarly to [2,4]:

$$J_{\mathcal{M}}(\mathbf{M}) = \sum_{(\mathbf{x}_i, \mathbf{x}_j) \in \mathcal{M}} \left(1 - (m_{i\emptyset} + m_{j\emptyset} - m_{i\emptyset} m_{j\emptyset}) - \sum_{A_k \subseteq \Omega, |A_k| = 1} m_{ik} m_{jk} \right),$$
(10)

$$J_{\mathcal{C}}(\mathbf{M}) = \sum_{(\mathbf{x}_i, \mathbf{x}_j) \in \mathcal{C}} \sum_{A_k \cap A_l \neq \emptyset} m_{ik} m_{jl},$$
(11)

$$J_{\mathcal{L}}(\mathbf{M}) = \sum_{i=1}^{n} \sum_{A_k \subseteq \Omega, A_k \neq \emptyset} b_{ik} \left(1 - \left(\sum_{A_k \cap A_l \neq \emptyset} \frac{|A_k \cap A_l|^{\frac{r}{2}}}{|A_l|^r} m_{il} \right) \right),$$
(12)

where b_{ik} denotes whether the ith instance belongs to the subset A_k or not:

$$b_{ik} = \begin{cases} 1 \text{ if } \mathbf{x}_i \text{ is constrained to subset } A_k, \\ 0 \text{ otherwise.} \end{cases}$$
(13)

It should be emphasized that in this study, unlike [2], each labeled object is constrained to only one subset. Indeed, it makes more coherent the set of constraints retrieved from the background knowledge. Constraints are gathered in three different sets such that $\mathcal{M}$ corresponds to the set of must-link con-straints, $\mathcal{C}$ to the set of cannot-link constraints and $\mathcal{L}$ denotes the labeled data constraints set. The $J_{\mathcal{M}}$ function returns the sum of the plausibilities that must-link constrained objects to belong to the same class. Similarly, $J_{\mathcal{C}}$ returns the sum of the plausibilities that cannot-link constrained objects are not in the same class. The $J_{\mathcal{L}}$ term calculates for each labeled object a weighted plausibility to belong to the label.

3.2 Optimization

The objective function is minimized as the ECM algorithm, i.e. by carrying out an iterative scheme where first $\mathbf{V}$ and $\mathbf{S}$ are fixed to optimize $\mathbf{M}$, second $\mathbf{M}$ and $\mathbf{S}$ are fixed to optimize $\mathbf{V}$ and finally $\mathbf{M}$ and $\mathbf{V}$ are fixed to optimize $\mathbf{S}$.

Centroids Optimization. It can be observed from (9) that the three penalty terms included in the objective function of the LPECM algorithm do not depend on the cluster centroids. Hence, the update scheme of $\mathbf{V}$ is identical to the ECM algorithm [14].

Masses Optimization. In order to obtain a quadratic objective function with linear constraints, we set parameter $\beta = 2$. A classical optimal approach can then be used to solve the problem [7]. The following equations present how to transform the objective function (9) in order to obtain a format accepted by most usual quadratic optimization function.

Let us define $\mathbf{m}_i^T = (m_{i\emptyset}, m_{i\omega_1}, \ldots, m_{i\Omega})$ the vector of masses for object $\mathbf{x}_i$. The first term of J_{LPECM} is then:

$$J_{ECM}(\mathbf{M}) = \sum_{i=1}^{n} \mathbf{m}_i^T \mathbf{\Phi}^i \mathbf{m}_i, \tag{14}$$

where $\mathbf{\Phi}^i = \left[\phi_{kl}^i\right]$ is a diagonal matrix of size $(2^c \times 2^c)$ associated to object $\mathbf{x}_i$ and defined such as:

$$\phi_{kl}^i = \begin{cases} \rho^2 & \text{if } A_k = A_l \text{ and } A_k = \emptyset, \\ d_{ik}^2 \, |A_k|^{\alpha} & \text{if } A_k = A_l \text{ and } A_k \neq \emptyset, \\ 0 & \text{otherwise.} \end{cases} \tag{15}$$

Penalty term used for must-link constraints can be rewritten as follows:

$$J_{\mathcal{M}}(\mathbf{M}) = n_{\mathcal{M}} + \sum_{(\mathbf{x}_i,\mathbf{x}_j)\in\mathcal{M}} \left(\mathbf{F}_{\mathcal{M}}^T \mathbf{m}_i + \mathbf{F}_{\mathcal{M}}^T \mathbf{m}_j\right) + \sum_{(\mathbf{x}_i,\mathbf{x}_j)\in\mathcal{M}} \mathbf{m}_i^T \mathbf{\Delta}^{\mathcal{M}} \mathbf{m}_j, \tag{16}$$

where $n_{\mathcal{M}}$ denotes the number of must-link constraints, $\mathbf{F}_{\mathcal{M}}$ is a vector of size 2^c and $\mathbf{\Delta}^{\mathcal{M}} = \left[\delta_{kl}^{\mathcal{M}}\right]$ corresponds to a matrix $(2^c \times 2^c)$ such that:

$$\mathbf{F}_{\mathcal{M}}^T = \underbrace{[-1, 0, \ldots, 0]}_{2^c} \quad \text{and} \quad \delta_{kl}^{\mathcal{M}} = \begin{cases} 1 & \text{if } A_k = \emptyset \text{ or } A_l = \emptyset, \\ -1 & \text{if } A_k = A_l \text{ and } |A_k| = |A_l| = 1, \\ 0 & \text{otherwise.} \end{cases}$$
$$\tag{17}$$

The penalty term associated to cannot-link constraints is:

$$J_{\mathcal{C}}(\mathbf{M}) = \sum_{(\mathbf{x}_i,\mathbf{x}_j)\in\mathcal{C}} \mathbf{m}_i^T \mathbf{\Delta}^{\mathcal{C}} \mathbf{m}_j, \tag{18}$$

where $\mathbf{\Delta}^{\mathcal{C}} = \left[\delta_{kl}^{\mathcal{C}}\right]$ is a matrix $(2^c \times 2^c)$ such that:

$$\delta_{kl}^{\mathcal{C}} = \begin{cases} 1 & \text{if } A_k \cap A_l \neq \emptyset, \\ 0 & \text{otherwise.} \end{cases} \tag{19}$$

Finally, the penalty term for the labeled data constraints is denoted as follows:

$$J_{\mathcal{L}}(\mathbf{M}) = n_{\mathcal{L}} - \sum_{i=1}^{n} \mathbf{F}_{\mathcal{L}}^T \mathbf{m}_i, \tag{20}$$

where $n_{\mathscr{L}}$ denotes the number of labeled data constraints and $\mathbf{F}_{\mathscr{L}}$ is a vector of size 2^c such that:

$$\mathbf{F}_{\mathscr{L}}^T = v_{ikl}c_{lk}, \quad \forall A_l \in \Omega, \tag{21}$$

$$c_{lk} = \frac{|A_k \cap A_l|^{\frac{r}{2}}}{|A_l|^r}, \tag{22}$$

$$v_{ikl} = \begin{cases} 1 & \text{if } (\mathbf{x}_i, A_k) \in \mathscr{L} \text{ and } A_k \cap A_l \neq \emptyset, \\ 0 & \text{otherwise.} \end{cases} \tag{23}$$

where expression $(\mathbf{x}_i, A_k) \in \mathscr{L}$ means that the labeled data constraint on object i is the subset A_k. Function $v_{ikl} = \{0, 1\}$ equals to 1 for subsets A_l that has an intersection with A_k knowing the constraint $\mathbf{x}_i \in A_k$.

Now, let us define $\mathbf{m}^T = (\mathbf{m}_1^T, \ldots, \mathbf{m}_n^T)$ the vector of size $n2^c$ containing the masses for each object and each subset, $\mathbf{H}$ a matrix of size $(n2^c \times n2^c)$ and $\mathbf{F}$ a vector of size $n2^c$ such that:

$$\mathbf{H} = \begin{pmatrix} \mathbf{\Phi}^1 & \mathbf{\Delta}_{12} & \cdots & \mathbf{\Delta}_{1n} \\ \mathbf{\Delta}_{21} & \mathbf{\Phi}^2 & \cdots & \\ \vdots & \vdots & \ddots & \vdots \\ \mathbf{\Delta}_{n1} & & \cdots & \mathbf{\Phi}^n \end{pmatrix}, \quad \text{where} \quad \mathbf{\Delta}_{ij} = \begin{cases} \mathbf{\Delta}^{\mathscr{M}}, & \text{if } (\mathbf{x}_i, \mathbf{x}_j) \in \mathscr{M}, \\ \mathbf{\Delta}^{\mathscr{C}}, & \text{else if } (\mathbf{x}_i, \mathbf{x}_j) \in \mathscr{C}, \\ 0, & \text{otherwise.} \end{cases}$$

$$\tag{24}$$

$$\mathbf{F}^T = \begin{pmatrix} \mathbf{F}_1 & \cdots & \mathbf{F}_i & \cdots & \mathbf{F}_n \end{pmatrix}, \quad \text{where} \quad \mathbf{F}_i = t_i \mathbf{F}_{\mathscr{M}} - b_i \mathbf{F}_{\mathscr{L}}, \tag{25}$$

$$t_i = \begin{cases} 1, & \text{if } \mathbf{x}_i \in \mathscr{M}, \\ 0, & \text{otherwise.} \end{cases}, \quad \text{and} \quad b_i = \begin{cases} 1, & \text{if } \mathbf{x}_i \in \mathscr{L}, \\ 0, & \text{otherwise.} \end{cases}. \tag{26}$$

Finally, the objective function (9) can be rewritten as follows:

$$J_{LPECM}(\mathbf{M}) = \mathbf{m}^T \mathbf{H} \mathbf{m} + \mathbf{F}^T \mathbf{m}. \tag{27}$$

3.3   Metric Optimization

It can be observed from (9), the three penalty terms of the LPECM algorithm objective function do not depend on the Mahalanobis distance. Since the set of metric $\mathbf{S}$ only appears in J_{ECM}, the update method is identical to the ECM algorithm [4]. The overall procedure of the LPECM algorithm is summarized in Algorithm 1.

4   Experiments

4.1   Experimental Protocols

Performances and time consumption of the LPECM algorithm have been tested on a toy data set and several classical data sets from UCI Machine Learning Repository [9]. For the Letters data set, we kept only the three letters {I,J,L} as done in [6]. As in [14], fixed parameters associated to the ECM algorithm

---

Algorithm 1. The LPECM algorithm with an adaptive metric

Require: c: Number of desired clusters; $\mathbf{X} = (\mathbf{x}_1, \ldots, \mathbf{x}_n)$ the data set; $\mathscr{C}$: Set of cannot-link constraints ; $\mathscr{M}$: Set of must-link constraints ; $\mathscr{L}$: Set of labeled data constraints ;

Ensure: credal partition matrix $\mathbf{M}$, centroids matrix $\mathbf{V}$, distance metric matrix $\mathbf{S}$

1: Initialization of $\mathbf{V}$;

2: **repeat**

3:     Calculate the new credal partition matrix $\mathbf{M}$ by solving the quadratic programming problem defined by (27) subject to (4);

4:     Calculate the new centroids matrix $\mathbf{V}$ by solving the linear system defined as in the ECM algorithm [14];

5:     Calculate the new metric matrix $\mathbf{S}$ and new associated distances using [4];

6: **until** No significant change in $\mathbf{V}$ between two successive iterations;

---

were set such as $\alpha = 1$, $\beta = 2$ and $\rho^2 = 100$. In order to balance the importance of the data structure, must-link constraints, cannot-link constraints and labeled data constraints respectively, we respectively set $\xi = \frac{1}{n2^c}$, $\gamma = \frac{1}{|\mathscr{M}|}$, $\eta = \frac{1}{|\mathscr{C}|}$ and $\delta = \frac{1}{|\mathscr{L}|}$ as coefficients.

Experiment on a data set consists of 20 simulations with a random selection of the constraints. For each simulation, five runs of the LPECM algorithm with random initialization of the centroids are performed. Then, in order to avoid local optimum, the clustering solution with the minimum value of the objective function is selected.

The accuracy of the obtained credal partition is measured with the Adjusted Rand Index (ARI) [12], which is the corrected-for-chance version of the Rand Index that compares a hard partition with the true partition of a data set. As a consequence, the credal partition generated by the LPECM algorithm is first transformed into a fuzzy partition using the pignistic transformation and then the maximum of probability on each object is retrieved to obtain a hard partition.

4.2   Toy Data Set

In order to show the interest of the LPECM algorithm, we started our experiments with a tiny synthetic data set composed of 15 objects and three classes. Figure 1 presents the hard credal partition obtained using the ECM algorithm. Big cross marks denote the centroid of each cluster. Centroids for subsets with higher cardinalities are not represented to ease the reading. As it can be observed, objects located between two clusters are assigned in subsets with cardinality equal to two. Notice also that, due to the stochastic initialization of the centroids, there may exist a small difference between the results obtained from every execution of the ECM algorithm. After the addition of background knowledge in the form of must-link constraints, cannot-link constraints and labeled data constraints and the execution of the LPECM algorithm with a Euclidean distance, it is interesting to observe that previous uncertainties have vanished.

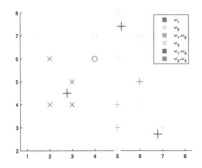
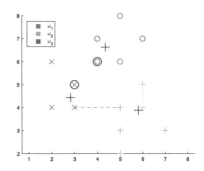

Fig. 1. Hard credal partition obtained on Toy data set with the ECM algorithm

Fig. 2. Hard credal partition obtained on Toy data set with the LPECM algorithm

Figure 2 presents the hard credal partition obtained. The magenta dashed line describes cannot-link constraints, the light green solid line represents must-link constraints and the circled point corresponds to the labeled data constraints .

Figure 3 illustrates, for the execution of the LPECM algorithm, the mass distribution for singletons with respect to the point numbers, allowing us a more distinct sight of the masses allocations. Table 1 displays the accuracy as well as time consumption for the ECM algorithm and the LPECM algorithm when first only the cannot-link constraint is incorporated, second when the cannot-link and the must-link constraint are introduced (Cannot-Must-Link line in Table 1), finally when all constraints are added (Cannot-Must-Labeled line in Table 1). Our results demonstrate that the combination of pairwise constraints and labeled data constraints improved the performance of the semi-clustering algorithm with tolerable time consumption. As expected, the more constraints are added, the better are the performance.

4.3   Real Data Sets

The LPECM algorithm has been tested on three known data sets from the UCI Machine Learning Repository namely Iris, Glass, and Wdbc and a derived Letters data set from UCI. Table 2 indicates for each data set its number of objects, its number of attributes and its number of classes.

For each data set, we randomly created 5%, 8%, and 10% of each type of constraints out of the whole objects, leading to a total of 15%, 24%, and 30% of constraints. As an example, Fig. 4 shows the hard credal partition obtained with the Iris data set after executing the LPECM algorithm with a Mahalanobis distance and 24% of constraints in total. As can be observed, all the constrained objects are clustered with certainty in a singleton. Ellipses represent the covariance matrices obtained for each cluster.

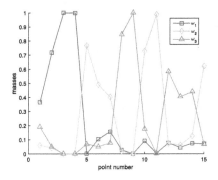

Table 1. Performance obtained on toy data set with the LPECM algorithm

	ARI	Time(s)
ECM	0.60	0.07
LPECM-Cannot	0.68	0.41
LPECM-C-Must	0.85	0.29
LPECM-C-M-labeled	1.00	0.22

Fig. 3. Mass curve obtained on Toy data set with the LPECM algorithm

Tables 3 and 4 illustrate for all data sets the accuracy results with a Euclidean and a Mahalanobis distance respectively when the different percentage of constraints are employed. Mean and standard deviation are calculated over 20 simulations. As it can be observed, incorporating constraints lead most of the time to significant improvement of the clustering solution. Using a Mahalanobis distance particularly help to achieve better accuracy than using a Euclidean distance. Indeed, the Mahalanobis distance corresponds to an adaptive metric giving more freedom than a Euclidean distance to respect the constraints while finding a coherent data structure.

Table 2. Description of the data sets from UCIMLR

Name	Objects	Attributes	Clusters
Iris	150	4	3
Letters	227	16	3
Wdbc	569	31	2
Glass	214	10	3

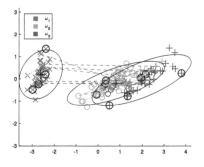

Fig. 4. Hard credal partition obtained on Iris data set with the LPECM algorithm

For the time consumption, as it can be observed from Fig. 5, (1) Adding constraints gives higher computation time than no constraints. (2) most of the time, the more constraints are added, the less time is needed to finish the computation.

Table 3. LPECM's performance (ARI) with Euclidean distance

	ECM	LPECM		
		5.00%	8.00%	10.00%
Iris	0.59 ± 0.00	0.70 ± 0.01	0.71 ± 0.00	0.70 ± 0.01
Letters	0.04 ± 0.01	0.09 ± 0.03	0.09 ± 0.04	0.10 ± 0.02
Wdbc	0.67 ± 0.00	0.71 ± 0.00	0.71 ± 0.01	0.71 ± 0.00
Glass	0.59 ± 0.07	0.60 ± 0.07	0.62 ± 0.06	0.65 ± 0.08

Table 4. LPECM's performance (ARI) with Mahalanobis distance

	ECM	LPECM		
		5.00%	8.00%	10.00%
Iris	0.67 ± 0.01	0.71 ± 0.05	0.82 ± 0.01	0.83 ± 0.04
Letters	0.08 ± 0.01	0.45 ± 0.03	0.47 ± 0.02	0.60 ± 0.05
Wdbc	0.73 ± 0.02	0.74 ± 0.03	0.75 ± 0.02	0.77 ± 0.05
Glass	0.56 ± 0.03	0.60 ± 0.03	0.65 ± 0.02	0.65 ± 0.03

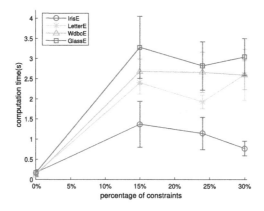

Fig. 5. Time consumption (CPU) of the LPECM algorithm with Euclidean distance

5   Conclusion

In this paper, we introduced a new algorithm named Labeled and Pairwise constraints Evidential C-Means (LPECM). It generates a credal partition and mixes three main types of instance-level constraints together, allowing us to retrieve more constraints from the background knowledge than other semi-supervised clustering algorithms. In addition, the framework of belief function employed in our algorithm allows us (1) to represent doubts for the labeled data constraints (2) to clearly express, with the credal partition as a result, the uncertainties about the class memberships of the objects. Experiments show that the LPECM algorithm does obtain better accuracy with the introduction of constraints, particularly with a Mahalanobis distance. Further investigations have to be performed to fine-tune parameters and to study the influence of the constraints on the clustering solution. The LPECM algorithm can also be applied for a real application to show the interest in gathering various types of constraints. In this framework, active learning schemes, which automatically retrieve few informative constraints with the help of an expert, are interesting to study. Finally, in order to scale and fast the LPECM algorithm, a new minimization process can be developed by relaxing some optimization constraints.

References

1. Antoine, V., Quost, B., Masson, M.H., Denœux, T.: CEVCLUS: evidential clustering with instance-level constraints for relational data. Soft Comput. - Fusion Found. Methodol. Appl. **18**(7), 1321–1335 (2014)
2. Antoine, V., Gravouil, K., Labroche, N.: On evidential clustering with partial supervision. In: Destercke, S., Denoeux, T., Cuzzolin, F., Martin, A. (eds.) BELIEF 2018. LNCS (LNAI), vol. 11069, pp. 14–21. Springer, Cham (2018). https://doi.org/10.1007/978-3-319-99383-6_3
3. Antoine, V., Labroche, N., Vu, V.V.: Evidential seed-based semi-supervised clustering. In: International Symposium on Soft Computing & Intelligent Systems, Kitakyushu, Japan (2014)
4. Antoine, V., Quost, B., Masson, M.H., Denœux, T.: CECM: constrained evidential C-means algorithm. Comput. Stat. Data Anal. **56**(4), 894–914 (2012)
5. Bezdek, J.C., Ehrlich, R., Full, W.: FCM: the fuzzy C-means clustering algorithm. Comput. Geosci. **10**(2–3), 191–203 (1984)
6. Bilenko, M., Basu, S., Mooney, R.: Integrating constraints and metric learning in semi-supervised clustering. In: Proceedings of the Twenty-First International Conference on Machine Learning. ACM New York, NY, USA (2004)
7. Coleman, T.F., Li, Y.: A reflective newton method for minimizing a quadratic function subject to bounds on some of the variables. SIAM J. Optim. **6**(4), 1040–1058 (1996)
8. Denœux, T.: Evidential clustering of large dissimilarity data. Knowl.-Based Syst. **106**(C), 179–195 (2016)
9. Dua, D., Graff, C.: UCI machine learning repository (2017). http://archive.ics.uci.edu/ml
10. Grira, N., Crucianu, M., Boujemaa, N.: Active semi-supervised fuzzy clustering. Pattern Recogn. **41**(5), 1834–1844 (2008)
11. Gustafson, D.E., Kessel, W.C.: Fuzzy clustering with a fuzzy covariance matrix. In: IEEE Conference on Decision & Control Including the Symposium on Adaptive Processes, New Orleans, LA (2007)
12. Hubert, L., Arabie, P.: Comparing partitions. J. Classif. **2**(1), 193–218 (1985)
13. Li, F., Li, S., Denoeux, T.: k-CEVCLUS: constrained evidential clustering of large dissimilarity data. Knowl.-Based Syst. **142**, 29–44 (2018)
14. Masson, M.H., Denœux, T.: ECM: an evidential version of the fuzzy C-means algorithm. Pattern Recogn. **41**(4), 1384–1397 (2008)
15. Shafer, G.: A Mathematical Theory of Evidence, vol. 42. Princeton University Press, Princeton (1976)

16. Smets, P., Kennes, R.: The transferable belief model. Artif. Intell. **66**, 191–234 (1994)
17. Vu, V.V., Do, H.Q., Dang, V.T., Do, N.T.: An efficient density-based clustering with side information and active learning: a case study for facial expression recognition task. Intell. Data Anal. **23**(1), 227–240 (2019)
18. Wagstaff, K., Cardie, C., Rogers, S., Schrœdl, S.: Constrained k-means clustering with background knowledge. In: Proceedings of the Eighteenth International Conference on Machine Learning (ICML), Williamstown, MA, USA, vol. 1, pp. 577–584 (2001)
19. Zhang, H., Lu, J.: Semi-supervised fuzzy clustering: a kernel-based approach. Knowl.-Based Syst. **22**(6), 477–481 (2009)

Hybrid Reasoning on a Bipolar Argumentation Framework

Tatsuki Kawasaki, Sosuke Moriguchi, and Kazuko Takahashi(✉)

Kwansei Gakuin University, 2-1 Gakuen, Sanda, Hyogo 669-1337, Japan
{dxk96093,ktaka}@kwansei.ac.jp, chiguri@acm.org

Abstract. We develop a method of reasoning using an incrementally constructed bipolar argumentation framework (BAF) aiming to apply computational argumentation to legal reasoning. A BAF that explains the judgment of a certain case is constructed based on the user's knowledge and recognition. More specifically, a set of effective laws are derived as the conclusions from evidential facts recognized by the user, in a bottom-up manner; conversely, the evidences required to derive a new conclusion are identified if certain conditions are added, in a top-down manner. The BAF is incrementally constructed by repeated exercise of this bidirectional reasoning. The method provides support for those who are not familiar with the law, so that they can understand the judgment process and identify strategies that might allow them to win their case.

Keywords: Argumentation · Bidirectional reasoning · Legal reasoning

1 Introduction

An argumentation framework (AF) is a powerful tool in the context of inconsistent knowledge [15,21]. There are several possible application areas of AFs, including law [4,20]. To date, research on applications has focused principally on AF updating to yield an acceptable set of facts when a new argument is presented, and strategies to win the argumentation when all of the dialog paths are known. However, in real legal cases, an AF representing a law in its entirety is usually incompletely grasped at the initial stage. Thus, it is more realistic to construct the AF incrementally; recognized facts are added in combination with AF reasoning.

For example, consider a case in which a person leased her house to another person, and the lessee then sub-leased a room to his sister; the lessor now wants to cancel the contract. (This is a simplified version of the case discussed in Satoh et al. [23].) The lessor decides to prosecute the lessee. The lessor knows that there was a lease, that they handed over the house to the lessee, and that the room was handed over by the lessee to the sublessee. However, if the lessor is not familiar with the law, she does not know what law might be applicable to her circumstances or what additional facts should be proven to make it effective. In addition, laws commonly include exceptions; that is, a law is effective if certain conditions are satisfied provided there is no exception.

© Springer Nature Switzerland AG 2019
N. Ben Amor et al. (Eds.): SUM 2019, LNAI 11940, pp. 79–92, 2019.
https://doi.org/10.1007/978-3-030-35514-2_7

For example, if there is no abuse of confidence, then the law of cancellation is not effective. Therefore, the lessor should check that there is "no abuse of confidence," as well as regarding facts that prove what must be proven. In addition, other facts may be needed to prove that there has been no abuse of confidence. Also, the presence of an exception may render another law effective. For those lacking a legal background, it can be difficult to grasp the entire structure of a particular law, which may be extensive and complicated. Thus, s/he often consults with, or even fully delegates the problem-solving process to, a lawyer. However, if the argumentation structure of the law was clear, s/he would be more likely to adequately understand the judgment process, obviating the need for a lawyer.

In this paper, we develop a bidirectional reasoning method using a bipolar argumentation framework (BAF) [2] that is applicable to legal reasoning. In a BAF, a general rule is represented as a support relation, and an exception as an attack relation. The facts of a case become arguments that are not attacked or supported by other arguments.

We explore the BAF in both a bottom-up and top-down manner, search for effective laws based on proven facts, and identify the facts required for application to other laws.

Beginning with the user-recognized facts of a specific case, laws that may be effective are searched for using a bottom-up process. Next, new conclusions are considered if specific conditions are satisfied. If such conclusions exist, the required facts are then identified in a top-down manner, so that the conditions are satisfied. If the existence of such facts can be proven, the facts are added as evidence, and the next round then begins. The procedure terminates if the user is satisfied with the conclusions obtained, or if no new conclusions are derived. By repeating this process, a user can simulate and scrutinize the judgment process to identify a strategy that may allow them to win the case.

This paper is organized as follows. In Sect. 2, we present the BAF, and the semantics thereof. In Sect. 3, we describe how the law is interpreted and represented using a BAF. In Sect. 4, we show the reasoning process of a BAF. In Sect. 5, we discuss related works. Finally, in Sect. 6, we present our conclusions and describe our planned future work.

2   Bipolar Argumentation Framework

A BAF is an extension of an AF in which the two relations of attack and support are defined over a set of arguments [2]. We define a support relation between a power set of arguments and a set of arguments; this differs from the common support relation of a BAF, so that it corresponds to a legal structure.

Definition 1 (bipolar argumentation framework). *A BAF is defined as a triple* $\langle AR, ATT, SUP \rangle$, *where* AR *is a finite set of arguments,* $ATT \subseteq AR \times AR$ *and* $SUP \subseteq (2^{AR} \setminus \{\emptyset\}) \times AR$. *If* $(B, A) \in ATT$, *then* B *attacks* A; *if* $(\mathbf{A}, A) \in SUP$, *then* $\mathbf{A}$ *supports* A.

A BAF can be regarded as a directed graph where the nodes and edges correspond to the arguments and the relations, respectively. Below, we represent a BAF graphically; a simple solid arrow indicates a support relation, and a straight arrow with a cutting edge indicates an attack relation. The dashed rectangle shows a set of arguments supporting a certain argument; it is sometimes omitted if the supporting set is a singleton.

Example 1. Figure 1 is a graphical representation of a BAF
$\langle \{a, b, c, d, e\}, \{(b, a), (e, d)\}, \{(\{c, d\}, a)\} \rangle$.

Fig. 1. Example of BAF.

Definition 2 (leaf). *An argument that is neither attacked nor supported by any other argument in a BAF is said to be* a leaf *of the BAF.*

For a BAF $\langle AR, ATT, SUP \rangle$, let $\rightarrow$ be a binary relation over AR as follows:

$$\rightarrow = ATT \cup \{(A, B) | \exists \mathbf{A} \subseteq AR, A \in \mathbf{A} \wedge (\mathbf{A}, B) \in SUP\}.$$

Definition 3 (acyclic). *A BAF $\langle AR, ATT, SUP \rangle$ is said to be* acyclic *if there is no $A \in AR$ such that $(A, A) \in \rightarrow^{+}$, where $\rightarrow^{+}$ is a transitive closure of $\rightarrow$.*

We define semantics for the BAF based on labeling [9]. Usually, labeling is a function from a set of arguments to $\{in, out, undec\}$, but $undec$ is unnecessary here because we consider only acyclic BAFs. An argument labeled in is considered an acceptable argument.

Definition 4 (labeling). *For a BAF $\langle AR, ATT, SUP \rangle$, a labeling $\mathcal{L}$ is a function from AR to $\{in, out\}$.*

Labeling of a set of arguments proceeds as follows: $\mathcal{L}(\mathbf{A}) = in$ if $\mathcal{L}(A) = in$ for all $A \in \mathbf{A}$; and $\mathcal{L}(\mathbf{A}) = out$ otherwise.

Definition 5 (complete labeling). *For a BAF baf $= \langle AR, ATT, SUP \rangle$, labeling $\mathcal{L}$ is* complete *iff the following conditions are satisfied: for any argument $A \in AR$, (i) $\mathcal{L}(A) = in$ if A is a leaf or $(\forall B \in AR; (B, A) \in ATT \Rightarrow \mathcal{L}(B) = out) \wedge (\exists \mathbf{A} \in 2^{AR}; (\mathbf{A}, A) \in SUP \wedge \mathcal{L}(\mathbf{A}) = in)$, (ii) $\mathcal{L}(A) = out$ otherwise.*

If an argument is both attacked and supported, the attack is taken to be stronger than the support. For any acyclic BAF, there is exactly one complete labeling.

3   Description of Legal Knowledge in a BAF

In this paper, we consider an application of the Japanese civil code.

We assume that the BAFs are acyclic and that each law features both general rules and exceptions. A law is effective if the conditions of the general rule are satisfied unless an exception holds. We construct a BAF in which each condition in a rule is represented by an argument; the general rules can be represented by support relations, and the exceptions by attack relations. Therefore, our interpretations of attack and support relations differ from those used in the other BAFs. First, a support relation is defined as a binary relation of a power set and a set of arguments, since if one of the conditions is not met, the law is ineffective. Second, an argument lacking support is labeled *out*, even if it is attacked by an argument labeled *out*, since a law is not defined only by its exceptions and any argument other than a leaf should have an argument that supports it. The correspondence between the "acceptance" criterion of our BAF and that of a logic program is shown in [17].

We assume that the entire set of laws can be represented by a BAF termed *a universal BAF*, denoted as follows:

$$ubaf = \langle UAR, UATT, USUP \rangle.$$

It is almost impossible for a person who is not an expert to understand all of the laws. Therefore, we construct a specific BAF for each incident; relevant evidential facts are disclosed, and applicable laws identified using the universal BAF.

Definition 6 (existence/absence argument). *For an argument A, an argument showing the existence of an evidential fact for A is termed* an existence argument *and is denoted by* $ex(A)$*; and an argument showing the absence of an evidential fact for A is termed* an absence argument *and is denoted by* $ab(A)$*. These arguments are abbreviated as ex/ab arguments, respectively.*

Definition 7 (consistent ex/ab arguments set). *For a set of ex/ab arguments S, if there does not exist an argument A that satisfies both $ex(A) \in S$ and $ab(A) \in S$, then S is said to be* consistent.

Example 2. Figure 2 shows a BAF for the house lease case shown in Sect. 1, together with the relevant ex/ab arguments.

In this Figure, $ex(a1)$, $ex(a2)$, and $ex(a4)$ are existence arguments for agreement_of_lease_contract, handover_to_lessee, and handover_to_ sublessee, respectively; $ab(b1)$ is an absence argument for fact_of_non_abuse_of_confidence; and no evidence is currently shown for the other leaves.

4   Reasoning Using the BAF

4.1   Outline

We employ a running example throughout this section.

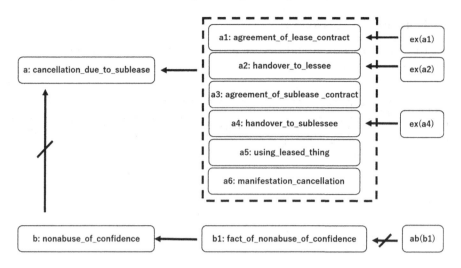

Fig. 2. Example of a BAF for a house-lease case.

Example 3. We assume the existence of the universal BAF *ubaf* = ⟨*UAR*, *UATT*, *USUP*⟩ shown in Fig. 3.

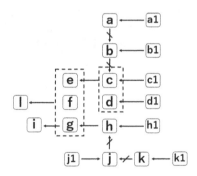

Fig. 3. Example of a universal BAF *ubaf*.

Let *Ex* be a set of ex/ab arguments that is currently recognized by a user. For either ex(*A*) or ab(*A*) of *Ex*, *A* ∈ *UAR*, and *A* is a leaf in *ubaf*.

Initially, a user recognizes a set of facts related to a certain incident. The reasoning proceeds by repeating two methods in turn. The first is used is to derive conclusions from the facts in a bottom-up manner, and the other is employed to find the evidence needed to draw a new conclusion if certain other conditions are met, this exercise proceeds in a top-down manner.

4.2   Bottom-Up Reasoning

In bottom-up reasoning, arguments are derived by following the support relations from an ex/ab argument. The algorithm is shown in Algorithm 1.

Algorithm 1. BUP: find conclusions

Let Ex be a set of ex/ab arguments and $AR = \{A|\mathrm{ex}(A) \in Ex\} \cup \{A|\mathrm{ab}(A) \in Ex\}$.
Find a pair of a set of arguments $\mathbf{A} \subseteq AR$ and an argument $A \in UAR \setminus AR$ such that $(\mathbf{A}, A) \in USUP$.
while there exists such a pair $(\mathbf{A}, A)$ **do**
    Set $AR = AR \cup \{A\}$.
end while
Set $SUP = USUP \cap (2^{AR} \times AR) \cup \{(\{\mathrm{ex}(A)\}, A)|\mathrm{ex}(A) \in Ex\}$.
Set $ATT = UATT \cap (AR \times AR) \cup \{(\mathrm{ab}(A), A)|\mathrm{ab}(A) \in Ex\}$. Set $AR = AR \cup Ex$.
Apply the complete labeling $\mathcal{L}$ to $baf = \langle AR, ATT, SUP \rangle$.
$Concl(Ex) = \{A \mid \mathcal{L}(A) = in \wedge \neg\exists(\mathbf{A}, B) \in SUP; A \in \mathbf{A} \subseteq AR\}$.
return $Concl(Ex)$.

The resulting set of conclusions is the set of arguments that are acceptable, and no more conclusions can be drawn from the currently known facts.

Example 4 (Cont'd). Let Ex be $\{\mathrm{ex}(a1), \mathrm{ex}(b1), \mathrm{ex}(c1), \mathrm{ex}(d1)\}$, and $ubaf$ be a BAF in Fig. 3. Then, the BAF can be constructed using the process shown in Fig. 4(a) and (b); finally, baf_1 is obtained, and $Concl(Ex) = \{a, e\}$ is derived as the set of conclusions. The complete labeling of the BAF baf_1 is shown in Fig. 4(c).

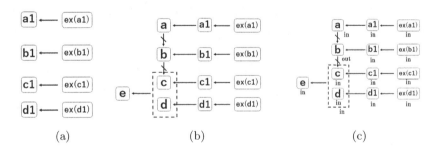

(a)	(b)	(c)

Fig. 4. The bottom-up reasoning used to construct baf_1.

4.3   Top-Down Reasoning

On the other hand, we can seek additional facts that must be proven if a new conclusion is to be derived. Here, we search for a new conclusion and a set of supports, and identify the facts required to derive the arguments of the set.

Definition 8 (differential support pair, differential supporting set of arguments, differentially supported argument). *For a BAF* $baf = \langle AR, ATT, SUP \rangle$, *if* $(\mathbf{A} \cap AR) \neq \emptyset \wedge (\mathbf{A} \cap AR) \neq \mathbf{A} \wedge (\mathbf{A}, A) \in USUP$, *then* $(\mathbf{A} \setminus AR, A)$ *is said to be* a differential support pair on *baf. In addition,* $\mathbf{A} \setminus AR$ *and* A *are said to be* a *differential supporting set of arguments on baf*, *and* a *differentially supported argument on baf*, *respectively.*

Intuitively, differential support pair means that A cannot be derived using the current BAF due to the lack of required conditions, but it can be derived if all of the arguments in $\mathbf{A} \setminus AR$ are accepted. In general, there may exist several differential support pairs on any BAF.

Example 5 (Cont'd). For baf_1, we find differential support pair $(\{f, g\}, l)$, because $\{e, f, g\} \cap AR = \{e\} \neq \emptyset$ and $(\{e, f, g\}, l) \in USUP$ (Fig. 5).

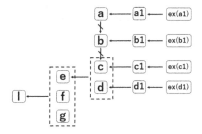

Fig. 5. A differential support pair on $baf_1 : (\{f, g\}, l)$.

For a BAF $baf = \langle AR, ATT, SUP \rangle$ and an argument $A \in AR$, we detect a set of facts that satisfies $\mathcal{L}(A) = in$. For an argument A, we check the conditions for labeling of the arguments that attack A and the sets of arguments that support A. This is achieved by repeatedly applying the following two algorithms: $PC(A)$ and $NC(A)$, which are shown in Algorithms 2 and 3, respectively. Note that there is no argument that both lacks support and is attacked.

Then, discovery of the required facts proceeds using the algorithm shown in Algorithm 4.

As a result, a set of ex/ab arguments is generated. An existence argument $ex(A)$ shows that the fact is required if $\mathcal{L}(\mathbf{A}) = in$ is to hold, whereas an absence argument $ab(A)$ shows that the evidence is an obstacle to prove $\mathcal{L}(\mathbf{A}) = in$.

Example 6 (Cont'd). For a differential supporting set of arguments $\{f, g\}$, we find $Sol(\{f, g\}) = PC(f) \cup PC(g)$.

(i) $PC(f) = \{ex(f)\}$.
(ii) $PC(g) = PC(h) = PC(h1) \cup NC(j)$ (Fig. 6). As for $PC(h1)$, we obtain $\{ex(h1)\}$. As for $NC(j)$, we have two alternatives: $NC(j1)$ and $PC(k)$ (Fig. 7).

Algorithm 2. $PC(A)$: find required arguments for $\mathcal{L}(A) = in$.

Let A be an argument in UAR.
if A is a leaf of $ubaf$ **then**
    $Sol(A) = \{\mathrm{ex}(A)\}$.
else
    Choose an arbitrary set of arguments $\mathbf{A}$ that satisfies $(\mathbf{A}, A) \in USUP$.
    $Sol(A) = \bigcup_{(B,A)\in UATT} NC(B) \cup \bigcup_{A_i \in \mathbf{A}} PC(A_i)$.
end if
return $Sol(A)$.

Algorithm 3. $NC(A)$: find required arguments for $\mathcal{L}(A) = out$.

Let A be an argument in UAR.
if A is a leaf of $ubaf$ **then**
    $Sol(A) = \{\mathrm{ab}(A)\}$.
else
    Choose an arbitrary argument B that satisfies $(B, A) \in UATT$.
    Let $\mathbf{A}_1, \ldots, \mathbf{A}_n$ be all sets of arguments such that $(\mathbf{A}_i, A) \in USUP(i = 1, \ldots, n)$.
    Choose an arbitrary argument $A_i \in \mathbf{A}_i$ $(i = 1, \ldots, n)$.
    Either $Sol(A) = PC(B)$
        or $Sol(A) = \bigcup_{i=1,\ldots,n} NC(A_i)$.
end if
return $Sol(A)$.

Assume that we choose the condition $NC(j1)$. Then, we find $\{\mathrm{ab}(j1)\}$ as $Sol(j1)$ (Fig. 8). Finally, we obtain a set of required facts $\{\mathrm{ex}(f), \mathrm{ex}(h1), \mathrm{ab}(j1)\}$ (Fig. 9).

4.4   Hybrid Reasoning

The algorithm used for hybrid reasoning is Algorithm 5.

As a result, the required facts are identified, and conclusions are derived from these facts.

Example 7 (Cont'd). For a set of required facts $\{\mathrm{ex}(f), \mathrm{ex}(h1), \mathrm{ab}(j1)\}$, assume that a user has confirmed the existence of f and $h1$, and the absence of $j1$. Then, we construct a new BAF baf_2 in a bottom-up manner from this set. Part of the labeling of baf_2 is shown in Fig. 10. Finally, we obtain a new conclusion set $Concl = \{a, i, l\}$.

Algorithm 4. TDN: find required facts

Let $baf = \langle AR, ATT, SUP \rangle$ be a BAF and $\mathbf{A}$ a differential supporting set of arguments on baf.
$Sol(\mathbf{A}) = \bigcup_{A \in \mathbf{A}} PC(A)$.
return $Sol(\mathbf{A})$.

Fig. 6. Top-down reasoning: Both $NC(j)$ and $PC(h1)$ are required.

Fig. 7. Top-down reasoning: Either $NC(j1)$ or $PC(k)$ is required.

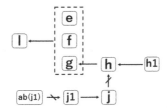

Fig. 8. Top-down reasoning: The situation when choosing $NC(j1)$.

The hybrid algorithm is nondeterministic at several steps and there are multiple possible solutions.

Example 8 (Cont'd). Assume that we choose the condition $PC(k)$ in Fig. 7. Then, we find $\{\text{ex}(k1)\}$ as $Sol(k1)$, and the set of required facts is $\{\text{ex}(f), \text{ex}(h1), \text{ex}(k1)\}$. In this case, we construct the different BAF baf'_2 shown in Fig. 11 after a second round of bottom-up reasoning. Strictly speaking, an argument j and the attack relations (k, j) and (j, h) do not appear in baf'_2 because a new argument is created by tracing only a support relation in BUP. However, it is reasonable to show the attack relation traced in the TDN, considering that the BAF is constructed based on the user's current knowledge. Note that these attacks do not affect the label $\mathcal{L}(h) = in$.

4.5   Correctness

We now prove the validity of hybrid reasoning.

In the proof, we use the height of an argument in *ubaf*, as defined in [17].

Definition 9. *For the acyclic universal BAF ubaf, the height of an argument A is defined as follows:*

- *If A is a leaf, then the height of A is 0.*
- *If there are some arguments B such that $(B, A) \in \rightarrow$, then the height of A is $h + 1$, where h is the maximum height of this B.*

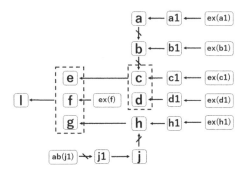

Fig. 9. Top-down reasoning: $Ex = \{\mathrm{ex}(f), \mathrm{ex}(h1), \mathrm{ab}(j1)\}$.

Algorithm 5. HR: hybrid reasoning

Let Ex be a set of ex/ab arguments in an initial state.
For Ex, obtain $Concl(Ex)$ and a new baf by BUP.
while a user does not attain a goal that satisfies him/her, and TDN returns a consistent solution with Ex **do**
    For a baf and an arbitrary **A** on baf, obtain $Sol(\mathbf{A})$ by TDN.
    for each ex(A) or ab(A) in $Sol(\mathbf{A})$ **do**
        Ask the user to confirm that existence or absence.
        if there exists a fact for A **then**
            Set $Ex = Ex \cup \{\mathrm{ex}(A)\}$.
        else
            Set $Ex = Ex \cup \{\mathrm{ab}(A)\}$.
        end if
        Get $Concl(Ex)$ and a new baf by BUP.
    end for
end while

It is easy to show that the heights of arguments are definable when $ubaf$ is acyclic.

Here, we prove two specifications, one for a BUP, and the other for a TDN. For a BUP, the built BAF includes arguments pertaining to the evidential facts that the user recognizes. Notably, the acceptability of such arguments is the same as that of the universal BAF.

Theorem 1. *Assume that $ubaf$ is acyclic. Let baf be built by BUP from Ex. When UEx is defined as $\{\mathrm{ex}(A)|\mathrm{ex}(A) \in Ex\} \cup \{\mathrm{ab}(A)|A$ is a leaf of $ubaf \wedge \mathrm{ex}(A) \notin Ex\}$, and $\mathcal{L}_U$ is a complete labeling for $\langle UAR \cup UEx, UATT \cup \{(\mathrm{ab}(A), A)|\mathrm{ab}(A) \in UEx\}, USUP \cup \{(\{\mathrm{ex}(A)\}, A)|\mathrm{ex}(A) \in UEx\}\rangle$, for any argument $A \in UAR$, $A \in AR \wedge \mathcal{L}(A) = \mathcal{L}_U(A)$, or $A \notin AR \wedge \mathcal{L}_U(A) = out$.*

Proof. We prove this by induction on the height of A. When A is a leaf, if $\mathrm{ex}(A) \in Ex$ (i.e., $\mathrm{ex}(A) \in UEx$), then $A \in AR$ and $\mathcal{L}(A) = \mathcal{L}_U(A) = in$. If $\mathrm{ex}(A) \notin Ex$ (i.e., $\mathrm{ab}(A) \in UEx$), then $\mathcal{L}_U(A) = out$. In this case, if $\mathrm{ab}(A) \in Ex$

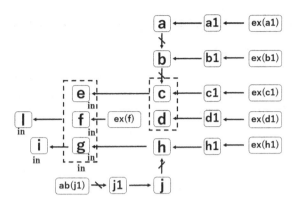

Fig. 10. The BAF obtained after the second round of bottom-up reasoning: baf_2.

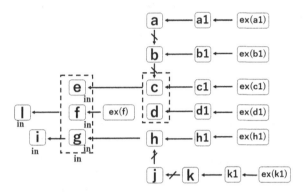

Fig. 11. The BAF obtained after the second round of bottom-up reasoning: baf'_2.

then $A \in AR$ but $\mathcal{L}(A) = out$, and, otherwise $A \notin AR$. Both cases satisfy the proposition.

Assume that A is not a leaf. If $\mathcal{L}_U(A) = in$, then there are some supports $(\mathbf{A}, A) \in USUP$ such that $\mathcal{L}_U(\mathbf{A}) = in$, and any attacks $(B, A) \in UATT$, $\mathcal{L}_U(B) = out$. From the induction hypothesis, for any $C \in \mathbf{A}$, $C \in AR$, and $\mathcal{L}(C) = in$; and for any attackers B of A, $\mathcal{L}(B) = out$ or $B \notin AR$. The definition of BUP immediately shows that $A \in AR$, and therefore $\mathcal{L}(A) = in = \mathcal{L}_U(A)$.

Assume that $\mathcal{L}_U(A) = out$. If $A \notin AR$, the proposition is satisfied. Otherwise, $A \in AR$, and from the definition of BUP, there are some supports $(\mathbf{A}, A)$ such that $\mathbf{A} \subseteq AR$, so A is not a leaf of baf. From $\mathcal{L}_U(A) = out$, there are some attacks $(B, A) \in UATT$ such that $\mathcal{L}_U(B) = in$, or for any supports $(\mathbf{A}, A) \in USUP$, $\mathcal{L}_U(A) = out$ (i.e., there exists $C \in \mathbf{A}$ such that $\mathcal{L}_U(C) = out$). From the induction hypothesis, there are some attacks $(B, A) \in UATT$ such that $\mathcal{L}(B) = in$, or for any supports $(\mathbf{A}, A) \in USUP$, there exists $C \in \mathbf{A}$ such that $C \notin AR$ or $\mathcal{L}(C) = out$. For the former case, $(B, A) \in ATT$, and therefore $\mathcal{L}(A) = out$. For the latter case, for any $(\mathbf{A}, A) \in SUP$, $\mathcal{L}(\mathbf{A}) = out$, and therefore $\mathcal{L}(A) = out$.

From the above, $A \in AR \wedge \mathcal{L}(A) = \mathcal{L}_U(A)$, or $A \notin AR \wedge \mathcal{L}_U(A) = out$. $\square$

For a TDN, the facts found by $PC(A)$ make the argument A acceptable.

Theorem 2. *Assume that ubaf is acyclic and that A is an argument in UAR. If $Ex \cup PC(A)$ is consistent and baf is built by BUP from $Ex \cup PC(A)$, then $A \in AR$, and the complete labeling $\mathcal{L}$ satisfies $\mathcal{L}(A) = in$. If $Ex \cup NC(A)$ is consistent and baf is built by BUP from $Ex \cup NC(A)$, then $A \notin AR$, or $A \in AR$ and the complete labeling $\mathcal{L}$ satisfies $\mathcal{L}(A) = out$.*

Proof. We prove this by induction on the height of A. For the former case, assume that $Ex \cup PC(A)$ is consistent. When A is a leaf (thus of height 0), $PC(A) = \{ex(A)\}$ (i.e., *baf* includes A and $ex(A)$), and therefore, $\mathcal{L}(A) = in$. Otherwise, for some $\mathbf{A}$ satisfying $(\mathbf{A}, A) \in USUP$, $PC(A) = \bigcup_{(B,A) \in UATT} NC(B) \cup \bigcup_{C \in \mathbf{A}} PC(C)$. For each B such that $(B, A) \in ATT$, $NC(B) \subseteq PC(A)$, and $Ex \cup NC(B)$ is thus consistent. As the height of B is less than that of A, from the induction hypothesis, $B \notin AR$ or $B \in AR$ but $\mathcal{L}(B) = out$. In a similar fashion, for each $C \in \mathbf{A}$, $C \in AR$ and $\mathcal{L}(C) = in$, and therefore $\mathcal{L}(\mathbf{A}) = in$. From the definitions of BUP and complete labeling, $A \in AR$ and $\mathcal{L}(A) = in$.

The proof for the case of $NC(A)$ is the same. $\square$

5   Related Works

Support relations play important roles in our approach. Such relations can be interpreted in several ways [12]. Cayrol et al. defined several types of indirect attacks by combining attacks with supports, and defined several types of extensions in BAF [10]. Boella et al. revised the semantics by introducing different meta-arguments and meta-supports [6]. Noueioua et al. developed a BAF that considered a support relation to be a "necessity" relation [18]. Čyras et al. considered that several semantics of a BAF could be captured using assumption-based argumentation [13]. Brewka et al. developed an abstract dialectical framework (ADF) as a generalization of Dung's AF [7,8]; a BAF was represented using an ADF. These works focus on acceptance of arguments. Here, we define a support relation and develop semantics that can represent a law.

Several authors have studied changes in AFs when arguments are added or deleted [14]. Cayrol et al. investigated changes in acceptable arguments when an argument was added to a current AF [11]. Baumann et al. developed a strategy for AF diagnosis and repair, and explored the computational complexity thereof [3]. Most research has focused on semantics, and changes in acceptable sets when arguments are added/deleted. The computational complexity associated with AF updating via argument addition/deletion is a significant issue [1]. Here, we propose the reasoning based on an incrementally constructed BAF, potentially broadening the applications of such frameworks. Complexity is not of concern; we do not need to consider all possibilities since solutions can be derived from a given universal BAF. However, it is possible to use efficient computational methods when executing our algorithm.

Our reasoning mechanism may be considered a form of hypothetical reasoning, or an abduction, which is a method used to search for the set of facts necessary to derive an observed conclusion [19]. In assumption-based argumentation, abduction is used to explain a conclusion supported by an argument [5]. Combinations of abduction and argumentation have been discussed in several works. Kakas et al. developed a method to determine the conditions that support arguments [16]. Sakama studied an abduction in argumentation framework [22] and proposed a method to search for the conditions explaining the justification state. This may include removal of an argument if it is not justified. Also, a computational method was developed by transforming an AF into a logic program. In our approach, we do not remove arguments; instead, we add absence arguments, which is equivalent to argument removal. It is reasonable to confirm the existence or absence of evidential facts when aiming to establish whether a certain law applies. The difference between the cited works and our method is that, in the previous works, observations are given and the facts that can explain those arguments are searched. In our case, potential conclusions justified by the observed facts are not specified; instead, bidirectional reasoning is performed repeatedly to assemble a knowledge set in an incremental manner. In addition, the purpose of our research is to support simulations. A minimal set of facts does not necessarily yield the best solution, unlike the cases of conventional hypothetical reasoning and common abduction.

6 Conclusion

In this paper, we developed a hybrid method featuring both bottom-up and top-down reasoning using an incrementally constructed BAF. The method can be applied to find a relevant law based on proven facts, and suggests facts that might make another law applicable. The proposed method can support those who are not familiar with a law through a simulation process, allowing a better understanding of the law to be achieved, in addition to identifying potential strategies for winning the case.

We are currently exploring reasoning processes that use three-valued representation, of which *undecided* is one possible representation. In future, we plan to implement visualization of our method.

Acknowledgment. This work was supported by JSPS KAKENHI Grant Number JP17H06103.

References

1. Alfano, G., Greco, S., Parisi, F.: A meta-argumentation approach for the efficient computation of stable and preferred extensions in dynamic bipolar argumentation frameworks. Intelligenza Artificiale **12**(2), 193–211 (2018)
2. Amgoud, L., Cayrol, C., Lagasquie-Schiex, M.C., Livet, P.: On bipolarity in argumentation frameworks. Int. J. Intell. Syst. **23**(10), 1062–1093 (2008)

3. Baumann, R., Ulbricht, M.: If nothing is accepted - repairing argumentation frameworks. In: Proceedings of KR 2010, pp. 108–117 (2018)
4. Bench-Capon, T., Prakken, H., Sartor, G.: Argumentation in legal reasoning. In: Simari, G., Rahwan, I. (eds.) Argumentation in Artificial Intelligence, pp. 363–382. Springer, Boston (2009). https://doi.org/10.1007/978-0-387-98197-0_18
5. Bondarenko, A., Dung, P.M., Kowalski, R., Toni, F.: An abstract, argumentation-theoretic approach to default reasoning. Artif. Intell. **93**, 63–101 (1997)
6. Boella, G., Gabbay, D.M., van der Torre, L., Villata, S.: Support in abstract argumentation. In: Proceedings of COMMA 2010, pp. 40–51 (2010)
7. Brewka, G., Woltran, S.: Abstract dialectical frameworks. In: Proceedings of KR 2010, pp. 102–111 (2010)
8. Brewka, G., Ellmauthaler, S., Strass, H., Wallner, J.P., Woltran, S.: Abstract dialectical frameworks revisited. In: Proceedings of IJCAI 2013, pp. 803–809 (2013)
9. Caminada, M.: On the issue of reinstatement in argumentation. In: Proceedings of JELIA 2006, 111–123 (2006)
10. Cayrol, C., Lagasquie-Schiex, M.: On the acceptability of arguments in bipolar argumentation frameworks. In: Proceedings of ECSQARU 2005, pp. 378–389 (2005)
11. Cayrol, C., de Saint-Cyr, F.D., Lagasquie-Schiex, M.: Change in abstract argumentation frameworks: adding an argument. J. Artif. Intell. Res. **28**, 49–84 (2010)
12. Cohen, A., Gottifredi, S., Garcia, A., Simari, G.: A survey of different approaches to support in argumentation systems. Knowl. Eng. Rev. **29**(5), 513–550 (2013)
13. Čyras, K., Schulz, C., Toni, F.: Capturing bipolar argumentation in non-flat assumption-based argumentation. In: Proceedings of PRIMA 2017, pp. 386–402 (2017)
14. Doutre, S., Jean-Guyb, M.: Constraints and changes: a survey of abstract argumentation dynamics. Argum. Comput. **9**(3), 223–248 (2018)
15. Dung, P.M.: On the acceptability of arguments and its fundamental role in non-monotonic reasoning, logic programming and n-person games. Artif. Intell. **77**, 321–357 (1995)
16. Kakas, A.C., Moraitis, P.: Argumentative agent deliberation, roles and context. Electron. Notes Theor. Comput. Sci. **70**, 39–53 (2002)
17. Kawasaki, T., Moriguchi, S., Takahashi, K.: Transformation from PROLEG to a bipolar argumentation framework. In: Proceedings of SAFA 2018, pp. 36–47 (2018)
18. Nouioua, F., Risch, V.: Argumentation framework with necessities. In: Proceedings of SUM 2011, pp. 163–176 (2011)
19. Poole, D.: Logical framework for default reasoning. Artif. Intell. **36**, 27–47 (1988)
20. Prakken, H., Sartor, G.: Law and logic: a review from an argumentation perspective. Artif. Intell. **36**, 214–245 (2015)
21. Rahwan, I., Simari, G. (eds.): Argumentation in Artificial Intelligence. Springer, Boston (2009). https://doi.org/10.1007/978-0-387-98197-0
22. Sakama, C.: Abduction in argumentation frameworks. J. Appl. Non-Class. Log. **28**, 218–239 (2018)
23. Satoh, K., et al.: PROLEG: an implementation of the presupposed ultimate fact theory of Japanese civil code by PROLOG technology. In: Onada, T., Bekki, D., McCready, E. (eds.) JSAI-isAI 2010. LNCS (LNAI), vol. 6797, pp. 153–164. Springer, Heidelberg (2011). https://doi.org/10.1007/978-3-642-25655-4_14

Active Preference Elicitation by Bayesian Updating on Optimality Polyhedra

Nadjet Bourdache$^{(\boxtimes)}$, Patrice Perny, and Olivier Spanjaard

Sorbonne Université, CNRS, LIP6, 75005 Paris, France
{nadjet.bourdache,patrice.perny,olivier.spanjaard}@lip6.fr

Abstract. We consider the problem of actively eliciting the preferences of a Decision Maker (DM) that may exhibit some versatility when answering preference queries. Given a set of multicriteria alternatives (choice set) and an aggregation function whose parameter values are unknown, we propose a new incremental elicitation method where the parameter space is partitioned into optimality polyhedra in the same way as in *stochastic multicriteria acceptability analysis*. Each polyhedron encompasses the subset of parameter values for which a given alternative is optimal (one optimality polyhedron, possibly empty, per alternative in the choice set). The uncertainty about the DM's judgment is modeled by a probability distribution over the polyhedra of the partition. At each step of the elicitation procedure, the distribution is revised in a Bayesian manner using preference queries whose choice is based on the *current solution strategy*, that we adapt to minimize the expected regret of the recommended alternative. We interleave the analysis of the set of alternatives with the elicitation of the parameters of the aggregation function (weighted sum or ordered weighted average). Numerical tests have been performed to evaluate the interest of the proposed approach.

Keywords: Incremental preference elicitation · Optimality polyhedra · Bayesian updating · Expected regrets

1 Introduction

Preference elicitation is an essential part of computer-aided multicriteria decision support. Indeed, criteria being often conflicting, the notion of optimality is subjective and fully depends on the Decision Maker's (DM) view on the relative importance attached to every criteria. Thus, the relevance of the recommendation depends on our ability to elicit this information and the way we model the uncertainty about the DM's preferences.

A standard way to compare feasible solutions in multicriteria decision problems is to use parameterized aggregation functions assigning a value (overall utility) to every solution. This function can be fitted to the DM preferences by eliciting the weighting coefficients that specify the importance of criteria in the aggregation. In many real cases, it is impractical but also useless to precisely

© Springer Nature Switzerland AG 2019
N. Ben Amor et al. (Eds.): SUM 2019, LNAI 11940, pp. 93–106, 2019.
https://doi.org/10.1007/978-3-030-35514-2_8

specify the parameters of the aggregation function. Given a decision model, exact choices can often be derived from a partial specification of weighting parameters. Dealing with partially specified parameters requires the development of solution methods that can determine an optimal or near optimal solution with such partial information. This is the aim of *incremental preference elicitation*, that consists on interleaving the elicitation with the exploration of the set of alternatives to adapt the elicitation process to the considered instance and to the DM's answers. Thus, the elicitation effort is focused on the useful part of the preference information. The purpose of incremental elicitation is not to learn precisely the values of the parameters of the aggregation function but to specify them sufficiently to be able to determine a relevant recommendation.

Incremental preference elicitation is the subject of several contributions in various contexts, see e.g. [3,4,7,16]. Starting from the entire set of possible parameter values, incremental elicitation methods are based on the reduction of the uncertainty about the parameter values by iteratively asking the DM to provide new preference information (e.g., with pairwise comparisons between alternatives). Any new information is translated into a hard constraint that allows to reduce the parameter space. In this way, preference data are collected until a necessarily optimal or near optimal solution can be determined, i.e., a solution that is optimal or near optimal for all the possible parameter values. These methods are very efficient because they allow a fast reduction of the parameter space. Nevertheless, they are very sensitive to possible mistakes of the DM in her answers. Indeed, in case of a wrong answer, the definitive reduction of the parameter space will exclude the wrong part of the set of possible parameter values, which is likely to exclude the optimal solution from the set of *possibly optimal* solutions (i.e., solutions that are optimal for at least one possible parameter value). Consequently, the relevance of the recommendation may be significantly impacted if there is no possible backtrack. A way to overcome this drawback is to use probabilistic approaches that allow to model the uncertainty about the DM's answers, and thus to give her the opportunity to contradict herself without impacting too much the quality of the recommendation. In such methods, the parameter space remains unchanged throughout the algorithm and the uncertainty about the real parameter values (which characterize the DM's preferences) is represented by a probability density function that is updated when new preference statements are collected.

This idea has been developed in the literature. In the context of incremental elicitation of utility values, Chajewska et al. [8] proposed to update a probability distribution over the DM's utility function to represent the belief about the utility value. The probability distribution is incrementally adjusted until the expected loss of the recommendation is sufficiently small. This method does not apply in our setting because we consider that the utility values of the alternatives on every criterion are known and that we elicit the values of the weighting coefficients of the aggregation function. Sauré and Vielma [15] introduced a method based on maintaining a confidence ellipsoid region using a multivariate Gaussian distribution over the parameter space. They use mixed integer programming to select a preference query that is the most likely to reduce the volume of the confidence region. In a recent work [5], the uncertainty about the parameter values is represented by a Gaussian distribution over the parameter space of rank-dependent aggregation functions. Preference queries are selected by minimizing expected regrets to update the density function using Bayesian linear regression. As the updating of a continuous density function is computationally cumbersome (especially when analytical results for the obtention of the posterior density function do not exist), data augmentation and sampling techniques are used to approximate the posterior density function. These methods are time consuming and require to make a tradeoff between computation time and accuracy of the approximation. In addition, the information provided by a continuous density function may be much richer than the information really needed by the algorithm to conclude. Indeed, it is generally sufficient to know that the true parameter values belong to a given restricted area of the parameter space to be able to identify an optimal solution without ambiguity. Thus, we introduce in this paper a new model-based incremental elicitation algorithm based on a discretization of the parameter space. We partition the parameter space into optimality polyhedra and we define a probability distribution over the partition. After each query, this distribution is updated using Bayes' rule.

The paper is organised as follows. Section 2 recalls some background on weighted sums and ordered weighted averages. We also introduce the optimality polyhedra we use in our method and we discuss our contribution with regard to related works relying on the optimality polyhedra. We present our incremental elicitation method in Sect. 3. Finally, some numerical tests showing the interest of the proposed approach are provided in Sect. 4.

2   Background and Notations

Let $\mathcal{X}$ be a set of n alternatives evaluated on p criteria. Any alternative of $\mathcal{X}$ is characterized by a performance vector $x = (x_1, \ldots, x_p)$, where $x_i \in [0, U]$ is the performance of the alternative on criterion i, and U is the maximum utility value. All utilities x_i are expressed on the same scale; the utility functions must be defined from the input data (criterion or attribute values), as proposed by, e.g., Grabisch and Labreuche [10]. To refine the Pareto dominance relation and to be able to better discriminate between alternatives in $\mathcal{X}$, we use a parametrized aggregation function denoted by f_w. The weighting vector w of the function defines how the components of x should be aggregated and thus makes it possible to model the decision behavior of the DM. In this paper, we consider two operators: the weighted sum (WS) and the ordered weighted average (OWA). We give some notations and recall some basic notions about this two aggregation functions in the following.

Weighted Sum. Let $x \in \mathbb{R}^p_+$ be a performance vector and $w \in \mathbb{R}^p_+$ be a weighting vector. The weighted sum is defined by:

$$\mathrm{WS}_w(x) = \sum_{i=1}^{p} w_i x_i \tag{1}$$

Ordered Weighted Average. Introduced by Yager [17], the OWA is a rank-dependent aggregation function, where the weights are not associated to the criteria but to the ranks in the ordered performance vector, giving more or less importance to good or bad performances. Let $x \in \mathbb{R}_+^p$ be a performance vector and $w \in \mathbb{R}_+^p$ be a weighting vector. The ordered weighted average is defined by:

$$\text{OWA}_w(x) = \sum_{i=1}^{p} w_i x_{(i)} \tag{2}$$

where $x_{(.)}$ is a permutation of vector x such that $x_{(1)} \leq \cdots \leq x_{(p)}$.

Example 1. *Let $x = (14, 9, 10)$, $y = (10, 12, 10)$ and $z = (9, 16, 6)$ be three performance vectors to compare, and assume that the weighting vector is $w = (\frac{1}{4}, \frac{1}{2}, \frac{1}{4})$. Applying Eq. (2), we obtain: $\text{OWA}_w(x) = 10.75 > \text{OWA}_w(y) = 10.5 > \text{OWA}_w(z) = 10$.*

Note that OWA includes the minimum ($w_1 = 1$ and $w_i = 0, \forall i \in [\![2, p]\!]$), the maximum ($w_p = 1$ and $w_i = 0, \forall i \in [\![1, p-1]\!]$), the arithmetic mean ($w_i = \frac{1}{p}, \forall i \in [\![1, p]\!]$) and all other order statistics as special cases.

If w is chosen with decreasing components (i.e., the greatest weight is assigned to the worst performance), the OWA function is *concave* and well-balanced performance vectors are favoured. We indeed have, for all $x \in \mathcal{X}$, $\text{OWA}_w((x_1, \ldots, x_i - \varepsilon, \ldots, x_j + \varepsilon, \ldots, x_p)) \geq \text{OWA}_w(x)$ for all i, j and $\varepsilon > 0$ such that $x_i - x_j \geq \varepsilon$. Depending on the choice of the weighting vector w, a concave OWA function allows to define a wide range of mean type aggregation operators between the minimum and the arithmetic mean. In the remainder of the paper, we only consider *concave* OWA functions. For the sake of brevity, we will say OWA for concave OWA.

Example 2. *Consider vectors x, y and z defined in Example 1 and assume that the weighting vector is now $w = (\frac{1}{2}, \frac{1}{3}, \frac{1}{6})$. We have: $\text{OWA}_w(x) = \frac{61}{6}$, $\text{OWA}_w(y) = \frac{62}{6}$ and $\text{OWA}_w(z) = \frac{52}{6}$. The alternative y, which corresponds to the most balanced performance vector, is the preferred one.*

Using f_w (defined with (1) or (2)) as an aggregation function, we call f_w-*optimal* an alternative x that maximizes $f_w(x)$. Eliciting the DM's preferences amounts to eliciting the weighting vector w. The rest of the section defines how we deal with the imprecise knowledge of the parameter values in the optimization process involved in the elicitation.

Optimality Polyhedra. We denote by W the set of all feasible weighting vectors. Note that, to limit the scale of this set, one can add the additional non restrictive normalisation constraint $\sum_{i=1}^{p} w_i = 1$. Thus, W is defined by $W = \{w \in \mathbb{R}_+^p \mid \sum_{i=1}^{p} w_i = 1 \text{ and } w_i \geq 0, \forall i\}$. In the case of a concave OWA, the additional constraint $w_1 \geq \cdots \geq w_p$ is enforced.

Starting from W and the set $\mathcal{X}$ of alternatives, we partition W into *optimality polyhedra*: the optimality polyhedron associated to an alternative x is the set of weighting vectors such that x is optimal. Note that the aggregation functions we use are linear in w (even though OWA is not linear in x because of the sorting of x before applying the aggregation operation).

This explains why the sets of the partition are convex polyhedra. Any preference statement of the form "Alternative x is preferred to alternative y" is indeed translated into a constraint $f_w(x) \geq f_w(y)$ which is linear in w.

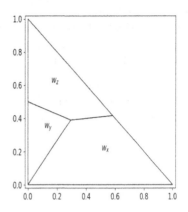

Fig. 1. Optimality polyhedra for x, y and z in Example 1 with WS.

More formally, the optimality polyhedron W_x associated to an alternative $x \in \mathcal{X}$ is defined by $W_x = \{w \in W | f_w(x) \geq f_w(y), \forall y \in \mathcal{X}\}$. Note that any empty set W_x (there is no $w \in W$ such that x is f_w-optimal) or not full dimensional set (i.e., $\forall w \in W_x, \exists y \in \mathcal{X}$ such that $f_w(x) = f_w(y)$) can be omitted. An example of such partition is given in Fig. 1 for the instance of Example 1, where the aggregation function is a weighted sum. Note that w_3 can be omitted thanks to the normalization constraint ($w_3 = 1 - w_1 - w_2$).

In order to represent the uncertainty about the exact values of parameters, a probability distribution is defined over the polyhedra of the partition. This distribution is updated using an incremental elicitation approach that will be described in the next section.

Related Works. The idea of partitioning the parameter space is closely related to Stochastic Multiobjective Acceptability Analysis (SMAA for short). The SMAA methodology has been introduced by Charnetski and Soland under the name of *multiple attribute decision making with partial information* [9]. Given a set of utility vectors and a set of linear constraints characterizing the feasible parameter space for a weighted sum (partial information elicited from the DM), they assume that the probability of optimality for each alternative is proportional to the hypervolume of its optimality polyhedron (the hypervolume reflects how likely an alternative is to be optimal). Lahdelma et al. [12] developed this idea in the case of imprecision or uncertainty in the input data (utilities of the alternatives according to the different criteria) by considering the criteria values as probability distributions. They defined the *acceptability index* for an alternative, that measures the variety of different valuations which allow for that alternative to be optimal, and is proportional to the *expected* volume of its optimality polyhedron. They also introduced a *confidence factor*, that measures if the input data is accurate enough for making an informed decision. The methodology has been adapted to the 2-additive Choquet integral model by Angilella et al. [2]. These works consider that the uncertainty comes from the criterion values or from the variation in the answers provided by *different* DMs. They also consider that some prior preference information is given and that there is no opportunity to ask the DM for new preference statements. Our work differentiates from these works in the following points:

- the criterion values are accurately known and only the parameter values of the aggregation function must be elicited;
- the uncertainty comes from possible errors in the DM's answers to preference queries;
- the elicitation process is incremental.

3   Incremental Elicitation Approach

Once the parameter space W is partitioned into *optimality polyhedra* as explained above, a prior density function is associated to the partition. This distribution informs us on how likely each solution is to be optimal. In the absence of a prior information about the DM's preferences, we define the prior distribution such that the probability of any polyhedron is proportional to its volume, as suggested by Charnetski and Soland [9]. The volume of W_x gives indeed a measure on the proportion of weighting vectors for which the alternative x is ranked first. More formally, the prior probability of x to be optimal is $P(x) = \frac{vol_{W_x}}{vol_W}$ where vol_W denotes the volume of a convex polyhedron W. We assume here a complete ignorance of the continuous probability distribution for w within each polyhedron. After each new preference statement, the probability distribution P is updated using Bayes' rule.

The choice of the next query to ask is a key point for the efficiency of the elicitation process in acquiring enough preferential information to make a recommendation with sufficient confidence.

Query Selection Strategy. In order to get the most informative possible query we use a strategy based on the minimization of expected regrets. Let us first introduce how we define expected regrets in our setting:

Definition 1. *Given two alternatives x and y, and a probability distribution P on $\mathcal{X}$, the pairwise expected regret PER is defined by:*

$$\text{PER}(x, y, \mathcal{X}, P) = \sum_{z \in \mathcal{X}} \max\{0, \text{PMR}(x, y, W_z)\} P(z)$$

where $P(z)$ represents the probability for z to be optimal and $\text{PMR}(x, y, \mathcal{W})$ is the pairwise maximum regret over a polyhedron $\mathcal{W}$, defined by:

$$\text{PMR}(x, y, \mathcal{W}) = \max_{w \in \mathcal{W}} \{f_w(y) - f_w(x)\}$$

In other words, the PER defines the expected worst utility loss incurred by recommending an alternative x instead of an alternative y, and $\text{PMR}(x, y, \mathcal{W})$ is the worst utility loss in recommending alternative x instead of alternative y

given that w belongs to $\mathcal{W}$. The use of the PMR within a polyhedron is justified by the complete ignorance about the probability distribution in the polyhedron, thereby, the worst case is considered.

Definition 2. *Given a set $\mathcal{X}$ of alternatives, the maximum expected regret of $x \in \mathcal{X}$ and the minimax expected regret over $\mathcal{X}$ are defined by:*

$$\text{MER}(x, \mathcal{X}, P) = \max_{y \in \mathcal{X}} \text{PER}(x, y, \mathcal{X}, P)$$

$$\text{MMER}(\mathcal{X}, P) = \min_{x \in \mathcal{X}} \text{MER}(x, \mathcal{X}, P)$$

In other words, the MER value defines the worst utility loss incurred by recommending an alternative $x \in \mathcal{X}$ and the MMER value defines the minimal MER value over $\mathcal{X}$.

The notion of regret expresses a measure of the interest of an alternative. At any step of the algorithm, the solution achieving the MMER value is a relevant recommendation because it minimizes the expected loss in the current state of knowledge. It also allows to determine an informative query to ask. Various query selection strategies based on regrets and expected regrets have indeed been introduced in the literature, see e.g. [6] in a deterministic context (*current solution strategy*) and [11] in a probabilistic context (a probability distribution is used to model the uncertainty about the parameter values). Adapting the current solution strategy to our probabilistic setting, we propose here a strategy that consists in asking the DM to compare the current recommendation $x^* = \arg \min_{x \in \mathcal{X}} \text{MER}(x, \mathcal{X}, P)$ to its best challenger defined by $y^* = \arg \max_{y \in \mathcal{X}} \text{PER}(x^*, y, P)$. The current probability distribution is then updated according to the DM's answer, as explained hereafter. The procedure can be iterated until the MMER value drops below a predefined threshold ε.

The approach proposed in this paper consists in interleaving preference queries and Bayesian updating of the probability distribution based on the DM's answers. The elicitation procedure is detailed in Algorithm 1. At each step i of the algorithm, we ask the DM to compare two alternatives $x^{(i)}$ and $y^{(i)}$. The answer is denoted by a_i, where $a_i = 1$ if $x^{(i)}$ is preferred to $y^{(i)}$ and $a_i = 0$ otherwise. From each answer a_i, the conditional probability $P(.|a_1, \ldots, a_{i-1})$ over the set of alternatives is updated in a Bayesian manner (Line 13 of Algorithm 1).

Bayesian Updating. We assume that answers a_i are independent binary random variables, i.e. $P(a_i|x^{(i)}, y^{(i)})$ only depends on the (unknown) weighting vector w and on the performance vectors of $x^{(i)}, y^{(i)}$. This is a standard assumption in Bayesian analysis of binary response data [1]. To alleviate the notations, we omit the conditioning statement in $P(a_i|x^{(i)}, y^{(i)})$, that we abbreviate by $P(a_i)$. Using Bayes' rule, the posterior probability of any alternative $z \in \mathcal{X}$ is given by:

$$P(z|a_1, \ldots, a_i) = \frac{P(a_1, \ldots, a_i|z)P(z)}{P(a_1, \ldots, a_i)} = \frac{P(a_i|z)P(a_1, \ldots, a_{i-1}|z)P(z)}{P(a_i)P(a_1, \ldots, a_{i-1})} \quad (3)$$

$$= \frac{P(a_i|z)P(z|a_1, \ldots, a_{i-1})}{P(a_i)} \quad (4)$$

Algorithm 1: Incremental Elicitation Procedure

Input: $\mathcal{X}$: set of alternatives, ε: acceptance threshold; W: parameter space.
Output: x^* : best recommendation in $\mathcal{X}$

1   $P(z) \leftarrow \frac{vol_{W_z}}{vol_W}, \forall z \in \mathcal{X}$

2   $i \leftarrow 0$

3 **repeat**

4     $i \leftarrow i + 1$

5     $x^{(i)} \leftarrow \arg\min_{x \in \mathcal{X}} \mathrm{MER}(x, \mathcal{X}, P(.|a_1, \ldots, a_{i-1}))$

6     $y^{(i)} \leftarrow \arg\max_{y \in \mathcal{X}} \mathrm{PER}(x^{(i)}, y, P(.|a_1, \ldots, a_{i-1}))$

7   Ask the DM if $x^{(i)}$ is preferred to $y^{(i)}$

8   **if** *the answer is yes* **then**

9       $a_i \leftarrow 1$

10   **else**

11       $a_i \leftarrow 0$

12   **for** $z \in \mathcal{X}$ **do**

13   Compute $P(z|a_1, \ldots, a_i)$ using Bayesian updating

14 **until** $\mathrm{MMER}(\mathcal{X}, P(.|a_1, \ldots, a_i)) \leq \varepsilon$;

15 **return** x^* selected in $\arg\min_{x \in \mathcal{X}} \mathrm{MER}(x, \mathcal{X}, P(.|a_1, \ldots, a_i))$

The likelihood function $P(a_i|z)$ is the conditional probability that the answer is a_i given that z is optimal. Let us denote by $W_{x^{(i)} \succsim y^{(i)}}$ the subset of W containing all vectors w such that $f_w(x^{(i)}) \geq f_w(y^{(i)})$; the likelihood function is defined as:

$$P(a_i = 1|z) = \begin{cases} \delta & \text{if } W_z \subseteq W_{x^{(i)} \succsim y^{(i)}} \\ 1 - \delta & \text{if } W_z \cap W_{x^{(i)} \succ y^{(i)}} = \emptyset \\ P(a_i = 1) & \text{otherwise} \end{cases}$$

where $\delta \in (\frac{1}{2}, 1]$ is a constant. The corresponding update of the probability masses follows the idea used by Nowak in noisy generalized binary search [14] and its effect is simple; the probability masses of polyhedra that are compatible with the preference statement are boosted relative to those that are not compatible, while the probability masses of the other polyhedra remain unchanged. The parameter δ controls the size of the boost, and can be seen as a lower bound on the probability of a correct answer. The three cases are depicted in Fig. 2.

In the third case (on the right of Fig. 2), due to the assumption of complete ignorance within a polyhedron, the new preference statement is not informative enough to update the probability of z to be optimal. Therefore, for all alternatives z such that W_z is cut by the constraint $f_w(x^{(i)}) \geq f_w(y^{(i)})$ no updating is performed and therefore $P(a_i|z) = P(a_i)$; consequently $P(z|a_1, \ldots, a_i) = P(z|a_1, \ldots, a_{i-1})$ by Eq. 4.

Regarding Eq. 4, note that, in practice, we do not need to determine $P(a_i)$. For any alternative $z \in \mathcal{X}$ such that W_z is not cut by the constraint, we have indeed $P(z|a_1, \ldots, a_i) \propto P(a_i|z)P(z|a_1, \ldots, a_{i-1})$. More precisely, $P(z|a_1, \ldots, a_i)$ is obtained by the following equation:

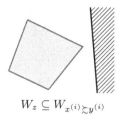
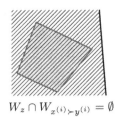
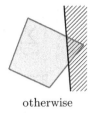

$$W_z \subseteq W_{x^{(i)} \succsim y^{(i)}} \qquad W_z \cap W_{x^{(i)} \succ y^{(i)}} = \emptyset \qquad \text{otherwise}$$

Fig. 2. The polyhedron is W_z. The non-hatched area is the half-space $W_{x^{(i)} \succsim y^{(i)}}$.

$$P(z|a_1, \ldots, a_i) = \sum_{y \in \mathcal{Y}} P(y|a_1, \ldots, a_{i-1}) \frac{P(a_i|z)P(z|a_1, \ldots, a_{i-1})}{\sum\limits_{y \in \mathcal{Y}} P(a_i|y)P(y|a_1, \ldots, a_{i-1})} \qquad (5)$$

where $\mathcal{Y}$ is the subset of alternatives whose optimality polyhedra are not cut by the constraint. The condition $\sum_{z \in \mathcal{X}} P(z|a_1, \ldots, a_i) = 1$ obviously holds.

If the optimal alternative x^* is unique, the proposition below states that, using Algorithm 1, the probability assigned to x^* cannot decrease if the DM always answers correctly.

Proposition 1. *Let us denote by x^* a uniquely optimal alternative. At any step i of Algorithm 1, if the answer to query i is correct, then:*

$$P(x^*|a_1, \ldots, a_i) \geq P(x^*|a_1, \ldots, a_{i-1})$$

Proof. Two cases can be distinguished:

Case 1. If $W_{x^*} \not\subseteq W_{x^{(i)} \succsim y^{(i)}}$ and $W_{x^*} \cap W_{x^{(i)} \succ y^{(i)}} \neq \emptyset$, then, as mentioned above, $P(x^*|a_1, \ldots, a_i) = P(x^*|a_1, \ldots, a_{i-1})$ by Eq. 4 because $P(a_i|x^*) = P(a_i)$.

Case 2. Otherwise, whatever the answer α of the DM, we have $P(a_i = \alpha|x^*) = \delta$ because the answer to query i is correct. By Eq. 5, it follows that:

$$P(x^*|a_1, \ldots, a_i) = \underbrace{\frac{\delta \sum_{y \in \mathcal{Y}} P(y|a_1, \ldots, a_{i-1})}{\sum_{y \in \mathcal{Y}} P(a_i = \alpha|y)P(y|a_1, \ldots, a_{i-1})}}_{\text{ratio } \rho} P(x^*|a_1, \ldots, a_{i-1})$$

We now show that $\rho \geq 1$ for $\delta > \frac{1}{2}$. Let us denote by $\mathcal{Y}_\delta$ the subset of alternatives $y \in \mathcal{Y}$ such that $P(a_i = \alpha|y) = \delta$. We have:

$$\sum_{y \in \mathcal{Y}} P(a_i = \alpha|y)P(y|a_1, \ldots, a_{i-1})$$

$$= \delta \sum_{y \in \mathcal{Y}_\delta} P(y|a_1, \ldots, a_{i-1}) + (1 - \delta) \sum_{y \in \mathcal{Y}_{1-\delta}} P(y|a_1, \ldots, a_{i-1})$$

because $\mathcal{Y} = \mathcal{Y}_\delta \cup \mathcal{Y}_{1-\delta}$ and $\mathcal{Y}_\delta \cap \mathcal{Y}_{1-\delta} = \emptyset$

$$\leq \delta \sum_{y \in \mathcal{Y}_\delta} P(y|a_1,\ldots,a_{i-1}) + \delta \sum_{y \in \mathcal{Y}_{1-\delta}} P(y|a_1,\ldots,a_{i-1})$$

because $\delta > \dfrac{1}{2}$ (the only case of equality is when $\mathcal{Y}_{1-\delta} = \emptyset$)

$$= \delta \sum_{y \in \mathcal{Y}} P(y|a_1,\ldots,a_{i-1})$$

Consequently, $\rho \geq 1$ and thus $P(x^*|a_1,\ldots,a_i) \geq P(x^*|a_1,\ldots,a_{i-1})$. ☐

Toward an Efficient Implementation. As mentioned above, in order to update the probability of an alternative z, we need to know the relative position of its optimality polyhedron W_z compared to the constraint induced by the new preference statement $f_w(x^{(i)}) \geq f_w(y^{(i)})$. In this purpose, we can consider the Linear Programs (LPs) $\mathrm{opt}\{f_w(x^{(i)}) - f_w(y^{(i)})|w \in W_z\}$, where $\mathrm{opt} = \min$ or $\max$.

If the optimal values of both LPs share the same sign, then we can conclude that the polyhedron is not cut by the constraint, otherwise it is cut. To limit the number of LPs that need to be solved (determining the positions of all the polyhedra would indeed require to solve $2n$ LPs), and thereby speed up the Bayesian updating, we propose to approximate the polyhedra by their outer Chebyshev balls (i.e., the smallest ball that contains the polyhedron). Let us denote by r the radius of the Chebyshev ball and by d the distance between the center of the ball and the hyperplane induced by the preference statement:

- if $d \geq r$ then the polyhedron is not cut by the constraint (see Fig. 3a). In order to know whether the polyhedron verifies the constraint or not, we just need to check whether the center of the ball verifies it or not. Thus, in this case, only two scalar products are required.
- if $d < r$ then an exact computation is required because the polyhedron can either be cut by the constraint (Fig. 3b) or not (Fig. 3c). In this way, the use of Chebyshev balls does not impact the results of the Bayesian updating but only speeds up the computations.

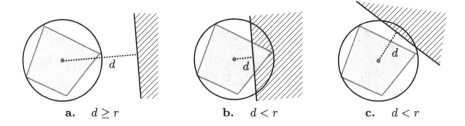

   a. $\ d \geq r$ **b.** $\ d < r$ **c.** $\ d < r$

Fig. 3. Example of an approximation of a polyhedron by an outer Chebyshev ball.

4   Experimental Results

Algorithm 1 has been implemented in Python using the *polytope* library to manage optimality polyhedra, and tested on randomly generated instances. We performed the tests on an Intel(R) Core(TM) i7-4790 CPU with 15 GB of RAM.

Random Generation of Instances. To evaluate the performances of Algorithm 1, we generated instances with 100 alternatives evaluated on 5 criteria, all possibly f_w-optimal (i.e., $W_x \neq \emptyset \; \forall x \in \mathcal{X}$). The generation of the performance vectors depends on the aggregation function (WS or OWA) that is considered:

- **WS instances.** An alternative x of the instance is generated as follows: a vector y of size 4 is uniformly drawn in $[0, 1]^4$, then x is obtained by setting $x_i = y_i - y_{i-1}$ for $i = 1, \ldots, 5$, where $y_0 = 0$ and $y_5 = 1$. The vectors thus generated all belong to the same hyperplane (because $\sum_{i=1}^{5} x_i = 1$ for all $x \in \mathcal{X}$) and the set of possibly *unique* WS-optimal alternatives is therefore significantly reduced (because the optimality polyhedra of many alternatives are not full dimensional). To avoid this issue, as suggested by Li [13], we apply the square root function on all components x_i for all $x \in \mathcal{X}$ in order to *concavify* the Pareto front. The set of performance vectors obtained is illustrated on the left of Fig. 4 in the bicriteria case.
- **OWA instances.** An alternative x is possibly OWA-optimal in a set $\mathcal{X}$ if its *Lorenz curve* $L(x)$ defined by $L_k(x) = \sum_{i=1}^{k} x_{(i)} (k \in [\![1, 5]\!])$ is possibly WS-optimal in $\{L(x) : x \in \mathcal{X}\}$. We say that a vector z is *Lorenz* if there exists a vector x such that $z = L(x)$. Given a Lorenz vector z, we denote by $L^{-1}(z)$ any vector x such that $L(x) = z$. For such a vector x, we have $x_{(i)} = z_i - z_{i-1}$ for all $i = 1, \ldots, 5$, where $z_0 = 0$. An alternative x of the instance is generated as follows: we first generate a point y in the polyhedron defined by the following linear constraints:

$$(\mathcal{P}) \begin{cases} y_{i+1} \geq y_i & \forall i \in [\![0, 4]\!] & (1) \\ (i+1)^2 y_{i+1} - i^2 y_i \geq i^2 y_i - (i-1)^2 y_{i-1} & \forall i \in [\![1, 4]\!] & (2) \\ \sum_{i=1}^{5} i^2 y_i = \sum_{i=1}^{5} i^2 & & (3) \\ y_0 = 0 \end{cases}$$

The set $\mathcal{L} = \{(i^2 y_i)_{i \in [\![1,5]\!]} : y \in \mathcal{P}\}$ contains vectors that are all Lorenz thanks to constraints (1) and (2). Furthermore, they belong to the same hyperplane due to constraint (3), and therefore they are all possibly WS-optimal. Consequently, all the alternatives in the set $\{L^{-1}(z) : z \in \mathcal{L}\}$ are possibly OWA-optimal. As above, to make them all possibly *unique* OWA-optimal, the square root function is applied on each component of vectors z in $\mathcal{L}$. The obtained set is $\mathcal{L}' = \{(i\sqrt{y_i})_{i \in [\![1,5]\!]} : y \in \mathcal{P}\}$. All the vectors in $\mathcal{X}' = \{L^{-1}(z) : z \in \mathcal{L}'\}$ are possibly unique OWA-optimal. Finally, to generate an alternative x in $\mathcal{X}'$, we randomly draw a convex combination $y = \sum_{j=1}^{m} \alpha_j \hat{y}^j$ of vertices $\hat{y}^1, \ldots, \hat{y}^m$ of $\mathcal{P}$. The obtained alternative is then defined by $x = L^{-1}((i\sqrt{y_i})_{i \in [\![1,5]\!]})$. The set of performance vectors obtained is illustrated on the right of Fig. 4 in the bicriteria case.

Finally, for both types of instances, a hidden vector w is generated to simulate the preferences of the DM.

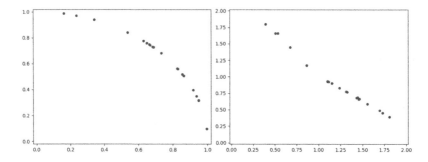

Fig. 4. Example of WS (left) and OWA (right) instances with $n = 20$ and $p = 2$

Simulation of the Interactions with the DM. To simulate the DM's answer to query i, we represent the intensity of preference between alternatives $x^{(i)}$ and $y^{(i)}$ by the variable $u^{(i)} = f_w(x^{(i)}) - f_w(y^{(i)}) + \varepsilon^{(i)}$ where $\varepsilon^{(i)} \sim \mathcal{N}(0, \sigma^2)$ is a Gaussian noise modelling the possible DM's error, with σ determining how wrong the DM can be. The DM states that $x^{(i)} \succsim y^{(i)}$ if and only if $u^{(i)} \geq 0$.

Analysis of the Results. We evaluated the efficiency of Algorithm 1 in terms of the actual rank of the recommended alternative. We considered different values for σ in order to test the tolerance to possible errors. More precisely, $\sigma = 0$ gives an error free model while $\sigma \in \{0.1, 0.2, 0.3\}$ models different rates of errors in the answers to queries. In the considered instances, these values led to, respectively, $3.6\%, 10\%$ and 22% of wrong answers for WS and to $3.2\%, 16\%$ and 25% of wrong answers for OWA. We set $\delta = 0.8$, which corresponds to a prior assumption of an error rate of 20%. Thus, the value of δ we used in the experiments is uncorrelated to the ones of σ. The computation time between two queries is less than 1 s in all cases. Results are averaged over 40 instances.

We first observed the evolution of the actual rank of the MMER alternative over queries (actual rank according to a hidden weighting vector representing the DM's preferences). Figure 5 (resp. Fig. 6) shows the curves obtained for WS (resp. OWA). We observe that the mean rank drops below 2 (out of 100 alternatives) after about 14 queries for WS with $\sigma < 0.3$, while the same happens for OWA whatever value of σ. We see that, in practice, the efficiency of the approach can be significantly impacted only when the error rate becomes greater than 20%.

We next compared the performance of Algorithm 1 with a deterministic approach described in [4], that consists in reducing the parameter space after each query (assuming that all answers are correct). The results are illustrated by the boxplots in Fig. 7 for WS, and in Fig. 8 for OWA. We can see that our probabilistic approach is more tolerant to errors than the deterministic approach. As the value of σ increases, the deterministic approach makes less and less relevant recommendations. The deterministic approach indeed recommends, in the

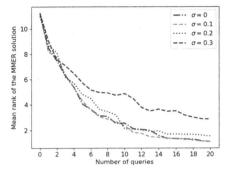

Fig. 5. Mean rank vs. queries (WS)

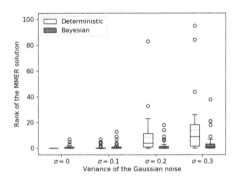

Fig. 6. Mean rank vs. queries (OWA)

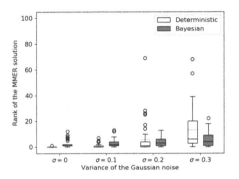

Fig. 7. Rank vs. error rate (WS)

Fig. 8. Rank vs. error rate (OWA)

worst case, alternatives that are ranked around 90 while it is less than 40 for Algorithm 1. More precisely, when $\sigma = 0.3$ (for both WS and OWA), in more than 75% of instances, Algorithm 1 recommends an alternative with a better rank than the mean rank obtained in the deterministic case.

5   Conclusion

We introduced in this paper a new model based incremental multicriteria elicitation method relying on a partition of the parameter space. The elements of the partition are the optimality polyhedra of the different alternatives, relatively to a weighted sum or an ordered weighted average. A probability distribution is defined over this partition, where each probability represents the likelihood that the true weighting vector belongs to the polyhedron. The approach is robust to possible mistakes in the DM's answers thanks to the incremental revision of probabilities in a Bayesian setting. We provide numerical tests showing the efficiency of the proposed algorithm in terms of number of queries, as well as the interest of using such a probabilistic approach compared to a deterministic approach. A short term research direction is to investigate if it is possible to further speed up the Bayesian updating by using outer Löwner-John ellipsoids instead of Chebyshev balls. The answer is not straightforward because, on the one hand, the use of ellipsoids indeed refines the approximation of the polyhedra, but on the other hand, this requires the use of matrix calculations to establish whether or not an ellipsoid is cut by the constraint induced by a preference statement. Another natural research direction is to extend our approach to more sophisticated aggregation functions admitting a linear representation, such as Weighted OWAs and other Choquet integrals, to improve our descriptive possibilities.

References

1. Albert, J.H., Chib, S.: Bayesian analysis of binary and polychotomous response data. J. Am. Stat. Assoc. **88**(422), 669–679 (1993)
2. Angilella, S., Corrente, S., Greco, S.: Stochastic multiobjective acceptability analysis for the choquet integral preference model and the scale construction problem. Eur. J. Oper. Res. **240**(1), 172–182 (2015)
3. Benabbou, N., Perny, P.: Incremental weight elicitation for multiobjective state space search. In: AAAI-15, pp. 1093–1099 (2015)
4. Bourdache, N., Perny, P.: Active preference elicitation based on generalized gini functions: application to the multiagent knapsack problem. In: AAAI 2019, pp. 7741–7748 (2019)
5. Bourdache, N., Perny, P., Spanjaard, O.: Incremental elicitation of rank-dependent aggregation functions based on Bayesian linear regression. In: IJCAI 2019, pp. 2023–2029 (2019)
6. Boutilier, C., Patrascu, R., Poupart, P., Schuurmans, D.: Constraint-based optimization and utility elicitation using the minimax decision criterion. Artif. Intell. **170**(8–9), 686–713 (2006)
7. Braziunas, D., Boutilier, C.: Minimax regret based elicitation of generalized additive utilities. In: Proceedings of UAI-07, pp. 25–32 (2007)
8. Chajewska, U., Koller, D., Parr, R.: Making rational decisions using adaptive utility elicitation. In: Proceedings of AAAI-00, pp. 363–369 (2000)
9. Charnetski, J.R., Soland, R.M.: Multiple-attribute decision making with partial information: the comparative hypervolume criterion. Nav. Res. Logist. Q. **25**(2), 279–288 (1978)
10. Grabisch, M., Labreuche, C.: A decade of application of the Choquet and Sugeno integrals in multi-criteria decision aid. Ann. OR **175**(1), 247–286 (2010)
11. Guo, S., Sanner, S.: Multiattribute Bayesian preference elicitation with pairwise comparison queries. In: NIPS, pp. 396–403 (2010)
12. Lahdelma, R., Hokkanen, J., Salminen, P.: SMAA - stochastic multiobjective acceptability analysis. Eur. J. Oper. Res. **106**(1), 137–143 (1998)
13. Li, D.: Convexification of a noninferior frontier. J. Optim. Theory Appl. **88**(1), 177–196 (1996)
14. Nowak, R.: Noisy generalized binary search. In: Bengio, Y., Schuurmans, D., Lafferty, J.D., Williams, C.K.I., Culotta, A. (eds.) Advances in Neural Information Processing Systems, vol. 22, pp. 1366–1374. Curran Associates, Inc. (2009)
15. Sauré, D., Vielma, J.P.: Ellipsoidal methods for adaptive choice-based conjoint analysis. Oper. Res. **67**, 315–338 (2019)
16. Wang, T., Boutilier, C.: Incremental utility elicitation with the minimax regret decision criterion. IJCAI **3**, 309–316 (2003)
17. Yager, R.R.: On ordered weighted averaging aggregation operators in multicriteria decision making. IEEE Trans. Syst. Man Cybern. **18**(1), 183–190 (1988)

Selecting Relevant Association Rules
From Imperfect Data

Cécile L'Héritier[1,2(✉)], Sébastien Harispe[1], Abdelhak Imoussaten[1],
Gilles Dusserre[1], and Benoît Roig[2]

[1] LGI2P, IMT Mines Ales, Univ Montpellier, Alès, France
{cecile.lheritier,sebastien.harispe,abdelhak.imoussaten,
gilles.dusserre}@mines-ales.fr
[2] EA7352 CHROME, Université de Nîmes, Nîmes, France
{cecile.lhritier,benoit.roig}@unimes.fr

Abstract. Association Rule Mining (ARM) in the context of imperfect data (*e.g.* imprecise data) has received little attention so far despite the prevalence of such data in a wide range of real-world applications. In this work, we present an ARM approach that can be used to handle imprecise data and derive imprecise rules. Based on evidence theory and Multiple Criteria Decision Analysis, the proposed approach relies on a selection procedure for identifying the most relevant rules while considering information characterizing their interestingness. The several measures of interestingness defined for comparing the rules as well as the selection procedure are presented. We also show how *a priori* knowledge about attribute values defined into domain taxonomies can be used to (i) ease the mining process, and to (ii) help identifying relevant rules for a domain of interest. Our approach is illustrated using a concrete simplified case study related to humanitarian projects analysis.

Keywords: Association rules · Imperfect data · Evidence theory · Multiple Criteria Decision Analysis (MCDA)

1 Introduction

Association rule mining (ARM) is a well-known data mining technique designed to extract interesting patterns in databases. It has been introduced in the context of market basket analysis [1], and has received a lot of attention since then [15]. An association rule is usually formally defined as an implication between an *antecedent* and a *consequent*, being conjunctions of attributes in a database, e.g. "People who have age-group between 20 and 30 and a monthly income greater than \$2k are likely to buy product X". Such rules are interesting for extracting simple intelligible knowledge from a database; they can also further be used in several applications, e.g. recommendation, customer or patient analysis. A large literature is dedicated to the study of ARM, and numerous algorithms have been defined for efficiently extracting rules handling a large range of data

© Springer Nature Switzerland AG 2019
N. Ben Amor et al. (Eds.): SUM 2019, LNAI 11940, pp. 107–121, 2019.
https://doi.org/10.1007/978-3-030-35514-2_9

types, e.g., nominal, ordinal, quantitative, sequential [15]. Nevertheless, only a few contributions of the literature study the case of ARM with imperfect data, e.g. [13,24], even if such data is central in numerous real-world applications.

In order to extend the body of work related to ARM with imperfect data, and to answer some of the limitations of existing contributions, this paper presents a novel ARM approach that can be used to handle imprecise data and derive imprecise rules. In this study, to simplify, the proposed approach focuses on a specific case where the *antecedent* and the *consequent* are composed of predefined disjoint sets of attributes forming a partition of the whole set of attributes. This particular case is relevant, for example in classification tasks in which the label value to predict can be defined as consequent of the rules of interest. To sum up, our goal is threefold: (i) to enrich the expressivity of existing proposed frameworks, (ii) to complement them with a richer procedure for selecting relevant rules, and (iii) to present simple way to incorporate domain knowledge to ease the mining process, and to help identifying relevant rules for a domain of interest. Based on the evidence theory framework and Multiple Criteria Decision Analysis, a selection procedure for identifying the most relevant rules while considering information characterizing their interestingness is proposed. The several measures of interestingness defined for comparing the rules, as well as the selection procedure, are presented. We also show how *a priori* knowledge in the form of taxonomies about consequent and antecedent (i.e. attribute values) can be used to focus on rules of interest for a domain. We also present an illustration using a simplified case study related to humanitarian projects analysis.

The paper is structured as follows: Sect. 2 formally introduces traditional ARM, the theoretical notions on which our approach is based, and formally defines the problem we are considering. It also introduces related work focusing on rule selection and ARM with imperfect data. The proposed approach is detailed in Sect. 3, and Sect. 4 presents the illustration. Finally, perspectives and concluding remarks are provided in Sect. 5.

2   Theoretical Background and Related Work

This section briefly presents some of the theoretical notions required to introduce our work. We next provide the problem statement of ARM with imperfect data, and our positioning w.r.t. existing contributions.

2.1   Theoretical Background

Association Rule Mining (ARM): In classical ARM [1], a database $\mathcal{D} = \{d_1, \ldots, d_m\}$ to be mined consists of m observations of a set of n attributes. The set of attribute indices is denoted by $N = \{1, \ldots, n\}$. Each attribute i takes its values in a discrete -boolean, nominal or numerical- finite scale denoted Θ_i. An association rule r denoted $r : X \to Y$ links an antecedent X with a consequent Y where $X \in \prod_{i \in I} \Theta_i$, $I \subset N$ and $Y \in \prod_{j \in J} \Theta_j$, $J \subseteq N \setminus I$.

The main challenge in ARM is to extract *interesting* rules from a large search space, *e.g.*, n and m are large. In this context, defining the *interestingness* of a rule is central.

Interestingness of Rules. Numerous works have studied notions related to the *interestingness* of a rule, [16,22,23]. No formal and widely accepted definition arose from those works, and discussing the numerous existing formulations is out of the scope of this paper. However, interestingness is generally regarded as a general concept covering several features of interest for a rule, e.g. *reliability* (how reliable is the rule?) and *conciseness* (is the rule complex?, i.e. based on numerous attribute-value pairs). Other aspects of a rule are also considered, e.g. *peculiarity*, *surprisingness*, or *actionability*, to name a few - the reader can refer to [12] for details. The literature also distinguishes objective and subjective measures, the latter being defined based on domain-dependent considerations. The two main (objective) measures used in the literature are *Support* and *Confidence* [2]. The *support* of a rule $r : X \rightarrow Y$ denoted $supp(X \rightarrow Y)$ is traditionally defined as the proportion of the realization of X and Y in $\mathcal{D}$, and the *confidence* denoted $conf(X \rightarrow Y)$ is defined as the proportion of the realization of Y when X is observed in $\mathcal{D}$. Given support and confidence thresholds, ARM usually aims at identifying rules exceeding those thresholds [2]. In classical ARM, support and confidence are quantified using probability theory framework. When ARM involves imperfect data, this quantification requires reformulating the problem in a theoretical framework suited for handling data imperfection. In this work, we focus on contributions based on evidence theory.

Evidence Theory has been introduced to represent imprecision and uncertainty [21]. We briefly introduce its main concepts. Let Θ be a finite set of elements being the most precise available information, referred to as the *frame of discernment*. A *mass function* $m : 2^{\Theta} \rightarrow [0,1]$ is a set function such that $\sum_{A \subseteq \Theta} m(A) = 1$. The quantity $m(A)$, $A \subseteq \Theta$ is interpreted as the portion of belief that is exactly committed to A and to nothing smaller. The subsets of Θ having a strictly positive mass are called *focal elements*, their set is denoted $\mathcal{F}$. The total belief committed to any $A \subseteq \Theta$ is measured by the *belief function*: $Bel : 2^{\Theta} \rightarrow [0,1]$ with $Bel(A) = \sum_{B \subseteq \Theta, B \subseteq A} m(B)$. In evidence theory, $Bel(\overline{A})$, where $\overline{A}$ denotes the complement of A in Θ, is characterized through the notion of *plausibility*: $Pl : 2^{\Theta} \rightarrow [0,1]$, with $Pl(A) = 1 - Bel(\overline{A}) = \sum_{B \subseteq \Theta, B \cap A \neq \emptyset} m(B)$.

In order to provide a complete generalization of the probability framework, conditioning has also been defined in evidence theory. Several expressions have been proposed, none of them leading to a full consensus [7,10]. In this paper, we will adopt the definition corresponding to the conditioning process stated by Fagin et al. [10], a natural extension of the Bayesian conditioning. We do not consider the definition proposed in Dempster [7] based on Dempster-Shafer combination rule, where a new information is interpreted as a modification of the initial belief function and used in a revision process [9]. Thus, for $A, B \subseteq \Theta$, such that $Bel(A) > 0$, we will further consider:

$$Bel(B|A) = \frac{Bel(A \cap B)}{Bel(A \cap B) + Pl(A \cap \overline{B})}, \quad Pl(B|A) = \frac{Pl(A \cap B)}{Pl(A \cap B) + Bel(A \cap \overline{B})}$$

2.2   Problem Statement and Related Work

Problem Statement. In classical ARM, where only precise information is considered, e.g., the value of attribute i is $X_i \in \Theta_i$, $i \in N$. In this paper, we consider observations as "the value of attribute i is in $A_i \subseteq \Theta_i$". The case $A_i \subset \Theta_i$ with $|A_i| > 1$ corresponds to imprecision, while $A_i = \Theta_i$ is considered when information is missing, i.e. it corresponds to the ignorance about the value of attribute i. In this setting, a rule r is defined as:

$$r : A \rightarrow B \text{ where } A = \prod_{i \in I} A_i, A_i \subseteq \Theta_i \text{ and } B = \prod_{j \in J} B_j, B_j \subseteq \Theta_j$$
$$\text{for all } I \subset N \text{ and } J \subseteq N \setminus I$$

As mentioned previously, in this paper we consider the case where antecedent A concerns only a subset $I_1 \subset N$ of attributes and consequent B concerns a subset $I_2 \subset N$ where I_1 and I_2 form partition of N, and $I_1 \neq \emptyset$. Thus:

$$r : A \rightarrow B \quad \text{where} \quad A = \prod_{i \in I_1} A_i, A_i \subseteq \Theta_i \quad \text{and} \quad B = \prod_{j \in I_2} B_j, B_j \subseteq \Theta_j \qquad (1)$$

We denote by $\mathcal{R}$ the set of rules defined by Formula (1). The problem addressed here is to reduce $\mathcal{R}$ by selecting only the relevant rules.

Related Work and Positioning. As stated in the introduction, our goal is threefold: (i) to enrich the expressivity of existing proposed frameworks dedicated to ARM with imperfect data, (ii) to complement them with a richer procedure for selecting relevant rules (rule pruning), and (iii) to present a simple way to incorporate domain knowledge to ease the mining process, and to help identifying relevant rules for a domain of interest.[1]

Rule Pruning. Most of the approaches use thresholds to select rules - only using support and confidence most often allows drastically reducing the number of rules in traditional ARM [1]. A post-mining step is generally performed to rank the remaining rules according to one specific interestingness measure -the measure used is generally selected according to the application domain and context-specific measure properties [23,27]. Nevertheless, processing this way does not

---

[1] Note that the simplification of the mining process here refers to a reduction of complexity in terms of the number of rules analysed, i.e. search space size. Algorithmic contributions and therefore complexity analyses regarding efficient implementations of the proposed approach are left for future work.

enable selecting rules when conflicting interestingness measures are used, e.g. maximizing both support and specificity of rules. This is the purpose of MCDA methods. Some works propose to take advantage of MCDA methods [3–6,17] in the context of ARM. Those works can be divided into two categories: (1) those incorporating the end-user's preferences using Analytic Hierarchy Process (AHP) and Electre II [6], or using Electre tri [3]; and (2) those that do not incorporate such information and use Data Envelopment Analysis (DEA) [5,26], or Choquet integral [17]. Our approach is hybrid and falls within the two categories. First, selection is made based only on database information as in Bouker et al. [4]. Second, if the set of selected rules is large, a trade-off based on end-user's preferences is used within an appropriate MCDA method. As our aim is to select a subset of interesting rules, Electre I [18] seems to be the most appropriate.

ARM and Imperfect Data. Several frameworks have been studied to deal with imperfect data in ARM. The assumptions entailed in the approaches based on probabilistic models do not preserve imprecision and might lead to unreliable inferences [13]. Uncertainty theories have also been investigated for imperfect data in ARM using fuzzy logic [14], or using possibility theory [8]. In the case of missing and incomplete data, evidential theory seems the appropriate setting to handle ARM problem [13,19,24,25]. Our approach is adopting this setting. In addition to studying a richer modelling that enables incorporating more information, we propose to combine it with a selection process taking advantage of an MCDA method, namely Electre I, to assess rules interestingness considering different viewpoints. Although some works previously mentioned tackle rule selection using MCDA, and few approaches have been addressing ARM problem using evidence theory, none of them is addressing both issues simultaneously.

We also present how to benefit from *a priori* knowledge about attribute values -organised into taxonomies- for improving the rule selection process, and reducing the increase of complexity induced by the proposed extension of modellings used so far in existing ARM approaches suited for imperfect data.

3   Proposed Approach

This section presents our ARM approach for imperfect data. We first introduce how rule interestingness is evaluated by presenting the selected measures and their formalization in the evidence theory framework. Then, the main steps of the proposed approach for selecting rules based on these measures are detailed.

3.1   Assessing Rule Interestingness from Imprecise Data

In this study, we focus on important objective measures of interestingness - subjective ones, involving further interactions with final user, are most often considered context-dependent and will not be considered in this paper. We propose to evaluate rules according to (i) their support, (ii) their confidence, as well as (iii) indirect evaluations used to criticize their potential relevance. In addition, since in our context rules are imprecise, and since very imprecise rules are most often considered useless, the (iv) degree of imprecision embedded in the mined rules is also evaluated. These four notions of interest considered in the study are defined below. For convenience, we consider that we are computing measures to evaluate a rule $r : A \rightarrow B$ where $A = \prod_{i \in I_1} A_i, A_i \subseteq \Theta_i$ and $B = \prod_{j \in I_2} B_j, B_j \subseteq \Theta_j$ with $I_1 \cup I_2 = N$. In our context, since we consider $n = |N|$ attributes, the set functions mass m, belief Bel and plausibility Pl are defined on subsets of $\Theta = \prod_{i \in N} \Theta_i$.

Support. A rule is said to be supported if observations of its realization are frequent [2]. In our context, the support of a rule relates to the masses of evidence associated to observations supporting the rule, either explicitly or implicitly. The belief function is thus used to express support:

$$supp(r : A \rightarrow B) = Bel(A \times B) \tag{2}$$

Note that the belief function is monotone, then, the rules composed of the most imprecise attribute values will necessarily be the most supported.

Confidence. A rule is said to be reliable if the relationship described by the rule is verified in a sufficiently great number of applicable cases [12]. The *Confidence* measure is traditionally evaluated as a conditional probability [1]. Its natural counterpart in evidence theory is given by the conditional belief, leading to the following expression:

$$conf(r : A \rightarrow B) = Bel(B \mid A) = \frac{Bel(A \times B)}{Bel(A \times B) + Pl(A \times \overline{B})} \tag{3}$$

The elements defining the consequent are conditioned to the elements composing the antecedent. Note that the belief and conditional belief functions have also been adopted to express support and confidence for ARM with imprecise data [13,24]. In those cases the modelling and domain definition were different, i.e. restricted to the cartesian products of the power-sets of attribute domains.

Indirect Measures of Potential Relevance. These measures will be introduced through an illustration. Consider humanitarian projects described by two attributes: the *transport means* with $\Theta_1 = \{truck, motorbike, helicopter\}$, and the final *coverage reached* in the project (proportion of beneficiaries), with $\Theta_2 = \{low, moderate, high\}$. To criticize the relevance of a rule $r : A \rightarrow B$, e.g. $r : \{truck\} \rightarrow \{high\}$, we propose to evaluate the following relations:

- $A \rightarrow \overline{B}$. In the example, if the rule $\{truck\} \rightarrow \{\overline{high}\}$ holds, it means that most often using *trucks* also leads to a *coverage* that is *not high*. Hence we consider that validating $A \rightarrow \overline{B}$ conveys a contradictory information w.r.t. to the rule $A \rightarrow B$ and tends to invalidate it.

- $\overline{A} \rightarrow B$. If the rule $\{\overline{truck}\} \rightarrow \{high\}$ holds, it means that in some cases, some of the *other means of transport* also allow to reach a *high coverage*. Such an information tends to decrease the interest of the rule $r : A \rightarrow B$ if we assume that B is not explained by multiple causes.
- $\overline{A} \rightarrow \overline{B}$. The rule $\{\overline{truck}\} \rightarrow \{\overline{high}\}$ means that when *trucks* are not used, a *low or moderate coverage* (not high) is obtained. We assume that most commonly, if $\{truck\} \rightarrow \{high\}$ is somehow assumed to be considered as valid, supporting $\{\overline{truck}\} \rightarrow \{\overline{high}\}$ will reinforce our interest over $\{truck\} \rightarrow \{high\}$.

In a probabilistic framework, only the relationship $\overline{A} \rightarrow \overline{B}$ would have to be studied, since the other ones do not provide additional information, i.e. $P(\overline{B}|A) = 1 - P(B|A)$, $P(B|\overline{A}) = 1 - P(\overline{B}|\overline{A})$, $P(A \times \overline{B}) = P(A)P(\overline{B}|A)$ and $P(\overline{A} \times B) = (1 - P(A))P(B|\overline{A})$. Thus, the potential relevance of a rule takes into consideration the confidence of the rule composed of the complements of the antecedent and the consequent, given by: $P(\overline{B}|\overline{A})$. Note that, in the literature, this measure is also referred to as *specificity*. When considering evidence theory, the information about the complement is provided by the plausibility function, such as $Bel(A) = 1 - Pl(\overline{A})$ and then $Bel(B|A) = 1 - Pl(\overline{B}|A)$. In this context, Table 1 introduces the relationships between the confidence of a rule (conditional belief) and the ones involving the complement of its antecedent and/or consequent.

Note that to criticize the relevance of a rule using the three rules involving its complements, we propose to consider their respective *support* and *confidence*: criticizing a rule on the basis of weakly supported rules would not be appropriate.

Table 1. Relationships between support and confidence of a rule $r : A \rightarrow B$ and rules involving its complements.

Rule	Support	Confidence	Depends on quantities
$A \rightarrow B$	$Bel(A \times B)$	$Bel(B \mid A)$	$Bel(A \times B)$ and $Pl(A \times \overline{B})$
$A \rightarrow \overline{B}$	$Bel(A \times \overline{B})$	$Bel(\overline{B} \mid A) = 1 - Pl(B \mid A)$	$Bel(A \times \overline{B})$ and $Pl(A \times B)$
$\overline{A} \rightarrow B$	$Bel(\overline{A} \times B)$	$Bel(B \mid \overline{A}) = 1 - Pl(\overline{B} \mid \overline{A})$	$Bel(\overline{A} \times B)$ and $Pl(\overline{A} \times \overline{B})$
$\overline{A} \rightarrow \overline{B}$	$Bel(\overline{A} \times \overline{B})$	$Bel(\overline{B} \mid \overline{A})$	$Bel(\overline{A} \times \overline{B})$ and $Pl(\overline{A} \times B)$

Specificity Using Information Content. Finally, we propose to incorporate the specificity of a rule. Let's consider the information "the value of attribute i is in the subset A_i". This information is more specific than the information "the value of attribute i is in the subset A_i'" where $A_i \subset A_i'$. Based on the notion of Information Content (IC) defined for comparing concept specificities in ontologies [20], we propose to quantify the specificity of a rule r by:

$$IC(r : A \rightarrow B) = 1 - \frac{\log |\{X : X \subseteq A \times B\}|}{|\Theta|} \tag{4}$$

$|X|$ denotes the number of elements in the set X and $\Theta = \prod_{i \in N} \Theta_i$.

3.2 Search Space Reduction

Let us remind the starting set $\mathcal{R}$ -see Formula (1)- of rules from which a small subset $\mathcal{R}^*$ of interesting rules should be selected:

$$\mathcal{R} = \{r:\ A \to B \mid A = \prod_{i \in I_1} A_i,\ A_i \subseteq \Theta_i,\ B = \prod_{j \in I_2} B_j,\ B_j \subseteq \Theta_j\}$$

We assume that I_1 and I_2 are fixed before starting the ARM process.

To simplify notations in the rest of the paper, we will denote by $r_{A,B}$ the rule $r:\ A \to B$ where A and B are as in the Formula (1). Two restrictions are proposed below:

1. All rules being supported are generalizations (supersets) of focal elements $\mathcal{F}$, i.e. $\mathcal{F} = \{X : X \subseteq \Theta, m(X) > 0\}$. Since support is a prerequisite for assessing rule validity, we further consider that the evaluation will be restricted to the set:

$$\mathcal{R}_r = \{r_{A,B} \in \mathcal{R} \mid \exists X \in \mathcal{F} \text{ st. } X \subseteq A \times B\}$$

2. The search space can also be reduced using prior knowledge defined into ontologies expressing taxonomies of attribute values. Since the ontology defines the concepts of interest for a domain, a restriction can be performed only considering the attribute values defined into taxonomies. Thus, for each $i \in N$, only a subset $\mathcal{O}_i$ of 2^{Θ_i} of the information of interest for a domain is considered. We can then define the following restriction:

$$\mathcal{R}_{r,t} = \{r_{A,B} \in \mathcal{R}_r \mid A = \prod_{i \in I_1} A_i,\ A_i \in \mathcal{O}_i,\ B = \prod_{j \in I_2} B_j,\ B_j \in \mathcal{O}_j\}$$

Table 2. Summary of interestingness measures considered in the selection process

$k \in K$	Measures	Formulae $\forall r \in \mathcal{R}_{r,t}\ r : A \to B$	Variation	Weight
1	Rule Support	$supp(r) = Bel(A \times B)$	Maximize	w_1
2	Rule Confidence	$conf(r) = Bel(B\|A)$	Maximize	w_2
3	Rule Specificity	$IC(r)$	Maximize	w_3
4	$A \to \overline{B}$	$Bel(A \times \overline{B})$	Minimize	w_4
5		$Bel(\overline{B}\|A)$	Minimize	w_5
6	$\overline{A} \to B$	$Bel(\overline{A} \times B)$	Minimize	w_6
7		$Bel(B\|\overline{A})$	Minimize	w_7
8	$\overline{A} \to \overline{B}$	$Bel(\overline{A} \times \overline{B})$	Maximize	w_8
9		$Bel(\overline{B}\|\overline{A})$	Maximize	w_9

3.3   Rules Selection Process

The proposed approach aims at selecting the most relevant rules $\mathcal{R}^*$ according to their evaluations on a set of interestingness measures listed in Table 2. We here consider that the evaluated rules are members of the restriction $\mathcal{R}_{r,t} \subseteq \mathcal{R}$, even if that condition could further be relaxed. We denote the set of interestingness measures by K ($|K| = 9$), and $g_k(r)$ the score of rule r for the measure $k \in K$. To simplify notations, we consider that $g_k(r)$ is to maximize[2] for all $k \in K$. A two-step pruning strategy is proposed.

Step 1: Dominance-Based Pruning. A reduction of the concurrent rules in $\mathcal{R}_{r,t}$ is carried out by focusing on non-dominated rules on the basis of the considered measures. A rule r_1 dominates a rule r_2, we write $r_2 \prec r_1$, iff r_1 is at least equal to r_2 on all measures and it exists a measure where r_1 is strictly superior to r_2. More formally,

$$r_2 \prec r_1 \text{ iff } g_k(r_2) \leq g_k(r_1), \forall k \in K \text{ and } \exists j \in K \text{ such that } g_j(r_2) < g_j(r_1).$$

The reduced set of rules can be stated as:

$$\mathcal{R}_{r,t,d} = \{r \in \mathcal{R}_{r,t} \mid \nexists r' \in \mathcal{R}_{r,t} : r \prec r'\}$$

Step 2: Pruning Using Electre I. When $\mathcal{R}_{r,t,d}$ remains too large to be manually analyzed, a subjective pruning procedure based on the selection procedure Electre I is applied. This MCDA method enables expressing subjectivity through parameters that can be given by decision makers [18]. We use it for finding the final set of rules $\mathcal{R}^* \subseteq \mathcal{R}_{r,t,d}$. Electre I builds an outranking relation between pairs of rules allowing to select a subset of the best rules: $\mathcal{R}^*$. This subset is such that (i) any rules excluded from $\mathcal{R}_{r,t,d}$ is outranked by at least one rule from $\mathcal{R}^*$, (ii) rules from $\mathcal{R}^*$ do not outrank each other. To do so, Electre I procedure (a) constructs outranking relationships through pairwise comparisons of rules, to further (b) exploit those relationships to build $\mathcal{R}^*$.

(a) Outranking relations: the relationship "*r outranks r'*" (rSr') means that r is at least as good as r' on the set of measures K. The outranking assertion rSr' holds if: (i) a sufficient coalition of measures supports it, and (ii) none of the measures is too strongly opposed to it. These conditions are respectively referred to as concordance $c(rSr')$ and discordance indices $d(rSr')$, such that:

$$c(rSr') = \sum_{\{k : g_k(r) \geq g_k(r')\}} w_k \text{ and } d(rSr') = \max_{\{k : g_k(r) < g_k(r')\}} [g_k(r') - g_k(r)],$$

with w_k the relative importance of measure k.
From these notations, we consider rSr' if $c(rSr') \geq \widehat{c}$ and $d(rSr') \leq \widehat{d}$; with $\widehat{c}$ and $\widehat{d}$, two thresholds defining when the outranking should be considered or not.

---

[2] Indeed all the measures used in our approach take values in the interval $[0, 1]$, then a measure k to minimize can be changed to a measure to maximize by considering $1 - g_k(r)$ instead of $g_k(r)$.

(b) Relations exploitation: a graph of outranking relationships is obtained from these pairwise comparisons. The kernel of this graph is our final reduced set of rules $\mathcal{R}^*$ to be considered, such that:

$$- \forall r' \in \mathcal{R}_{r,t,d} \setminus \mathcal{R}^*, \exists r \in \mathcal{R}^* : rSr', \text{ and}$$
$$- \forall (r, r') \in \mathcal{R}^* \times \mathcal{R}^*, \neg(rSr'). \tag{5}$$

The set of model parameters that have to be defined for applying the subjective reduction based on Electre I are: weights $w_k, \forall k \in K$, and the concordance and discordance thresholds, $\widehat{c}, \widehat{d}$.[3] The choice of parameter values will be further discussed in the illustration Sect. 4.

4   Illustration

As an illustration, we consider the context of humanitarian projects carried out for answering to emergency situations. A dataset of observations describes these emergency situations according to four attributes: (1) the *type of disaster* faced, (2) the *season*, (3) the *environment* in which it occurred, and (4) an evaluation of the situation w.r.t. the *human cost*. We further refer to these attributes using

Table 3. Database of observations expressed using precise, imprecise or missing values.

	Disaster type	Season	Environment	Human cost
d_1	{earthquake}	{autumn}	{rural}	{medium}
d_2	{tsunami}	{autumn}	{urban}	{medium}
d_3	{epidemic}	-	{urban}	{veryHigh}
d_4	{earthquake, epidemic, tsunami}	{spring}	-	{high, veryHigh}
d_5	{epidemic}	{spring}	{urban}	{high}
d_6	{epidemic}	{spring, summer}	-	{high, veryHigh}
d_7	{epidemic}	{spring, summer}	{urban}	{high, veryHigh}
d_8	{epidemic}	{spring, summer}	{urban}	{veryHigh}
d_9	{earthquake, epidemic, tsunami}	{summer}	{rural}	{high}
d_{10}	{epidemic}	{summer}	{urban}	{high}
d_{11}	{epidemic}	{summer}	{urban}	{veryHigh}
d_{12}	{earthquake}	{winter}	{rural}	{high, medium, veryHigh}
d_{13}	{earthquake}	{winter}	{rural}	{low}
d_{14}	{earthquake, epidemic, tsunami}	{winter}	{rural}	{high}

---

[3] Evaluating support and confidence of $\overline{A} \rightarrow B$ and $\overline{A} \rightarrow \overline{B}$ can lead to undefined values, e.g. evaluating $\overline{A} \rightarrow B$, we have $Bel(\overline{A} \times B) = 0$ when $\overline{A}$ has never been observed, leading to $Bel(B|\overline{A})$ being undefined. However, pruning using dominance and Electre I requires the same measures to be defined. Undefined values are thus substituted by an arbitrary value that neither favor nor penalize the evaluation of the rule $A \rightarrow B$. The median of $Bel(\overline{A} \times B)$ (resp. $Bel(\overline{A} \times \overline{B})$) has been chosen. Note that $A \rightarrow \overline{B}$ is not concerned since evaluating $A \rightarrow B$ implies evidence on A.

their number, considering that they respectively take discrete values in: $\Theta_1 = \{tsunami, earthquake, epidemic, conflict, pop.displacement\}$, $\Theta_2 = \{spring, summer, autumn, winter\}$, $\Theta_3 = \{urban, rural\}$, $\Theta_4 = \{low, medium, high, veryHigh\}$. Besides, for each attribute, prior knowledge is defined into ontologies determining the values of interest. In this specific case study, the purpose of association rules is to highlight the influence of a situation contextual features on its evaluation according to the *Human Cost*, a useful information for project planning. Thus the searched rules $r : A \rightarrow B$ will imply the attributes in the following set $I_1 = \{1, 2, 3\}$ in the *antecedent* and in $I_2 = \{4\}$ for the *consequent*.

Table 4. Set of non-dominated rules, $\mathcal{R}_{r,t,d}$.

	Disaster Type	Season	Environment	Human cost
r_0 :	$\{earthquake\}$	$\wedge \{autumn\}$	$\wedge \{rural\}$	$\rightarrow \{medium\}$
r_1 :	$\{earthquake, tsunami\}$	$\wedge \{autumn\}$	$\wedge \Theta_3$	$\rightarrow \{medium\}$
r_2 :	$\{tsunami\}$	$\wedge \{autumn\}$	$\wedge \{urban\}$	$\rightarrow \{medium\}$
r_3 :	$\{earthquake, epidemic, tsunami\} \wedge \Theta_2$		$\wedge \Theta_3$	$\rightarrow \Theta_4$
r_4 :	$\{earthquake, epidemic, tsunami\} \wedge \Theta_2$		$\wedge \Theta_3$	$\rightarrow \{high, medium, veryHigh\}$
r_5 :	$\{earthquake, epidemic, tsunami\} \wedge \Theta_2$		$\wedge \Theta_3$	$\rightarrow \{high, veryHigh\}$
r_6 :	$\{epidemic\}$	$\wedge \Theta_2$	$\wedge \Theta_3$	$\rightarrow \{high, veryHigh\}$
r_7 :	$\{epidemic\}$	$\wedge \Theta_2$	$\wedge \{urban\}$	$\rightarrow \{veryHigh\}$
r_8 :	$\{earthquake\}$	$\wedge \{autumn, winter\} \wedge \{rural\}$		$\rightarrow \{medium\}$
r_9 :	$\{earthquake, tsunami\}$	$\wedge \{autumn, winter\} \wedge \Theta_3$		$\rightarrow \{low, medium\}$
r_{10} :	$\{earthquake, tsunami\}$	$\wedge \{autumn, winter\} \wedge \Theta_3$		$\rightarrow \{medium\}$
r_{11} :	$\{earthquake, epidemic, tsunami\} \wedge \{spring, summer\} \wedge \Theta_3$			$\rightarrow \{high, veryHigh\}$
r_{12} :	$\{epidemic\}$	$\wedge \{spring, summer\} \wedge \Theta_3$		$\rightarrow \{high, veryHigh\}$
r_{13} :	$\{epidemic\}$	$\wedge \{spring, summer\} \wedge \{urban\}$		$\rightarrow \{high, veryHigh\}$
r_{14} :	$\{epidemic\}$	$\wedge \{spring, summer\} \wedge \{urban\}$		$\rightarrow \{veryHigh\}$
r_{15} :	$\{epidemic\}$	$\wedge \{summer\}$	$\wedge \{urban\}$	$\rightarrow \{high, veryHigh\}$
r_{16} :	$\{epidemic\}$	$\wedge \{summer\}$	$\wedge \{urban\}$	$\rightarrow \{veryHigh\}$
r_{17} :	$\{earthquake\}$	$\wedge \{winter\}$	$\wedge \{rural\}$	$\rightarrow \{low\}$

Among the observations of 14 projects given in Table 3, some attribute values are expressed with imprecision, e.g. *Human cost* values may be unclear such that *"human Cost is High or VeryHigh"*. When values are missing the total ignorance is considered. In this setting, the size of the initial studied space $\mathcal{R}$ is $\prod_{i=1}^{4} 2^{|\Theta_i \setminus \emptyset|} = 20925$. Using the restrictions focusing on rules with non-null support, and involving attribute values of interest defined into ontologies (cf. Sect. 3), we obtain a reduced search space $\mathcal{R}_{r,t}$ composed of 484 rules.

The rule evaluation and selection process is further applied to $\mathcal{R}_{r,t}$ using the 9 interestingness measures proposed in Table 2. Using dominance-based pruning (Step 1/2), a set of 18 non-dominated rules $\mathcal{R}_{r,t,d}$ is identified among the 484 rules initially considered. These rules are listed in Table 4, and indexed from r_0 to r_{17}. Pruning using Electre I is then applied over the set of non-dominated rules $\mathcal{R}_{r,t,d}$ (Step 2/2). Different sets of selected rules -i.e. $\mathcal{R}^*$- are given in Table 5 for different sets of model parameters. The results being sensitive to parameter values, we propose to discuss different parameter settings. We remind that these parameters are: $\forall k \in K$, w_k the weights of interestingness measures, and $\hat{c}$ and $\hat{d}$ the concordance and discordance thresholds. They represent end-user's preferences. They can be given directly; the weights w_k can also be elicited using Simos, a well-known weighting procedure [11].

Table 5. Final sets of rules ($\mathcal{R}^*$) obtained with Electre I pruning using four parameter settings (a to e).

	w_1	w_2	w_3	w_4	w_5	w_6	w_7	w_8	w_9	$\hat{d}$	$\mathcal{R}^*$
a	0.27	0.15	0.1	0.08	0.08	0.08	0.08	0.08	0.08	0.3	$\{r_1, r_3, r_6, r_9, r_{11}\}$
b	0.18	0.18	0.18	0.1	0.1	0.1	0.1	0.03	0.03	0.3	$\{r_1, r_3, r_6\}$
										0.2	$\{r_0, r_1, r_2, r_3, r_6, r_{13}, r_{16}, r_{17}\}$
c	0.12	0.2	0.2	0.08	0.08	0.08	0.08	0.08	0.08	0.3	$\{r_1, r_3, r_6\}$
										0.2	$\{r_0, r_1, r_2, r_3, r_6, r_{13}, r_{16}, r_{17}\}$
d	0.15	0.25	0.25	0.05	0.05	0.05	0.05	0.075	0.075	0.3	$\{r_1, r_3, r_6, r_{17}\}$
										0.2	$\{r_0, r_1, r_2, r_3, r_6, r_{13}, r_{16}, r_{17}\}$
e	0.33	0.33	0.34	0	0	0	0	0	0	0.3	$\mathcal{R}_{r,t,d} \setminus \{r_8, r_{10}, r_{16}, r_{17}\}$

Different sets of parameters, with $\hat{c} = 0.7$

Among the considered interestingness measures, according to the literature, we assume that *support, confidence* and *IC* are the most significant ones w.r.t. rule interest. They have to be associated to the most important weights. Conversely, we assume that the other measures -about rule complements- are secondary and will provide additional information for comparing and criticizing the relevance of rules. In the first set of parameters *(a)* (cf. Table 5), the weight given to *support* and *confidence* is maximized to represent 60% of the votes required for the outranking (to exceed $\hat{c} = 0.7$). This setting will tend to favor the rules having a high degree of imprecision, being well supported and then reliable, since $Bel(B|A) \geq Bel(A \times B)$. For example, in this setting the rules r_3, r_6, r_{11}, see Tables 4 and 5, are among the selected rules in $\mathcal{R}^*$; e.g. with r_3 involving the total imprecision on three attributes.

When restricting $\hat{d}$ to 0.2 with the parameter settings *(b), (c), (d)*, it increases the size of the kernel, while still discarding more than half of the rules among the set of non-dominated ones. With parameters *(d)* and $\hat{d} = 0.3$, highest importance is given to *confidence* and *IC*, providing these 2 measures with 71% of the voting power to reach the outranking condition $\hat{c} = 0.7$. Thus, a rule with a better score on *confidence*, *IC* and on some of the other measures -except *support*- can be selected while having a low support. This is illustrated with the selection of r_{17} for example. Lastly, the parameter setting (e) is equivalent to considering only the three main measures with equal importance. Here, it enables to discard only 4 extra rules in comparison to dominance relationships. This is explained by the fact that the absence of dominance between rules is more frequent.

Finally, the parameter settings *(b), (c)* or *(d)* with $\hat{d} = 0.2$, favoring the *support, confidence* and *IC* over the other measures tend to provide interesting results. This setting enables the selection of both precise and imprecise rules of interest w.r.t. the initial set of observations, such as r_{16} and r_{13}. In the initial dataset -see Table 3- the imprecise information $\{spring, summer\}$ for the *season* or $\{high, veryHigh\}$ for the *Human Cost* are frequently observed. Indeed, selecting the imprecise rule $r_{13} : \{epidemic\} \wedge \{spring, summer\} \wedge \{urban\} \rightarrow \{high, veryHigh\}$ in $\mathcal{R}^*$ is not surprising. As an interpretation of this rule, we say that the analysis of the database tends to relate the occurrence of epidemics in urban areas to a specific season, spring or summer, and human cost. In particular, the rule seems valid at least for one the conjunction "summer and high human cost", "summer and a very High human cost", "spring and high" or "spring and veryHigh". In this illustration, different sets of parameters and their results on rule selection have been presented. However, these parameters have to be set by the end-user.

To further discuss these results, it is interesting to note that all the selected measures for rules comparison, except the *IC*, are based on observations frequency. In order to counterbalance the preponderance of this factor, it might be relevant to add subjective measures and not only data-driven ones. Subjective interestingness measures have been studied in the literature. Relying on these works, we could include here measures based for example on user expected rules or expected conjunction of attribute values. Furthermore, investigating the dependencies among frequency based measures, and considering them in the selection process will be valuable. Nevertheless, considering additional measures (especially data-driven), as the ones proposed for classical ARM, is not necessarily straightforward within the evidence theory framework. It indeed implies to define their right expression and meaning in this framework.

5   Conclusion and Perspectives

Mining association rules from imperfect data is a key challenge for real-world applications dealing with imperfect data, e.g., imprecise, missing data, etc. The ARM approach introduced in this paper enables to deal with imprecise data and derive imprecise rules under specific conditions (e.g. fixing both antecedent and consequent). Relying on evidence theory and Multiple Criteria Decision Analysis, this new framework enriches expressivity of existing works while providing a novel selection procedure for identifying most interesting rules according to several viewpoints. To this aim, several interestingness measures have been proposed, and used in a two-step selection procedure based on dominance relationships and Electre I. A restriction using *a priori* knowledge has also been proposed to focus and ease the mining process by incorporating symbolic knowledge defined into domain ontologies. To further improve the approach, additional measures of interestingness could be added. Future work related to subjective measures (e.g., user-oriented) would be particularly relevant to enrich the set of frequency-based measures that are currently involved in the approach. Studying the interactions between the measures would also be of interest. Finally, only an illustration using a simplified case study related to humanitarian projects analysis has been presented in this paper. Thorough algorithmic complexity and performance evaluations of the approach have to be discussed. Difficult challenges related to algorithmic complexity and efficiency issues of the procedure also have to be addressed in order to mine rules involving numerous attributes.

References

1. Agrawal, R., Imieliński, T., Swami, A.: Mining association rules between sets of items in large databases. In: ACM SIGMOD Record, vol. 22, pp. 207–216. ACM (1993)
2. Agrawal, R., Srikant, R., et al.: Fast algorithms for mining association rules. In: Proceedings of 20th International Conference on Very Large Data Bases, VLDB, vol. 1215, pp. 487–499 (1994)
3. Ait-Mlouk, A., Gharnati, F., Agouti, T.: Multi-agent-based modeling for extracting relevant association rules using a multi-criteria analysis approach. Vietnam J. Comput. Sci. **3**(4), 235–245 (2016)
4. Bouker, S., Saidi, R., Yahia, S.B., Nguifo, E.M.: Ranking and selecting association rules based on dominance relationship. In: 2012 IEEE 24th International Conference on Tools with Artificial Intelligence, vol. 1, pp. 658–665. IEEE (2012)
5. Chen, M.C.: Ranking discovered rules from data mining with multiple criteria by data envelopment analysis. Expert Syst. Appl. **33**(4), 1110–1116 (2007)
6. Choi, D.H., Ahn, B.S., Kim, S.H.: Prioritization of association rules in data mining: multiple criteria decision approach. Expert Syst. Appl. **29**(4), 867–878 (2005)
7. Dempster, A.P.: Upper and lower probabilities induced by a multivalued mapping. Ann. Math. Stat. **38**, 325–339 (1967)
8. Djouadi, Y., Redaoui, S., Amroun, K.: Mining association rules under imprecision and vagueness: towards a possibilistic approach. In: 2007 IEEE International Fuzzy Systems Conference, pp. 1–6. IEEE (2007)
9. Dubois, D., Denoeux, T.: Conditioning in dempster-shafer theory: prediction vs. revision. In: Denoeux, T., Masson, M.H. (eds.) Belief Functions: Theory and Applications, pp. 385–392. Springer, Heidelberg (2012). https://doi.org/10.1007/978-3-642-29461-7_45
10. Fagin, R., Halpern, J.Y.: A new approach to updating beliefs. In: Proceedings of the Sixth Annual Conference on Uncertainty in Artificial Intelligence, UAI 1990, pp. 347–374. Elsevier Science Inc., New York, NY, USA (1991). http://dl.acm.org/citation.cfm?id=647233.760137
11. Figueira, J., Roy, B.: Determining the weights of criteria in the electre type methods with a revised simos' procedure. Eur. J. Oper. Res. **139**(2), 317–326 (2002)
12. Geng, L., Hamilton, H.J.: Interestingness measures for data mining: a survey. ACM Comput. Surv. **38**(3), 9-es (2006)
13. Hewawasam, K., Premaratne, K., Subasingha, S., Shyu, M.L.: Rule mining and classification in imperfect databases. In: 2005 7th International Conference on Information Fusion, vol. 1, p. 8. IEEE (2005)
14. Hong, T.P., Lin, K.Y., Wang, S.L.: Fuzzy data mining for interesting generalized association rules. Fuzzy Sets Syst. **138**(2), 255–269 (2003)
15. Kotsiantis, S., Kanellopoulos, D.: Association rules mining: a recent overview. GESTS Int. Trans. Comput. Sci. Eng. **32**(1), 71–82 (2006)
16. Liu, B., Hsu, W., Chen, S., Ma, Y.: Analyzing the subjective interestigness of association rules. IEEE Intell. Syst. **15**(5), 47–55 (2000). https://doi.org/10.1109/5254.889106

17. Nguyen Le, T.T., Huynh, H.X., Guillet, F.: Finding the most interesting association rules by aggregating objective interestingness measures. In: Richards, D., Kang, B.-H. (eds.) PKAW 2008. LNCS (LNAI), vol. 5465, pp. 40–49. Springer, Heidelberg (2009). https://doi.org/10.1007/978-3-642-01715-5_4

18. Roy, B.: Classement et choix en présence de points de vue multiples. Revue française d'informatique et de recherche opérationnelle 2(8), 57–75 (1968)

19. Samet, A., Lefèvre, E., Yahia, S.B.: Evidential data mining: precise support and confidence. J. Intell. Inf. Syst. 47(1), 135–163 (2016)

20. Seco, N., Veale, T., Hayes, J.: An intrinsic information content metric for semantic similarity in wordNet. In: Ecai, vol. 16, p. 1089 (2004)

21. Shafer, G.: A Mathematical Theory of Evidence, vol. 42. Princeton University Press, Princeton (1976)

22. Silberschatz, A., Tuzhilin, A.: What makes patterns interesting in knowledge discovery systems. IEEE Trans. Knowl. Data Eng. 8(6), 970–974 (1996)

23. Tan, P.N., Kumar, V., Srivastava, J.: Selecting the right interestingness measure for association patterns. In: Proceedings of the Eighth ACM SIGKDD International Conference on Knowledge Discovery and Data Mining, pp. 32–41. ACM (2002)

24. Tobji, M.B., Yaghlane, B.B., Mellouli, K.: A new algorithm for mining frequent itemsets from evidential databases. Proc. IPMU 8, 1535–1542 (2008)

25. Bach Tobji, M.A., Ben Yaghlane, B., Mellouli, K.: Frequent itemset mining from databases including one evidential attribute. In: Greco, S., Lukasiewicz, T. (eds.) SUM 2008. LNCS (LNAI), vol. 5291, pp. 19–32. Springer, Heidelberg (2008). https://doi.org/10.1007/978-3-540-87993-0_4

26. Toloo, M., Sohrabi, B., Nalchigar, S.: A new method for ranking discovered rules from data mining by dea. Expert Syst. Appl. 36(4), 8503–8508 (2009)

27. Vaillant, B., Lenca, P., Lallich, S.: A clustering of interestingness measures. In: Suzuki, E., Arikawa, S. (eds.) DS 2004. LNCS (LNAI), vol. 3245, pp. 290–297. Springer, Heidelberg (2004). https://doi.org/10.1007/978-3-540-30214-8_23

Evidential Classification of Incomplete Data via Imprecise Relabelling: Application to Plastic Sorting

Lucie Jacquin[1](✉), Abdelhak Imoussaten[1](✉), François Trousset[1](✉),
Jacky Montmain[1](✉), and Didier Perrin[2](✉)

[1] LGI2P, IMT Mines Ales, Univ Montpellier, Ales, France
{lucie.jacquin,abdelhak.imoussaten,francois.trousset,
jacky.montmain}@mines-ales.fr
[2] C2MA, IMT Mines Ales, Univ Montpellier, Ales, France
didier.perrin@mines-ales.fr

Abstract. Besides ecological issues, the recycling of plastics involves economic incentives that encourage industrial firms to invest in the field. Some of them have focused on the waste sorting phase by designing optical devices able to discriminate on-line between plastic categories. To achieve both ecological and economic objectives, sorting errors must be minimized to avoid serious recycling problems and significant quality degradation of the final recycled product. Even with the most recent acquisition technologies based on spectral imaging, plastic recognition remains a tough task due to the presence of imprecision and uncertainty, e.g. variability in measurement due to atmospheric disturbances, ageing of plastics, black or dark-coloured materials etc. The enhancement of recent sorting techniques based on classification algorithms has led to quite good performance results, however the remaining errors have serious consequences for such applications. In this article, we propose an imprecise classification algorithm to minimize the sorting errors of standard classifiers when dealing with incomplete data, by both integrating the processing of classification doubt and hesitation in the decision process and improving the classification performances. To this end, we propose a relabelling procedure that enables better representation of the imprecision of the learning data, and we introduce the belief functions framework to represent the posterior probability provided by a classifier. Finally, the performances of our approach compared to existing imprecise classifiers is illustrated on the sorting problem of four plastic categories from mid-wavelength infra-red spectra acquired in an industrial context.

Keywords: Machine learning · Imprecise classification · Reliable classification · Belief functions · Plastic separation

1 Introduction

Plastic recycling is a promising alternative to landfills for dealing with the fastest growing waste stream in the world [8]. However, for physiochemical reasons related to non-miscibility between plastics, most plastics must be recycled

© Springer Nature Switzerland AG 2019
N. Ben Amor et al. (Eds.): SUM 2019, LNAI 11940, pp. 122–135, 2019.
https://doi.org/10.1007/978-3-030-35514-2_10

separately. Plastic category identification is therefore a major challenge in the recycling process. With the emergence of hyperspectral imaging, some industrial firms have designed sorting devices able to discriminate between several categories of plastics based on their absorption or transmittance spectra. The sorting process is generally performed using supervised classification, which has been well developed with the emergence of computer sciences and data science [18,22,38]. The classification performance might be affected by several issues such as noise or overlapping regions in the feature space [21,34]. The latter problem occurs when samples from different classes share very similar characteristics. We are particularly faced with these problems when attempting to classify industrially acquired spectra. Indeed, in an industrial context, the acquisition process is subject to technical and financial constraints to ensure throughput and financial competitiveness. For this reason one cannot expect the same quality of data as for equivalent laboratory measures. Several issues imply the presence of imprecision and uncertainty in the acquired spectra: (i) the available spectral range might be insufficient; (ii) the plastic categories to be recycled are chemically close; (iii) atmospheric perturbations may cause noise; (iv) plastic ageing and plastic additives are known to change spectral information; (v) impurities like dust deposits or remains of tags will also produce spectral noise. As in solving many other decision-making problems, classification errors may have serious consequences, *e.g.*, medical diagnosis applications. Regarding plastic sorting, identification errors will cause serious recycling difficulties and significant degradation of the secondary raw material performances and thus quality degradation of the recycled products. Usually, the problem of plastic identification is treated using standard classification algorithms that are designed to produce point predictions, *i.e.*, a single plastic category. In cases of imperfect data, standard classifiers become confused and inevitably commit errors. This brings us to consider alternative representations of the information that take into account imprecision and uncertainty to achieve more accurate classification. Modern theories of uncertainty such as fuzzy subsets [35], possibility theory [14], imprecise probabilities [33] or belief functions [26,30] offer better representations of the data-imperfection of information. Several classification algorithms have been proposed in these frameworks. Most of them are extensions of standard algorithms. We can cite the fuzzy version of the well known k-means algorithm [15], fuzzy and evidential versions of k-Nearest Neighbour (k-NN) [10,19] or some fuzzy and evidential revisions of neural network algorithms [4,11].

In this paper we consider the case where the original imperfections come from data features only. Available training example labels are precise and considered trustworthy, *e.g.*, based on laboratory measures and expertise. In order to better represent all available information, we think that labels should conform with the feature imprecision. If an object of class θ_1 has its vector of features x in the overlapping region θ_1 and θ_2, then the example should be relabelled by the set $\{\theta_1, \theta_2\}$. In order to achieve such representation we propose to relabel each training example in accordance with their discriminatory nature. New labels are therefore subsets of the original set of classes. This imprecise relabelling would better represent the learning data by mapping overlaps in the feature space. The resulting imprecise label can be naturally treated in the belief functions theory context. Indeed, belief functions theory [26] is an interesting framework for representing imprecise and uncertain data by allowing the allocation of a probability mass for imprecise data. Thus, imprecision and ignorance is better captured in this framework compared to the probability framework where equiprobability and imprecision are confused. The recent growing interest in this theory has allowed techniques to be developed for resolving a diverse range of problems such as estimation [12,17], standard classification [10,32], or even hierarchical classification [1,23].

Our proposed approach, called Evidential CLAssification of incomplete data via Imprecise Relabelling (ECLAIR), is based on a relabelling procedure of the training examples that enables better representation of the missing information about some data features. Then a classifier is trained on the relabelled data producing a posterior mass function. With imprecise relabelling we try to quantify, using a mass function, the extend to which a subsets of classes is reliable and relevant as output for a new data. In other words, we look for the set of classes which any more precise subset output would lead inevitably to an error. The resulting classification algorithm can enhance the classification accuracy as well as cope with difficult examples by allowing less precise but more reliable classification output which will optimize the recycling process.

The remainder of this paper is organized as follows: Sect. 2 sets down the main notations and provides a reminder on supervised classification and elements of belief functions theory; in Sect. 3 we present the proposed approach; Sect. 4 briefly describes the related works; Sect. 5 presents results of experimentation on the sorting problem of four plastics.

2   Theoretical Background

Classification is a technique allowing to assign objects to categories from the observations of several of their characteristics. A classifier is a function that maps an object represented by its values of characteristics on a finite set of variables, to a category represented by a value of a categorical variable. More precisely, let us consider a set of n categories represented by a set $\Theta = \{\theta_1, \theta_2, \ldots, \theta_n\}$, also refereed as a set of labels or classes. In the framework of belief functions Θ is called a frame of discernment. Each θ_j, $j \in \{1, ..., n\}$ denotes a singleton which represents the lowest level of discernible information in Θ. Let us denote by $X_1, X_2, \ldots, X_p$, p variables where the taken values represent the characteristics, also called attributes or features, of the objects, to be classified. In the rest of the paper we refer to Θ as a set of classes and to $(X_1, X_2, \ldots, X_p)$ as a vector of features where $\forall i \in \{1, \ldots, p\}$, X_i refers both to the name of the feature and to the space of the values taken by the feature, i.e., $X_i \subseteq \mathbb{R}$. For an object x belonging to $\mathcal{X} = \prod_{i=1}^{p} X_i \subseteq \mathbb{R}^p$, let $\theta(x) \in \Theta$ denote the unknown label that should be associated to x.

In this article, we focus on a supervised classification problem. The specificity of the considered data, referred to as *incomplete data*, is that some features of some examples are missing due to technological aspects. Therefore, only part of the data of these examples is obtained. The proposed classification approach, qualified as *imprecise*, integrates the incompleteness of the data in its process to predict subsets of classes comprising the true class when standard counterpart classifier would have predicted the wrong class. To this aim we diverted standard probabilistic classifiers from their natural use for computing probability on sets of classes. Such uncertain resulting information is then captured by belief functions. The following subsections, briefly recalls the notions discussed.

2.1   Supervised Classification

To determine $\theta(x)$ in a supervised classification manner, a standard classifier $\delta_\Theta : \mathcal{X} \to \Theta$ is trained on a set of examples $(x_i, \theta_i)_{1 \leq i \leq N}$ such that for all $1 \leq i \leq N$, x_i belongs to $\mathcal{X}$ and θ_i to Θ. By standard classifier we mean a classifier that assigns to x a single class $\theta(x) = \theta_j$, $j \in \{1, \ldots, n\}$. In some cases when the input data is too voluminous or redundant, it may be appropriate to perform some extraction features before the training of δ_Θ. By reducing the dimension of $\mathcal{X}$, and thus, working with a reduced feature space $\mathcal{X}' \subseteq \mathbb{R}^{p'}$ with $p' < p$, the extraction such as Principal Component Analysis (PCA), Linear Discriminant Analysis (LDA) or Independent Component Analysis (ICA) facilitates the learning and may enhance the classification performance. When feature extraction is designed taking into account the labels of the training examples it is termed as supervised feature extraction. For instance LDA also known as Fisher discriminant analysis reduces the number of features to $n-1$ by looking for a linear combination of the variables maximizing the within-groups and minimizing between-groups variance.

2.2   Probabilistic Classifier and Decision Rule

When δ_Θ can also provide for x a *posterior probability* distribution $p(.|x) : \Theta \to [0, 1]$, it is called a probabilistic classifier. Many classifier algorithms base their decision only on $p(.|x)$ as follows: $\theta(x) = arg \max_{j=1,\ldots,n} p(\theta_j|x)$. For more sophisticated decisions, one can use the decision rule technique classically used in decision theory. Let $\mathcal{A} = \{a_1, a_2, \ldots, a_m\}$ be a finite set of actions that can be taken. In the case of a standard classifier, an action $a \in \mathcal{A}$ corresponds to assign a class $\theta \in \Theta$ to an object x. In such case, we simplify by setting $\mathcal{A} = \Theta$. In order to compare decisions in $\mathcal{A}$ or to compare the classifier δ_Θ to another decision rule, two functions are introduced: *loss function* and risk function. A loss function $L : \mathcal{A} \times \Theta \to \mathbb{R}$ is considered to quantify the loss $L(a, \theta)$ incurred when choosing the action $a \in \mathcal{A}$ while the true state of nature is $\theta \in \Theta$. A *risk function* $r_{\delta_\Theta} : \mathcal{A} \to \mathbb{R}$ is defined as the following expectation: $r_{\delta_\Theta}(a) = E_{p(.|x)}(L(a, \theta))$. In the case of discrete and finite probability distribution, we have $r_{\delta_\Theta}(\theta_j) = \sum_{k=1}^{n} L(\theta_j, \theta_k) \, p(\theta_k|x)$, $j \in \{1, \ldots, n\}$. Thus, considering the decision rule δ_Θ, the class θ_j minimizing the risk $r_{\delta_\Theta}(\theta_j)$ over Θ should be chosen.

2.3   Elements of Belief Functions Theory

Due to the additivity constraint inherent to the definition of a probability distribution, one cannot built a probability distribution when measures, observations, etc. are imprecise. Belief functions theory, as an extension of probability theory, allows masses to be assigned to imprecise data. Two levels are considered when introducing belief functions: credal and pignistic levels. At the credal level, beliefs are captured and quantified by belief functions, while at the pignistic level or decision level, beliefs are quantified using probability distributions.

Credal Level. A *mass function*, also called *basic belief assignment* (bba), is a set function $m : 2^\Theta \to [0,1]$ satisfying $\sum_{A \subseteq \Theta} m(A) = 1$. For a set $A \subseteq \Theta$, the quantity $m(A)$ is interpreted as a measure of evidence committed exactly to the set A and not to any more specific subsets of A. The elements $A \in 2^\Theta$ such that $m(A) > 0$ are called focal elements and they form a set denoted $\mathbb{F}$. $(m, \mathbb{F})$ is called body of evidence. The total belief committed to A is measured by the sum of all masses of A's subsets. This is expressed by the *belief function* $Bel : 2^\Theta \to [0,1]$, $Bel(A) = \sum_{B \subseteq \Theta, B \subseteq A} m(B)$. Furthermore the *plausibility* of A, $Pl : 2^\Theta \to [0,1]$, quantifies the maximum amount of support that could be allocated to A, $Pl(A) = \sum_{B \subseteq \Theta, B \cap A \neq \emptyset} m(B)$.

Pignistic Level. In the transferable belief model [29], the decision is made in the pignistic level. The evidential information is transferred into a probabilistic framework by means of the pignistic probability distribution $betP_m$, for $\theta \in \Theta$, $betP_m(\theta) = \sum_{A \subseteq \Theta, A \ni \theta} m(A)/|A|$, where $|A|$ denotes the number of elements in A.

Decision Rule. The risk associated with a decision rule is adaptable for the evidential framework [9,13,27]. In the case of imprecise data, the set of actions $\mathcal{A}$ is $2^\Theta \setminus \{\emptyset\}$. In order to decide between the elements of $\mathcal{A}$ according to the chosen loss function L, it is possible to adopt different strategies. Two strategies are proposed in the literature: the optimistic strategy by minimizing $\underline{r}_{\delta_\Theta}$ or the pessimistic strategy by minimizing $\overline{r}_{\delta_\Theta}$ which are defined as follows:

$$\underline{r}(A) = \sum_{B \subseteq \Theta} m(B) \min_{\theta \in B} L(A, \theta), \quad \overline{r}(A) = \sum_{B \subseteq \Theta} m(B) \max_{\theta \in B} L(A, \theta). \tag{1}$$

3   Problem Statement and Proposed Approach

3.1   Imprecise Supervised Classification

For a new example x, the output of an imprecise classifier is a set of classes, all its elements are candidates for the true class θ and the missing information prevent more precise output. In this case a possible output of the classifier is the information: "$\theta \in A$", $A \subseteq \Theta$. To perform an imprecise classification, two cases need to be distinguished related to the training examples: (*case 1*) learning examples are precisely labelled, i.e., only a single class is assigned to each example; (*case 2*) one or more classes are assigned to each training example. In the first case described in the Subsect. 2.1, standard classifiers give a single class as prediction to a new object x but some recent classifiers [6,7,36] give a set of classes as prediction of x. Some of these recent classifiers base their algorithm on the posterior probability provided by standard classifiers. More precisely, if we denote by $\mathbb{P}(.|x)$ the probability measure associated to the posterior probability distribution $p(.|x)$, $\mathbb{P}(A|x) = \sum_{\theta \in A} p(\theta|x)$, $A \subseteq \Theta$ is used to determine the relevant subset of classes to be assigned to x. In the second case when the imprecision or doubt is explicitly expressed by the labels, [2,5,37], a classifier $\delta_{2\Theta} : \mathcal{X} \to 2^{\Theta} \setminus \{\emptyset\}$ is trained on a set of examples $(x_i, A_i)_{1 \leq i \leq N}$ such that for all $1 \leq i \leq N$, x_i belongs to $\mathcal{X}$ and $\emptyset \neq A_i \subseteq \Theta$. This case is refereed in our paper as *imprecise supervised classification*.

3.2   Problem Statement

Let us consider the supervised classification problem where the available training examples that are precisely labelled (*case 1*) $(x_i, \theta_i)_{1 \leq i \leq N}$, $x_i \in \mathcal{X}$ and $\theta_i \in \Theta$ are such that (i) the labels $\theta_{i=1,...,N}$ are trusted. They may derive from expertise on other features $x^*_{i=1,...,N}$ which contain more complete information than $x_{i=1,...,N}$, (ii) this loss of information induces overlapping on some examples. In other words, $\exists i, j \in \{1, ..., N\}$ such that the characteristics of x_i are very close to those of x_j but $\theta_i \neq \theta_j$. When a standard classifier is trained on such data, it will commit inevitable errors. The problem that we handle in this paper is how to improve the learning step to better consider this type of data and get better performances and reliable predictions.

3.3   The Imprecise Classification Approach

The proposed approach of imprecise classification is constituted by three steps: (i) the **relabelling step** which consists in analysing the training example in order to add to the class that is initially associated to an example the classes associated to the other examples having characteristics very close. Thus a new set of examples is built: $(x_i, A_i)_{1 \leq i \leq N}$ such that for all $1 \leq i \leq N$, x_i belongs to $\mathcal{X}$ and $\emptyset \neq A_i \subseteq \Theta$; (ii) the **training step** which consists on the training of probabilistic classifier $\delta_{2\Theta} : \mathcal{X} \to 2^{\Theta} \setminus \{\emptyset\}$. The classifier $\delta_{2\Theta}$ provides for a new object $x \in \mathcal{X}$ a posterior probability distribution on 2^{Θ} which is also a mass function denoted $m(.|x)$. The trained classifier ignores the existence of inclusion or intersection between subsets of classes. This unawareness of relations between the labels may seem counter intuitive, but is compatible with the purpose of finding a potentially imprecise label associated to a new incoming example; (iii) the **decision step** which consists of proposing a loss function adapted for the case of imprecise classification that calculates the prediction that minimize the risk function associated to the classifier $\delta_{2^{\Theta}}$. Figure 1 illustrates the global process and the steps of relabelling, classification and decision are presented in detail below.

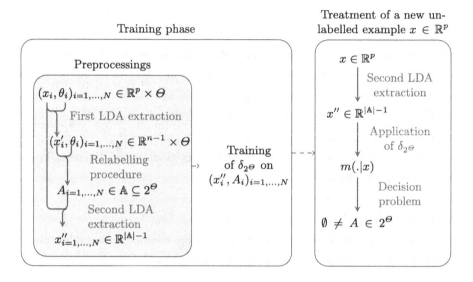

Fig. 1. Steps of evidential classification of incomplete data

Relabelling Procedure. First we perform LDA extraction on the training examples (cf Fig. 1) in order to reduce complexity. The resulting features are $x_i' \in \mathbb{R}^{n-1}, i = 1, ..., N$ where $n = |\Theta|$. Then we consider a set of C standard classifiers $\delta_{\Theta}^1, ..., \delta_{\Theta}^C$ where on each classifier $\delta_{\Theta}^c : \mathbb{R}^{n-1} \to \Theta$, $c \in \{1, ..., C\}$ we compute leave-one-out (LOO) cross validation predictions for the training data $(x_i', \theta_i)_{i=1,...,N}$.

The relabelling of the example (x_i', θ_i) is based on a vote procedure of the LOO predictions of the C classifiers. The vote procedure is the following: when more than 50% majority of the classifiers predict a class θ_{maj_i}, the example is relabelled as the union $A_i = \{\theta_i, \theta_{maj_i}\}$. Note that when $\theta_{maj_i} = \theta_i$ the original label remains, i.e., $A_i = \theta_i$. If none of the predicted classes from the C classifiers gets the majority, then the ignorance is expressed for this example by relabelling it as $A_i = \Theta$. Note that the new labels are consistent with the original classes that were trusted. The fact that several (C) classifiers are used to express the imprecision permits a better objectivity on the real imprecision of the features, i,e, the example is difficult not only for a single classifier. We denote by $\mathbb{A} \subseteq 2^\Theta$ the set of the new training labels $A_i, i = 1, ..., N$.

Note that we limited the new labels A_i to have at most two elements except when expressing ignorance $A_i = \Theta$. This is done for avoiding too unbalanced training sets, but more general relabelling could be considered. Once all the training examples are relabelled, a classifier δ_{2^Θ} can be trained.

Learning δ_{2^Θ}. As indicated throughout this paper, δ_{2^Θ} is learnt using the new labels ignoring the relations that might exist between the elements of $\mathbb{A}$. Reinforcing the idea of independence of treatment between the classes, LDA is applied to the relabelled training set $(x_i, A_i)_{i=1,...,N}$. This results to the reduction of the space dimension from p to $|\mathbb{A}| - 1$ which better expresses the repartition of relabelled training examples. For the training example $i \in \{1, ..., N\}$, let $x_i'' \in \mathbb{R}^{|\mathbb{A}|-1}$ be the new projection of x_i on this $|\mathbb{A}| - 1$ dimension space. The classifier δ_{2^Θ} is finally taught on $(x_i'', A_i)_{i=1,...,N}$.

Decision Problem. As recalled in Subsects. 2.2 and 2.3, the decision to assign a new object x to a single class or a set of classes usually relies on the minimisation of the risk function which is associated to a loss function $L : 2^\Theta \setminus \{\emptyset\} \times \Theta \to \mathbb{R}$. As mentioned in the introduction to this paper, the application of our work concerns situations where errors may have serious consequences. It would then be legitimate to consider the pessimistic strategy by minimizing $\overline{r}_{\delta_\Theta}$. Furthermore, in the definition of $\overline{r}_{\delta_\Theta}$, Eq. (1), the quantity $\max_{\theta \in B} L(A, \theta)$ concerns the loss incurred by choosing $A \subseteq \Theta$, when the true nature is comprised in $B \subseteq \Theta$. On the basis of this fact, we proposed a new definition of the loss function, $L(A, B)$, $A, B \subseteq \Theta$, which directly takes into account the relations between A and B. This is actually a generalisation of the definition proposed in [7] that is based on F-measure, recall and precision for imprecise classification. Let us consider $A, B \in 2^\Theta \setminus \{\emptyset\}$, where $A = \theta(x)$ is the prediction for the object x and B is its state of nature. Recall is defined as the proportion of relevant classes included in the prediction $\theta(x)$. We define the recall of A and B as:

$$R(A, B) = \frac{|A \cap B|}{|B|}. \tag{2}$$

Precision is defined as the proportion of classes in the prediction that are relevant. We define the precision of A and B as:

$$P(A, B) = \frac{|A \cap B|}{|A|}. \tag{3}$$

Considering these two definition, the F-measure can be defined as follows:

$$F_\beta(A, B) = \frac{(1 + \beta^2)PR}{\beta^2 P + R} = \frac{(1 + \beta^2)|A \cap B|}{\beta^2|B| + |A|}. \tag{4}$$

Note that $\beta = 0$, induce $F_\beta(A, B) = P(A, B)$, whereas when $\beta \to \infty$, $F_\beta(A, B) \underset{\beta \to \infty}{\to} P(A, B)$. Let us comment on some situations according to the "true set" B and the predicted set A. The worse scenario of prediction is when there is no intersection between A and B. This would always be sanctioned by $F_\beta(A, B) = 0$. On the contrary, when $A = B$, $F_\beta(A, B) = 1$ for every β. Between those extreme cases, the errors of generalisation i.e., $B \subset A$, are controlled by the precision while the errors of specialisation i.e., $A \subset B$, are controlled by the recall. Finally, the loss function $L_\beta : 2^\Theta \setminus \{\emptyset\} \times 2^\Theta \setminus \{\emptyset\} \to \mathbb{R}$ is extended:

$$L_\beta(A, B) = 1 - F_\beta(A, B). \tag{5}$$

For an example x to be classified, whose mass function $m(.|x)$ has been calculated by δ_{2^Θ}, we predict the set A minimizing the following risk function:

$$\text{Risk}_\beta(A) = \sum_{B \subseteq \Theta} m(B) L_\beta(A, B). \tag{6}$$

4   Related Works

Regarding relabelling procedures, much research has been carried out to identify suspect examples with the intention to suppress or relabel them into a concurrent more appropriate class [16, 20]. This is generally done to enhance the performance. Other approaches consist in relabelling into imprecise classes. This has been done to test the evidential classification approach on imprecise labelled data in [37]. But, as already stated, our relabelling serves a different purpose, better mapping overlaps in the feature space. Concerning the imprecise classification, several works have been dedicated to tackle this problem. Instead of the term "imprecise classification" that is adopted in our article, authors use terms like "nondeterministic classification" [7], "reliable classification" [24], "indeterminate classification" [6, 36], "set-valued classification" [28, 31] or "conformal prediction" [3] (see [24] for a short state of the art). In [36], the Naive Credal Classifier (NCC) is proposed as the extension of Naive Bayes Classifier (NBC) to sets of probability distributions. In [24] the authors propose an approach that starts from the outputs of a binary classification [25] using classifier that are trained to distinguish aleatoric and epistemic uncertainty. The outputs are epistemic uncertainty, aleatoric uncertainty and two preference degrees in favor of the two concurrent classes. [24] generalizes this approach to the multi-class and providing set of classes as output. Closer to our approach are approaches of [5] and [7]. The approach in [7] is based on a posterior probability distribution provided by a probabilistic classifier. The advantage of such approach and ours is that any standard probabilistic classifier may be used to perform an imprecise classification. Our approach distinguishes itself by the relabelling step and by the way probabilities are allowed on sets of classes. To the best of our knowledge existing works algorithms do not train a probabilistic classifier on partially labelled data to quantify the body of evidence. Although we insisted for the use of standard probabilistic classifier δ_{2^Θ} unaware of relations between the sets, it is possible to run our procedure with an evidential classifier as the evidential k-NN [5].

5   Illustration

5.1   Settings

We performed experiments on the classification problem of four plastic categories designated plastics A, B, C and D on the basis of industrially acquired spectra. The total of 11540 available data examples is summarized in Table 1. Each plastic example was identified by experts on the basis of laboratory measure of attenuated total reflectance spectra (ATR) which is considered as a reliable source of information for plastic category's determination. As a consequence, original training classes are trusted and were not questioned. However data provided by the industrial devices may be challenged. These data consist in spectra composed of the reflectance intensity of 256 different wavelengths. Therefore and for the enumerated reasons in Sect. 1, the features are subject to ambiguity. Prior to experiments, all the feature vectors, i.e., spectra, were corrected by the standard normal variate technique to avoid light scattering and spectral noise effects. We implemented our approach and compared it to the approaches in [5] and [7]. The implementation is made using R packages, using existing functions for the application of the following 8 classifiers naive Bayes classifier: (nbayes), k-Nearest Neighbour (k-NN), decision tree (tree), random forest (rf), linear discriminant analysis (lda), partial least squares discriminant analysis (pls-da), support vector machine (svm) and neural networks (nnet).[1]

Table 1. Number of spectra of each original class in learning and testing bases.

Classes	Category A	Category B	Category C	Category D
Learning base	1416	1412	1425	1434
Testing base	1469	1458	1454	1472

5.2   Results

In order to apply our procedure, we must primary choose a set of classifiers to perform the relabelling. These classifiers are not necessarily probabilistic but producing point prediction. Thus, for every experimentation, our algorithm ECLAIR was performed with the ensemble relabelling using 7 classifiers: nbayes, k-NN, tree, rf, lda, svm, nnet[2]. Then, we are able to perform the ECLAIR imprecise version of a selected probabilistic classifier. Figure 2, shows the recall and precision scores of the probabilistic classifier nbayes to show the role of β. We see the same influence of β as mentioned in [7]. Indeed, (cf Subsect. 3.3), with small

---

[1] Experiments concerning these learning algorithm rely on the following functions (and R packages) : naiveBayes (e1071), knn3 (caret), rpart (rpart), randomForest (randomForest), lda (MASS), plsda (caret), svm (e1071), nnet (nnet).

[2] In order to limit unbalanced classes, we choose to exclude form the learning base examples which new labels count less than 20 examples.

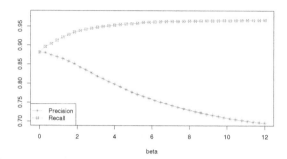

Fig. 2. Recall and precision of ECLAIR using nbayes, i.e. $\delta_{2\ominus}$ is nbayes, against β.

values of β we have good precision, traducing the relevance of prediction, i.e., the size of the predicted set is reasonable; while high values of β give good recall, meaning reliability, i.e., better chance to have true class included in the predictions. The choice of β should then result form a compromise between relevance and reliability requirement.

Table 2. Precision P of ECLAIR compared with *nondeterministic* with βs chosen such that recalls equal to 0.90.

	nbayes	k-NN	tree	rf	lda	pls-da	svm	evidential k-NN
Nondeterministic	86.70	86.94	**85.00**	86.52	**83.41**	85.35	88.20	86.58
ECLAIR	**87.78**	**87.89**	83.88	**87.45**	82.94	**86.33**	**88.31**	**86.69**

In order to evaluate the performances of ECLAIR, we compared our results to the classifier proposed in [7] that is called here *nondeterministic* classifier. As *nondeterministic* classifier and ECLAIR are set up for a parameter β, we decided to set βs such that global recalls equal to 0.90, and compare global precisions on a fair basis. For even more neutrality regarding the features used in both approach, we furnish to the *nondeterministic* classifier, the same reduced features x_i'', $i = 1, ..., N$, that those used by ECLAIR in the training phase (see Fig. 1). The 7 first columns of Table 2 shows the so obtained precisions for 7 classifiers. These results show the competitiveness of our approach for most of the classifiers, especially nbayes, k-NN, rf and pls-da. However, these results are only partial since they do not show the general trend for different βs that are generally in favour of our approach. Therefore we present more complete results for nbayes and svm in Fig. 3, showing evaluation of precision score against recall score for several values of β varying in $[0, 6]$. On the same figure, we also present the results of *nondeterministic* classifier with different input feature (in black): raw features, i.e., $x_i \in \mathbb{R}^p$, LDA reduced features, i.e., $x_i' \in \mathbb{R}^{n-1}$ and the same features as those used for ECLAIR, i.e., $x_i'' \in \mathbb{R}^{|\mathbb{A}|-1}$ (see Fig. 1 for more details). Doing so, we show that the good performances of ECLAIR are not only

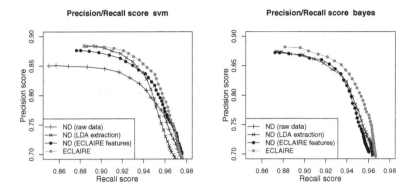

Fig. 3. Precision vs recall of Nondeterministic (ND) and ECLAIR

attributable to extraction phase. To facilitate the understanding of the results plotted in Fig. 3, one should understand that the best performances are those illustrated by points on the top right of the plots, i.e., higher precision and recall scores. We observe that ECLAIR generally makes a better compromise between the recall and precision scores for the used classifiers. Regarding the special case when ECLAIR is performed with an evidential classifier performing example imprecise labelled training (see Sect. 4), the comparison is less straightforward. We considered the evidential k-NN [10] for imprecise labels by minimizing the error suggested in [39]. Using this evidential k-NN as a classifier $\delta_{2\Theta}$ in ECLAIR procedure is straightforward. Concerning the application of *nondeterministic* classifier, we decided to keep the same parameter and turn the classifier into probabilistic by applying the pignistic transformation to the mass output of the k-NN classifier (see column of Table 2). ECLAIR obtains a slightly better results.

6   Conclusion

In this article, a method of evidential classification of incomplete data via imprecise relabelling was proposed. For any probabilistic classifier, our approach proposes an adaptation to get more cautious output. The benefit of our approach was illustrated on the problem of sorting plastics and showed competitive performances. Our algorithm is generic it can be applied in any other context where incomplete data on the features are presents. In future works we plan to exploit our procedure to provide cautious decision-making for the problem of plastic sorting. This application requires high reliability of the decision for preserving the physiochemical properties of the recycle product. At the same time, the decision shall ensure reasonable relevance to guarantee financial interest, indeed the more one plastic category is finely sorted the more benefice the industrial gets. We also plan to strengthen our approach evaluation by confronting it with other state of the art imprecise classifiers and by preforming experiments on several datasets from machine learning repositories.

References

1. Alshamaa, D., Chehade, F.M., Honeine, P.: A hierarchical classification method using belief functions. Signal Process. **148**, 68–77 (2018)
2. Ambroise, C., Denoeux, T., Govaert, G., Smets, P.: Learning from an imprecise teacher: probabilistic and evidential approaches. Appl. Stoch. Models Data Anal. **1**, 100–105 (2001)
3. Balasubramanian, V., Ho, S.S., Vovk, V.: Conformal Prediction for Reliable Machine Learning: Theory, Adaptations and Applications. Newnes, Oxford (2014)
4. Buckley, J.J., Hayashi, Y.: Fuzzy neural networks: a survey. Fuzzy Sets Syst. **66**(1), 1–13 (1994)
5. Côme, E., Oukhellou, L., Denoeux, T., Aknin, P.: Learning from partially supervised data using mixture models and belief functions. Pattern Recogn. **42**(3), 334–348 (2009)
6. Corani, G., Zaffalon, M.: Learning reliable classifiers from small or incomplete data sets: the naive credal classifier 2. J. Mach. Learn. Res. **9**(Apr), 581–621 (2008)
7. Coz, J.J.D., Díez, J., Bahamonde, A.: Learning nondeterministic classifiers. J. Mach. Learn. Res. **10**(Oct), 2273–2293 (2009)
8. Cucchiella, F., D'Adamo, I., Koh, S.L., Rosa, P.: Recycling of weees: an economic assessment of present and future e-waste streams. Renew. Sustain. Energy Rev. **51**, 263–272 (2015)
9. Dempster, A.P.: Upper and lower probabilities induced by a multivalued mapping. In: Yager, R.R., Liu, L. (eds.) Classic Works of the Dempster-Shafer Theory of Belief Functions. Studies in Fuzziness and Soft Computing, vol. 219. Springer, Heidelberg (2008). https://doi.org/10.1007/978-3-540-44792-4_3
10. Denoeux, T.: A k-nearest neighbor classification rule based on dempster-shafer theory. IEEE Trans. Syst. Man Cybern. **25**(5), 804–813 (1995)
11. Denoeux, T.: A neural network classifier based on dempster-shafer theory. IEEE Trans. Syst. Man Cybern. Part A Syst. Hum. **30**(2), 131–150 (2000)
12. Denoeux, T.: Maximum likelihood estimation from uncertain data in the belief function framework. IEEE Trans. Knowl. Data Eng. **25**(1), 119–130 (2013)
13. Denoeux, T.: Logistic regression, neural networks and dempster-shafer theory: a new perspective. Knowl.-Based Syst. **176**, 54–67 (2019)
14. Dubois, D., Prade, H.: Possibility theory. In: Meyers, R. (ed.) Computational Complexity. Springer, New York (2012). https://doi.org/10.1007/978-1-4614-1800-9
15. Dunn, J.C.: A fuzzy relative of the ISODATA process and its use in detecting compact well-separated clusters. J. Cybern. **3**, 32–57 (1973)
16. Kanj, S., Abdallah, F., Denoeux, T., Tout, K.: Editing training data for multi-label classification with the k-nearest neighbor rule. Pattern Anal. Appl. **19**(1), 145–161 (2016)
17. Kanjanatarakul, O., Kaewsompong, N., Sriboonchitta, S., Denoeux, T.: Estimation and prediction using belief functions: Application to stochastic frontier analysis. In: Huynh, V.N., Kreinovich, V., Sriboonchitta, S., Suriya, K. (eds.) Econometrics of Risk. Studies in Computational Intelligence, vol. 583. Springer, Cham (2015). https://doi.org/10.1007/978-3-319-13449-9_12
18. Kassouf, A., Maalouly, J., Rutledge, D.N., Chebib, H., Ducruet, V.: Rapid discrimination of plastic packaging materials using mir spectroscopy coupled with independent components analysis (ICA). Waste Manage. **34**(11), 2131–2138 (2014)
19. Keller, J.M., Gray, M.R., Givens, J.A.: A fuzzy k-nearest neighbor algorithm. IEEE Trans. Syst. Man Cybern. **4**, 580–585 (1985)

20. Lallich, S., Muhlenbach, F., Zighed, D.A.: Improving classification by removing or relabeling mislabeled instances. In: Hacid, M.-S., Raś, Z.W., Zighed, D.A., Kodratoff, Y. (eds.) ISMIS 2002. LNCS (LNAI), vol. 2366, pp. 5–15. Springer, Heidelberg (2002). https://doi.org/10.1007/3-540-48050-1_3

21. Lee, H.K., Kim, S.B.: An overlap-sensitive margin classifier for imbalanced and overlapping data. Expert Syst. Appl. **98**, 72–83 (2018)

22. Leitner, R., Mairer, H., Kercek, A.: Real-time classification of polymers with NIR spectral imaging and blob analysis. Real-Time Imaging **9**(4), 245–251 (2003)

23. Naeini, M.P., Moshiri, B., Araabi, B.N., Sadeghi, M.: Learning by abstraction: hierarchical classification model using evidential theoretic approach and Bayesian ensemble model. Neurocomputing **130**, 73–82 (2014)

24. Nguyen, V.L., Destercke, S., Masson, M.H., Hüllermeier, E.: Reliable multi-class classification based on pairwise epistemic and aleatoric uncertainty. In: International Joint Conference on Artificial Intelligence (2018)

25. Senge, R., et al.: Reliable classification: learning classifiers that distinguish aleatoric and epistemic uncertainty. Inf. Sci. **255**, 16–29 (2014)

26. Shafer, G.: A Mathematical Theory of Evidence, vol. 42. Princeton University Press, Princeton (1976)

27. Shafer, G.: Constructive probability. Synthese **48**(1), 1–60 (1981)

28. Shafer, G., Vovk, V.: A tutorial on conformal prediction. J. Mach. Learn. Res. **9**(Mar), 371–421 (2008)

29. Smets, P.: Non-Standard Logics for Automated Reasoning. Academic Press, London (1988)

30. Smets, P., Kennes, R.: The transferable belief model. Artif. Intell. **66**(2), 191–234 (1994)

31. Soullard, Y., Destercke, S., Thouvenin, I.: Co-training with credal models. In: Schwenker, F., Abbas, H.M., El Gayar, N., Trentin, E. (eds.) ANNPR 2016. LNCS (LNAI), vol. 9896, pp. 92–104. Springer, Cham (2016). https://doi.org/10.1007/978-3-319-46182-3_8

32. Sutton-Charani, N., Imoussaten, A., Harispe, S., Montmain, J.: Evidential bagging: combining heterogeneous classifiers in the belief functions framework. In: Medina, J., et al. (eds.) IPMU 2018. CCIS, vol. 853, pp. 297–309. Springer, Cham (2018). https://doi.org/10.1007/978-3-319-91473-2_26

33. Walley, P.: Towards a unified theory of imprecise probability. Int. J. Approximate Reasoning **24**(2–3), 125–148 (2000)

34. Xiong, H., Li, M., Jiang, T., Zhao, S.: Classification algorithm based on nb for class overlapping problem. Appl. Math **7**(2L), 409–415 (2013)

35. Zadeh, L.A.: Fuzzy sets as a basis for a theory of possibility. Fuzzy Sets Syst. **1**(1), 3–28 (1978)

36. Zaffalon, M.: The naive credal classifier. J. Stat. Plan. Infer. **105**(1), 5–21 (2002)

37. Zhang, J., Subasingha, S., Premaratne, K., Shyu, M.L., Kubat, M., Hewawasam, K.: A novel belief theoretic association rule mining based classifier for handling class label ambiguities. In: Proceeidngs of Workshop Foundations of Data Mining (FDM 2004), International Conferenece on Data Mining (ICDM 2004) (2004)

38. Zheng, Y., Bai, J., Xu, J., Li, X., Zhang, Y.: A discrimination model in waste plastics sorting using nir hyperspectral imaging system. Waste Manage. **72**, 87–98 (2018)

39. Zouhal, L.M., Denoeux, T.: An evidence-theoretic k-NN rule with parameter optimization. IEEE Trans. Syst. Man Cybern. Part C (Appl. Rev.) **28**(2), 263–271 (1998)

An Analogical Interpolation Method for Enlarging a Training Dataset

Myriam Bounhas[1,2(✉)] and Henri Prade[3]

[1] Emirates College of Technology, Abu Dhabi, UAE
myriam_bounhas@yahoo.fr
[2] LARODEC Lab., ISG de Tunis, Tunis, Tunisia
[3] IRIT, Université Paul Sabatier, 118 route de Narbonne,
31062 Toulouse cedex 09, France
prade@irit.fr

Abstract. In classification problems, it happens that the training set remains scarce. Given a data set, described in terms of discrete, ordered attribute values, we propose an interpolation-based approach in order to predict *new* examples useful for enlarging the original data set. The proposed approach relies on the use of continuous analogical proportions that are statements of the form "a is to x as x is to c". The prediction is made on the basis of pairs of examples (a, c) present in the data set, for which one can find a value for x for each attribute value as well as for the corresponding class label of the example thus created. The first option that we consider is to select x as the *midpoint* between a and c, attribute by attribute. To extend the search space, we may also choose x as any randomly selected value *between* the values of a and c. We first propose a basic algorithm implementing these two interpolation definitions, then we extend it to two improved algorithms. In the former, we only consider the nearest neighbor pairs (a, c) to x for prediction, while, in the latter, we further restrict the search to those pairs (a, c) having the same class label. The experimental results, for classical ML classifiers applied to the enlarged data sets built by the proposed algorithms, show the effectiveness of analogical interpolation methods for enlarging data sets.

1 Introduction

Analogical proportions are statements of the form "a is to b as c is to d". In the *Nicomachean Ethics*, Aristotle makes an explicit parallel between such statements and geometric proportions of the form "$\frac{a}{b} = \frac{c}{d}$", where a, b, c, d are numbers. It also parallels arithmetic proportions, or difference proportions, which are of the form "$a - b = c - d$". The logical modeling of an analogical proportion as a quaternary connective between four Boolean items appears to be a logical counterpart of such numerical proportions [15]. It has been extended to items described by vectors of Boolean, nominal or numerical values [2].

A particular case of such statements, named *continuous* analogical proportions, is obtained when the two central components are equal, namely they are

© Springer Nature Switzerland AG 2019
N. Ben Amor et al. (Eds.): SUM 2019, LNAI 11940, pp. 136–152, 2019.
https://doi.org/10.1007/978-3-030-35514-2_11

statements of the form "a is to b as b is to c". In case of numerical proportions, if we assume that b is unknown, it can be expressed in terms of a and c as $b = \sqrt{a \cdot c}$ in the geometric case, and as $\frac{a+c}{2}$ in the arithmetic case. Note that similar inequalities hold in both cases: $\min(a,c) \leq \sqrt{a \cdot c} \leq \max(a,c)$ and $\min(a,c) \leq \frac{a+c}{2} \leq \max(a,c)$. This means that the continuous analogical proportion induces a kind of interpolation between a and c in the numerical case by involving an intermediary value that can be obtained from a and c.

General analogical proportions when d is unknown provides an extrapolation mechanism, which with numbers yields $d = \frac{b \cdot c}{a}$ and $d = b + c - a$ in the geometric and arithmetic cases respectively. We recognize the expression of the well-known Rule of Three in the first expression. Analogical proportions-based inference [2] offers a similar extrapolation device relying on the parallel between (a,b) and (c,d) stated by "a is to b as c is to d".

The analogical proportions-based extrapolation has been successfully applied to classification problems. It may be used either directly as a new classification paradigm [2,12], or as a way of completing a training set on which classical classification methods are applied once this set has been completed [1,4]. This paper investigates the effectiveness of the simpler option of using only continuous analogical proportions that involve pairs instead of triples of items, in order to enlarge a training set.

The paper is organized as follows. Section 2 provides a short background on analogical proportions and more particularly on continuous ones. Then Sect. 3 surveys related work on analogical interpolation or extrapolation. Section 4 presents different variants of algorithms for completing a training set based on the idea of continuous analogical proportions. Section 5 reports the results of the use of different classical classification techniques on the corresponding enlarged training sets for various benchmarks.

2 Background: Continuous Analogical Proportion

The statement "a is to b as c is to d", here denoted $a : b :: c : d$, expresses that "a differs from b as c differs from d, and b differs from a as d differs from c". The logical counterpart of the latter statement, where a, b, c, d are Boolean variables, is given by:

$$a : b :: c : d = (\neg a \wedge b \equiv \neg c \wedge d) \wedge (\neg b \wedge a \equiv \neg d \wedge c)$$

See [13,16] for justifications. This expression is true for only 6 patterns of values for $abcd$, namely $\{0000, 0011, 0101, 1111, 1100, 1010\}$. This extends to nominal values where $a : b :: c : d$ holds true if and only if $abcd$ is one of the following patterns $ssss$, $stst$, or $sstt$, where s and t are two possible distinct values of items a, b, c and d.

Regarding continuous analogical proportions, it can be easily checked that the unique solutions of equations $1 : x :: x : 1$ and $0 : x :: x : 0$ are respectively $x = 1$ and $x = 0$, while $1 : x :: x : 0$ or $0 : x :: x : 1$ have no solution in the

Boolean case. This somewhat trivializes continuous analogical proportions in the Boolean case. The situation for nominal values is the same.

The case of numerical values is richer. a, b, c, d are now supposed to be normalized values in the real interval $[0, 1]$. The reader is referred to [6] for a general discussion of multiple-valued logic extensions of analogical proportions. They can be associated with the following expression:

$$a : b :: c : d = \begin{cases} 1 - \mid (a - b) - (c - d) \mid, \\ \quad \text{if } a \geq b \text{ and } c \geq d, \text{ or } a \leq b \text{ and } c \leq d \\ 1 - \max(\mid a - b \mid, \mid c - d \mid), \\ \quad \text{if } a \leq b \text{ and } c \geq d, \text{ or } a \geq b \text{ and } c \leq d \end{cases} \quad (1)$$

It coincides with $a : b :: c : d$ on $\{0, 1\}$. As can be seen, $a : b :: c : d$ is equal to 1 if and only if $(a - b) = (c - d)$. For instance, $0.2 : 0.5 :: 0.6 : 0.9$, or $0.2 : 0.5 :: 0.2 : 0.5$ holds true. Because $|a - b| = |(1 - a) - (1 - b)|$, it is easy to check that the code independence property: $a : b :: c : d = (1 - a) : (1 - b) :: (1 - c) : (1 - d)$ holds (0 and 1 play symmetric roles, and it is the same to encode an attribute positively or negatively).

Then the corresponding expression for continuous analogical proportions is [16]:

$$a : b :: b : c = \begin{cases} 1 - \mid a + c - 2b \mid, \\ \quad \text{if } a \geq b \text{ and } b \geq c, \text{ or } a \leq b \text{ and } b \leq c \\ 1 - \max(\mid a - b \mid, \mid b - c \mid), \\ \quad \text{if } a \leq b \text{ and } b \geq c, \text{ or } a \geq b \text{ and } b \leq c \end{cases} \quad (2)$$

As can be seen $a : b :: b : c = 1$ if and only if $b = (a + c)/2$ (which includes the case $a = b = c$). The proportions $0 : \frac{1}{2} :: \frac{1}{2} : 1$ or $0.3 : 0.6 :: 0.6 : 0.9$ are examples of continuous analogical proportions. Moreover, $1 : 3 :: 3 : 5$ is an example of continuous analogical proportion between nominal ordered grades. Thus this extension captures the idea of betweenness implicit in statements of the form "a is to b as b is to c". Note that we have $0 : 1 :: 1 : 0 = 0$ and $1 : 0 :: 0 : 1 = 0$, as expected.

Analogical proportions extend to vectors in a component-wise manner. Let $\boldsymbol{a} = (a_1, \ldots, a_m)$, where each a_i belongs to $\{0, 1\}$ (Boolean case), or to a finite set with more than 2 elements (nominal case), or to $[0, 1]$ (numerical case). $\boldsymbol{b}, \boldsymbol{c}, \boldsymbol{d}$ are defined similarly. Then $\boldsymbol{a} : \boldsymbol{b} :: \boldsymbol{c} : \boldsymbol{d}$ has a truth value which is just $\min_{i=1}^{m} a_i : b_i :: c_i : d_i$.

In this paper, we deal with classification. So each vector $\boldsymbol{a}$ in a training set is associated with its class $cl(\boldsymbol{a})$. Thus saying that the continuous analogical proportion $\boldsymbol{a} : \boldsymbol{x} :: \boldsymbol{x} : \boldsymbol{c}$ holds true amounts to say:

$$\boldsymbol{a} : \boldsymbol{x} :: \boldsymbol{x} : \boldsymbol{c} = 1 \text{ iff}$$
$$a_j : x_j :: x_j : c_j = 1 \text{ for each attribute } j \text{ and } cl(\boldsymbol{a}) : cl(\boldsymbol{x}) :: cl(\boldsymbol{x}) : cl(\boldsymbol{c}) = 1 \quad (3)$$

Moreover, since continuous analogical proportions are trivial for a Boolean or a nominal variable, we shall also use a more liberal extension of betweenness for the vectorial case [10] in this paper. Namely, we shall say x is *between* a and c defined as:

$$between(a, x, c) = 1 \text{ iff } a_j \leq x_j \leq c_j \text{ or } c_j \leq x_j \leq a_j \text{ for each attribute } j. \quad (4)$$

Then we can define the set Between(a, c) of vectors between two vectors a and c. For instance, we have Between$(01000, 11010) = \{01000, 11000, 01010, 11010\}$. Note that in case of Boolean values, the betweenness condition can also be written as $\forall i = 1, \cdots, m, (a_i \wedge c_i \rightarrow x_i) \wedge (x_i \rightarrow a_i \vee c_i) = 1$.

3   Related Work

The idea of generating, or completing, a third example from two examples can be encountered in different settings. An option, quite different from interpolation, is the "feature knock out" method [23], where a third example is built by modifying a randomly chosen feature of the first example with that of the second one. A somewhat related idea can be found in a recent proposal [3] which introduces a measure of oddness with respect to a class that is computed on the basis of pairs made of two nearest neighbors in the same class; this amounts to replace the two neighbors by a fictitious representative of the class.

Reasoning with a system of fuzzy if-then rules provides an interpolation mechanism [14], which, from these rules and an input "in-between" their condition parts, yields a new conclusion "in-between" their conclusion parts, by taking advantage of membership functions that can be seen as defining fuzzy "neighborhoods".

Moreover, several approaches based on the use of interpolation and analogical proportions have been developed in the past decade. In [17], the problem considered is to complete a set of parallel if-then rules, represented by a set of condition variables associated to a conclusion variable. The values of the variables are assumed to belong to finite sets of ordered labels. The basic idea is to apply analogical proportion inference in order to induce missing rules from an initial set of rules, when an analogical proportions hold between the variable labels of several parallel rules. Although this approach may seem close to the analogical interpolation-based approach proposed in this paper, our goal is not to predict just the conclusion part of an incomplete rule, but rather a whole example including its attribute-based description and its class. Moreover, we restrict our study to the use of pairs of examples for this prediction, while in [17] the authors use both pairs or triples of rules for completing rules. An extended version of the above-mentioned work has been presented in [22] where the authors also propose a more cautious method that makes explicit the basic assumptions under which rule conclusions are produced from analogical proportions. Along the same line, see also [21] on interpolation between default rules.

Let us also mention the general approach proposed by Schockaert and Prade [20] to interpolative and extrapolative reasoning from incomplete generic knowledge represented by sets of symbolic rules, handled in a purely qualitative manner, where labels are represented in conceptual spaces. This work is an extended version of [19] in which only interpolative inference is considered. The same authors present an illustrative case study in [18] in the music domain. In the context of natural language modeling, Derrac and Schockaert [5] have proposed a data-driven approach that exploits betweenness and a fortiori inference to derive semantic relations within conceptual spaces.

Besides, some previous works have considered, discussed and experimented the idea of an analogical proportion-based enlargement of a training set, based on triples of examples. In [1], the authors proposed an approach to generate synthetic data to tune a handwritten character classifier. Couceiro et al. [4] presented a way to extend a Boolean sample set for classification using the notion of "analogy preserving" functions that generate examples on the basis of triples of examples in the training set. The authors only tested their approach on Boolean data.

In a more recent work, Lieber et al. [10] have extended the paradigm of classical Case-Based Reasoning to link the current case to either pairs of known cases by performing a restricted form of interpolation, or to triples of known cases by exploiting extrapolation, taking advantage of betweenness and analogical proportion relations.

Lastly, in the context of deep learning, Goodfellow et al. [7] invented the idea of a generative adversarial network (GAN) as a class of machine learning systems. Given a training set, two neural networks, contesting with each other in a game, are learnt in order to generate new data with the same statistics as the training set. More recently, Inoue [9] presented a data augmentation technique for image classification that mix two randomly picked images to train a classifier.

4   Analogical Interpolation-Based Predictor (AIP)

Analogical proportions have been recently applied to classification problems and have shown their efficiency for classifying a variety of datasets [2]. In this paper, we aim to investigate if continuous analogical proportions could be useful for a prediction purpose, namely enlarging a training set with made examples, and if standard classification methods applied to this enlarged set can compete with the direct application of analogical proportions-based inference for classification. As said before, the basic idea of the paper is to apply an interpolation method for predicting *new* examples not present in the original data set which is just enlarged.

In the following, we describe the basic principle of our predicting approach.

4.1   Basic Procedure

Consider a set E of n classified examples i.e., $E = \{(\boldsymbol{x^1}, y^1), ..., (\boldsymbol{x^i}, y^i), ..., (\boldsymbol{x^n}, y^n)\}$ such that the class label $y^i = cl(\boldsymbol{x^i})$ is known for each $i \in 1, ..., n$. The goal is to predict a new set of examples $S = \{(\boldsymbol{x^k}, y^k) \notin E\}$ by interpolating examples from the set E. The new set S will serve for enlarging E.

The basic idea is to find pairs of examples $(a, c) \in E^2$ with known labels such that the analogical proportion (3) is solvable attribute by attribute i.e., there exists x such that $a_j : x_j :: x_j : c_j = 1$ for each attribute $j = 1, ..., m$, and the class equation has $cl(x)$ as a solution, i.e., $cl(a) : cl(x) :: cl(x) : cl(c) = 1$.

As mentioned before in Sect. 2, the solution for the previous equation $a_j : x_j :: x_j : c_j = 1$ in the numerical case is just the midpoint $x_j = (a_j + c_j)/2$ for each attribute $j = 1, ..., m$. We are interested in the case of ordered nominal values in this paper. Moreover, we assume that the distances between any two successive values in such an ordered set of values are the same. Let $V = \{v_1, \cdots, v_k\}$ be an ordered set of nominal values, then, v_i will be regarded as the midpoint of v_{i-j} and v_{i+j} with $j \geq 1$, provided that both v_{i-j} and v_{i+j} exist. For instance, if $V = \{1, \cdots, 5\}$, the analogical proportions $1 : 3 :: 3 : 5$ or $2 : 3 :: 3 : 4$ hold, while $2 : x :: x : 5 = 1$ has no solution. So it is clear that some pairs (a, c) will not lead to any solution since we restrict the search space to the pairs for which the midpoint (attribute by attribute) exists.

This condition may be too restrictive especially for datasets with high number of attributes which may reduce the set of predicted examples. In case of success, the predicted example $x = \{x_1, ..., x_j, ...x_m\}$ will be assigned to the predicted class label $cl(x)$ and saved in a *candidate* set.

Since different voting pairs may predict the same example x more than once (x may be the midpoint of more than one pair (a, c)), a candidate example may have different class labels. Then has to perform a *vote* on class labels for each candidate example classified differently in the candidate set. This leads to the final predicted set of examples where each example is classified uniquely.

This process can be described by the following procedure:

1. Find solvable pairs (a, c) such that Eq. 3 has a unique *non null* solution x.
2. In case of ties (an example x is predicted with different class labels), apply voting on all its predicted class labels and assign to x the success label.
3. Add x to the set of predicted examples (together with $cl(x)$).

In the next section, we first present a basic algorithm applying the process described above, then we propose two options that may help to improve the search space for the voting pairs.

4.2 Algorithms

The simplest way is to systematically consider *all* pairs $(a, c) \in E^2$, for which Eq. 3 is solvable, as candidate pairs for prediction. Algorithm 1 implements a basic *Analogical Interpolation-based Predictor*, denoted AIP_{std}, without applying any filter on the voting pairs.

Considering all pairs (a, c) for prediction may seem unreasonable especially when the domain of attribute values is large since this may blur prediction results. A first improvement of Algorithm 1 is to restrict the search for pairs to those that are among the *nearest neighbors* (in terms of Hamming distance) to the example to be predicted.

Algorithm 1. Analogical Interpolation-based Predictor (AIP_{std})

Input: A set E of classified instances
CandidatesSet $= \emptyset$
S $= \emptyset$
for each pair (a, c) in E^2 **do**
　if $cl(a) : cl(x) :: cl(x) : cl(c) = 1$ has solution l **then**
　　if $a : x :: x : c = 1$ has solution b **then**
　　　$cl(b) = l$
　　　CandidatesSet.add(b)
　　end if
　end if
end for
S = VoteOnclasses(CandidatesSet)
Comp(E)= $E + S$
return (Comp(E))

Let us consider two different pairs (a, c) and $(d, e) \in E^2$. We assume that $a : x :: x : c = 1$ produces as solution an example b and $d : x :: x : e = 1$ produces an other example $b' \neq b$. If b' is closest to (d, e) than b is to (a, c) in terms of Hamming distance, it is more reasonable to consider *only* the pair (d, e) for prediction. This means that example b' will be predicted while b will be rejected. We denote AIP_{NN} this second improved Algorithm 2 in the following.

Algorithm 3 (that we denote $AIP_{NN,SC}$) is exactly the same as Algorithm 2 in all respects, except that we look for only pairs (a, c) belonging to the same class in this case. Note that the two algorithms only differ for non binary classification problems, since $s : x :: x : t = 1$ has no solution in $\{0, 1\}$ for $s \neq t$.

4.3 Another Option

As can be seen in the next section, searching for the best pairs (described in Algorithms 2 and 3) limits the number of accepted voting pairs. Moreover, there is a second constraint to be satisfied, that is limiting the solutions of Eq. 3 to the values of x that are the midpoint of a and c which is hard to be satisfied in the ordered nominal setting. To relax this last constraint, we may think to use the "betweenness" definition given in Eq. 4. In this definition, the equation $between(a, x, c) = 1$ has, as a solution, *any* x such that x is *between* a and c for each attribute $j \in 1, ..., m$. This last option is implemented by the algorithm denoted AIP_{Btw} which is exactly the same as Algorithm 3 except that we use the definition (4) to solve the analogical interpolation.

Algorithm 2. Analogical Interpolation-based Predictor using Nearest Neighbors pairs for prediction (AIP_{NN})

Input: A set E of classified instances
CandidatesSet $= \emptyset$
PredictedSet $= \emptyset$
Min_{HD} = NbrAttribute
for each pair (a, c) in E^2 **do**
  if $cl(a) : cl(x) :: cl(x) : cl(c) = 1$ has solution l **then**
    if $a : x :: x : c = 1$ has solution b **then**
      $cl(b) = l$
      HD = Max(HammingDistance(b,a), HammingDistance(b,c))
      if $HD < Min_{HD}$ **then**
        Min_{HD} = HD
        CandidateSet.clean()
        CandidatesSet.add(b)
      else if HD $= Min_{HD}$ **then**
        CandidatesSet.add(b)
      end if
    end if
  end if
end for
S = VoteOnclasses(CandidatesSet)
Comp(E)= $E + S$
return (Comp(E))

---

5   Experimentations and Discussion

In this section, we aim to evaluate the efficiency of the proposed algorithms for predicting new examples. For this purpose, we first run different standard ML classifiers on the original dataset, then we apply each AI-Predictor to generate a new set of predicted examples that is used to enlarge the original data set. This leads us to four different enlarged datasets, one for each proposed algorithm. Finally, we re-evaluate again ML classifiers on each of these completed datasets. For both original and enlarged datasets, we apply the testing protocol presented in the next sub-section.

In this experimentation, we tested with the following standard ML classifiers:

- **IBk:** a k-NN classifier, we use the Manhattan distance and we tune the classifier on different values of the parameter $k = 1, 2, ..., 11$.
- **C4.5:** generating a pruned or unpruned C4.5 decision tree. We tune the classifier with different confidence factors used for pruning $C = 0.1, 0.2, ..., 0.5$.
- **JRip:** propositional rule learner, Repeated Incremental Pruning to Produce Error Reduction (RIPPER). We tune the classifier for different values of optimization runs $O = 2, 4, ...10$ and we apply pruning.

Algorithm 3. Analogical Interpolation-based Predictor using Nearest Neighbors pairs in the same class for prediction ($AIP_{NN,SC}$)

Input: A set E of classified instances
CandidatesSet = $\emptyset$
PredictedSet = $\emptyset$
Min_{HD} = NbrAttribute
for each pair (a, c) in E^2 **do**
  if $cl(a) = cl(c)$ **then**
    if $a : x :: x : c = 1$ has solution b **then**
      $cl(b) = cl(a)$ //or $cl(c)$
      HD = Max(HammingDistance(b,a), HammingDistance(b,c))
      if $HD < Min_{HD}$ **then**
        Min_{HD} = HD
        CandidateSet.clean()
        CandidatesSet.add(b)
      else if HD $= Min_{HD}$ **then**
        CandidatesSet.add(b)
      end if
    end if
  end if
end for
S = VoteOnclasses(CandidatesSet)
Comp(E)= $E + S$
return (Comp(E))

5.1 Datasets for Experiments

The experimental study is based on several datasets taken from the U.C.I. machine learning repository [11]. A brief description of these data sets is given in Table 1.

To apply the analogical interpolation, we have chosen to deal only with ordered nominal datasets in this study (the extension to the numerical case is the topic of a future work). Table 1 includes 10 datasets with ordered nominal or Boolean attribute values. In terms of classes, we deal with a maximum number of 5 classes.

- Balance, Car, Hayes-Roth and Nursery are multiple classes datasets.
- Monk1, Monk2, Monk3, Breast Cancer, Voting and W. B. Cancer datasets are binary class problems. Monk3 has noise added (in the sample set only). Voting data set contains only binary attributes and has missing attribute values. As a missing value, in this dataset, simply means that this value is not "yes" nor "no", we replace each missing value by a third value other than 0 and 1. These data sets are described in Table 1.

5.2 Testing Protocol

To test ML classifiers, we apply a standard 10 fold cross-validation technique. As usual, the final accuracy is obtained by averaging the 10 different accuracies (computed as the ratio of the number of correct predictions to the total number of test examples) for each fold. However, each ML classifier requires a parameter to be tuned before performing this cross-validation.

Table 1. Description of datasets

Datasets	Instances	Nominal Att.	Binary Att.	Classes
Balance	625	4	0	3
Car	743	6	0	4
Monk1	432	4	2	2
Monk2	432	4	2	2
Monk3	432	4	2	2
Breast Cancer	286	6	3	2
Voting	435	0	16	2
Hayes-Roth	132	5	0	3
W. B. Cancer	699	9	0	2
Nursery	1102	8	0	5

In order to do that, we *randomly* choose a fold (as recommended by [8]), we keep only the corresponding training set (i.e. which represents 90% of the full dataset). On this training set, we again perform a 10-fold cross-validation with diverse values of the parameters. We then select the parameter values providing the best accuracy. These tuned parameters are then used to perform the initial cross-validation. As expected, these tuned parameters change with the target dataset. To be sure that our results are stable enough, we run each algorithm (with the previous procedure) 10 times so we have 10 different parameter optimizations. The displayed parameter p is the average value over the 10 different values (one for each run). The results shown in Table 2 are the average values obtained from 10 rounds of this complete process.

5.3   Experimental Results

In the following, we first provide a comparative study of the overall accuracies for ML classifiers obtained with original and enlarged datasets. This study aims to check if examples predicted by the *AIP* are of *good quality* (namely labeled with the suitable class). In such case, the efficiency of ML classifiers should be improved when applied to enlarged datasets. Then we also report the main characteristics of these predicted datasets. Finally, we compare ML classification results with enlarged datasets to the ones obtained by directly applying Analogy-based Classification [2] to the original datasets. In this last study, we wonder if using ML classifiers with enlarged datasets may perform similarly as Analogy-based Classification [2] to the original datasets while maintaining a reduced complexity.

Results of ML-Classifiers. Accuracy results for IBk, C4.5 and JRIP are obtained by using the free implementation of Weka software to the enlarged datasets obtained from *AI*-Predictors. To run IBk, C4.5 and JRIP, we first optimize the corresponding parameter for each classifier, using the meta CVParameterSelection class provided by Weka using a cross-validation applied to the training set only. This enables us to select the best value of the parameter for each dataset, then we train and test the classifier using this selected value of this parameter.

Table 2 provides classification results of ML classifiers obtained with a 10-fold cross validation and for the best/optimized value of the tuned parameter (denoted p in this table).

Results in the previous table show that:

Table 2. Results for ML classifiers obtained with the enlarged datasets

Datasets		KNN		C4.5		JRIP	
		Accuracy	p	*Accuracy*	p	*Accuracy*	p
Balance	$AIP_{NN,SC}$	**85.7 ± 2.13**	1	**74.15 ± 2.42**	0.5	**76.05 ± 2.85**	9
	AIP_{NN}	85.31 ± 3.24	1	73.73 ± 4.12	0.5	75.09 ± 3.23	6
	AIP_{Std}	78.16 ± 1.15	3	65.92 ± 2.73	0.5	68.45 ± 3.73	6
	AIP_{Btw}	83.04 ± 3.42	3	75.44 ± 3.89	0.5	75.21 ± 4.64	7
	Orig.	84.05 ± 2.6	11	63.79 ± 4.33	0.3	72.74 ± 3.48	6
Car	$AIP_{NN,SC}$	91.4 ± 1.84	1	92.78 ± 1.28	0.4	88.6 ± 2.82	8
	AIP_{NN}	91.5 ± 1.95	1	**93.14 ± 1.95**	0.5	**89.13 ± 2.55**	8
	AIP_{Std}	91.51 ± 1.91	3	92.26 ± 1.85	0.3	89.09 ± 1.93	6
	AIP_{Btw}	86.74 ± 2.71	4	88.74 ± 1.99	0.4	85.61 ± 2.38	8
	Orig.	**92.38 ± 2.51**	1	90.84 ± 3.61	0.5	86.58 ± 3.67	8
Monk1	$AIP_{NN,SC}$	94.58 ± 2.7	5	94.11 ± 2.88	0.2	**93.75 ± 2.48**	2
	AIP_{NN}	94.82 ± 2.37	3	94.53 ± 2.35	0.1	93.62 ± 1.9	2
	AIP_{Std}	87.07 ± 4.48	3	87.35 ± 2.49	0.1	83.21 ± 4.34	6
	AIP_{Btw}	85.34 ± 3.91	3	88.15 ± 4.78	0.3	89.46 ± 3.66	4
	Orig.	**98.37 ± 2.78**	2	**99.36 ± 0.64**	0.4	90.99 ± 13.15	2
Monk2	$AIP_{NN,SC}$	82.41 ± 4.77	1	72.44 ± 0.19	0.1	71.91 ± 3.32	5
	AIP_{NN}	**82.49 ± 7.56**	1	72.44 ± 0.19	0.1	71.87 ± 3.8	3
	AIP_{Std}	76.12 ± 4.28	3	77.03 ± 0.0	0.1	76.6 ± 0.43	4
	AIP_{Btw}	80.86 ± 0.79	3	**80.79 ± 0.78**	0.1	**80.56 ± 0.82**	3
	Orig.	65.29 ± 1.74	11	67.13 ± 0.61	0.1	64.64 ± 3.69	4
Monk3	$AIP_{NN,SC}$	98.38 ± 1.31	3	98.41 ± 1.31	0.1	98.24 ± 1.49	2
	AIP_{NN}	98.38 ± 1.41	3	98.41 ± 1.41	0.1	98.27 ± 1.42	2
	AIP_{Std}	92.91 ± 2.47	3	93.58 ± 3.09	0.1	92.09 ± 2.63	4
	AIP_{Btw}	97.75 ± 1.76	3	97.71 ± 1.76	0.1	97.87 ± 1.79	2
	Orig.	**99.14 ± 1.49**	1	**99.82 ± 0.18**	0.2	**98.95 ± 1.48**	2

(continued)

Table 2. (*continued*)

Datasets		KNN		C4.5		JRIP	
		Accuracy	*p*	*Accuracy*	*p*	*Accuracy*	*p*
Breast Cancer	$AIP_{NN,SC}$	75.57 ± 8.31	4	74.01 ± 7.29	0.2	71.9 ± 8.6	4
	AIP_{NN}	75.59 ± 4.95	5	73.68 ± 6.85	0.2	71.0 ± 7.49	5
	AIP_{Std}	$\mathbf{83.0 \pm 3.19}$	6	$\mathbf{82.47 \pm 3.93}$	0.1	$\mathbf{80.3 \pm 7.01}$	3
	AIP_{Btw}	75.86 ± 5.27	4	75.94 ± 5.99	0.2	72.61 ± 5.84	4
	Orig.	72.81 ± 7.65	9	71.58 ± 6.55	0.2	70.11 ± 8.59	2
Voting	$AIP_{NN,SC}$	93.32 ± 3.58	4	95.65 ± 2.67	0.2	95.62 ± 2.85	3
	AIP_{NN}	93.05 ± 3.17	3	95.79 ± 3.59	0.3	95.67 ± 3.07	3
	AIP_{Std}	$\mathbf{93.89 \pm 2.31}$	2	96.12 ± 2.02	0.3	$\mathbf{96.1 \pm 2.04}$	3
	AIP_{Btw}	93.22 ± 3.84	2	95.45 ± 2.37	0.2	95.73 ± 2.13	2
	Orig.	92.5 ± 3.59	4	$\mathbf{96.38 \pm 2.63}$	0.2	95.84 ± 2.39	4
Hayes-Roth	$AIP_{NN,SC}$	$\mathbf{74.62 \pm 8.84}$	1	74.4 ± 9.63	0.2	84.79 ± 7.65	4
	AIP_{NN}	73.91 ± 8.0	1	74.13 ± 7.65	0.2	85.12 ± 6.58	5
	AIP_{Std}	60.45 ± 11.59	3	70.62 ± 9.3	0.4	78.78 ± 9.67	4
	AIP_{Btw}	69.87 ± 7.77	1	$\mathbf{80.43 \pm 12.53}$	0.1	$\mathbf{88.52 \pm 8.8}$	2
	Orig.	61.41 ± 10.31	3	68.2 ± 6.66	0.2	83.26 ± 9.04	4
W. B. Cancer	$AIP_{NN,SC}$	95.92 ± 1.69	1	94.38 ± 3.38	0.4	94.57 ± 2.15	5
	AIP_{NN}	96.12 ± 2.47	1	94.05 ± 2.82	0.3	94.5 ± 2.31	4
	AIP_{Std}	$\mathbf{96.82 \pm 1.22}$	3	$\mathbf{97.37 \pm 1.23}$	0.5	$\mathbf{96.56 \pm 2.19}$	5
	AIP_{Btw}	95.99 ± 1.17	2	94.43 ± 1.49	0.4	94.44 ± 2.16	5
	Orig.	96.7 ± 1.73	3	94.79 ± 3.19	0.2	95.87 ± 2.9	4
Nursery	$AIP_{NN,SC}$	98.23 ± 0.96	1	98.69 ± 0.56	0.4	97.78 ± 1.12	6
	AIP_{NN}	$\mathbf{98.25 \pm 0.78}$	1	$\mathbf{98.74 \pm 0.64}$	0.5	$\mathbf{97.83 \pm 1.25}$	5
	AIP_{Std}	97.73 ± 0.88	1	98.0 ± 0.96	0.5	97.74 ± 0.99	5
	AIP_{Btw}	95.9 ± 0.97	3	96.51 ± 1.34	0.4	95.78 ± 1.5	6
	Orig.	97.45 ± 1.34	3	97.7 ± 1.36	0.5	95.58 ± 2.04	4
Average	$AIP_{NN,SC}$	**89,01**	–	86,90	–	87,32	–
	AIP_{NN}	88,94	–	86,86	–	87,21	–
	AIP_{Std}	85,76	–	86,07	–	85,89	–
	AIP_{Btw}	86,45	–	**87,35**	-	**87,57**	–
	Orig.	86,01	–	84,96	–	85,46	–

- The accuracy results have been improved when applying ML classifiers on the new predicted data instead of the original data. This is noticed for all datasets except for Monk1 and Monk3 datasets. The highest improvement percentage was noticed with the IBk classifier for the dataset Monk2 (17%), Hayes-Roth (13%) and Breast Cancer (11%).
- Regarding the two artificial datasets Monk1 and Monk3, it is known in the original dataset, that only two attributes among 6 are involved to define the class label for each example. We may think that using the *midpoint* value for each attribute as well as the class label, applied in the proposed analogical

interpolation which treat equally *all* attributes, is not compatible with this kind of classification.

- The good improvement observed for Monk2 dataset confirms our previous intuition since, contrary to Monk1 and Mon3, in Monk2 all attributes are involved in defining the class label in this dataset.
- The standard Algorithm 1 outperforms other algorithms in case of Cancer and Breast Cancer datasets. It is important to note that only these two datasets include attributes with large range of values (with maximum of 10 different values for Cancer and 13 different values for Breast Cancer). Moreover the number of attributes is also high if compared to other datasets. We expect that, in case ordered nominal data is represented by a large scale, using only nearest neighbor pairs for prediction seems too restrictive and leads to a local search for new examples.
- There is no particular algorithm that provides the best results for all datasets.
- We computed the average accuracy for each proposed algorithm and for each ML classifier over all datasets. Results are given at the end of Table 2. We can note that IBk classifier performs the best accuracy when using the enlarged data built from the $AIP_{NN,SC}$ Algorithm. While C4.5 and JRIP perform better when applied to the dataset built from AIP_{Btw} Algorithm.
- Overall, the IBK classifier shows the highest classification accuracy over all datasets.

In this first study, the improved results of ML classifiers when applied to enlarged datasets show the ability of the proposed algorithms (especially, $AIP_{NN,SC}$ and AIP_{Btw}) to predict examples that are labeled with the suitable class.

Characteristics of the Predicted Datasets. To have a better understanding of the previous shown results, in this subsection we aim to investigate more the new predicted datasets. For this end, we compute the number of predicted examples for each dataset and the proportion of these examples that are assigned to the correct/suitable class label. This proportion is computed on the basis of the predicted examples that are compatible with the original set. For this new experimentation, we only consider examples predicted by Algorithm $AIP_{NN,SC}$ (and AIP_{Std} for some datasets). We save these additional results in Table 3. From these results, we can see that:

- In seven among ten datasets, the proportion of predicted examples that are successfully classified is 100%. This means that *all* predicted examples that match the original set are assigned to the correct class label and thus are *fully* compatible with the original set (see for example Monk2, Breast Cancer, Hayes Roth and Nursery).
- Predicting accurate examples in these datasets may explain why ML classifiers show high classification improvement when applied to the new enlarged dataset.
- Although $AIP_{NN,SC}$ Algorithm succeeds to predict accurate examples, the number of predicted examples is very reduced for some datasets such as for

Breast Cancer, Voting and Cancer. This due to the fact that we restrict the search for only nearest neighbors pairs belonging to the same class in this Algorithm. It is important to note that these datasets contains large number of attributes which make the process of pairs filter more constraining.

- As can be seen in Table 3, the size of the predicted sets is considerably increased, for these three datasets, when applying AIP_{Std} Algorithm which is less constraining than $AIP_{NN,SC}$ (520 examples instead of 46 are predicted for Cancer dataset). In Table 2, we also noticed that, only for these three cited datasets, IBK performs considerably better when applied to the datasets built from the standard algorithm AIP_{Std} (producing larger sets). It is clear that in case the predicted set is very reduced, the enlarged dataset remains similar to the original set that's why the improvement percentage of ML classifiers cannot be clearly noticed in the case of datasets predicted from $AIP_{NN,SC}$ Algorithm.

- Lastly for some datasets such as Monk1 and Monk3, the proportion of predicted examples that are compatible with the original set is low if compared to other datasets. As explained before, in the original sets, the classification function involves only 2 among 6 attributes which seems incompatible with continuous analogical interpolation assuming that all attributes as well as class label are the midpoint of the attributes and the class label of the pair used for prediction.

Table 3. Nbr. of predicted examples, proportion of predicted examples that are compatible with the original set

Datasets	Nbr. predicted	Prop. of success
Balance	529	85.82
Car	630	93.44
Monk1	288	87.5
Monk2	221	100
Monk3	320	96.25
Breast Cancer-$AIP_{NN,SC}$	14	100
Breast Cancer-AIP_{Std}	152	83.78
Voting-$AIP_{NN,SC}$	38	100
Voting-AIP_{Std}	95	100
Hayes-Roth	27	100
Cancer-$AIP_{NN,SC}$	46	100
Cancer-AIP_{Std}	520	100
Nursery	883	99.89

Comparison with AP-Classifier [2]. Finally, we provide a comparative study of ML classifiers results, reported in Sect. 5.3, to the results obtained with a direct application of analogical proportions for a classification purpose [2]. Note that in [2], analogical proportions-based extrapolation has been directly applied to define a new classification paradigm while in this paper we exploit analogical proportions-based interpolation to enlarge datasets on which classical ML classifiers are applied. Classification accuracies of analogical proportions-based classifiers [2] are given in Table 4 and compared to the *best* result of each ML classifier applied to the enlarged datasets. Results in Table 4 shows that AP-Classifier outperforms classic ML classifiers on five datasets especially on the three Monks datasets. However enlarged datasets, using analogical interpolation, helped to reduce the gap between AP-Classifier and other ML classifiers once they were applied to these enlarged data. On the other side, ML classifiers provides better accuracies on four other datasets (see for example the Breast cancer (resp. Hayes-Roth) dataset for which the IBK (resp. JRIP) is largely better than AP-Classifier).

Table 4. Results for ML classifiers obtained with the enlarged datasets and comparison with AP-Classifier [2]

Datasets	AP-Classifier [2]		KNN		C4.5		JRIP	
	Accuracy	*p*	*Accuracy*	*p*	*Accuracy*	*p*	*Accuracy*	*p*
Balance	**86.35 ± 2.27**	11	85.7 ± 2.13	1	74.15 ± 2.42	0.5	76.05 ± 2.85	9
Car	**94.16 ± 4.11**	11	91.5 ± 1.95	1	93.14 ± 1.95	0.5	89.13 ± 2.55	8
Monk1	**99.77 ± 0.71**	7	94.82 ± 2.37	3	94.53 ± 2.35	0.1	93.75 ± 2.48	2
Monk2	**99.77 ± 0.7**	11	82.49 ± 7.56	1	80.79 ± 0.78	0.1	80.56 ± 0.82	3
Monk3	**99.63 ± 0.7**	9	98.38 ± 1.41	3	98.41 ± 1.41	0.1	98.27 ± 1.42	2
Breast Cancer	73.68 ± 6.36	10	**83.0 ± 3.19**	6	82.47 ± 3.93	0.1	80.3 ± 7.01	3
Voting	94.73 ± 3.72	7	93.89 ± 2.31	2	**96.12 ± 2.02**	0.3	96.1 ± 2.04	3
Hayes-Roth	79.29 ± 9.3	7	74.62 ± 8.84	1	80.43 ± 12.53	0.1	**88.52 ± 8.8**	2
W. B. Cancer	97.01 ± 3.35	4	96.82 ± 1.22	3	**97.37 ± 1.23**	0.5	96.56 ± 2.19	5

This comparison firstly shows the interest of analogical proportions as a classification tool for some datasets and secondly as way for enlarging datasets for other cases. Identifying on which dataset each of these methods may be better applied should be deeply investigated in future.

In terms of complexity, the proposed Analogical Interpolation approaches (which are quadratic due to the use of pairs of examples) if combined with the IBK classifier for example (which is linear), leads to a improved classifier. This latter shows better classification accuracy and enjoining reduced complexity if compared to the AP-classifier having cubic complexity (that may be computationally costly for large datasets [2]).

6   Conclusion

This paper has studied the idea of enlarging a training set using analogical proportions as in [4], with two main differences: we only consider pairs of examples by using continuous analogical proportions which contribute to reduce the complexity to be quadratic instead of cubic, and we test with ordered nominal datasets instead of Boolean one.

On the one hand the results obtained by classical machine learning methods on the enlarged training set generally improve those obtained by applying these methods to the original training sets. On the other hand, these results, obtained with a smaller level of complexity, are often not so far from those obtained by directly applying the analogical proportion-based classification method on the original training set [2].

References

1. Bayoudh, S., Mouchère, H., Miclet, L., Anquetil, E.: Learning a classifier with very few examples: analogy based and knowledge based generation of new examples for character recognition. In: Kok, J.N., Koronacki, J., Mantaras, R.L., Matwin, S., Mladenič, D., Skowron, A. (eds.) ECML 2007. LNCS (LNAI), vol. 4701, pp. 527–534. Springer, Heidelberg (2007). https://doi.org/10.1007/978-3-540-74958-5_49
2. Bounhas, M., Prade, H., Richard, G.: Analogy-based classifiers for nominal or numerical data. Int. J. Approximate Reasoning **91**, 36–55 (2017)
3. Bounhas, M., Prade, H., Richard, G.: Oddness-based classification: a new way of exploiting neighbors. Int. J. Intell. Syst. **33**(12), 2379–2401 (2018)
4. Couceiro, M., Hug, N., Prade, H., Richard, G.: Analogy-preserving functions: a way to extend Boolean samples. In: Proceedings 26th International Joint Conference on Artificial Intelligence, IJCAI 2017, Melbourne, 19–25 August, pp. 1575–1581 (2017)
5. Derrac, J., Schockaert, S.: Inducing semantic relations from conceptual spaces: a data-driven approach to plausible reasoning. Artif. Intell. **228**, 66–94 (2015)
6. Dubois, D., Prade, H., Richard, G.: Multiple-valued extensions of analogical proportions. Fuzzy Sets Syst. **292**, 193–202 (2016)
7. Goodfellow, I., et al.: Generative adversarial nets. In: Ghahramani, Z., Welling, M., Cortes, C., Lawrence, N.D., Weinberger, K.Q. (eds.) Advances in Neural Information Processing Systems 27, pp. 2672–2680. Curran Associates, Inc. (2014)
8. Hsu, C., Chang, C., Lin, C.: A practical guide to support vector classification. Technical report, Department of Computer Science, National Taiwan University (2010)
9. Inoue, H.: Data augmentation by pairing samples for images classification. CoRR abs/1801.02929 (2018). http://arxiv.org/abs/1801.02929
10. Lieber, J., Nauer, E., Prade, H., Richard, G.: Making the best of cases by approximation, interpolation and extrapolation. In: Cox, M.T., Funk, P., Begum, S. (eds.) ICCBR 2018. LNCS (LNAI), vol. 11156, pp. 580–596. Springer, Cham (2018). https://doi.org/10.1007/978-3-030-01081-2_38
11. Mertz, J., Murphy, P.: UCI repository of machine learning databases (2000). ftp://ftp.ics.uci.edu/pub/machine-learning-databases

12. Miclet, L., Bayoudh, S., Delhay, A.: Analogical dissimilarity: definition, algorithms and two experiments in machine learning. JAIR **32**, 793–824 (2008)
13. Miclet, L., Prade, H.: Handling analogical proportions in classical logic and fuzzy logics settings. In: Sossai, C., Chemello, G. (eds.) ECSQARU 2009. LNCS (LNAI), vol. 5590, pp. 638–650. Springer, Heidelberg (2009). https://doi.org/10.1007/978-3-642-02906-6_55
14. Perfilieva, I., Dubois, D., Prade, H., Esteva, F., Godo, L., Hodáková, P.: Interpolation of fuzzy data: analytical approach and overview. Fuzzy Sets Syst. **192**, 134–158 (2012)
15. Prade, H., Richard, G.: From analogical proportion to logical proportions. Logica Universalis **7**(4), 441–505 (2013)
16. Prade, H., Richard, G.: Analogical proportions: from equality to inequality. Int. J. Approximate Reasoning **101**, 234–254 (2018)
17. Prade, H., Schockaert, S.: Completing rule bases in symbolic domains by analogy making. In: Galichet, S., Montero, J., Mauris, G. (eds.) Proceedings 7th Conference European Society for Fuzzy Logic and Technology (EUSFLAT), Aix-les-Bains, 18–22 July, pp. 928–934. Atlantis Press (2011)
18. Schockaert, S., Prade, H.: Interpolation and extrapolation in conceptual spaces: a case study in the music domain. In: Rudolph, S., Gutierrez, C. (eds.) RR 2011. LNCS, vol. 6902, pp. 217–231. Springer, Heidelberg (2011). https://doi.org/10.1007/978-3-642-23580-1_16
19. Schockaert, S., Prade, H.: Qualitative reasoning about incomplete categorization rules based on interpolation and extrapolation in conceptual spaces. In: Benferhat, S., Grant, J. (eds.) SUM 2011. LNCS (LNAI), vol. 6929, pp. 303–316. Springer, Heidelberg (2011). https://doi.org/10.1007/978-3-642-23963-2_24
20. Schockaert, S., Prade, H.: Interpolative and extrapolative reasoning in propositional theories using qualitative knowledge about conceptual spaces. Artif. Intell. **202**, 86–131 (2013)
21. Schockaert, S., Prade, H.: Interpolative reasoning with default rules. In: Rossi, F. (ed.) IJCAI 2013, Proceedings 23rd International Joint Conference on Artificial Intelligence, Beijing, 3–9 August, pp. 1090–1096 (2013)
22. Schockaert, S., Prade, H.: Completing symbolic rule bases using betweenness and analogical proportion. In: Prade, H., Richard, G. (eds.) Computational Approaches to Analogical Reasoning: Current Trends. SCI, vol. 548, pp. 195–215. Springer, Heidelberg (2014). https://doi.org/10.1007/978-3-642-54516-0_8
23. Wolf, L., Martin, I.: Regularization through feature knock out. MIT Computer Science and Artificial Intelligence Laboratory (CBCL Memo 242) (2004)

Towards a Reconciliation Between Reasoning and Learning - A Position Paper

Didier Dubois and Henri Prade[(⊠)]

IRIT - CNRS, 118 route de Narbonne, 31062 Toulouse Cedex 09, France
{dubois,prade}@irit.fr

Abstract. The paper first examines the contours of artificial intelligence (AI) at its beginnings, more than sixty years ago, and points out the important place that machine learning already had at that time. The ambition of AI of making machines capable of performing any information processing task that the human mind can do, means that AI should cover the two modes of human thinking: the instinctive (reactive) one and the deliberative one. This also corresponds to the difference between mastering a skill without being able to articulate it and holding some pieces of knowledge that one can use to explain and teach. In case a function-based representation applies to a considered AI problem, the respective merits of learning a universal approximation of the function vs. a rule-based representation are discussed, with a view to better draw the contours of AI. Moreover, the paper reviews the relative positions of knowledge and data in reasoning and learning, and advocates the need for bridging the two tasks. The paper is also a plea for a unified view of the various facets of AI as a science.

1   Introduction

What is artificial intelligence (AI) about? What are the research topics that belong to AI? What are the topics that stand outside? In other words, what are the contours of AI? Answers to these questions may have evolved with time, as did the issue of the proper way (if any) of doing AI. Indeed over time, AI has been successively dominated by logical approaches (until the mid 1990's) giving birth to the so-called "symbolic AI", then by (Bayesian) probabilistic approaches, and since recently by another type of numerical approach, artificial neural networks. This state of facts has contributed to developing antagonistic feelings between different schools of thought, including claims of supremacy of some methods over others, rather than fostering attempts to understand the

A preliminary version of this paper was presented at the 2018 IJCAI-ECAI workshop "Learning and Reasoning: Principles & Applications to Everyday Spatial and Temporal Knowledge", Stockholm, July 13–14.

potential complementarity of approaches. Moreover, when some breakthrough takes place in some sector of AI such as expert systems in the 1980's, or fuzzy logic in the 1990's (outside mainstream AI), or yet deep learning [51] nowadays, it is presented through its technological achievements rather than its actual scientific results. So we may even - provocatively - wonder: Is AI a science, or just a bunch of engineering tools? In fact, AI has developed over more than sixty years in several directions, and many different tools have been proposed for a variety of purposes. This increasing diversity, rather than being a valuable asset, may be harmful for an understanding of AI as a whole, all the more so as most AI researchers are highly specialized in some area and are largely ignoring the rest of the field.

Besides, beyond the phantasms and fears teased by the phrase 'artificial intelligence', the meaning of words such as 'intelligence', 'learning', or 'reasoning' has a large spectrum and may refer to quite different facets of human mind activities, which contributes to blur the meaning of what we claim when we are using the acronym AI. Starting with 'intelligence', it is useful to remember the dichotomy popularized in [44] between two modes of thinking: "System 1" which is fast, instinctive and emotional, while "System 2" is slower, more deliberative, and more logical. See [76] for an illustration of similar ideas in the area of radiological diagnosis, where "super-experts" provide correct diagnosis, even on difficult cases, without any deliberation, while "ordinary experts" may hesitate, deliberate on the difficult cases and finally make a wrong diagnosis. Yet, a "super-expert" is able to explain what went wrong to an "ordinary expert" and what important features should have been noticed in the difficult cases.

Darwiche [21] has recently pointed out that what is achieved by deep leaning corresponds to tasks that do not require much deliberation, at least for a top expert, and is far from covering all that may be expected from AI. In other words, the system is mastering skills rather than being also able to elaborate knowledge for thinking and communicating about its skills. This is the difference between an excellent driver (without teaching capability) and a driving instructor.

The intended purpose of this paper is to advocate in favor of a unified view of AI both in terms of problems and in terms of methods. The paper is organized as follows. First, in Sect. 2 a reminder on the history of the early years of AI emphasizes the idea that the diversity of AI has been there from its inception. Then Sect. 3 first discusses relations between a function-based view and a rule-based view of problems, in relation with "modeling versus explaining" concerns. The main paradigms of AI are then restated and the need for a variety of approaches ranging from logic to probability and beyond is highlighted. Section 4 reviews the roles of knowledge and data both in reasoning and in machine learning. Then, Sect. 5 points out problems where bridging reasoning and learning might be fruitful. Section 6 calls for a unified view of AI, a necessary condition for letting it become a mature science.

2   A Short Reminder of the Beginnings of AI

To have a better understanding of AI, it may be useful to have a historical view of the emergence of the main ideas underling it [53,54,64]. We only focus here on its beginnings. Still it is worth mentioning that exactly three hundreds years before the expression 'artificial intelligence' was coined, the English philosopher Thomas Hobbes of Malmesbury (1588–1679) described human thinking as a symbolic manipulation of terms similar to mathematical calculation [39]. Indeed, he wrote *"Per Ratiocinationem autem intelligo computationem."* (or in English one year later *"By ratiocination I mean computation."*) The text continues with *"Now to compute, is either to collect the sum of many things that are added together, or to know what remains when one thing is taken out of another. Ratiocination, therefore, is the same with addition and subtraction;"* One page after one reads: *"We must not therefore think that computation, that is, ratiocination, has place only in numbers, as if man were distinguished from other living creatures (which is said to have been the opinion of Pythagoras) by nothing but the faculty of numbering; for magnitude, body, motion, time, degrees of quality, action, conception, proportion, speech and names (in which all the kinds of philosophy consist) are capable of addition and subtraction."* Such a description appears retrospectively quite consonant with what AI programs are trying to do!

In the late 1940's with the advent of cybernetics [96], the introduction of artificial neural networks [56][1], the principle of synaptic plasticity [37] and the concept of computing machines [91] lead to the idea of thinking machines with learning capabilities. In 1950, the idea of machine intelligence appeared in a famous paper by Turing [92], while Shannon [89] was investigating the possibility of a program playing chess, and the young Zadeh [97] was already suggesting multiple-valued logic as a tool for the conception of thinking machines.

As it is well-known, the official birthday act of AI corresponds to a research program whose application for getting a financial support, was written in the summer of 1955, and entitled "A proposal for the Dartmouth summer research project on artificial intelligence" (thus putting the name of the new field in the title!); it was signed by the two fathers of AI, John McCarthy (1927–2011), and Marvin Minsky (1927–2016), and their two mentors Nathaniel Rochester (1919–2001) (who designed the IBM 701 computer and was also interested in neural network computational machines), and Claude Shannon (1916–2001) [55] (in 1950 he was already the founder of digital circuit design theory based on Boolean logic, the founder of information theory, but also the designer of an electromechanical mouse (Theseus) capable of searching through the corridors of a maze until reaching a target and of acquiring and using knowledge from past experience). Then a series of meetings was organized at Dartmouth College (Hanover, New Hampshire, USA) during the summer of 1956. At that time, McCarthy was already interested in symbolic logic representations, while Minsky

---

[1] One would notice the word 'logical' in the title of this pioneering paper.

had already built a neural network learning machine (he was also a friend of Rosenblatt [79] the inventor of perceptrons).

The interests of the six other participants can be roughly divided into reasoning and learning concerns, they were on the one hand Simon (1916–2001), Newell (1927–1992) [63] (together authors with John Clifford Shaw (1922–1991) of a program *The Logic Theorist* able to prove theorems in mathematical logic), and More [60] (a logician interested in natural deduction at that time), and on the other hand Samuel (1901–1990) [81] (author of programs for checkers, and later chess games), Selfridge (1926–2008) [84] (one of the fathers of pattern recognition), and Solomonoff (1926–2009) [90] (already author of a theory of probabilistic induction).

Interestingly enough, as it can be seen, these ten participants, with different backgrounds ranging from psychology to electrical engineering, physics and mathematics, were already the carriers of a large variety of research directions that are still present in modern AI, from machine learning to knowledge representation and reasoning.

3   Representing Functions and Beyond

There are two modes of representation of knowledge, that can be called respectively functional and logical. The first mode consists in building a large, often numerical, function that produces a result when triggered by some input. The second mode consists of separate, possibly related, chunks of explicit knowledge, expressed in some language. The current dominant machine learning paradigm (up to noticeable exceptions) has adopted the functional approach[2], which ensures impressive successes in tasks requiring reactiveness, at the cost of losing explanatory power. Indeed, we can argue that what is learnt is know-how or skills, rather than knowledge. The other, logical, mode of representation, is much more adapted to the encoding of articulated knowledge, reasoning from it, and to the production of explanations via deliberation, but its connection to learning from data is for the most part still in infancy.

A simple starting point for discussing relationships between learning and reasoning is to compare the machineries of a classifier and a rule-based expert system, for diagnosis for instance. In both cases, a function-based view may apply. On the one hand, from a set of examples (of inputs and outputs of the function, such as pairs (symptoms, disease)) one can easily predict the disease corresponding to a new case via its input symptoms, after learning some function (e.g., using neural nets). On the other hand, one may have a set of expert rules stating that if the values of the inputs are such and such, the global evaluation should be in some subset. Such rules are mimicking the function. If collected from an expert, rules may turn out to be much less successful than the function learned from data. Clearly, the first view may provide better approximations and does not require the elicitation of expert rules, which is costly. However, the explanatory power will be poor in any case, because it will not be possible

---

[2] Still this function-based approach is often cast in a probabilistic modeling paradigm.

to answer "why not" questions and to articulate explanations based on causal relations. On the contrary, if causal knowledge is explicitly represented in the knowledge base, it has at least the merit of offering a basis for explanations (in a way that should be cognitively more appropriate for the end-user). It is moreover well-known that causal information cannot easily be extracted from data: only correlations can be laid bare if no extra information is added [66].

The fuzzy set literature offers early examples of the replacement of an automatic control law by a set of rules. Indeed Zadeh [98] proposed to use fuzzy expert rules for controlling complex non linear dynamic systems that might be difficult to model using a classical automatic control approach, while skilled humans can do the job. This was rapidly shown to be successful [52]. The fact of using fuzzy rules, rather than standard Boolean if-then rules, had the advantage of providing a basis for an interpolation mechanism, when an input was firing several rules to some degree. Although the approach was numerical and quite far from the symbolic logic-based AI mainstream trend in those times, it was perceived as an AI-inspired approach, since it was relying on the representation of expert know-how by chunks of knowledge, rather than on the derivation of a control law from the modeling of the physical system to be controlled (i.e., the classical control engineering paradigm). After some time, it was soon recognized that fuzzy rules could be learnt rather than obtained from experts, while keeping excellent results thanks to the property of universal approximation possessed by sets of fuzzy rules. Mathematical models of such fuzzy rules are in fact closely related to neural network radial basis functions. But, fuzzy rules thus obtained by learning may become hardly intelligible. This research trend, known under the names of 'soft computing' or 'computational intelligence', thus often drifted away from an important AI concern, the explainability power; see [27] for a discussion.

The long term ambition of AI is to make machines capable of performing any information processing task the human mind can perform. This certainly includes recognition, identification, decision and diagnosis tasks (including sophisticated ones). They are "System 1" tasks (using Kahneman terminology) as long as we do not need to explain and reason about obtained results. But there are other problems that are not fully of this kind, even if machine learning may also play a role in their solving. Consider for instance the solving of quadratic equations. Even if we could predict, in a bounded domain, by machine learning techniques, whether an equation has zero, one or two solutions and what are their values (with a good approximation) from a large amount of examples, the solving of such equations by discovering their analytical solution(s), via factorization through symbolic calculations, seems to be a more powerful way of handling of the problem (the machine could then teach students).

AI problems cannot always be viewed in terms of the function-based view mentioned above. There are cases where we do not have a function, only a one-to-many mapping, e.g., when finding all the solutions (if any) of a set of constraints. Apart from solving combinatorial problems, tasks such as reasoning about static or dynamical situations, or building action plans, or explaining results, communicating explanations pertaining to machine decisions in a meaningful way to an end-user, or analyzing arguments and determining their possible weakness, or understanding what is going on in a text, a dialog in natural langage, in an image, a video, or finding relevant information and summarizing it are examples that may require capabilities beyond pure machine learning. This is why AI, over the years, has developed general representation settings and methods capable of handling large classes of situations, while mastering computation complexity. Thus, at least five general paradigms have emerged in AI:

- **Knowledge representation** with symbolic or numerical structured settings for representing knowledge or preferences, such as logical languages, graphical representations like Bayesian networks, or domain ontologies describing taxonomy of concepts. Dedicated settings have been also developed for the representation of temporal or spatial information, of uncertain pieces of information, or of independence relations.
- **Reasoning and decision** Different types of reasoning tasks, beyond classical deduction, have been formalized such as: non monotonic reasoning for dealing with exception-tolerant rules in the presence of incomplete information, or reasoning from inconsistent information, or belief revision, belief updating, information fusion in the presence of conflicts, or formal argumentation handling pros and cons, or yet reasoning directly from data (case-based reasoning, analogical reasoning, interpolation, extrapolation). Models for qualitative (or quantitative) decision from compact representations have been proposed for decision under uncertainty, multiple criteria, or group decisions.
- **General algorithms for problem solving** This covers a panoply of generic tools ranging from heuristic ordered search methods, general problem solver techniques, methods for handling constraints satisfaction problems, to efficient algorithms for classical logic inference (e.g., SAT methods), or for deduction in modal and other non-classical logics.
- **Learning** The word 'learning' also covers different problems, from the classification of new items based on a set of examples (and counter-examples), the induction of general laws describing concepts, the synthesis of a function by regression, the clustering of similar data (separating dissimilar data into different clusters) and the labelling of clusters, to reinforcement learning and to the discovery of regularities in data bases and data mining. Moreover, each of these problems can often be solved by a variety of methods.
- **Multiple agent AI** Under this umbrella, there are quite different problems such as: the cooperation between human or artificial agents and the organization of tasks for achieving collective goals, the modeling of BDI agents (Belief, Desire, Intention), possibly in situations of dialogue (where, e.g., agents, which have different information items at their disposal, do not pursue the same goals, and try to guess the intentions of the other ones), or the study of the emergence of collective behaviors from the behaviors of elementary agents.

4   Reasoning with Knowledge or with Data

In the above research areas, knowledge and data are often handled separately. In fact, AI traditionally deals with knowledge rather than with data, with the important exception of machine learning, whose aim can sometimes be viewed as changing data into knowledge. Indeed, basic knowledge is obtained from data by induction, while prior background knowledge may help learning machineries. These remarks suggest that the joint handling of knowledge and data is a general issue, and that combining reasoning and learning methods should be seriously considered.

Rule-based systems, or ontologies expressed by means of description logics, or yet Bayesian networks, represent background knowledge that is useful to make prediction from facts and data. In these reasoning tasks, knowledge as well as data is often pervaded with uncertainty. This has been extensively investigated.

Data, provided that they are reliable, are positive in nature since their existence manifests the *actual* possibility of what is observed or reported. This contrasts with knowledge that delimit the extent of what is *potentially* possible by specifying what is impossible (which has thus a negative flavor). This is why reasoning from both knowledge and data goes much beyond the application of generic knowledge to factual data as in expert systems, and even the separate treatment of knowledge and data in description logics via 'TBox' and 'ABox' [4]. It is is a complex issue, which has received little attention until now [93].

As pointed out in [71], reasoning directly with data has been much less studied. The idea of similarity naturally applies to data and gives birth to specific forms of reasoning such as case-based reasoning [45], case-based decision [35], or even case-based argumentation. "Betweenness" and similarity are at the basis of interpolation mechanisms, while analogical reasoning, which may be both a matter of similarity and dissimilarity, provides a mechanism for extrapolation. A well-known way of handling similarity and interpolation is to use fuzzy rules (where fuzzy set membership degrees capture the idea of similarity w.r.t. the core value(s) of the fuzzy set) [67]. Besides, analogical reasoning, based on analogical proportions (i.e., statements of the form "a is to b as c is to d", where items a, b, c, d are represented in terms of Boolean, nominal or numerical variables), which can be logically represented [28,58,72], provides an extrapolation mechanism that from three items a, b, c described by complete vectors, amounts to inferring the missing value(s) in incomplete vector d, providing that a, b, c, d makes an analogical proportion component-wise on the known part of d; this was successfully applied to classification [14,18,57], and more recently to preference learning [13,32].

Lastly, the ideas of interpolation and extrapolation closely related to analogical proportion-based inference seem to be of crucial importance in many numerical domains. They can be applied to symbolic settings in the case of propositional categorization rules, using relations of betweenness and parallelism respectively, under a conceptual spaces semantics [83]; see [82] for an illustration.

5   Issues in Learning: Incomplete Data and Representation Formats

The need for reasoning from incomplete, uncertain, vague, or inconsistent information, has led to the development of new approaches beyond logic and probability. Incompleteness is a well-known phenomenon in classical logic. However, many reasoning problems exceed the capabilities of classical logic (initially developed in relation with the foundations of mathematics where statements are true or false, and there is no uncertainty in principle). As for probability theory, single probability distributions, often modeled by Bayesian networks are not fully appropriate for handling incomplete information nor epistemic uncertainty. There are different, but related, frameworks for modeling ill-known probabilities that were developed in the last 50 years by the Artificial Intelligence community at large [95]: belief functions and evidence theory (which may be viewed as a randomization of the set-based approach to incomplete information), imprecise probability theory [3,94] (which uses convex families of probability functions) and quantitative possibility theory (which is the simplest model since one of the lower and the upper probability bounds is trivial).

The traditional approach for going from data to knowledge is to resort to statistical inferential methods. However, these methods used to assume data that are precise and in sufficient quantity. The recent concern with big data seems to even strengthen the relevance of probability theory and statistics. However there are a number of circumstances where data is missing or is of poor quality, especially if one tries to collect information for building machines or algorithms supposed to face very complex or unexpected situations (e.g., autonomous vehicles in crowded areas). The concern of Artificial Intelligence for reasoning about partial knowledge has led to a questioning of traditional statistical methods when data is of poor quality [19,38,42,43].

Besides, the fact that we may have to work with incomplete relational data and that knowledge may also be uncertain has motivated the development of a new probabilistic programming language first called "Probabilistic Similarity Logic", and then "Probabilistic Soft Logic" (PSL, for short) where each ground atom in a rule has a truth value in $[0, 1]$. It uses the Łukasiewicz t-norm and co-t-norm to handle the fuzzy logical connectives [5,33,34]. We are close to representation concerns of fuzzy answer set programs [61]. Besides, there is a need for combining symbolic reasoning with the subsymbolic vector representation of neural networks in order to use gradient descent for training the neural network to infer facts from an incomplete knowledge base, using similarity between vectors [16,17,78].

Machine learning may find some advantages to use advanced representation formats as target languages, such as weighted logics [26] (Markov logic, probabilistic logic programs, multi-valued logics, possibilistic logic, etc.). For instance, qualitative possibility theory extends classical logic by attaching lower bounds of necessity degrees and captures nonmonotonic reasoning, while generalized possibilistic logic [30] is more powerful and can capture answer-set programming, or reason about the ignorance of an agent. Can such kinds of qualitative uncertainty modeling, or yet fuzzy or uncertain description logics, uncertainty representation formalisms, weighted logics, be used more extensively in machine learning? Various answers and proposals can be found in [48–50, 86, 88]. This also raises the question of extending version space learning [59] to such new representation schemes [41, 73, 75].

If-then rules, in classical logic formats, are a popular representation format in relational learning [80]. Association rules have logical and statistical bases ; they are rules with exceptions completed by confidence and support degrees [1, 36]. But, other types of rules may be of interest. Mining genuine default rules that obey Kraus, Lehmann and Magidor postulates [47] for nonmonotonic reasoning relies on the discovery of big-stepped probabilities [8] in a database [9]. Multiple threshold rules, i.e., rules describing how a global evaluation depends on multiple criteria evaluations on linearly ordered scales, such as, e.g., selection rules of the form "if $x_1 \geq a_1$ and $\cdots$ and $x_n \geq a_n$ then $y \geq b$" play a central role in ordinal classification [46] and can be represented by Sugeno integrals or their extensions [15, 74]. Gradual rules, i.e., statements of the form "the more x is A, the more y is B", where A, and B are fuzzy sets, are another representation format of interest [65, 87]. Other types of fuzzy rules may provide a rule-based interpretation [20] for neural nets, which may be also related to non-monotonic inference [7, 22]. All these examples indicates the variety of rules that makes sense and be considered both in reasoning and in learning.

Another trend of research has been also motivated by the extraction of symbolic knowledge from neural networks [22] under the form of nonmonotonic rules. The goal of a neuro-symbolic integration has been pursued with the proposal of a connectionist modal logic, where extended modal logic programs are translated into neural network ensembles, thus providing a neural net view of, e.g., the muddy children problem [24]. Following a similar line of thought, the same authors translate a logic program encoding an argumentation network, which is then 'turned into a neural network for arguments [23]. A more recent series of works [25, 85, 86] propose another form of integration between logic and neural nets using a so-called "Real Logic", implemented in deep Tensor Neural Networks, for integrating deductive reasoning and machine learning. The semantics of the logical constants is in terms of vectors of real numbers, and first order logic formulas have degrees of truth in [0, 1] handled with Łukasiewicz multiple-valued logic connectives. Somewhat related is a work on ontology reasoning [40] where the goal is to generate a neural network with binary outputs that, given a database storing tuples of the form (subject, predicate, object), is able, for any input literal, to decide the entailment problem for a logic program describing the ontology. Others look for an exact representation of a binarized neural network as a Boolean formula [62].

The use of degrees of truth multiple-valued logic raises the question of the exact meaning of these degrees. In relation with this kind of work, some have advocated a non-probabilistic view of uncertainty [11], but strangely enough without any reference to the other uncertainty representation frameworks! Maybe more promising is the line of research initiated a long time ago by

Pinkas [68,69] where the idea of penalty logic (related to belief functions [31]) has been developed in relation with neural networks, where penalty weights reflect priorities attached to logical constraints to be satisfied by a neural network [70]. Penalty logics and Markov logic [77] are also closely related to possibilistic logic [30].

Another intriguing question would be to explore possible relations between spikes neurons [12], which are physiologically more plausible than classical artificial neural networks, and fire when conjunctions of thresholds are reached, with Sugeno integrals (then viewed as a System 1-like black box) and their logical counterparts [29] (corresponding to a System 2-like representation).

6   Conclusion

Knowledge representation and reasoning on the one hand, and machine learning on the other hand, have been developed largely as independent research trends in artificial intelligence in the last three decades. Yet, reasoning and learning are two basic capabilities of the human mind that do interact. Similarly the two corresponding AI research areas may benefit from mutual exchanges. Current learning methods derive know-how from data in the form of complex functions involving many tuning parameters, but they should also aim at producing articulated knowledge, so that repositories, storing interpretable chunks of information, could be fed from data. More precisely, a number of logical-like formalisms, whose explanatory capabilities could be exploited, have been developed in the last 30 years (non-monotonic logics, modal logics, logic programming, probabilistic and possibilistic logics, many-valued logics, etc.) that could be used as target languages for learning techniques, without restricting to first-order logic, nor to Bayes nets.

Interfacing classifiers with human users may require some ability to provide high level explanations about recommendations or decisions that are understandable by an end-user. Reasoning methods should handle knowledge and information extracted from data. The joint use of (supervised or unsupervised) machine learning techniques and of inference machineries raises new issues. There is a number of other points, worth mentioning, which have not be addressed in the above discussions:

- *Teachability* A related issue is more generally how to move from machine learning models to knowledge communicated to humans, about the way the machine proceeds when solving problems.
- *Using prior knowledge* Another issue is a more systematic exploitation of symbolic background knowledge in machine learning devices. Can prior causal knowledge help exploiting data and getting rid of spurious correlations? Can an argumentation-based view of learning be developed?
- *Representation learning* Data representation impacts the performance of machine learning algorithms [10]. In that respect, what may be, for instance, the role of vector space embeddings, or conceptual spaces?

– *Unification of learning paradigms* Would it be possible to bridge learning paradigms from transduction to inductive logic programming? Even including formal concept analysis, or rough set theory?

This paper has especially advocated the interest of a cooperation between two basic areas of AI: knowledge representation and reasoning on the one hand and machine learning on the other hand, reflecting the natural cooperation between two modes, respectively reactive and deliberative, of human intelligence. It is also a plea for maintaining a unified view of AI, all facets of which have been present from the very beginning, as recalled in Sect. 2 of this paper. It is time that AI comes of age as a genuine science, which means ending unproductive rivalries between different approaches, and fostering a better shared understanding of the basics of AI through open-minded studies bridging sub-areas in a constructive way. In the same spirit, a plea for a unified view of computer science can be found in [6]. Mixing, bridging, hybridizing advanced ideas in knowledge representation, reasoning, and machine learning or data mining should renew basic research in AI and contribute in the long term to a more unified view of AI methodology. The interested reader may follow the work in progress of the group "Amel" [2] aiming at a better mutual understanding of research trends in knowledge representation, reasoning and machine learning, and how they could cooperate.

Acknowledgements. The authors thank Emiliano Lorini, Dominique Longin, Gilles Richard, Steven Schockaert, Mathieu Serrurier for useful exchanges on some of the issues surveyed in this paper. This work was partially supported by ANR-11-LABX-0040-CIMI (Centre International de Mathématiques et d'Informatique) within the program ANR-11-IDEX-0002-02, project ISIPA.

References

1. Agrawal, R., Imielinski, T., Swami, A.N.: Mining association rules between sets of items in large databases. In: Buneman, P., Jajodia, S. (eds.) Proceedings 1993 ACM SIGMOD International Conference on Management of Data, Washington, DC, 26–28 May 1993, pp. 207–216. ACM Press (1993)
2. Amel, K.R.: From shallow to deep interactions between knowledge representation, reasoning and machine learning. In: BenAmor, N., Theobald, M. (eds.) Proceedings 13th International Conference Scala Uncertainity Mgmt (SUM 2019), Compiègne, LNCS, 16–18 December 2019. Springer, Heidelberg (2019)
3. Augustin, T., Coolen, F.P.A., De Cooman, G., Troffaes, M.C.M.: Introduction to Imprecise Probabilities. Wiley, Hoboken (2014)
4. Baader, F., Horrocks, I., Lutz, C., Sattler, U.: An Introduction to Description Logic. Cambridge University Press, Cambridge (2017)
5. Bach, S.H., Broecheler, M., Huang, B., Getoor, L.: Hinge-loss Markov random fields and probabilistic soft logic. J. Mach. Learn. Res. **18**, 109:1–109:67 (2017)
6. Bajcsy, R., Reynolds, C.W.: Computer science: the science of and about information and computation. Commun. ACM **45**(3), 94–98 (2002)
7. Balkenius, C., Gärdenfors, P.: Nonmonotonic inferences in neural networks. In: Proceedings 2nd International Conference on Principle of Knowledge Representation and Reasoning (KR 1991), Cambridge, MA, pp. 32–39 (1991)

8. Benferhat, S., Dubois, D., Prade, H.: Possibilistic and standard probabilistic semantics of conditional knowledge bases. J. Log. Comput. **9**(6), 873–895 (1999)
9. Benferhat, S., Dubois, D., Lagrue, S., Prade, H.: A big-stepped probability approach for discovering default rules. Int. J. Uncert. Fuzz. Knowl.-based Syst. **11**(Suppl.–1), 1–14 (2003)
10. Bengio, Y., Courville, A., Vincent, P.: Representation learning: a review and new perspectives. IEEE Trans. Pattern Anal. Mach. Intell. **35**(8), 1798–1828 (2013)
11. Besold, T.R., Garcez, A.D.A., Stenning, K., van der Torre, L., van Lambalgen, M.: Reasoning in non-probabilistic uncertainty: logic programming and neural-symbolic computing as examples. Minds Mach. **27**(1), 37–77 (2017)
12. Bichler, O., Querlioz, D., Thorpe, S.J., Bourgoin, J.-P., Gamrat, C.: Extraction of temporally correlated features from dynamic vision sensors with spike-timing-dependent plasticity. Neural Netw. **32**, 339–348 (2012)
13. Bounhas, M., Pirlot, M., Prade, H., Sobrie, O.: Comparison of analogy-based methods for predicting preferences. In: BenAmor, N., Theobald, M. (eds.) Proceedings 13th International Conference on Scala Uncertainity Mgmt (SUM 2019), Compiègne, LNCS, 16–18 December. Springer, Heidelberg (2019)
14. Bounhas, M., Prade, H., Richard, G.: Analogy-based classifiers for nominal or numerical data. Int. J. Approx. Reasoning **91**, 36–55 (2017)
15. Brabant, Q., Couceiro, M., Dubois, D., Prade, H., Rico, A.: Extracting decision rules from qualitative data via sugeno utility functionals. In: Medina, J., Ojeda-Aciego, M., Verdegay, J.L., Pelta, D.A., Cabrera, I.P., Bouchon-Meunier, B., Yager, R.R. (eds.) IPMU 2018. CCIS, vol. 853, pp. 253–265. Springer, Cham (2018). https://doi.org/10.1007/978-3-319-91473-2_22
16. Cohen, W.W.: TensorLog: a differentiable deductive database. CoRR, abs/1605.06523 (2016)
17. Cohen, W.W., Yang, F., Mazaitis, K.: TensorLog: deep learning meets probabilistic DBs. CoRR, abs/1707.05390 (2017)
18. Couceiro, M., Hug, N., Prade, H., Richard, G.: Analogy-preserving functions: a way to extend Boolean samples. In: Proceedings 26th International Joint Conference on Artificial Intelligence, (IJCAI 2017), Melbourne, 19–25 August, pp. 1575–1581 (2017)
19. Couso, I., Dubois, D.: A general framework for maximizing likelihood under incomplete data. Int. J. Approx. Reasoning **93**, 238–260 (2018)
20. d'Alché-Buc, F., Andrés, V., Nadal, J.-P.: Rule extraction with fuzzy neural network. Int. J. Neural Syst. **5**(1), 1–11 (1994)
21. Darwiche, A.: Human-level intelligence or animal-like abilities?. CoRR, abs/1707.04327 (2017)
22. d'Avila Garcez, A.S., Broda, K., Gabbay, D.M.: Symbolic knowledge extraction from trained neural networks: a sound approach. Artif. Intell. **125**(1–2), 155–207 (2001)
23. d'Avila Garcez, A.S., Gabbay, D.M., Lamb, L.C.: Value-based argumentation frameworks as neural-symbolic learning systems. J. Logic Comput. **15**(6), 1041–1058 (2005)
24. d'Avila Garcez, A.S., Lamb, L.C., Gabbay, D.M.: Connectionist modal logic: representing modalities in neural networks. Theor. Comput. Sci. **371**(1–2), 34–53 (2007)
25. Donadello, I., Serafini, L., Garcez, A.D.A.: Logic tensor networks for semantic image interpretation. In: Sierra, C. (ed) Proceedings 26th International Joint Conference on Artificial Intelligence (IJCAI 2017), Melbourne, 19–25 August 2017, pp. 1596–1602 (2017)

26. Dubois, D., Godo, L., Prade, H.: Weighted logics for artificial intelligence - an introductory discussion. Int. J. Approx. Reasoning **55**(9), 1819–1829 (2014)
27. Dubois, D., Prade, H.: Soft computing, fuzzy logic, and artificial intelligence. Soft Comput. **2**(1), 7–11 (1998)
28. Dubois, D., Prade, H., Richard, G.: Multiple-valued extensions of analogical proportions. Fuzzy Sets Syst. **292**, 193–202 (2016)
29. Dubois, D., Prade, H., Rico, A.: The logical encoding of Sugeno integrals. Fuzzy Sets Syst. **241**, 61–75 (2014)
30. Dubois, D., Prade, H., Schockaert, S.: Generalized possibilistic logic: foundations and applications to qualitative reasoning about uncertainty. Artif. Intell. **252**, 139–174 (2017)
31. Dupin de Saint-Cyr, F., Lang, J., Schiex, T.: Penalty logic and its link with Dempster-Shafer theory. In: de Mántaras, R.L., Poole, D. (eds.) Proceedings 10th Annual Conference on Uncertainty in Artificial Intelligence (UAI 1994), Seattle, 29–31 July, pp. 204–211 (1994)
32. Fahandar, M.A., Hüllermeier, E.: Learning to rank based on analogical reasoning. In: Proceedings 32th National Conference on Artificial Intelligence (AAAI 2018), New Orleans, 2–7 February 2018 (2018)
33. Fakhraei, S., Raschid, L., Getoor, L.: Drug-target interaction prediction for drug repurposing with probabilistic similarity logic. In: SIGKDD 12th International Workshop on Data Mining in Bioinformatics (BIOKDD). ACM (2013)
34. Farnadi, G., Bach, S.H., Moens, M.F., Getoor, L., De Cock, M.: Extending PSL with fuzzy quantifiers. In: Papers from the 2014 AAAI Workshop Statistical Relational Artificial Intelligence, Québec City, 27 July, pp. WS-14-13, 35–37 (2014)
35. Gilboa, I., Schmeidler, D.: Case-based decision theory. Q. J. Econ. **110**, 605–639 (1995)
36. Hájek, P., Havránek, T.: Mechanising Hypothesis Formation - Mathematical Foundations for a General Theory. Springer, Heidelberg (1978). https://doi.org/10.1007/978-3-642-66943-9
37. Hebb, D.O.: The Organization of Behaviour. Wiley, Hoboken (1949)
38. Heitjan, D., Rubin, D.: Ignorability and coarse ckata. Ann. Statist. **19**, 2244–2253 (1991)
39. Hobbes, T.: Elements of philosophy, the first section, concerning body. In: Molesworth, W. (ed.) The English works of Thomas Hobbes of Malmesbury, vol. 1. John Bohn, London, 1839. English translation of "Elementa Philosophiae I. De Corpore" (1655)
40. Hohenecker, P., Lukasiewicz, T.: Ontology reasoning with deep neural networks. CoRR, abs/1808.07980 (2018)
41. Hüllermeier, E.: Inducing fuzzy concepts through extended version space learning. In: Bilgiç, T., De Baets, B., Kaynak, O. (eds.) IFSA 2003. LNCS, vol. 2715, pp. 677–684. Springer, Heidelberg (2003). https://doi.org/10.1007/3-540-44967-1_81
42. Hüllermeier, E.: Learning from imprecise and fuzzy observations: data disambiguation through generalized loss minimization. Int. J. Approx. Reasoning **55**(7), 1519–1534 (2014)
43. Jaeger, M.: Ignorability in statistical and probabilistic inference. JAIR **24**, 889–917 (2005)
44. Kahneman, D.: Thinking, Fast and Slow. Farrar, Straus and Giroux, New York (2011)
45. Kolodner, J.L.: Case-Based Reasoning. Morgan Kaufmann, Burlington (1993)
46. Kotlowski, W., Slowinski, R.: On nonparametric ordinal classification with monotonicity constraints. IEEE Trans. Knowl. Data Eng. **25**(11), 2576–2589 (2013)

47. Kraus, S., Lehmann, D., Magidor, M.: Nonmonotonic reasoning, preferential models and cumulative logics. Artif. Intell. **44**, 167–207 (1990)
48. Kuzelka, O., Davis, J., Schockaert, S.: Encoding Markov logic networks in possibilistic logic. In: Meila, M., Heskes, T. (eds.) Proceedings 31st Conference on Uncertainty in Artificial Intelligence (UAI 2015), Amsterdam, 12–16 July 2015, pp. 454–463. AUAI Press (2015)
49. Kuzelka, O., Davis, J., Schockaert, S.: Learning possibilistic logic theories from default rules. In: Kambhampati, S. (ed.) Proceedings 25th International Joint Conference on Artificial Intelligence (IJCAI 2016), New York, 9–15 July 2016, pp. 1167–1173 (2016)
50. Kuzelka, O., Davis, J., Schockaert, S.: Induction of interpretable possibilistic logic theories from relational data. In: Sierra, C. (ed.) Proceedings 26th International Joint Conference on Artificial Intelligence, IJCAI 2017, Melbourne, 19–25 August 2017, pp. 1153–1159 (2017)
51. LeCun, Y., Bengio, Y., Hinton, G.E.: Deep learning. Nature **521**(7553), 436–444 (2015)
52. Mamdani, E.H., Assilian, S.: An experiment in linguistic synthesis with a fuzzy logic controller. Int. J. Man-Mach. Stu. **7**, 1–13 (1975)
53. Marquis, P., Papini, O., Prade, H.: Eléments pour une histoire de l'intelligence artificielle. In: Panorama de l'Intelligence Artificielle. Ses Bases Méthodologiques, ses Développements, vol. I, pp. 1–39. Cépaduès (2014)
54. Marquis, P., Papini, O., Prade, H.: Some elements for a prehistory of Artificial Intelligence in the last four centuries. In: Proceedings 21st Europoen Conference on Artificial Intelligence (ECAI 2014), Prague, pp. 609–614. IOS Press (2014)
55. McCarthy, J., Minsky, M., Roch-ester, N., Shannon, C.E.: A proposal for the Dartmouth summer research project on artificial intelligence, august 31, 1955. AI Mag. **27**(4), 12–14 (2006)
56. McCulloch, W.S., Pitts, W.: A logical calculus of ideas immanent in nervous activity. Bull. Math. Biophys. **5**, 115–133 (1943)
57. Miclet, L., Bayoudh, S., Delhay, A.: Analogical dissimilarity: definition, algorithms and two experiments in machine learning. JAIR **32**, 793–824 (2008)
58. Miclet, L., Prade, H.: Handling analogical proportions in classical logic and fuzzy logics settings. In: Sossai, C., Chemello, G. (eds.) ECSQARU 2009. LNCS (LNAI), vol. 5590, pp. 638–650. Springer, Heidelberg (2009). https://doi.org/10.1007/978-3-642-02906-6_55
59. Mitchell, T.: Version spaces: an approach to concept learning. Ph.D. thesis, Stanford (1979)
60. More, T.: On the construction of Venn diagrams. J. Symb. Logic **24**(4), 303–304 (1959)
61. Mushthofa, M., Schockaert, S., De Cock, M.: Solving disjunctive fuzzy answer set programs. In: Calimeri, F., Ianni, G., Truszczynski, M. (eds.) LPNMR 2015. LNCS (LNAI), vol. 9345, pp. 453–466. Springer, Cham (2015). https://doi.org/10.1007/978-3-319-23264-5_38
62. Narodytska, N.: Formal analysis of deep binarized neural networks. In: Lang, J. (ed.) Proceedings 27th International Joint Conference Artificial Intelligence (IJCAI 2018), Stockholm, 13–19 July 2018, pp. 5692–5696 (2018)
63. Newell, A., Simon, H.A.: The logic theory machine. a complex information processing system. In: Proceedings IRE Transactions on Information Theory(IT-2), The Rand Corporation, Santa Monica, Ca, 1956. Report P-868, 15 June 1956, pp. 61-79, September 1956

64. Nilsson, N.J.: The Quest for Artificial Intelligence : A History of Ideas andAchievements. Cambridge University Press, Cambridge (2010)
65. Nin, J., Laurent, A., Poncelet, P.: Speed up gradual rule mining from stream data! A B-tree and OWA-based approach. J. Intell. Inf. Syst. **35**(3), 447–463 (2010)
66. Pearl, J.: Causality, vol. 2000, 2nd edn. Cambridge University Press, Cambridge (2009)
67. Perfilieva, I., Dubois, D., Prade, H., Esteva, F., Godo, L., Hodáková, P.: Interpolation of fuzzy data: analytical approach and overview. Fuzzy Sets Syst. **192**, 134–158 (2012)
68. Pinkas, G.: Propositional non-monotonic reasoning and inconsistency in symmetric neural networks. In: Mylopoulos, J., Reiter, R. (eds.) Proceedings 12th International Joint Conference on Artificial Intelligence, Sydney, 24–30 August 1991, pp. 525–531. Morgan Kaufmann (1991)
69. Pinkas, G.: Reasoning, nonmonotonicity and learning in connectionist networks that capture propositional knowledge. Artif. Intell. **77**(2), 203–247 (1995)
70. Pinkas, G., Cohen, S.: High-order networks that learn to satisfy logic constraints. FLAP J. Appl. Logics IfCoLoG J. Logics Appl. **6**(4), 653–694 (2019)
71. Prade, H.: Reasoning with data - a new challenge for AI? In: Schockaert, S., Senellart, P. (eds.) SUM 2016. LNCS (LNAI), vol. 9858, pp. 274–288. Springer, Cham (2016). https://doi.org/10.1007/978-3-319-45856-4_19
72. Prade, H., Richard, G.: From analogical proportion to logical proportions. Logica Universalis **7**(4), 441–505 (2013)
73. Prade, H., Rico, A., Serrurier, M.: Elicitation of sugeno integrals: a version space learning perspective. In: Rauch, J., Raś, Z.W., Berka, P., Elomaa, T. (eds.) ISMIS 2009. LNCS (LNAI), vol. 5722, pp. 392–401. Springer, Heidelberg (2009). https://doi.org/10.1007/978-3-642-04125-9_42
74. Prade, H., Rico, A., Serrurier, M., Raufaste, E.: Elicitating sugeno integrals: methodology and a case study. In: Sossai, C., Chemello, G. (eds.) ECSQARU 2009. LNCS (LNAI), vol. 5590, pp. 712–723. Springer, Heidelberg (2009). https://doi.org/10.1007/978-3-642-02906-6_61
75. Prade, H., Serrurier, M.: Bipolar version space learning. Int. J. Intell. Syst. **23**, 1135–1152 (2008)
76. Raufaste, E.: Les Mécanismes Cognitifs du Diagnostic Médical : Optimisation et Expertise. Presses Universitaires de France (PUF), Paris (2001)
77. Richardson, M., Domingos, P.M.: Markov logic networks. Mach. Learn. **62**(1–2), 107–136 (2006)
78. Rocktäschel, T., Riedel, S.: End-to-end differentiable proving. In: Guyon, I., et al. (eds.) Proceedings 31st Annual Conference on Neural Information Processing Systems (NIPS 2017), Long Beach, 4–9 December 2017, pp. 3791–3803 (2017)
79. Rosenblatt, F.: The perceptron: a probabilistic model for information storage and organization in the brain. Psychol. Rev. **65**(6), 386–408 (1958)
80. Rückert, U., De Raedt, L.: An experimental evaluation of simplicity in rule learning. Artif. Intell. **172**(1), 19–28 (2008)
81. Samuel, A.: Some studies in machine learning using the game of checkers. IBM J. **3**, 210–229 (1959)
82. Schockaert, S., Prade, H.: Interpolation and extrapolation in conceptual spaces: a case study in the music domain. In: Rudolph, S., Gutierrez, C. (eds.) RR 2011. LNCS, vol. 6902, pp. 217–231. Springer, Heidelberg (2011). https://doi.org/10.1007/978-3-642-23580-1_16

83. Schockaert, S., Prade, H.: Interpolative and extrapolative reasoning in propositional theories using qualitative knowledge about conceptual spaces. Artif. Intell. **202**, 86–131 (2013)
84. Selfridge, O.G.: Pandemonium: a paradigm for learning. In: Blake, D.V., Uttley, A.M. (ed) Symposium on Mechanisation of Thought Processes, London, 24–27 November 1959, vol. 1958, pp. 511–529 (1959)
85. Serafini, L., Garcez, A.S.A.: Logic tensor networks: deep learning and logical reasoning from data and knowledge. In: Besold, T.R., Lamb, L.C., Serafini, L., Tabor, W. (eds.) Proceedings 11th International Workshop on Neural-Symbolic Learning and Reasoning (NeSy 2016), New York City, 16–17 July 2016, vol. 1768 of CEUR Workshop Proceedings (2016)
86. Serafini, L., Donadello, I., Garcez, A.S.A.: Learning and reasoning in logic tensor networks: theory and application to semantic image interpretation. In: Seffah, A., Penzenstadler, B., Alves, C., Peng, X. (eds.) Proceedings Symposium on Applied Computing (SAC 2017), Marrakech, 3–7 April 2017, pp. 125–130. ACM (2017)
87. Serrurier, M., Dubois, D., Prade, H., Sudkamp, T.: Learning fuzzy rules with their implication operators. Data Knowl. Eng. **60**(1), 71–89 (2007)
88. Serrurier, M., Prade, H.: Introducing possibilistic logic in ILP for dealing with exceptions. Artif. Intell. **171**(16–17), 939–950 (2007)
89. Shannon, C.E.: Programming a computer for playing chess. Philos. Mag. (7th series) **XLI** (314), 256–275 (1950)
90. Solomonoff, R.J.: An inductive inference machine. Tech. Res. Group, New York City (1956)
91. Turing, A.M.: Intelligent machinery. Technical report, National Physical Laboratory, London, 1948. Also. In: Machine Intelligence, vol. 5, pp. 3–23. Edinburgh University Press (1969)
92. Turing, A.M.: Computing machinery and intelligence. Mind **59**, 433–460 (1950)
93. Ughetto, L., Dubois, D., Prade, H.: Implicative and conjunctive fuzzy rules - a tool for reasoning from knowledge and examples. In: Hendler, J., Subramanian, D. (eds.) Proceedings 16th National Confernce on Artificial Intelligence, Orlando, 18–22 July 1999, pp. 214–219 (1999)
94. Walley, P.: Statistical Reasoning with Imprecise Probabilities. Chapman and Hall, London (1991)
95. Walley, P.: Measures of uncertainty in expert systems. Artif. Intell. **83**(1), 1–58 (1996)
96. Wiener, N.: Cybernetics or Control and Communication in the Animal and the Machine. Wiley, Hoboken (1949)
97. Zadeh, L.A.: Thinking machines - a new field in electrical engineering. Columbia Eng. Q. **3**, 12–13 (1950)
98. Zadeh, L.A.: Outline of a new approach to the analysis of complex systems and decision processes. IEEE Trans. Syst. Man Cybern. **3**(1), 28–44 (1973)

CP-Nets, π-pref Nets, and Pareto Dominance

Nic Wilson[1(✉)], Didier Dubois[2], and Henri Prade[2]

[1] Insight Centre for Data Analytics, School of Computer Science and IT,
University College Cork, Cork, Ireland
nic.wilson@insight-centre.org
[2] IRIT-CNRS Université Paul Sabatier, Toulouse, France

Abstract. Two approaches have been proposed for the graphical handling of qualitative conditional preferences between solutions described in terms of a finite set of features: Conditional Preference networks (CP-nets for short) and more recently, Possibilistic Preference networks (π-pref nets for short). The latter agree with Pareto dominance, in the sense that if a solution violates a subset of preferences violated by another one, the former solution is preferred to the latter one. Although such an agreement might be considered as a basic requirement, it was only conjectured to hold as well for CP-nets. This non-trivial result is established in the paper. Moreover it has important consequences for showing that π-pref nets can at least approximately mimic CP-nets by adding explicit constraints between symbolic weights encoding the *ceteris paribus* preferences, in case of Boolean features. We further show that dominance with respect to the extended π-pref nets is polynomial.

1 Introduction

Ceteris Paribus Conditional Preference Networks (CP-nets, for short) [5,6] were introduced in order to provide a convenient tool for the elicitation of multidimensional preferences and accordingly compare the relative merits of solutions to a problem. They are based on three assumptions: only ordinal information is required; the preference statements deal with the values of single decision variables in the context of fixed values for other variables that influence them; preferences are provided all else being equal (*ceteris paribus*). CP-nets were inspired by Bayesian networks (they use a dependency graph, most of the time a directed acyclic one, whose vertices are variables) but differ from them by being qualitative, by their use of the ceteris paribus assumption, and by the fact that the variables in a CP-net are decision variables rather than random variables. In the most common form of CP-nets, each preference statement in the preference graph translates into a strict preference between two solutions (i.e., value assignment to all decision variables) differing on a single variable (referred to as a worsening flip) and the dominance relation between solutions is the transitive closure of this worsening flip relation.

© Springer Nature Switzerland AG 2019
N. Ben Amor et al. (Eds.): SUM 2019, LNAI 11940, pp. 169–183, 2019.
https://doi.org/10.1007/978-3-030-35514-2_13

Another kind of conditional preference network, called π-pref nets, has been more recently introduced [1], and is directly inspired by the counterpart of Bayesian networks in possibility theory, called possibilistic networks [3]. A Π-pref net shares with CP-nets its directed acyclic graphical structure between decision variables, and conditional preference statements attached to each variable in the contexts defined by assignments of its parent variables in the graph. The preference for one value against another is captured by assigning degrees of possibility (here interpreted as utilities) to these values. When the only existing preferences are those expressed by the conditional statements (there are no preference statements across contexts or variables), it has been shown that the dominance relation between solutions is obtained by comparing vectors of symbolic utility values (one per variables) using Pareto-dominance.

Some results comparing the preference relations between solutions obtained from CP-nets and π-pref nets with Boolean decision variables are given in [1]. This is made easy by the fact that CP-nets and π-pref nets share the graph structure and the conditional preference tables. It was shown that the two obtained dominance relations between solutions cannot conflict with each other (there is no preference reversal between them), and that ceteris paribus information can be added to π-pref nets in the form of preference statements between specific products of symbolic weights. One pending question was to show that the dominance relation between solutions obtained from a CP-net refines the preference relation obtained from the corresponding π-pref net. In the case of Boolean variables, the π-pref net ordering can be viewed as a form of Pareto ordering: each assignment of a decision variables is either good (= in agreement with the preference statement) or bad. The pending question comes down to prove a monotonicity condition for the preference relation on solutions, stating that as soon as a solution contains more (in the sense of inclusion) good variable assignments than another solution, it should be strictly preferred by the CP-net. Strangely enough this natural question has hardly been addressed in the literature so far (see [2] for some discussion). The aim of this paper is to solve this problem, and more generally to compare the orderings of solutions using the two preference modeling structures.

We further show that dominance with respect to extended π-pref nets can be computed in polynomial time, using linear programming; it thus forms a polynomial upper approximation for the CP-net dominance relation.

The paper is structured as follows: In Sect. 2 we define a condition, that we call *local dominance*, that is shown to be a sufficient condition for dominance in a CP-net. The follow two sections, Sects. 3 and 4, make use of this sufficient condition in producing results that show that a form of Pareto ordering is a lower bound for a lower bound for CP-net dominance. Section 5 then uses the results of Sect. 4 to show that π-pref nets dominance is a lower bound for CP-net dominance. We also show there that the extended π-pref nets dominance, which is an upper bound for CP-net dominance, can be computed in polynomial time. Section 6 concludes.

2   A Sufficient Condition for Dominance in a CP-Net

We start by recalling the definition of CP-nets and a characterization of the corresponding dominance relation between solutions.

2.1   Defining CP-nets

We consider a finite set of variables $\mathcal{V}$. Each variable $X \in \mathcal{V}$ has an associated finite domain $\mathrm{Dom}(X)$. An outcome (also called a solution) is a complete assignment to the variables in $\mathcal{V}$, i.e., a function w that, for each variable $X \in \mathcal{V}$, $w(X) \in \mathrm{Dom}(X)$.

A CP-net Σ over set of variables $\mathcal{V}$ is a pair $\langle G, P \rangle$. The first component G is a directed graph with vertices in $\mathcal{V}$, and we say that CP-net Σ is *acyclic* if G is acyclic. For variable $X \in \mathcal{V}$, let $\mathcal{U}_X$ be the set of parents of X in G, i.e., the set of variables Y such that (Y, X) is an edge in G. The second component P of Σ consists of a collection of partial orders $\{>_u^X : X \in \mathcal{V}, u \in \mathrm{Dom}(\mathcal{U}_X)\}$, called conditional preference tables; for each variable $X \in \mathcal{V}$ and each assignment u to the parents $\mathcal{U}_X$ of X, relation $>_u^X$ is a strict partial order (i.e., a transitive and irreflexive relation) on $\mathrm{Dom}(X)$. We make the assumption that for each variable X there exists at least one assignment u to $\mathcal{U}_X$ such that $>_u^X$ is non-empty (i.e., for each $X \in \mathcal{V}$ there exists some $x, x' \in \mathrm{Dom}(X)$ and some u such that $x >_u^X x'$).

Let w be an outcome and, for variable $X \in \mathcal{V}$, let $u = w(\mathcal{U}_X)$ be the projection of w to the parents set of X. If $x >_u^X x'$ then we shall write, for simplicity, (with the understanding that x and x' are elements of $\mathrm{Dom}(X)$):

$$x > x' \text{ given } w \text{ [with respect to } \Sigma\text{]}.$$

Note that if v is any outcome whose projection to the parents set of X is also u then $[x > x' \text{ given } v]$ if and only if $[x > x' \text{ given } w]$; the values of $w(Y)$ and $u(Y)$ may differ for variables $Y \notin \mathcal{U}_X \cup \{X\}$, but the preference between x and x' in the context u does not depend on Y.

We say that Σ is *locally totally ordered* if each associated strict partial order $>_u^X$ is a strict total order, so that for each pair of different elements x and x' of $\mathrm{Dom}(X)$, we have either $x >_u^X x'$ or $x' >_u^X x$. We say that Σ is *Boolean* if for each $X \in \mathcal{V}$, each domain has exactly two elements: $|\mathrm{Dom}(X)| = 2$.[1]

The Dominance Relation Associated with a CP-Net. Given a CP-net Σ over variables $\mathcal{V}$, we say that w' is a *worsening flip* from w w.r.t. Σ, if w' and w are outcomes that differ on exactly one variable $X \in \mathcal{V}$ (so that $w'(X) \neq w(X)$ and for all $Y \in \mathcal{V} \setminus \{X\}$, $w'(Y) = w(Y)$), and $w(X) >_u^X w'(X)$, where u is the projection of w (or w') to the parent set $\mathcal{U}_X$ of X.

The set of direct consequences of CP-net Σ are the set of pairs (w, w'), where w' is a worsening flip from w w.r.t. Σ, forming an irreflexive relation:

---

[1] If a variable has only one element in its domain, it is a constant, and we could remove it if we wished.

Definition 1. *The worsening flip relation $>^{\Sigma}_{wf}$ is defined by $w >^{\Sigma}_{wf} w'$ if and only if $w(X) >^{X}_{u} w'(X)$ where $u = w(\mathcal{U}_X) = w'(\mathcal{U}_X)$. Let the binary relation $>^{\Sigma}_{cp}$ on outcomes denote the transitive closure of $>^{\Sigma}_{wf}$. If $w >^{\Sigma}_{cp} w'$ we say that w [cp-]dominates w' [with respect to Σ].*

The relation $>^{\Sigma}_{wf}$ is well-defined due to the ceteris paribus assumption. A sequence of outcomes $w_1, \ldots, w_k$ is said to be a worsening flipping sequence [with respect to CP-net Σ] from w_1 to w_k if, for each $i = 1, \ldots, k - 1$, w_{i+1} is a worsening flip from w_i. Thus, w cp-dominates w' if and only if there is a worsening flipping sequence from w to w'.

2.2   Some Simple Conditions for CP-Dominance

For outcomes w and v we define $\Delta(w, v)$ to be the set of variables on which they differ, i.e., $\{X \in \mathcal{V} : w(X) \neq v(X)\}$. The following lemma gives two simple sufficient conditions for w to dominate v with respect to a CP-net. In Case (i), for each variable X in $\Delta(w, v)$, there is a worsening flip from w, changing $w(X)$ to $v(X)$. In Case (ii) cflip from v changing $v(X)$ to $w(X)$.

Lemma 1. *Consider an acyclic CP-net Σ and two different outcomes w and v. Then w cp-dominates v w.r.t. CP-net Σ if either*

(i) for all $X \in \Delta(w, v)$, $w(X) > v(X)$ given w; or
(ii) for all $X \in \Delta(w, v)$, $w(X) > v(X)$ given v.

Proof. Let $k = |\Delta(w, v)|$, which is greater than zero because $w \neq v$. Let us label the elements of $\Delta(w, v)$ as $X_1, \ldots, X_k$ in such a way that if $i < j$ then X_i is not an ancestor of X_j with respect to the CP-net directed graph; this is possible because of the acyclicity assumption on Σ. To prove (i), beginning with outcome w, we flip variables of w to v in the order $X_1, \ldots, X_k$, so that we first change $w(X_1)$ to $v(X_1)$, and then change $w(X_2)$ to $v(X_2)$, and so on. The choice of variable ordering means that when we flip variable X_i the assignment to the parents $\mathcal{U}_{X_i}$ of X_i is just $w(\mathcal{U}_{X_i})$. It can be seen that this is a sequence of worsening flips from w to v, and thus, w cp-dominates v w.r.t. Σ.

Part (ii) is very similar, except that we start with v, and iteratively change X_i from $v(X_i)$ to $w(X_i)$ in the order $i = 1, \ldots, k$. The assumption behind part (ii) implies that we obtain an improving flipping sequence from v to w.   □

Lemma 1 can be used to prove a more general form of itself.

Proposition 1. *Consider an acyclic CP-net Σ and two different outcomes w and v. Assume that for each $X \in \Delta(w, v)$ either $w(X) > v(X)$ given w w.r.t. Σ, or $w(X) > v(X)$ given v w.r.t. Σ. Then w cp-dominates v w.r.t. Σ.*

Proof. Define outcome u by $u(X) = v(X)$ if X is such that $w(X) > v(X)$ given w (so $X \in \Delta(w, v)$), and $u(X) = w(X)$ otherwise. Then $\Delta(w, v)$ is the disjoint union of $\Delta(w, u)$ and $\Delta(u, v)$.

For all $X \in \Delta(w, u)$, $w(X) > u(X)$ given w, because $u(X) = v(X)$ and $w(X) > v(X)$ given w. Lemma 1 implies that w cp-dominates u w.r.t. Σ.

For all $X \in \Delta(u, v)$, $u(X) > v(X)$ given v, since $u(X) = w(X)$ and $u(X) > v(X)$ given w. Lemma 1 implies that u cp-dominates v w.r.t. Σ. Thus, w cp-dominates v w.r.t. Σ. □

2.3 The Local Dominance Relation

The conditions of Proposition 1 involve what might be called a local dominance condition.

Definition 2. *Given an acyclic CP-net Σ we say that outcome w locally dominates outcome v [w.r.t. CP-net Σ], written $w >_{LD}^{\Sigma} v$, if for each $X \in \Delta(w, v)$ either $w(X) > v(X)$ given w w.r.t. Σ; or $w(X) > v(X)$ given v w.r.t. Σ.*

Proposition 1 above implies that if w locally dominates v then w cp-dominates v, so that $w >_{LD}^{\Sigma} v$ implies $w >_{cp}^{\Sigma} v$. In fact, we even have the following result.

Proposition 2. *Given an acyclic CP-net Σ, binary relation $>_{cp}^{\Sigma}$ is the transitive closure of $>_{LD}^{\Sigma}$.*

Proof. Let $\succ$ be the transitive closure of $>_{LD}^{\Sigma}$. Since, by Proposition 1, $>_{LD}^{\Sigma}$ is a subset of $>_{cp}^{\Sigma}$, and the latter is transitive, we have that $\succ$ is a subset of $>_{cp}^{\Sigma}$.

Suppose that w' is a worsening flip from w w.r.t. Σ. Then, $w(X) > w'(X)$ given w and $\Delta(w, w') = \{X\}$, which implies that w locally dominates w'. This shows that $>_{LD}^{\Sigma}$, and thus, $\succ$, contains the worsening flip relation $>_{wf}^{\Sigma}$ induced by Σ. Being transitive, $\succ$ contains the transitive closure $>_{cp}^{\Sigma}$ of $>_{wf}^{\Sigma}$. We have therefore shown that $>_{cp}^{\Sigma}$ equals $\succ$, the transitive closure of $>_{LD}^{\Sigma}$. □

3 Pareto Ordering for CP-Nets in the General Case

A Pareto Ordering between outcomes comes down to saying that w dominates w' if $\forall X \in V, w(X)$ is at least as good an assignment as $w'(X)$ (and better for some X). However, it is not so easy to define Pareto dominance between outcomes in a CP-net when variables are not Boolean. It is often impossible to compare $w(X)$ and $w'(X)$ directly as there is generally no relation $>_{u}^{X}$ that compares them. To perform this kind of comparison in the general case of a dependency graph, we must in some way map the various preference relations $>_{u}^{X}$ on $\mathrm{Dom}(X)$ to some common scale, either totally (using a scoring function) or partially on some landmark values (mapping the best choices or the worst choices). We define a somewhat extreme Pareto-like relation, using the latter idea, below. As mentioned in Sect. 1, and discussed in detail in Sect. 4, the more natural form of Pareto dominance applies only for the case of Boolean CP-nets.

3.1   A Variant of Pareto Dominance for CP-nets

We define relation $>_{sp}^{\Sigma}$ on the set of outcomes, which can be viewed as being based on a strong variant of the Pareto condition (with sp standing for *strong Pareto*).

Definition 3 (Fully dominating and fully dominated). *For outcome w, we say that x is* fully dominating *in X given w [w.r.t. Σ] if $x \in \mathrm{Dom}(X)$, and for all $x' \in \mathrm{Dom}(X) \setminus \{x\}$ we have $x > x'$ given w w.r.t. Σ.*

Similarly, we say that x is fully dominated *in X given w [w.r.t. Σ] if $x \in \mathrm{Dom}(X)$, and $|\mathrm{Dom}(X)| > 1$ and for all $x' \in \mathrm{Dom}(X) \setminus \{x\}$ we have $x' > x$ given w w.r.t. Σ.*

Thus, if x is fully dominating in X given w then x is not fully dominated in X given w. Also, there can at most one element $x \in \mathrm{Dom}(X)$ that is fully dominating in X given w, and at most one that is fully dominated in X given w.

We define irreflexive relation $>_{sp}^{\Sigma}$ by, for different outcomes w and v, $w >_{sp}^{\Sigma} v$ if and only if for all $X \in \mathcal{V}$ either $v(X)$ is fully dominated in X given v w.r.t. Σ; or $w(X)$ is fully dominating in X given w w.r.t. Σ.

In the case in which the local relations $>_u^X$ are total orders, then the definitions can be simplified. Consider any outcome w, and value x in $\mathrm{Dom}(X)$, and let u be the projection of w to the parent set of X. Let x_u^* and x_{u*} be the best and the worst element (respectively) in $\mathrm{Dom}(X)$ for relation $>_u^X$. Then x is fully dominating in X given w if and only if $x = x_u^*$, and x is fully dominated in X given w if and only if $|\mathrm{Dom}(X)| > 1$ and $x = x_{u*}$. Another way of defining the $>_{sp}^{\Sigma}$ relation then consists, for each relation $>_u^X$, of mapping $\mathrm{Dom}(X)$ to a three-valued totally ordered scale $L = \{1, I, 0\}$ with $1 > I > 0$ using a kind of qualitative scoring function $f_u^X : \mathrm{Dom}(X) \to L$ defined by $f_u^X(x_u^*) = 1$, $f_u^X(x_{u*}) = 0$, and $f_u^X(x) = I$ otherwise. Note that relation $w >_{sp}^{\Sigma} w'$ expresses a very strong form of Pareto-dominance, since it requires that not only $w \neq w'$ and $f_u^X(w(X)) \geq f_u^X(w'(X))$, but also that either $f_u^X(w(X)) = 1$ or $f_u^X(w(X)) = 0, \forall X \in \mathcal{V}$.

Proposition 3. *Relation $>_{sp}^{\Sigma}$ is transitive, and is contained in $>_{LD}^{\Sigma}$, i.e., $w >_{sp}^{\Sigma} v$ implies $w >_{LD}^{\Sigma} v$, and thus $>_{sp}^{\Sigma} \subseteq >_{LD}^{\Sigma} \subseteq >_{cp}^{\Sigma}$. Furthermore, we have $>_{sp}^{\Sigma}$ and $>_{cp}^{\Sigma}$ are equal (i.e., are the same relation) if and only if $>_{sp}^{\Sigma}$ and $>_{LD}^{\Sigma}$ are equal.*

Proof. We will prove transitivity of $>_{sp}^{\Sigma}$ by showing that if $w_1 >_{sp}^{\Sigma} w_2$ and $w_2 >_{sp}^{\Sigma} w_3$ then $w_1 >_{sp}^{\Sigma} w_3$. Consider any $X \in \mathcal{V}$ such that $w_3(X)$ is not fully dominated in X given w_3. Since $w_2 >_{sp}^{\Sigma} w_3$, we have that $w_2(X)$ is fully dominating in X given w_2, and so $w_2(X)$ is not fully dominated in X given w_2. Since $w_1 >_{sp}^{\Sigma} w_2$, we have that $w_1(X)$ is fully dominating in X given w_1. Thus, for all $X \in \mathcal{V}$, if $w_3(X)$ is not fully dominated in X given w_3 then $w_1(X)$ is fully dominating in X given w_1, and hence, $w_1 >_{sp}^{\Sigma} w_3$, proving transitivity.

Now, suppose that $w_1 >_{sp}^{\Sigma} w_2$, and consider any $X \in \mathcal{V}$. Either (i) $w_2(X)$ is fully dominated in X given w_2, and thus, $w_1(X) > w_2(X)$ given w_2; or (ii)

$w_1(X)$ is fully dominating in X given w_1, and thus, $w_1(X) > w_2(X)$ given w_1; therefore we have $w_1 >^{\Sigma}_{LD} w_2$.

Clearly if $>^{\Sigma}_{sp}$ and $>^{\Sigma}_{cp}$ are equal then the inclusions $>^{\Sigma}_{sp} \subseteq >^{\Sigma}_{LD} \subseteq >^{\Sigma}_{cp}$ imply that $>^{\Sigma}_{sp}$ and $>^{\Sigma}_{LD}$ are equal. Conversely, assume that $>^{\Sigma}_{sp}$ and $>^{\Sigma}_{LD}$ are equal. We then have that $>^{\Sigma}_{LD}$ is transitive (since $>^{\Sigma}_{sp}$ is transitive), and thus it is equal to its transitive closure, which equals $>^{\Sigma}_{cp}$ by Proposition 2. □

3.2 Necessary and Sufficient Conditions for Equality of $>^{\Sigma}_{sp}$ and $>^{\Sigma}_{cp}$

We will show that $>^{\Sigma}_{sp}$ and $>^{\Sigma}_{cp}$ are only equal under extremely special conditions, including that the CP-net is unconditional and that each domain has at most two elements. We use a series of lemmas to prove the result.

The first lemma follows easily using the transitivity of $>^{\Sigma}_{sp}$.

Lemma 2. *Given CP-net Σ, then we have $>^{\Sigma}_{sp}$ equals $>^{\Sigma}_{cp}$ if and only if for all pairs (w, w') such that w' is a worsening flip from w we have $w >^{\Sigma}_{sp} w'$.*

Proof. We need to prove that $>^{\Sigma}_{sp}$ equals $>^{\Sigma}_{cp}$ if and only if $>^{\Sigma}_{sp}$ contains the worsening flip relation $>^{\Sigma}_{wf}$ induced by Σ. Since $>^{\Sigma}_{cp}$ is the transitive closure of $>^{\Sigma}_{wf}$, if $>^{\Sigma}_{sp}$ equals $>^{\Sigma}_{cp}$ then $>^{\Sigma}_{sp}$ contains $>^{\Sigma}_{wf}$.

Regarding the converse, assume that $>^{\Sigma}_{sp}$ contains $>^{\Sigma}_{wf}$. Since, by Proposition 3, $>^{\Sigma}_{sp}$ is transitive, then $>^{\Sigma}_{sp}$ contains the transitive closure $>^{\Sigma}_{cp}$ of $>^{\Sigma}_{wf}$. Proposition 3 implies that $>^{\Sigma}_{sp}$ is a subset of $>^{\Sigma}_{cp}$, so $>^{\Sigma}_{sp}$ equals $>^{\Sigma}_{cp}$. □

The definition of $>^{\Sigma}_{sp}$ leads to the following characterisation. Suppose that w' is a worsening flip from w w.r.t. CP-net Σ, with X being the variable on which they differ. Then $w >^{\Sigma}_{sp} w'$ if and only if (a) either $w(X)$ is fully dominating in X given w w.r.t. Σ, or $w'(X)$ is fully dominated in X given w w.r.t. Σ; and (b) for all $Y \in \mathcal{V} \setminus \{X\}$,

(i) if Y is not a child of X then $w(Y)$ is either fully dominated or fully dominating in Y given w w.r.t. Σ; and

(ii) if Y is a child of X then $w(Y)$ is either fully dominating in Y given w w.r.t. Σ or fully dominated in Y given w' w.r.t. Σ.

The above considerations lead to the following result.

Lemma 3. *Consider any $X \in \mathcal{V}$, and any assignment u to the parents of X, and any values $x, x' \in \mathrm{Dom}(X)$ such that $x >^{X}_{u} x'$. Assume that $w >^{\Sigma}_{sp} w'$ whenever (w, w') is an associated worsening flip, i.e., if $w(X) = x$ and $w'(X) = x'$, and w and w' agree on all other variables, and w extends u. Let (v, v') be one such associated worsening flip.*

If variable Z is not a child of X and z is any element of $\mathrm{Dom}(Z)$ then z is either fully dominated or fully dominating in X given v w.r.t. Σ. We have $|\mathrm{Dom}(Z)| \leq 2$.

If variable Y is a child of X and y is any element of $\mathrm{Dom}(Y)$ then y is either fully dominating given v or fully dominated given v'. We have $|\mathrm{Dom}(Y)| \leq 2$.

Note that the condition $|\mathrm{Dom}(Z)| \leq 2$ follows since there can be at most one fully dominated and at most one fully dominating element in X given v.

Lemma 3 implies that for any variable X, every other variable has at most two values, which immediately implies that every domain has at most two elements:

Lemma 4. *Suppose that $>_{sp}^{\Sigma}$ and $>_{cp}^{\Sigma}$ are equal. Then each domain has at most two values.*

Definition 4 (True parents and being unconditional). *Let Y be a variable and let X be an element of its parent set $\mathcal{U}_Y$. We say that X is not a true parent of Y if for all assignments u and u' to $\mathcal{U}_Y$ that differ only on the value of X, if $y >_u^Y y'$ then $y >_{u'}^Y y'$. We say that Y is unconditional in Σ if it has no true parents.*

If X is not a true parent of Y then $>_u^Y$ does not depend on X. For any CP-net Σ we can generate an equivalent CP-net (i.e., that generates the same ordering on outcomes) such that every parent of every variable is a true parent.

Lemma 5. *Suppose that $>_{sp}^{\Sigma}$ and $>_{cp}^{\Sigma}$ are equal. Assume that every parent of variable Y is unconditional, and let X be one such parent. Suppose that u is some assignment to the parents of Y, and that u' is another assignment that differs from u only on the value of X. If $y >_u^Y y'$ then $y >_{u'}^Y y'$.*

Proof. Suppose that $y >_u^Y y'$. Let v be any outcome extending u and let v' be any outcome extending u'. Lemma 4 implies that X has at most two values. If X had only one value then it is trivially not a true parent of Y, so we can assume that $\mathrm{Dom}(X) = 2$. X is unconditional so it has no parents. Our definition of a CP-net implies that the relation $>^X$ is non-empty, so we have $x_1 >^X x_2$, for some labelling x_1 and x_2 of the values of X. We first consider the case in which $u(X) = x_1$. Now, y' is not fully dominating given u and so, by Lemma 3, y' is fully dominated given u', which implies $y >_{u'}^Y y'$.

We now consider the other case in which $u(X) = x_2$. Then, y is not fully dominated given u, and so, by Lemma 3, y is fully dominating given u', and thus, also $y >_{u'}^Y y'$.   □

Lemma 5 implies that X is not a true parent of Y. Since X was an arbitrary parent of Y, it then implies that Y has no true parent, so is unconditional. Applying this result inductively then implies that every variable in $\mathcal{V}$ is unconditional with respect to Σ. Along with Lemmas 3 and 4, this leads to the following result.

Proposition 4. *Given CP-net Σ, then we have $>_{sp}^{\Sigma}$ equals $>_{cp}^{\Sigma}$ if and only if Σ be a Boolean locally totally ordered CP-net such that each variable X is unconditional in Σ.*

4   Pareto Ordering for the Boolean Case

As discussed earlier, there is a natural way of defining a Pareto ordering for the case of Boolean locally totally ordered CP-nets. Basically, if variables are Boolean, each of its values is either fully dominating or fully dominated in each context. So, the relation $>_{sp}^{\Sigma}$ becomes a full-fledged Pareto ordering. In this section we analyse the relationship between this Pareto ordering and the CP-net ordering.

Let Σ be a Boolean locally totally ordered CP-net. Consider any outcome w. We say that variable X is *bad for* w if there is an improving flip of variable X from w to another outcome w'. Define F_w to be the set of variables which are bad for w.

The definition of F_w and of the local dominance relation (see Sect. 2.3) immediately leads to the following expression of $>_{LD}^{\Sigma}$ in the Boolean locally totally ordered case.

Lemma 6. *Let Σ be a Boolean locally totally ordered CP-net. Then, for different outcomes w and v, we have $w >_{LD}^{\Sigma} v$ if and only if $F_w \cap \Delta(w, v) \subseteq F_v$.*

Proof. $w >_{LD}^{\Sigma} v$ if and only if for each $X \in \Delta(w, v)$ either $w(X) > v(X)$ given w, or $w(X) > v(X)$ given v. For $X \in \Delta(w, v)$, we have $w(X) > v(X)$ given w if and only if $X \notin F_w$; and we have $w(X) > v(X)$ given v if and only if $X \in F_v$. Thus, $w >_{LD}^{\Sigma} v$ if and only if for each $X \in \Delta(w, v)$ [$X \in F_w \Rightarrow X \in F_v$], which is if and only if $F_w \cap \Delta(w, v) \subseteq F_v$.   □

We define the irreflexive binary relation $>_{par}^{\Sigma}$ on outcomes as follows.

Definition 5. *For different outcomes w and v, $w >_{par}^{\Sigma} v$ if and only if $F_w \subseteq F_v$, i.e., every variable that is bad for w is also bad for v.*

This can be viewed as a kind of Pareto ordering, and equals the strong Pareto relation $>_{sp}^{\Sigma}$ (see Sect. 3.1) for the Boolean locally totally ordered case.

Lemma 7. *Let Σ be a Boolean locally totally ordered CP-net. Let w and v be outcomes. Then $w >_{sp}^{\Sigma} v$ if and only if $w >_{par}^{\Sigma} v$.*

Proof. For different w and v, $w >_{sp}^{\Sigma} v$ if and only if for all $X \in \mathcal{V}$ either $v(X)$ is fully dominated in X given v, or $w(X)$ is fully dominating in X given w.

Suppose that $w >_{sp}^{\Sigma} v$ and consider any $X \in \mathcal{V}$. If $X \in F_w$ then $w(X)$ is not fully dominating in X given w, and so $v(X)$ is fully dominated in X given v, which implies that $X \in F_v$. We have shown that $F_w \subseteq F_v$.

Conversely, assume that $F_w \subseteq F_v$, and consider any $X \in \mathcal{V}$. such that $v(X)$ is not fully dominated in X given v. Because Σ is a Boolean locally totally ordered CP-net this implies that X is not bad for v. Since $F_w \subseteq F_v$, this implies that X is not bad for w, and so, $w(X)$ is fully dominating in X given w. This proves that $w >_{sp}^{\Sigma} v$.   □

The CP-net relation contains the Pareto relation, with the local dominance relation being between the two.

Theorem 1. *Let Σ be a Boolean locally totally ordered CP-net. Relation $>_{par}^{\Sigma}$ is transitive, and is contained in $>_{LD}^{\Sigma}$, i.e., $w >_{par}^{\Sigma} v$ implies $w >_{LD}^{\Sigma} v$, and thus $>_{par}^{\Sigma} \subseteq >_{LD}^{\Sigma} \subseteq >_{cp}^{\Sigma}$. Furthermore, we have $>_{par}^{\Sigma}$ and $>_{cp}^{\Sigma}$ are equal (i.e., are the same relation) if and only if $>_{par}^{\Sigma}$ and $>_{LD}^{\Sigma}$ are equal, which happens only if every variable of the CP-net is unconditional.*

Proof. Theorem 1 follows immediately from Propositions 3 and 4 and Lemma 7. $\square$

As a consequence, we get that CP-nets are in agreement with Pareto ordering in the case of Boolean locally totally ordered variables: for any variable X and any configuration u of its parents, consider the mapping $f_u^X : \text{Dom}(X) \to \{0, 1\}$ such that $f_u^X(x^*) = 1$ and $f_u^X(x_{u*}) = 0$. For any two distinct outcomes w and w', we have that $\forall X \in \mathcal{V}, f_{w(U_X)}^X(w(X)) \geq f_{w'(U_X)}^X(w'(X))$ if and only $F_w \subseteq F_{w'}$, which is Pareto-ordering $>_{par}^{\Sigma}$.

We emphasise the following part of the theorem:

Corollary 1. *Let Σ be a Boolean locally totally ordered CP-net, $w >_{par}^{\Sigma} w'$ implies $w >_{cp}^{\Sigma} w'$.*

As shown in the previous section, it does not seem straightforward to extend this Pareto ordering in a natural way to non-Boolean variables without using scaling functions that map all partial orders $(\text{Dom}(X), >_u^X), u \in \mathcal{U}_X$ to a common value scale, unless the variables are all preferentially independent from one another. In this case, $\mathcal{U}_X = \emptyset, \forall X$, and $>_u^X = >^X, \forall X \in \text{Dom}(X)$. We could then define the Pareto dominance relation $>_{par}^{\Sigma}$ on outcomes as $w >_{par}^{\Sigma} w'$ if and only if $w \neq w'$ and $w(X) >^X w'(X)$ or $w(X) = w'(X)$ for all $X \in \mathcal{V}$.

5 π-pref Nets

Possibility theory [8] is a theory of uncertainty devoted to the representation of incomplete information. It is maxitive (addition is replaced by maximum) in contrast with probability theory. It ranges from purely ordinal to purely numerical representations. Possibility theory can be used for representing preferences [9]. It relies on the idea of a possibility distribution π, i.e., a mapping from a universe of discourse Ω to the unit interval $[0, 1]$. Possibility degrees $\pi(w)$ estimate to what extent the solution w is not unsatisfactory. π-pref nets are based on possibilistic networks [3], using conditional possibilities of the form $\pi(x|u) = \frac{\Pi(x \wedge u)}{\Pi(u)}$, for $u \in \text{Dom}(\mathcal{U}_X)$, where $\Pi(\varphi) = \max_{w \models \varphi} \pi(w)$. The use of product-based conditioning rather than min-based conditioning leads to possibilistic nets that are more similar to Bayesian nets.

The ceteris paribus assumption of CP-nets is replaced in possibilistic networks by a chain rule like in Bayesian networks. It enables one to compute, using an aggregation function, the degree of possibility of solutions. However it is supposed that these numerical values are unknown and represented by symbolic weights. Only ordering between symbolic values or products thereof can be expressed. The dominance relation between solutions is obtained by comparing products of symbolic utility values computed for them from the conditional preference tables.

Definition 6 *([1]). A Boolean possibilistic preference network (π-pref net) is a preference network, where $|Dom(X)| = 2, \forall X \in \mathcal{V}$, and each preference statement $x >_u^X x'$ is associated to a conditional possibility distribution such that $\pi(x|u) = 1 > \pi(x'|u)) = \alpha_X^u$, and α_X^u is a non-instantiated variable on $[0, 1)$ we call a* symbolic weight. *One may also have indifference statements $x \sim_u^X x'$, expressed by $\pi(x|u) = \pi(x'|u) = 1$.*

π-pref nets induce a partial ordering between solutions based on the comparison of their degrees of possibility in the sense of a joint possibility distribution computed using the product-based chain rule: $\pi(x_i, \ldots, x_n) = \prod_{i=1,\ldots,n} \pi(x_i|u_i)$. The preferences between solutions are of the form $w \succ_\pi w'$ if and only if $\pi(w) > \pi(w')$ *for all instantiations of the symbolic weights.*

5.1 π-pref Nets Vs CP-Nets

Let us compare preference relations between solutions induced by both CP-nets and π-pref nets. It has been shown [2] that the ordering between solutions induced by a π-pref net corresponds to the Pareto ordering between the vectors $w = (\theta_1(w), \ldots, \theta_n(w))$ where $\theta_i(w) = \pi(w(X_i)|w(\mathcal{U}_{X_i})), i = 1, \ldots, n$.

As symbolic weights are not comparable across variables, it is easy to see that the only way to have $\pi(w) \geq \pi(w')$ is to have $\theta_k(w) \geq \theta_k(w')$ in each component k of w and w'. Otherwise the products will be incomparable due to the presence of distinct symbolic variables on each side. So, if $w \neq w'$,

$$w \succ_\pi w' \text{ if and only if } \theta_k(w) \geq \theta_k(w'), k = 1, \ldots, n \text{ and } \exists i : \theta_i(w) > \theta_i(w').$$

It is then known that the π-pref net ordering between solutions induced by the preference tables is refined by comparing the sets F_w of bad variables for w:

$$w \succ_\pi w' \Rightarrow F_w \subset F_{w'}$$

since if two solutions contain variables having bad assignments in the sense of the preference tables, the corresponding symbolic values may differ if the contexts for assigning a value to this variable differ. It has been shown that if the weights α_X^u reflecting the satisfaction level due to assigning the bad value to X_i in the context u_i do not depend on this context, then we have an equivalence in the above implication:

$$\text{If } \forall X \in \mathcal{V}, \alpha_X^u = \alpha_X, \forall u_i \in \text{Dom}(\mathcal{U}_X), \text{ then } w \succ_\pi w' \iff w >_{par}^\Sigma w'.$$

As a consequence, using Corollary 1, it is clear that $w \succ_\pi w'$ implies $w >_{cp}^\Sigma w'$ so that the CP-net preference ordering refines the one induced by the corresponding Boolean π-pref net. It suggests that we can try to add ceteris paribus constraints to a π-pref net and so as to capture the preferences expressed by a CP-net.

In the following, we highlight local constraints between each node and its children that enable ceteris paribus to be simulated. Ceteris paribus constraints are of the form $w >_{cp}^{\Sigma} w'$ where w and w' differ by one flip. For each such statement (one per variable), we add the constraint on possibility degrees $\pi(w) > \pi(w')$. Using the chain rule, it corresponds to comparing products of symbolic weights. Let $\text{Dom}(\mathcal{U}_X) = \times_{X_i \in \mathcal{U}_X} \text{Dom}(X_i)$ denote the Cartesian product of domains of variables in $\mathcal{U}_X$, $\alpha_X^u = \pi(x^-|u)$, where x^- is bad for X and $\gamma_Y^{u'} = \pi(y^-|u')$. Suppose a CP-net and a π-pref net built from the same preference statements. It has been shown in [2] that the worsening flip constraints are all induced by the conditions: $\forall \ X \in \mathcal{V}$ s.t. X has children $\mathcal{C}h(X) \neq \emptyset$:

$$\max_{u \in \text{Dom}(\mathcal{U}_X)} \alpha_X^u < \prod_{Y \in \mathcal{C}h(X)} \min_{u' \in \text{Dom}(\mathcal{U}_Y)} \gamma_Y^{u'}$$

Let $\succ_{\pi}^{+}$ be the resulting preference ordering built from the preference tables and applying constraints of the above format between symbolic weights, then, it is clear that $w \succ_{cp} w' \Rightarrow w \succ_{\pi}^{+} w'$: relation $\succ_{\pi}^{+}$ is a bracketing from above of the CP-net ordering.

5.2   Relation $\succ_{\pi}^{+}$ as a Polynomial Upper Bound for CP-Net Dominance

In this section we give a characterisation of the relation $\succ_{\pi}^{+}$ in terms of deduction of linear constraints, which implies that determining dominance with respect to $\succ_{\pi}^{+}$ is polynomial. It is thus a polynomial upper bound for CP-net dominance.

We list all the different symbolic weights (not including 1) as $\alpha_1, \ldots, \alpha_m$, and let α represent the whole vector of symbolic weights $[\alpha_1, \ldots, \alpha_m]$.

Let a *weights vector* z be a vector of m real numbers $[z_1, \ldots, z_m]$ (with each z_i in $\{-1, 0, 1\}$). For each such weights vector z, we associate the product $\alpha_1^{z_1} \cdots \alpha_m^{z_m}$, which we abbreviate as $R_\alpha[z]$.

A comparison between products of symbolic weights can be encoded as a statement $R_\alpha[z] > 1$. For example, a comparison $\alpha_1 > \alpha_2 \alpha_3$ is equivalent to $R_\alpha[z] > 1$ where $z = [1, -1, -1, 0, 0, \ldots]$, since $R_\alpha[z] = \alpha_1^1 \alpha_2^{-1} \alpha_3^{-1}$ and so $R_\alpha[z] > 1 \iff \alpha_1^1 \alpha_2^{-1} \alpha_3^{-1} > 1 \iff \alpha_1 > \alpha_2 \alpha_3$. In this way, every ceteris paribus statement corresponds to a set of statements $R_\alpha[z] > 1$ for different vectors z.

For each $i = 1, \ldots, m$, define the vector $z^{(i)}$ as $z_i^{(i)} = -1$ and for all $j \neq i$, $z_j^{(i)} = 0$. $R_\alpha[z^{(i)}] > 1$ expresses that $\alpha_i^{-1} > 1$, i.e., $\alpha_i < 1$. For a CP-net Σ let $\check{Z}(\Sigma)$ be the set of weights vectors associated with symbolic weights comparisons for each ceteris paribus statement, plus for each $i = 1, \ldots, m$, the element $z^{(i)}$.

Similarly, every solution is associated with a product of symbolic weights, so a comparison $w > w'$ between solutions w and w' corresponds to a statement pertaining to a weights vector z'. The definitions lead easily to the following characterisation of this form of dominance.

Proposition 5. *Consider any CP-net Σ with associated set of weights vectors $Z(\Sigma)$, and let w and w' be two solutions, where the comparison $w > w'$ has associated vector z'. We have that $w \succ^+_\pi w'$ if and only if $\{R_\alpha[z] > 1 : z \in Z(\Sigma)\}$ implies $R_\alpha[z'] > 1$, i.e., if one replaces the values of symbolic weights α_i by any real values such that $R_\alpha[z] > 1$ holds for each $z \in Z(\Sigma)$ then $R_\alpha[z'] > 1$ also holds.*

We can write $\log R_\alpha[z]$ as $z_1\lambda_1 + \cdots + z_m\lambda_m = z \cdot \lambda$, where λ is the vector $(\lambda_1, \ldots, \lambda_m)$ and each $\lambda_i = \log \alpha_i$. Thus, $R_\alpha[z] > 1 \iff \log R_\alpha[z] > 0 \iff z \cdot \lambda > 0$. By Proposition 5 this implies that $w \succ^+_\pi w'$ if and only if for vectors λ, $\{z \cdot \lambda > 0 : z \in Z(\Sigma)\}$ implies $z' \cdot \lambda > 0$.

Using a standard result from convex sets, this leads to the following result, which gives a somewhat simpler characterisation that shows that dominance is polynomial. It also suggests potential links with Generalized Additive Iindependent (GAI) value function approximations of CP-nets [4,7].

Theorem 2. *Consider any CP-net with associated set of weights vectors $Z(\Sigma)$, and let w and w' be two different solutions, where $w > w'$ has associated vector z'. We have that $w \succ^+_\pi w'$ if and only if there exist non-negative real numbers r_z for each $z \in Z(\Sigma)$ such that $\sum_{z \in Z(\Sigma)} r_z z = z'$. Hence, whether or not $w \succ^+_\pi w'$ holds can be checked in polynomial time.*

Proof. As argued above, $w \succ^+_\pi w'$ holds if and only if for vectors λ, the set of inequalities $\{z \cdot \lambda > 0 : z \in Z(\Sigma)\}$ implies $z' \cdot \lambda > 0$. We need to show that this holds if and only if there exist non-negative real numbers r_z for each $z \in Z(\Sigma)$ such that $\sum_{z \in Z(\Sigma)} r_z z = z'$. Firstly, let us assume that there exist non-negative real numbers r_z for each $z \in Z(\Sigma)$ such that $\sum_{z \in Z(\Sigma)} r_z z = z'$. Consider any vector λ such that $z \cdot \lambda > 0$ for all $z \in Z(\Sigma)$. Then $z' \cdot \lambda = \sum_{z \in Z(\Sigma)} r_z z \cdot \lambda$ which is greater than zero since each r_z is non-negative, and at least some $r_z > 0$ (else z' is the zero vector, which would contradict $w \neq w'$).

Conversely, let us assume that there do not exist non-negative real numbers r_z for each $z \in Z(\Sigma)$ such that $\sum_{z \in Z(\Sigma)} r_z z = z'$. To prove that the set of inequalities $\{z \cdot \lambda > 0 : z \in Z(\Sigma)\}$ does not imply $z' \cdot \lambda > 0$, we will show that there exists a vector λ with $z \cdot \lambda > 0$ for all $z \in Z(\Sigma)$ but $z' \cdot \lambda \leq 0$. Let C be the set of vectors of the form $\sum_{z \in Z(\Sigma)} r_z z$ over all choices of non-negative reals r_z. Now, C is a convex and closed set, which by the hypothesis does not intersect with $\{z'\}$ (i.e., does not contain z'). Since $\{z'\}$ is closed and compact we can use a hyperplane separation theorem to show that there exists a vector λ and real numbers $c_1 < c_2$ such that for all $x \in C$, $x \cdot \lambda > c_2$ and $z' \cdot \lambda < c_1$. Because C is closed under strictly positive scalar multiplication (i.e., $x \in C$ implies $rx \in C$ for all real $r > 0$) we must have $c_2 \leq 0$, and $x \cdot \lambda \geq 0$ for all $x \in C$, and in particular $z \cdot \lambda \geq 0$ for all $z \in Z(\Sigma)$. Also, $z' \cdot \lambda < c_1 < c_2 \leq 0$ so $z' \cdot \lambda \leq 0$, as required.

The last part follows since linear programming is polynomial. $\square$

6   Summary and Discussion

In this paper we have compared CP-nets and π-pref nets, two qualitative counterparts of Bayes nets for the representation of conditional preferences. We have studied them from the point of view of their rationality, namely whether they respect Pareto dominance between multiple Boolean variable solutions to a decision problem expressed by such graphical models. While π-pref nets naturally respect this property, strangely enough, it was previously unknown whether the preference ordering induced by CP-nets respects it or not. For more general (non-Boolean) variables, it seems difficult to extend this notion of Pareto-dominance for a CP-net in an entirely natural way. Besides, it was shown previously that the ordering induced by π-pref nets is weaker than the one induced by CP-nets, but ceteris paribus constraints can be added to a π-pref net in the form of constraints between products of symbolic variables. Here we show the polynomial nature of this encoding. Thus we get a bracketing of the CP-net preference ordering by bounds which are apparently easier to compute than standard CP-net preferences. Further research includes constructing an example that explicitly proves that the upper approximation of the CP-net ordering is not tight; moreover the case of non-Boolean variables deserves further investigation.

Acknowledgements. This material is based upon works supported by the Science Foundation Ireland under Grants No. 12/RC/2289 and No. 12/RC/2289-P2 which are co-funded under the European Regional Development Fund.

References

1. Ben Amor, N., Dubois, D., Gouider, H., Prade, H.: Possibilistic preference networks. Inf. Sci. **460–461**, 401–415 (2018)
2. Ben Amor, N., Dubois, D., Gouider, H., Prade, H.: Expressivity of possibilistic preference networks with constraints. In: Moral, S., Pivert, O., Sánchez, D., Marín, N. (eds.) SUM 2017. LNCS (LNAI), vol. 10564, pp. 163–177. Springer, Cham (2017). https://doi.org/10.1007/978-3-319-67582-4_12
3. Benferhat, S., Dubois, D., Garcia, L., Prade, H.: On the transformation between possibilistic logic bases and possibilistic causal networks. Int. J. Approx. Reasoning **29**(2), 135–173 (2002)
4. Boutilier, C., Bacchus, F., Brafman, R.I.: UCP-networks: a directed graphical representation of conditional utilities. In: Proceedings of the 17th Conference on Uncertainty in AI, Seattle, Washington, USA, pp. 56–64 (2001)
5. Boutilier, C., Brafman, R.I., Hoos, H.H., Poole, D.: Reasoning with conditional ceteris paribus preference statements. In: Proceedings of the 15th Conference on Uncertainty in AI, Stockholm, Sweden, pp. 71–80 (1999)
6. Boutilier, C., Brafman, R.I., Domshlak, C., Hoos, H.H., Poole, D.: CP-nets: a tool for representing and reasoning with conditional ceteris paribus preference statements. J. Artif. Intell. Res. **21**, 135–191 (2004)
7. Brafman, R.I., Domshlak, C., Kogan, T.: Compact value-function representations for qualitative preferences. In: Proceedings of the 20th Conference on Uncertainty in AI, Banff, Canada, pp. 51–59 (2004)

8. Dubois, D., Prade, H.: Possibility Theory: An Approach to ComputerizedProcessing of Uncertainty. Plenum Press (1988)
9. Dubois, D., Prade, H.: Possibility theory as a basis for preference propagation in automated reasoning. In: Proceedings of the 1st IEEE International Conference on Fuzzy Systems, San Diego, CA, pp. 821–832 (1992)

Measuring Inconsistency Through Subformula Forgetting

Yakoub Salhi[✉]

CRIL - CNRS & Université d'Artois, Lens, France
salhi@cril.fr

Abstract. In this paper, we introduce a new approach for defining inconsistency measures. The key idea consists in forgetting subformula occurrences in order to restore consistency. Thus, our approach can be seen as a generalization of the approach based on forgetting only propositional variables. We here introduce rationality postulates of inconsistency measuring that take into account in a syntactic way the internal structure of the formulas. We also describe different inconsistency measures that are based on forgetting subformula occurrences.

1 Introduction

In this work, we are interested in quantifying conflicts for better analyzing the nature of the inconsistency in a knowledge base. Plenty of proposals for inconsistency measures have been defined in the literature (e.g. see [3,7,9,14,15]), and it has been shown that they can be applied in different domains, such as e-commerce protocols [4], integrity constraints [6], databases [13], multi-agent systems [10], spatio-temporal qualitative reasoning [5].

In the literature, an inconsistency measure is defined as a function that associates a non negative value to each knowledge base. In particular, the authors in [9] have proposed different rationality postulates for defining inconsistency measures that allow capturing important aspects related to inconsistency in the case of classical propositional logic. Furthermore, objections to some of them and many new postulates have also been proposed in [1]. The main advantage of the approach based on rationality postulates for defining inconsistency measures is its flexibility in the sense that the appropriate measure in a given context can be chosen through the desired properties from the existing postulates.

In [11,12], the authors have proposed a general framework for reasoning under inconsistency by forgetting propositional variables to restore consistency. Using the variable forgetting approach of this framework, an inconsistency measure has been proposed in [2]. The main idea consists in quantifying the amount of inconsistency as the minimum number of variable occurrences that have to be forgotten to restore consistency. We here propose a new approach for defining inconsistency measures that can be seen as a generalization of the previous approach. Indeed, our main idea consists in measuring the amount of inconsistency by considering sets of subformula occurrences that we need to forget to restore

N. Ben Amor et al. (Eds.): SUM 2019, LNAI 11940, pp. 184–191, 2019.
https://doi.org/10.1007/978-3-030-35514-2_14

consistency. To the best of our knowledge, we here provide the first approach that takes into account in a syntactic way the internal structure of the formulas.

In this work, we propose rationality postulates for measuring inconsistency that are based on reasoning about subformula occurrences. In particular, the postulate stating that forgetting any subformula occurrence does not increase the amount of inconsistency. Finally, we propose several inconsistency measures that are based on forgetting subformula occurrences. These measures are defined by considering the number of modified formulas and the size of the forgotten subformula occurrences to restore consistency. For instance, one of the proposed inconsistency measure quantifies the amount of inconsistency as the minimum size of the subformula occurrences that have to be forgotten to obtain consistency. It is worth mentioning that we show that two of the described inconsistency measures correspond to two measures existing in the literature: that introduced in [2] based on forgetting variables and that introduced in [7] based on consistent subsets.

2 Preliminaries

2.1 Classical Propositional Logic

We here consider that every piece of information is represented using classical propositional logic. We use Prop to denote the set of propositional variables. The set of propositional formulas is denoted Form. We use the letters p, q, r, s to denote the propositional variables, and the Greek letters ϕ, ψ and χ to denote the propositional formulas. Moreover, given a syntactic object o, we use $\mathcal{P}(o)$ to denote the set of propositional variables occurring in o. Given a set of variables S such that $\mathcal{P}(\phi) \subseteq S$, we use $Mod(\phi, S)$ to denote the set of all the models of ϕ defined over S.

Given a formula ϕ, the *size of a formula* ϕ, denoted $s(\phi)$, is inductively defined as follows: $s(p) = s(\bot) = s(\top) = 1$; $s(\neg\psi) = 1 + s(\psi)$; $s(\psi \otimes \chi) = 1 + s(\psi) + s(\chi)$ for $\otimes = \wedge, \vee, \rightarrow$. In other words, the size of a formula is defined as the number of the occurrences of propositional variables, constants and logical connectives that appear in it.

Similarly, the set of the subformulas of ϕ, denoted $SF(\phi)$, is inductively defined as follows: $SF(p) = \{p\}$; $SF(\bot) = \{\bot\}$; $SF(\top) = \{\top\}$; $SF(\neg\psi) = \{\neg\psi\} \cup SF(\psi)$; $SF(\psi \otimes \chi) = \{\psi \otimes \chi\} \cup SF(\psi) \cup SF(\chi)$ for $\otimes = \wedge, \vee, \rightarrow$.

Given a formula ϕ and $\psi \in SF(\phi)$, we use $O(\phi, \psi)$ to denote the number of the occurrences of ψ in ϕ. Moreover, we consider that *the occurrences of a subformula are ordered starting from the left*. For example, consider the formula $\phi = (p \wedge q) \rightarrow (\neg r \vee q)$. Then, $SF(\phi) = \{\phi, p \wedge q, \neg r \vee q, \neg r, p, q, r\}$. Further, $O(\phi, p) = 1$ and $O(\phi, q) = 2$. The first occurrence of q is that occurring in the subformula $p \wedge q$, while the second is that occurring in the subformula $\neg r \vee q$.

The *polarity* of a subformula occurrence within a formula that has a polarity (*positive* or *negative*) is defined as follows:

- ϕ is a positive (resp. negative) subformula occurrence of the positive (resp. negative) formula ϕ;

- if χ is a positive (resp. negative) subformula occurrence of ϕ, then χ is also a positive (resp. negative) subformula occurrence of $\phi \otimes \psi$, $\psi \otimes \phi$, $\psi \to \phi$ for every formula ψ and for $\otimes = \wedge, \vee$;
- if χ is a positive (resp. negative) subformula occurrence of ϕ, then χ is a negative (resp. positive) subformula occurrence of $\neg\phi$ and $\phi \to \psi$ for every formula ψ.

Consider, for instance, the formula $p \to (p \vee q)$ with the negative polarity. Then, the left-hand p is a positive subformula occurrence and the right-hand occurrence is negative.

A *knowledge base* is a finite set of propositional formulas. A knowledge base K is inconsistent if its associated formula $\bigwedge_{\phi \in K} \phi$ ($\top$ if $K = \emptyset$) is inconsistent, written $K \vdash \bot$, otherwise it is consistent, written $K \nvdash \bot$. We use $\mathcal{K}_{\mathsf{Form}}$ to denote the set of knowledge bases. Moreover, we use $SF(K)$ to denote the set $\bigcup_{\phi \in K} SF(\phi)$.

From now on, *we consider that the polarity of the formulas occurring in any knowledge base are negative*, the same results can be obtained by symmetrically considering the positive polarity.

Given a knowledge base K, a subset $K' \subseteq K$ is said to be a *minimal inconsistent subset* (MIS) of K if (i) $K' \vdash \bot$ and (ii) $\forall \phi \in K'$, $K' \setminus \{\phi\} \nvdash \bot$. Moreover, K' is said to be a *maximal consistent subset* (MCS) of K if (i) $K' \nvdash \bot$ and (ii) $\forall \phi \in K \setminus K'$, $K' \cup \{\phi\} \vdash \bot$. We use $MISes(K)$ and $MCSes(K)$ to denote respectively the set of all the MISes and the set of all the MCSes of K.

2.2   Substitution

Given a formula ϕ and two subformula occurrences ψ and χ in ϕ. We say that ψ and χ are *disjoint* if one does not occur in the other.

Given two propositional formulas ϕ and ψ, $\chi \in SF(\phi)$ and $i \in 1..O(\phi, \chi)$, we use $\phi[(\chi, i)/\psi]$ to denote the result of substituting the formula ψ for the ith occurrence of χ in ϕ. Further, we use $\phi[\chi/\psi]$, $\phi[(\chi)^+/\psi]$ and $\phi[(\chi)^-/\psi]$ to denote the result of substituting the formula ψ for respectively all the occurrences of χ, all the positive occurrences of χ and all the negative occurrences of χ in ϕ. Similarly, given the formulas $\phi, \psi_1, \ldots, \psi_l, \chi_1, \ldots \chi_l$ and the expressions $e_1, \ldots, e_l$ such that each e_i has one of the forms (χ_i, j), χ_i, $(\chi_i)^+$ and $(\chi_i)^-$, $\phi[e_1, \ldots, e_l/\psi_1, \ldots, \psi_l]$ is the result of *simultaneously* substituting $\psi_1, \ldots, \psi_l$ for the subformula occurrences corresponding to the expressions $e_1, \ldots, e_l$ respectively. It is worth mentioning that the subformula occurrences corresponding to the expressions $e_1, \ldots, e_l$ should be pairwise disjoint in ϕ.

For instance, consider the formula $\phi = (p \wedge q) \to (p \vee q)$ with the negative polarity. Then, $\phi[(p)^+, (q, 2)/(p \wedge \neg q), r]$ corresponds to the formula $((p \wedge \neg q) \wedge q) \to (p \vee r)$. Indeed, there is a unique positive occurrence of p which is on the left-hand side of the implication and it is replaced with $p \wedge \neg q$; and the second occurrence of q is on the right-hand side of the implication and it is replaced with r.

3   Inconsistency Measure

In the literature, an inconsistency measure is defined as a function that associates
a non negative value to each knowledge base (e.g. [3,7,9,14,15]). It is used to
quantify the amount of inconsistency in a knowledge base. The different works
on inconsistency measures use postulate-based approaches to capture important
aspects related to inconsistency. In particular, in the recent work [3], the authors
have proposed the following formal definition of inconsistency measure that we
consider in this work.

Definition 1 (Inconsistency Measure).   *An* inconsistency measure *is a
function* $I : \mathcal{K}_{\mathsf{Form}} \rightarrow \mathbb{R}^{+}_{\infty}$ *that satisfies the two following properties: (i)* $\forall K \in \mathcal{K}_{\mathsf{Form}}$, $I(K) = 0$ *iff* K *is consistent (Consistency); and (ii)* $\forall K, K' \in \mathcal{K}_{\mathsf{Form}}$, *if* $K \subseteq K'$ *then* $I(K) \leqslant I(K')$ *(Monotonicity). The set* $\mathbb{R}^{+}_{\infty}$ *corresponds to the set of positive real numbers augmented with a greatest element denoted* ∞.

The postulate (*Consistency*) means that an inconsistency measure must
allow distinguishing between consistent and inconsistent knowledge bases, and
(*Monotonicity*) means that the amount of inconsistency does not decrease by
adding new formulas to a knowledge base. Many other postulates have been
introduced in the literature to characterize particular aspects related to inconsistency (e.g. see [1,9,15]).

Let us now describe some simple inconsistency measures from the literature:

- $I_M(K) = |MISes(K)|$ ([8])
- $I_d^{hit}(K) = |K| - max\{|K'| \mid K' \in MCSes(K)\}$ ([7])
- $I_{hs}(K) = min\{|S| \mid S \subseteq M \text{ and } \forall \phi \in K, \exists \mathcal{B} \in S \text{ s.t. } \mathcal{B} \models \phi\} - 1$ with $M = \bigcup_{\phi \in K} Mod(\phi, \mathcal{P}(K))$ and $min\{\} = \infty$ ([14])
- $I_{forget}(K) = min\{n \mid \bigwedge_{\phi \in K} \phi[(p_1, i_1), \dots (p_n, i_n) / C_1, \dots, C_n], p_1, \dots, p_n \in$ Prop, $C_1, \dots, C_n \in \{\top, \bot\}\}$ ([2])

The measure I_M quantifies the amount of inconsistency through minimal inconsistent subsets: more MISes brings more conflicts; I_d^{hit} consider the dual of the
size of the greatest MCSes; I_{hs} is defined through an explicit use of the Boolean
semantics: the amount of inconsistency is related to the minimum number of
models that satisfy all the formulas in the considered knowledge base; and I_{forget}
defines the amount of inconsistency as the minimum number of variables that
we have to forget to restore consistency. It is worth mentioning that we consider
here the reformulation of I_{forget} proposed in [15].

4   Subformula-Based Rationality Postulates

In this section, we propose rationality postulates for measuring inconsistency
that are based on reasoning about forgetting subformula occurrences. In the same
way as in the case of the inconsistency measure I_{forget}, we use the constants $\top$
and $\bot$ to forget subformula occurrences.

The rationality postulates that we consider are defined as follows $\forall K \in \mathcal{K}_{\mathsf{Form}}$
and $\forall \phi \in \mathsf{Form}$ with $\phi \notin K$:

- $(ForgetNegOcc)$:
  1. $\forall \psi \in SF(\phi)$ and $\forall i \in 1..O(\phi, \psi)$ with the ith occurrence of ψ in ϕ is negative, $I(K \cup \{\phi[(\psi, i)/\top]\}) \leqslant I(K \cup \phi)$;
  2. $\forall \psi \in SF(\phi)$ and $\forall i \in 1..O(\phi, \psi)$ with the ith occurrence of ψ in ϕ is negative and $\phi[(\psi, i)/\bot] \notin K$, $I(K \cup \phi) \leqslant I(K \cup \{\phi[(\psi, i)/\bot]\})$.
- $(ForgetPosOcc)$:
  1. $\forall \psi \in SF(\phi)$ and $\forall i \in 1..O(\phi, \psi)$ with the ith occurrence of ψ in ϕ is positive and $\phi[(\psi, i)/\top] \notin K$, $I(K \cup \phi) \leqslant I(K \cup \{\phi[(\psi, i)/\top]\})$;
  2. $(ForgetPosOcc_\bot)$ $\forall \psi \in SF(\phi)$ and $\forall i \in 1..O(\phi, \psi)$ with the ith occurrence of ψ in ϕ is positive, $I(K \cup \{\phi[(\psi, i)/\bot]\}) \leqslant I(K \cup \phi)$.

The first property of $(ForgetNegOcc)$ expresses the fact that a negative subformula occurrence becomes useless to produce inconsistency if it is replaced with $\top$. Regarding the second property, it is worth mentioning that the condition $\phi[(\psi, i)/\bot] \notin K$ is only used to prevent formula deletion. The postulate $(ForgetPosOcc)$ is simply the counterpart in the case of positive subformula occurrences of $(ForgetNegOcc)$.

In a sense, the next proposition shows that the previous postulates can be seen as restrictions of the postulate $(Dominance)$, introduced in [9], in the case of consistent formulas. Let us recall that $(Dominance)$ is defined as follows:

- $\forall K \in \mathcal{K}_{\mathsf{Form}}$ and $\forall \phi, \psi \in \mathsf{Form}$ with $\phi \nvdash \bot$ and $\phi \vdash \psi$, $I(K \cup \{\phi\}) \geqslant I(K \cup \{\psi\})$.

Proposition 1. *The following two properties are satisfied for $\forall \phi \in \mathsf{Form}$ with negative polarity and $\forall \psi \in SF(\phi)$ and $\forall i \in 1..O(\phi, \psi)$: (i) if the ith occurrence of ψ in ϕ is negative, then $\phi[(\psi, i)/\bot] \vdash \phi$ and $\phi \vdash \phi[(\psi, i)/\top]$; (ii) if the ith occurrence of ψ in ϕ is positive, then $\phi[(\psi, i)/\top] \vdash \phi$ and $\phi \vdash \phi[(\psi, i)/\bot]$.*

Proof. We here consider only the case of $\phi[(\psi, i)/\bot] \vdash \phi$ when the considered occurrence is negative and the case of $\phi \vdash \phi[(\psi, i)/\bot]$ when the considered occurrence is positive, the other case being similar. The proof is by mutual induction on the value of $s(\phi)$. If $s(\phi) = 1$, then ϕ is a propositional variable or a constant, and as a consequence, $\phi[(\psi, i)/\bot] = \bot$ holds in the case where the ith occurrence of ψ in ϕ is negative. Thus, we obtain $\phi[(\psi, i)/\bot] = \bot \vdash \phi$. Moreover, there is no positive subformula occurrence in this case. Assume now that $s(\phi) > 1$. Then, ϕ has one of the following forms $\neg \phi'$ $\phi_1 \wedge \phi_2$, $\phi_1 \vee \phi_2$ and $\phi_1 \rightarrow \phi_2$. Consider first the case $\phi = \neg \phi'$ the proof is trivial in the case $\psi = \phi$. If the ith occurrence of ψ in ϕ is negative, then it is positive in ϕ', and using the induction hypothesis, $\phi' \vdash \phi'[(\psi, i)/\bot]$ holds. Thus, we obtain $\neg \phi'[(\psi, i)/\bot] = \phi[(\psi, i)/\bot] \vdash \neg \phi' = \phi$. The case where the ith occurrence of ψ in ϕ is positive is similar. The proof in the remaining cases can be obtained by simple application of the induction hypothesis, except the case $\phi_1 \rightarrow \phi_2$, which is similar to that of $\neg \phi'$.

For instance, a direct consequence of Proposition 1 is the fact that I_{hs} satisfies $(ForgetNegOcc)$ and $(ForgetPosOcc)$. However, I_M does not satisfy these postulates. Indeed, consider $K = \{p \wedge \neg p, p \wedge q, p \wedge r\}$. We clearly have $I_M(K) = 1$ since there is a single MIS, which is $\{p \wedge \neg p\}$, but $I_M(\{\top \wedge \neg p, p \wedge q, p \wedge r\}) = 2$ since there are two MISes $\{\top \wedge \neg p, p \wedge q\}$ and $\{\top \wedge \neg p, p \wedge r\}$.

We now introduce a rationality postulate, named $(ForgetSubformula)$, that is based on forgetting all the occurrences of a subformula. Before that, let us introduce a notational convention. Given a knowledge base K and a subformula $\psi \in SF(\phi)$ with $\phi \in K$, $K[\psi \downarrow]$ denotes $\bigcup_{\phi \in K} \phi[(\psi)^-, (\psi)^+/\top, \bot]$. In other words, $K[\psi \downarrow]$ is used to denote that all the occurrences of ψ are forgotten to restore consistency.

The postulate $(ForgetSubformula)$ is defined as follows: $\forall K \in \mathcal{K}_{\mathsf{Form}}$ and $\forall \psi \in SF(K), I(K[\psi \downarrow]) \leqslant I(K)$. It is clearly weaker than the previous postulates and expresses simply that the amount of inconsistency does not decrease by forgetting any subformula. This postulate can be used instead of $(ForgetNegOcc)$ and $(ForgetPosOcc)$ in the case where no distinction is made between the occurrences of any subformula.

5   Forgetting Based Inconsistency Measures

In this section, we define several inconsistency measures that are based on forgetting subformula occurrences. We show in particular that two of these measures correspond to I_d^{hit} and I_{forget} described previously.

The first inconsistency measure, denoted $I_{osf}^{\#}$, is defined as the minimum number of subformula occurrences that have to be forgotten to restore consistency. It is formally defined as follows:

$$I_{osf}^{\#}(\{\phi_1, \ldots, \phi_n\}) = min\{\textstyle\sum_{i=1}^{n} l_i \mid \{\phi_1[(\psi_1^1, j_1^1), \ldots, (\psi_{l_1}^1, j_{l_1}^1)/ C_1^1, \ldots, C_{l_1}^1]\} \cup$$
$$\cdots \cup \{\phi_n[(\psi_1^n, j_1^n), \ldots, (\psi_{l_n}^n, j_{l_n}^n)/C_1^n, \ldots, C_{l_n}^n]\} \nvdash \bot \text{ with } C_i^j \in \{\top, \bot\}\}.$$

The second inconsistency measure, denoted I_{osf}^{s}, is defined in the same way as $I_{osf}^{\#}$, but it takes into account the sizes of the forgotten subformula occurrences: $I_{osf}^{s}(\{\phi_1, \ldots, \phi_n\}) = min\{\sum_{i=1}^{n} \sum_{k=1}^{l_i} s(\psi_k^i) \mid \{\phi_1[(\psi_1^1, j_1^1), \ldots, (\psi_{l_1}^1, j_{l_1}^1)/ C_1^1, \ldots, C_{l_1}^1]\} \cup \cdots \cup \{\phi_n[(\psi_1^n, j_1^n), \ldots, (\psi_{l_n}^n, j_{l_n}^n)/C_1^n, \ldots, C_{l_n}^n]\} \nvdash \bot \text{ with } C_i^j \in \{\top, \bot\}\}$. The measure I_{osf}^{s} relates the effort needed to restore consistency to the size of the considered subformula occurrences instead of their number as in $I_{osf}^{\#}$.

The third inconsistency measure, denoted $I_{osf}^{s,1}$, takes also into account the sizes of the forgotten subformula occurrences, with the additional requirement that there is at most one forgotten occurrence in every formula in the knowledge base: $I_{osf}^{s,1}(\{\phi_1, \ldots, \phi_n\}) = min\{\sum_{i=1}^{n} s(\psi_i) \mid \{\phi_1[(\psi_1, j_1)/C_1]\} \cup \cdots \cup \{\phi_n[(\psi_n, j_n)/C_n]\} \nvdash \bot \text{ with } C_1, \ldots, C_n \in \{\top, \bot\}\}$. The measure $I_{osf}^{s,1}$ captures the fact that if we need to forget two disjoint subformula occurrences ψ and ψ' in the same formula ϕ to restore consistency, then we have to forget the smallest subformula occurrence in ϕ containing both ψ and ψ'. This measure allows considering the relationship between occurrences forgotten in the same piece of information.

For the sake of illustration, consider the base $K = \{p \wedge q, \neg p \wedge \neg q\}$. Then, we clearly have $I_{osf}^{s}(K) = 2$ since we only need to forget the first occurrences of p and q to restore consistency. However, $I_{osf}^{s,1}(K) = 3$ since we need to forget the entire formula $p \wedge q$ to forget its subformulas p and q. Compared to I_{osf}^{s} in this case, we also consider in $I_{osf}^{s,1}$ the fact that the first occurrences of p and q are related with conjunction.

The three following inconsistency measures can be seen as variants of the previous ones by considering subformulas instead of subformula occurrences. These measures can be used in the contexts where no distinction is made between the occurrences of a subformula with regard to the amount of inconsistency. For instance, the inconsistency measure denoted $I_{sf}^{\#}$ is defined as the minimum number of subformulas that have to be forgotten to restore consistency. Thus, forgetting any subformula once or more does not change the amount of inconsistency.

$$I_{sf}^{\#}(K) = min\{m \in \mathbb{N} \mid K[\psi_1 \downarrow] \cdots [\psi_m \downarrow] \not\vdash \bot\}$$

$$I_{sf}^{s}(K) = min\{\textstyle\sum_{i=1}^{m} s(\psi_i) \mid K[\psi_1 \downarrow] \cdots [\psi_m \downarrow] \not\vdash \bot$$
$$with\ s(\psi_1) = \ldots = s(\psi_m) = 1\}$$

$$I_{sf}^{s,1}(\{\phi_1, \ldots, \phi_n\}) = min\{\textstyle\sum_{\chi \in \bigcup_{i=1}^{n} \{\psi_i\}} s(\chi) \mid \{\phi_1[(\psi_1, j_1)/C_1]\} \cup \cdots \cup$$
$$\{\phi_n[(\psi_n, j_n)/C_n]\} \not\vdash \bot\ with\ C_1, \ldots, C_n \in \{\top, \bot\}\}$$

One can easily see that all the previous measures satisfy the two postulates (*Consistency*) and (*Monotonicity*), and as a consequence, they are inconsistency measures with respect to Definition 1. Further, from their definitions, it is clear that they also satisfy the rationality postulates (*ForgetNegOcc*) and (*ForgetPosOcc*).

In the following proposition, we have the fact that $I_{osf}^{\#}$ and I_d^{hit} are the same, and in addition $I_{osf}^{s}(K) = I_{forget}(K)$ for every constant free knowledge base K.

Proposition 2. *The following properties are satisfied:*

1. $I_{osf}^{\#}(K) = I_d^{hit}(K)$;
2. $I_{osf}^{s}(\{\phi_1, \ldots, \phi_n\}) = min\{\textstyle\sum_{i=1}^{n} l_i \mid \{\phi_1[(\psi_1^1, j_1^1), \ldots, (\psi_{l_1}^1, j_{l_1}^1)/C_1^1, \ldots, C_{l_1}^1]\}$
   $\cup \cdots \cup \{\phi_n[(\psi_1^n, j_1^n), \ldots, (\psi_{l_n}^n, j_{l_n}^n)/C_1^n, \ldots, C_{l_n}^n]\} \not\vdash \bot$
   with $C_1^1, \ldots, C_{l_n}^n \in \{\top, \bot\}$ *and* $s(\psi_1^1) = \ldots = s(\psi_{l_n}^n) = 1\}$.

6   Conclusion and Perspectives

We have proposed an approach for measuring inconsistency that takes into account in a syntactic way the internal structure of the formulas, which is based on forgetting subformula occurrences to restore consistency. As a future work, we intend to investigate the possibility to consider more rationality postulates that consider the internal structure in a syntactic way. The aim of such postulates is to capture other interesting links between inconsistency and the notion of subformula occurrence. We also plan to propose inconsistency measures that combine the subformula forgetting based approach with other syntactic approaches, such as those based on minimal inconsistent subsets.

References

1. Besnard, P.: Revisiting postulates for inconsistency measures. In: Fermé, E., Leite, J. (eds.) JELIA 2014. LNCS (LNAI), vol. 8761, pp. 383–396. Springer, Cham (2014). https://doi.org/10.1007/978-3-319-11558-0_27

2. Besnard, P.: Forgetting-based inconsistency measure. In: Schockaert, S., Senellart, P. (eds.) SUM 2016. LNCS (LNAI), vol. 9858, pp. 331–337. Springer, Cham (2016). https://doi.org/10.1007/978-3-319-45856-4_23

3. Bona, G.D., Grant, J., Hunter, A., Konieczny, S.: Towards a unified framework for syntactic inconsistency measures. In: Proceedings of the Thirty-Second AAAI Conference on Artificial Intelligence, New Orleans, Louisiana, USA (2018)

4. Chen, Q., Zhang, C., Zhang, S.: A verification model for electronic transaction protocols. In: Yu, J.X., Lin, X., Lu, H., Zhang, Y. (eds.) APWeb 2004. LNCS, vol. 3007, pp. 824–833. Springer, Heidelberg (2004). https://doi.org/10.1007/978-3-540-24655-8_90

5. Condotta, J., Raddaoui, B., Salhi, Y.: Quantifying conflicts for spatial and temporal information. In: Principles of Knowledge Representation and Reasoning: Proceedings of the Fifteenth International Conference, KR 2016, Cape Town, South Africa, 25–29 April 2016, pp. 443–452 (2016)

6. Grant, J., Hunter, A.: Measuring inconsistency in knowledgebases. J. Intell. Inf. Syst. **27**(2), 159–184 (2006)

7. Grant, J., Hunter, A.: Distance-based measures of inconsistency. In: van der Gaag, L.C. (ed.) ECSQARU 2013. LNCS (LNAI), vol. 7958, pp. 230–241. Springer, Heidelberg (2013). https://doi.org/10.1007/978-3-642-39091-3_20

8. Hunter, A., Konieczny, S.: Measuring inconsistency through minimal inconsistent sets. In: Principles of Knowledge Representation and Reasoning: Proceedings of the Eleventh International Conference, KR 2008, Sydney, Australia, 16–19 September 2008, pp. 358–366. AAAI Press (2008)

9. Hunter, A., Konieczny, S.: On the measure of conflicts: shapley inconsistency values. Artif. Intell. **174**(14), 1007–1026 (2010)

10. Hunter, A., Parsons, S., Wooldridge, M.: Measuring inconsistency in multi-agent systems. Kunstliche Intelligenz **28**, 169–178 (2014)

11. Lang, J., Marquis, P.: Resolving inconsistencies by variable forgetting. In: Proceedings of the Eights International Conference on Principles and Knowledge Representation and Reasoning (KR-02), Toulouse, France, 22–25 April 2002, pp. 239–250 (2002)

12. Lang, J., Marquis, P.: Reasoning under inconsistency: a forgetting-based approach. Artif. Intell. **174**(12–13), 799–823 (2010)

13. Martinez, M.V., Pugliese, A., Simari, G.I., Subrahmanian, V.S., Prade, H.: How dirty is your relational database? An axiomatic approach. In: Mellouli, K. (ed.) ECSQARU 2007. LNCS (LNAI), vol. 4724, pp. 103–114. Springer, Heidelberg (2007). https://doi.org/10.1007/978-3-540-75256-1_12

14. Thimm, M.: On the expressivity of inconsistency measures. Artif. Intell. **234**, 120–151 (2016)

15. Thimm, M.: On the evaluation of inconsistency measures. In: Grant, J., Martinez, M.V. (eds.) Measuring Inconsistency in Information, Volume 73 of Studies in Logic. College Publications, February 2018

Explaining Hierarchical Multi-linear Models

Christophe Labreuche$^{(\boxtimes)}$

Thales Research & Technology, Palaiseau, France
christophe.labreuche@thalesgroup.com

Abstract. We are interested in the explanation of the solution to a hierarchical multi-criteria decision aiding problem. We extend a previous approach in which the explanation amounts to identifying the most influential criteria in a decision. This is based on an influence index which extends the Shapley value on trees. The contribution of this paper is twofold. First, we show that the computation of the influence grows linearly and not exponentially with the depth of the tree for the multi-linear model. Secondly, we are interested in the case where the values of the alternatives are imprecise on the criteria. The influence indices become thus imprecise. An efficient computation approach is proposed for the multi-linear model.

1 Introduction

One of the major challenges of Artificial Intelligence (AI) methods is to explain their predictions and make them transparent for the user. The explanations can take very different forms depending on the area. For instance, in Computer Vision, one is interested in identifying the salient factors explaining the classification of an image [12]. In Machine Learning, one might look for the smallest modification to make on an instance to change its class (counter-factual example) [16]. In Constraint Programming, the aim is to find the simplest way to repair a set of inconsistent constraints [8]. And so on. There is thus a variety of explanation methods applicable to a wide range of AI methods.

Many decision problems involve multiple attributes to be taken into account. Multi-Criteria Decision Aiding (MCDA) aims at representing the preferences of a decision maker regarding options on the basis of multiple and conflicting criteria. In real applications, one shall use elaborate decision models able to capture complex expertise. A few models have been shown to have this ability, such as the Choquet integral [3], the multi-linear model [11] or the Generalized Additive Independence (GAI) model [1,6]. The main asset of these models is their ability to represent interacting criteria. The multi-linear model is especially important as it is the most natural multi-dimensional interpolation model. It is very smooth and does not have discontinuity of the Gradient that the Choquet integral has. The following example illustrates applications in which such models are important.

Example 1 (Example 1 in [9]). The DM is a Tactical Operator of an aircraft aiming at Maritime Patrol. It consists in monitoring a maritime area and in particular looking for illegal activity. The DM is helped by an automated system that evaluates in real time a Priority Level (PL) associated to each ship in this area. The higher the PL the

© Springer Nature Switzerland AG 2019
N. Ben Amor et al. (Eds.): SUM 2019, LNAI 11940, pp. 192–206, 2019.
https://doi.org/10.1007/978-3-030-35514-2_15

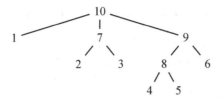

Fig. 1. Hierarchy of criteria for Example 2.

more suspicious a ship and the more urgent it is to intercept it. The PL is used to raise the attention of the DM on some specific ships. The computation of the PL depends on several criteria: 1. Incoherence between Automatic Identification System (AIS) data and radar detection; 2. Suspicion of drug smuggling on the ship; 3. Suspicion of human smuggling on the ship; 4. Current speed (since fast boats are often used to avoid being easily intercepted); 5. Maximum speed since the first detection of the ship (it represents the urgency for the potential interception); 6. Proximity of the ship to the shore (since smuggling ships often aim at reaching the shore as fast as possible). ■

In the previous example, as in most real-applications, the criteria are not considered in a flat way but are organized as a tree. The criteria are indeed organized hierarchically with several nested aggregation functions. The hierarchical structure shall represent the natural decomposition of the decision reasoning into points of view and sub-points of view. In the previous example, the six criteria are organized as in Fig. 1. The tree of the DM contains four aggregation nodes: 7. Suspicion of illegal activity; 8. Kinematics; 9. Capability to escape interception; 10. Overall PL.

The ability to explain the evaluation is very important in Example 2. If the PL of a ship suddenly increases over time, the tactical operator needs to understand where this comes from. This latter is under stress and time pressure. He is thus looking for an explanation highlighting the most influencing attributes in the evolution of the PL. This type of explanation has been recently widely studied under the name of *feature attribution*. The aim is to attribute to each feature its level of contribution. Among the many concepts that have been proposed, the Shapley value has been widely used in Machine Learning [4, 10].

The Shapley value has also been recently as an explanation means in MCDA [9]. In this reference, a new explanation approach for hierarchical MCDA models has been introduced. The idea is to highlight the criteria that contribute most to the decision. In Example 2, consider two ships represented by two alternatives x and y taking the following values on the six attributes $x = (x_1, x_2, x_3, x_4, x_5, x_6) = (+, -, -, -, -, -)$ and $y = (+, +, +, +, +, +)$ (where values '+' and '−' indicate a high and low value respectively). The type of explanation that is sought can typically be that the nodes contributing the most to the preference of y over x are nodes 8 (Kinematics) and 9 (Capability to escape interception) and not 2 (Suspicion of drug smuggling on the ship) or 3 (Suspicion of human smuggling on the ship). This helps the user to further analyze the values of criteria 8 and 9 (and not criteria 2 or 3). To this end, an indicator measuring the degree to which a node contributes to the preference between two alternatives has been defined in Ref. [9]. It is a generalization of the Shapley value on trees.

The contribution of this paper is to further develop this approach in two directions.

We are interested in the practical computation of the influence indicator. The main drawback of the Shapley value is that it has an exponential complexity in the number of nodes. It has been shown in Ref. [9] that the influence index for a node can be equivalently be computed on a subtree. The first contribution of this paper is to rewrite the influence index so as to improve the computational complexity. It cannot be further reduced without making assumptions on the utility model. An illustration of the influence indicator to the Choquet integral has been proposed in Ref. [9]. We consider in this paper another important class of aggregation model, based on the multi-linear extension. One of the main result of this paper shows that for the multi-linear model, the computations can be performed independently on each aggregation node, making the computation of the influence index much more tractable (see Sect. 5.2).

In practice, the values of the alternatives on the attributes are imprecise (second direction of this work). In Example 2, one needs to assess the PL of faraway ships for which the values of some attributes are not precisely known. In particular, the attributes related to the intent of the ship cannot readily be determined. Other attributes such as the heading of a ship cannot be assigned to a precise value as it is a fluctuating variable. The imprecision of the values of the attributes can also come from some disagreement among experts opinions (for attributes corresponding to a subjective judgment). For numerical attributes, the imprecise value can take the form of an interval. So far, there is no explanation approach able to capture imprecise values of the alternatives. In Example 2, the values of a ship on numerical attributes such as the maximum speed or the proximity of the shore might be given as an interval of confidence. The imprecisions on the value of the alternatives on the attributes propagate to the influence degrees in a very complex manner. We show that when the aggregation models are multi-linear models, the computation of the bounds on the influence degree can be easily obtained (see Sect. 4).

2   Preference Model and Notations

2.1   MCDA Model

We are given a set of criteria $N = \{1, \ldots, n\}$, each criterion $i \in N$ being associated with an attribute X_i, either discrete or continuous. The alternatives are characterized by a value on each attribute and are thus associated to an element in $X = X_1 \times \cdots \times X_n$. We assume that the preferences of the DM over the alternatives are represented by a utility model $U : X \to \mathbb{R}$.

The hierarchy of criteria is represented by a rooted tree T, defined by the set of nodes M_T (i.e. the set of criteria and aggregation nodes), and the children $\mathrm{Ch}_T(l)$ of node l (i.e. the nodes that are aggregated at each node l) [5]. We also denote by $N_T \subseteq M_T$ the set of leaves of tree T (i.e. the criteria), by $s_T \in M_T$ the root of tree T (i.e. the top aggregation node), by $\mathrm{Ch}_T(l)$ the children of node l in T, by $\mathrm{Desc}_T(l)$ the set of descendants of l, and by $\mathrm{Leaf}_T(l)$ the leaves at or below $l \in M_T$. A hierarchical model on criteria N is such that $N_T = N$.

The preference model is composed of an aggregation function H_l at each node $l \in M_T \setminus N_T$ and a partial utility function u_i for each criterion $i \in N_T$ (criteria). For $x \in X$, we can compute $U(x)$ recursively from a function v_i^U defined at each node $i \in M_T$:

- $v_i^U(x) = u_i(x_i)$ for every leaf $i \in N_T$,
- $v_l^U(x) = H_l\big((v_k^U(x))_{k \in \mathrm{Ch}_T(l)}\big)$ for every aggregation node $l \in M_T \setminus N_T$,
- $U(x) = v_{s_T}^U(x)$ is the overall utility.

Example 2 (Example 2 cont.). We have

$$v_i^U(x) = u_i(x_i) \text{ for } i \in \{1, 2, 3, 4, 5, 6\},$$
$$v_7^U(x) = H_7(v_2^U(x), v_3^U(x)), \; v_8^U(x) = H_8(v_4^U(x), v_5^U(x)),$$
$$v_9^U(x) = H_9(v_6^U(x), v_8^U(x)), \; U(x) = v_{10}^U(x) = H_{10}(v_1^U(x), v_7^U(x), v_9^U(x)). \qquad \blacksquare$$

2.2 Shapley Value

In Cooperative Game Theory, a *game* on N is a set function $v : 2^N \to \mathbb{R}$ such that $v(\varnothing) = 0$, N is the set of players, and $v(S)$ (for $S \subseteq N$) is the amount of wealth produced by S when they cooperate. It is a non-normalized capacity. The Shapley value is a fair share of the global wealth $v(N)$ produced by all players together, among themselves [14]:

$$\phi_i^{\mathrm{Sh}}(N, v) := \sum_{S \subseteq N \setminus i} \frac{(n - |S| - 1)! |S|!}{n!} \big[v(S \cup \{i\}) - v(S) \big]. \tag{1}$$

It can also be written as an average over the permutation on N:

$$\phi_i^{\mathrm{Sh}}(N, v) := \frac{1}{2^n} \sum_{\pi \in \Pi(N)} \big[v(S_\pi(i)) - v(S_\pi(i) \setminus \{i\}) \big], \tag{2}$$

where $S_\pi(\pi(k)) := \{\pi(1), \dots, \pi(k)\}$ and $\Pi(N)$ is the set of permutations on N.

2.3 Influence Index

Consider two alternatives x and y in X. One wishes to explain the reasons of the difference of preference between x and y. The explanation proposed in Ref. [9] takes the form of an index measuring the degree to which each node in M_T contributes to the difference of preference between x and y. An influence index denoted by $I_i(x, y; U, T)$ is computed for each node $i \in M_T$ for utility model U on the hierarchy T of criteria. The influence index is some kind of Shapley value applied to the game $v(S) = U(y_S, x_{N \setminus S})$ for all $S \subseteq N$, where $(y_S, x_{N \setminus S})$ denotes an alternative taking the values of y in S and the values of x in $N \setminus S$. As for the Shapley value, it is defined from permutations on N. Its expression is defined by [9]:

$$I_i(x, y, T, U) = \begin{cases} \frac{1}{|\Pi(T)|} \sum_{\pi \in \Pi(T)} \delta_\pi^{x,y,T,U}(i) & \text{if } i \in N_T, \\ \sum_{k \in \mathrm{Leaf}_T(i)} I_k(x, y, T, U) & \text{else,} \end{cases} \tag{3}$$

where $\delta_\pi^{x,y,T,U}(i) := U(y_{S_\pi(i)}, x_{N \setminus S_\pi(i)}) - U(y_{S_\pi(i) \setminus \{i\}}, x_{(N \setminus S_\pi(i)) \cup \{i\}})$. In (3), the set of admissible orderings $\Pi(T)$ is defined as the set of orderings of elements of N for which all elements of a subtree of T are consecutive. More precisely, $\pi \in \Pi(T)$ iff, for every $l \in M_T \setminus N$, indices $\pi^{-1}(\text{Leaf}_T(l))$ are consecutive.

2.4 Influence Index of the Restricted Tree

The complexity of computing I_i is equal to $|\Pi(T)|$, which is far too large. It has been shown in Ref. [9] that one can reduce this complexity by taking profit of some symmetries among permutations in $\Pi(T)$. The symmetries can be seen considering subtrees of T. We consider a subtree T' of T having the same root as T, taking a subset of nodes of T and having the same edges than T between nodes that are kept.

Definition of $U_{T'}$: Given $((u_i)_{i \in N_T}, (H_i)_{i \in M_T \setminus N_T})$ and a subtree T' of T, we can define $((u'_i)_{i \in N_{T'}}, (H'_i)_{i \in M_{T'} \setminus N_{T'}})$ by $u'_i = u_i$ for $i \in N_{T'} \cap N_T$, $u'_i(x_i) = x_i$ for $i \in N_{T'} \setminus N_T$ and $H'_i = H_i$ for $i \in M_{T'} \setminus N_{T'}$. The overall utility on the subtree is denoted by $U_{T'}$. We set $X_i = \mathbb{R}$ for every $i \in M_T \setminus N_T$. Then for $x \in X$, $U(x) = U_{T'}(x^{T'})$ where $x^{T'} \in X_{T'}$ is defined by $x_i^{T'} = x_i$ if $i \in N_{T'} \cap N_T$ and $x_i^{T'} = v_i^U(x)$ otherwise.

Definition of $T_{[j]}$: A particular subtree is when a node $j \in M_T$ of T becomes a leaf, and thus all descendants of j are encapsulated and represented by j. We define the restricted tree $T_{[j]}$ by $M_{T_{[j]}} := (M_T \setminus \text{Desc}_T(j)) \cup \{j\}$, $N_{T_{[j]}} := (N_T \setminus \text{Leaf}_T(j)) \cup \{j\}$, $s_{T_{[j]}} := s_T$, and $\text{Ch}_{T_{[j]}}(l) = \text{Ch}_T(l)$ for all $l \in M_{T_{[j]}} \setminus N_{T_{[j]}}$.

Definition of $T_{[J]}$: For $J = \{j_1, \ldots, j_p\}$, we set $T_{[J]} := \left(((T)_{[j_1]})_{[j_2]} \cdots \right)_{[j_p]}$.

Let us thus consider I_i for some fixed $i \in N$. The path from s_T to i in T consists of the nodes $r_0 = s_T, r_1, \ldots, r_t = i$. Let $J = \bigcup_{l=1}^{t-1} \text{Ch}_T(r_{l-1}) \setminus \{r_l\}$. Then we have [9]

$$I_i(x, y, T, U) = I_i(x^{T_{[J]}}, y^{T_{[J]}}, T_{[J]}, U_{T_{[J]}}). \tag{4}$$

The influence index can be equivalently be computed on the restricted tree $T_{[J]}$.

3   Generic Complexity Reduction of $I_i(x, y; U, T)$

Our aim is to implement the influence index in practice. The influence index contains an exponential number of terms. It is thus very challenging to perform its exact computation. A complexity analysis is performed in Sect. 3.1. An alternative expression of the influence index, reducing its computational complexity is proposed in Sect. 3.2.

3.1   Complexity Analysis

By Sect. 2.3, the expression of the influence index is given by (3). Hence the complexity of $I_i(x, y; U, T)$ depends on the number of permutations $\Pi(T)$. For $j \in M_T \setminus N$, we denote by $T_{|j}$ the subtree of T starting at node j, defined by $M_{T_{|j}} := \text{Desc}_T(j)$, $N_{T_{|j}} := \text{Leaf}_T(j)$, $s_{T_{|j}} := j$, and $\text{Ch}_{T_{|j}}(l) = \text{Ch}_T(l)$ for all $l \in M_{T_{|j}} \setminus N_{T_{|j}}$. Then the cardinality of $\Pi(T)$ can be recursively computed thanks to the next result.

Lemma 1. $|\Pi(T)| = |\mathrm{Ch}_T(s_T)|! \times \displaystyle\prod_{j \in \mathrm{Ch}_T(s_T)} |\Pi(T_{|j})|.$

The proofs of this result and the others are omitted for space limitation.

Lemma 3 provides a recursive formula to compute the number of compatible permutations in a tree T, that is the complexity of $I_i(x, y, T, U)$.

By (4), the extended Owen value of node i for tree T can be computed equivalently on tree $T_{[J]}$. The implementation of these formulae requires to enumerate over the permutations $\Pi(T_{[J]})$. This helps to drastically reduce the complexity.

Example 3. For T of Fig. 2(left), $i = 1$, we obtain $J = \{2, 10, 14\}$. Figure 2(right) shows $T_{[J]}$. ∎

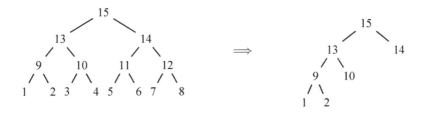

Fig. 2. Trees T (left) and $T_{[J]}$ (right), $J = \{2, 10, 14\}$.

In order to demonstrate the gain obtained by using $T_{[J]}$ instead of T, let us take the example of uniform trees, denoted by $T^{\mathrm{Un}}_{d,p}$ (with $d, p \in \mathbb{N}_*$) where each aggregation node contains exactly p children and each leaf is exactly at depth d of the root. Figure 2(left) illustrates $T^{\mathrm{Un}}_{3,2}$. The next lemma gives the expression of the number of permutations associated to the uniform tree $T^{\mathrm{Un}}_{d,p}$.

Lemma 2. $n = \left|N_{T^{\mathrm{Un}}_{d,p}}\right| = p^d, \ \left|\Pi(T^{\mathrm{Un}}_{d,p})\right| = (p!)^{\sum_{k=0}^{d-1} p^k}, \ and \ \left|\Pi\left((T^{\mathrm{Un}}_{d,p})_{[J]}\right)\right| = (p!)^d.$

Table 1 below shows a clear benefit of using $T_{[J]}$ instead of T in the computation of the influence index: the ratio amounts to orders of magnitude when n increases.

3.2 Alternative Expression of $I_i(x, y; U, T)$

Expression (3) takes the form of an average over permutations. The number of terms in the sum in (3) is equal to $C(N) := 2^{|N|-1}$. We give in this section an equivalent new expression taking profit of relation (4).

Consider I_i for some fixed $i \in N$. We set $V_l := \mathrm{Ch}_T(r_{l-1})$ for all $l \in \{1, \ldots, t\}$, $V'_l := V_l \setminus \{r_l\}$ – see Fig. 3.

Expression (3) can be turned into a sum over coalitions, which reduces a little bit the computation complexity:

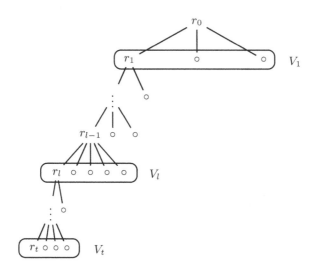

Fig. 3. Illustration of notation r_l and V_l.

Theorem 1. *We have*

$$I_i(x, y, U, T_{[J]}) = \sum_{S_1 \subseteq V_1'} \cdots \sum_{S_t \subseteq V_t'} \frac{\prod_{l=1}^t \left(|S_l|! \times (|V_l| - |S_l| - 1)! \right)}{\prod_{l=1}^t |V_l|!} \tag{5}$$
$$\times \left[U(y_i, [y_{S_{1..t}} x]_{V_{1..t}'}) - U(x_i, [y_{S_{1..t}} x]_{V_{1..t}'}) \right],$$

where $S_{l..j} = S_l \cup \cdots \cup S_j$, $V_{l..j}' = V_l' \cup \cdots \cup V_j'$, $x_k = v_k^U(x)$, $y_k = v_k^U(y)$ *and* $[y_S x]_T$ *(for* $S \subseteq T$*) denotes an alternative taking the value of* y *in* S *and the value of* x *in* $T \setminus S$.

The computation complexity of (5) is given by the next result.

Lemma 3. *The number of terms in (5) is of order* $C(T) := \prod_{l=1}^t 2^{|V_l'|}$.

The last two column in Table 1 presents the log of the number of operations in the expression of the influence index written over coalitions rather than on permutations. The complexity of computing the influence index reducing (resp. not reducing) to the restricted tree is denoted by $C(T_{d,p}^{\mathrm{Un}})$ (resp. $C(N_{T_{d,p}^{\mathrm{Un}}})$).

We obtain significant improvements on the computation time. In the second part of the paper, we will aim at drastically reducing this complexity – going from exponential complexity to polynomial or even linear – by taking an appropriate family of hierarchical aggregation models.

Table 1. Logarithm of the number of permutations and subsets for uniform trees $T_{d,p}^{\mathrm{Un}}$.

d	p	n	Expression (3)			Expression (5)	
			$\log_{10}|\varPi(N)|$	$\log_{10}|\varPi(T)|$	$\log_{10}\varPi(T_{[J]})$	$\log_{10} C(N_{T_{d,p}^{\mathrm{Un}}})$	$\log_{10} C(T_{d,p}^{\mathrm{Un}})$
3	3	27	28.036	10.115	2.334	8.128	1.806
3	4	64	89.1	28.984	4.14	19.266	2.709
3	5	125	209.27	64.454	6.237	37.629	3.612
3	6	216	412.0	122.86	8.571	65.022	4.515
4	3	81	120.76	31.126	3.112	24.383	2.408
4	4	256	506.93	117.31	5.520	77.063	3.612
4	5	625	1477.7	324.35	8.316	188.14	4.816
4	6	1296	3473.0	740.04	11.429	390.13	6.021
5	3	243	475.76	94.156	3.89	73.15	3.010
5	4	1024	2639.7	470.65	6.901	308.2	4.515
5	5	3125	9566.3	1623.84	10.395	940.7	6.021
5	6	7776	26879	4443.15	14.286	2340.9	7.526

4   Computation of the Influence Index with Imprecise Values

In many practical situations, the values of the alternatives are imprecise. We have justified this in the introduction, in particular for Example 2. For the sake of simplicity, the imprecision of the two alternatives on which the explanation is computed are given as intervals: $\widehat{x} = [\underline{x}, \overline{x}]$ and $\widehat{y} = [\underline{y}, \overline{y}]$, with $\underline{x}, \overline{x}, \underline{y}, \overline{y} \in X$. The problem is to define the influence index between $\widehat{x}$ and $\widehat{y}$.

The idea is to propagate the imprecisions on the values of x and y on the computation of the influence index. The influence of node i in the comparison between $\widehat{x}$ and $\widehat{y}$ is a closed interval defined by

$$\widehat{I}_i(\widehat{x}, \widehat{y}, T, U) = \left[\underline{I}_i(\widehat{x}, \widehat{y}, T, U), \overline{I}_i(\widehat{x}, \widehat{y}, T, U)\right],$$

where

$$\underline{I}_i(\widehat{x}, \widehat{y}, T, U) = \min_{x \in \widehat{x}} \min_{y \in \widehat{y}} I_i(x, y, T, U),$$
$$\overline{I}_i(\widehat{x}, \widehat{y}, T, U) = \max_{x \in \widehat{x}} \max_{y \in \widehat{y}} I_i(x, y, T, U).$$

We have

$$\underline{I_i}(\widehat{x},\widehat{y},T,U) = \min_{x\in\widehat{x}}\min_{y\in\widehat{y}} I_i(x,y,T,U)$$

$$= \min_{x\in\widehat{x}}\min_{y\in\widehat{y}} \frac{1}{|\Pi(T)|} \sum_{\pi\in\Pi(T)} \left[U(y_i, y_{S_\pi(i)\setminus\{i\}}, x_{-S_\pi(i)}) - U(x_i, y_{S_\pi(i)\setminus\{i\}}, x_{-S_\pi(i)}) \right]$$

$$= \min_{x_{-i}\in\widehat{x}_{-i}}\min_{y_{-i}\in\widehat{y}_{-i}} \frac{1}{|\Pi(T)|} \sum_{\pi\in\Pi(T)} \left[U(\underline{y}_i, y_{S_\pi(i)\setminus\{i\}}, x_{-S_\pi(i)}) \right.$$

$$\left. - U(\overline{x}_i, y_{S_\pi(i)\setminus\{i\}}, x_{-S_\pi(i)}) \right],$$

and

$$\overline{I_i}(\widehat{x},\widehat{y},T,U) = \max_{x\in\widehat{x}}\max_{y\in\widehat{y}} I_i(x,y,T,U)$$

$$= \max_{x\in\widehat{x}}\max_{y\in\widehat{y}} \frac{1}{|\Pi(T)|} \sum_{\pi\in\Pi(T)} \left[U(y_i, y_{S_\pi(i)\setminus\{i\}}, x_{-S_\pi(i)}) - U(x_i, y_{S_\pi(i)\setminus\{i\}}, x_{-S_\pi(i)}) \right]$$

$$= \max_{x_{-i}\in\widehat{x}_{-i}}\max_{y_{-i}\in\widehat{y}_{-i}} \frac{1}{|\Pi(T)|} \sum_{\pi\in\Pi(T)} \left[U(\overline{y}_i, y_{S_\pi(i)\setminus\{i\}}, x_{-S_\pi(i)}) \right.$$

$$\left. - U(\underline{x}_i, y_{S_\pi(i)\setminus\{i\}}, x_{-S_\pi(i)}) \right].$$

In the general case, computing $\underline{I_i}(\widehat{x},\widehat{y},T,U)$ or $\overline{I_i}(\widehat{x},\widehat{y},T,U)$ is difficult. We will show in the next section that these computations become tractable for the multi-linear model.

5   Case of the Multi-linear Model

Section 3.2 has provided an improved expression of the influence index reducing its computation complexity. However, it is still exponential in the number of criteria and the depth of the tree. We cannot further reduce the computation complexity without making assumptions on the utility model U. For applications requiring real-time computations of the explanations and/or presenting a large tree of criteria, we need to restrict ourselves to classes of models U having specific properties allowing to break the exponential complexity of the computation. This can be easily obtained considering very simple aggregation models. For example, if all aggregation models in the tree are simple weighted sums

$$v_l^U(x) = H_l\big((v_k^U(x))_{k\in\mathrm{Ch}_T(l)}\big) = \sum_{k\in\mathrm{Ch}_T(l)} w_l(k)\, v_k^U(x), \qquad (6)$$

where $w_l(k)$ is the weight of node k at aggregation ode l, then one can easily show that

$$I_i(x,y;T;U) = (u_i(y_i) - u_i(x_i)) \prod_{l=0}^{t-1} w_{r_l}(r_{l+1}). \qquad (7)$$

Even though the complexity of computing $I_i(x, y; T; U)$ is linear in the depth of the tree, the underlying model is very simple and far from being able to capture real-life preferences.

We are thus looking for a decision model realizing a good compromise between a high representation power (in particular being able to capture interaction among attributes) and a low computation time for the influence indices. We explore in this paper the multi-linear model and believe that it realizes such good compromise. Section 5.1 describes the multi-linear model. Section 5.2 shows that the expression of the influence index for the multi-linear model can be drastically simplified in terms of computational complexity. Section 5.3 shows that when the values of the alternatives are uncertain, the computation of the influence is also tractable for the multi-linear model.

5.1 Multi-linear Model

Consider an aggregation node $l \in M_T \setminus N$, which children are $\mathrm{Ch}_T(l)$. For the sake of simplicity, we assume that the components that are aggregated by H_l are simply denoted by the vector $a = (a_1, \ldots, a_{n_l})$, with $n_l = |\mathrm{Ch}_T(l)|$.

There exists many aggregation functions [2,7]. The simplest one is the weighted sum (see (6)):

$$\mathrm{WS}(a) = \sum_{i=1}^{n_l} w_l(i)\, a_i,$$

where $w_l(i)$ is the weight assigned to node i. This model assumes the independence among the criteria.

Without loss of generality, we can assume that the score lies in interval $[0, 1]$ where 0 (resp. 1) means the criterion is not satisfied at all (resp. completely satisfied). In order to represent interaction among criteria, the idea is to assign weights not only to single criteria but also to subsets of criteria. A *capacity* (also called *fuzzy measure* [15]) is a set function $v_l : 2^{n_l} \to [0, 1]$ such that $v_l(\emptyset) = 0$, $v_l(\{1, \ldots, n_l\}) = 1$ and $v_l(S) \leq v_l(T)$ whenever $S \subseteq T$ [3]. Term $v_l(S)$ represents the aggregated score of an option being very well-satisfied on criteria S (with score 1) and very ill-satisfied on the other criteria (with score 0).

The *Möbius transform* of v_l, denoted by $m_l : 2^{n_l} \to \mathbb{R}$, is given by [13]

$$m_l(A) = \sum_{B \subseteq A} (-1)^{|A \setminus B|} v_l(B).$$

A capacity is said to be 2-additive if the Möbius coefficients are zero for all subsets of three or more terms. Two classical aggregation functions can be obtained given the Möbius coefficients m_l. The first one is the Choquet integral [3]

$$\mathrm{C}_{m_l}(a) = \sum_{T \in \mathcal{S}_l} m_l(T) \times \min_{m \in T} a_m,$$

whereas the second one is the multi-linear model

$$\mathrm{M}_{m_l}(a) = \sum_{T \in \mathcal{S}_l} m_l(T) \times \prod_{m \in T} a_m, \tag{8}$$

where $\mathcal{S}_j$ is the subset of $\{1, \ldots, n_l\}$ on which the Möbius coefficients are non-null.

The next example illustrates the multi-linear model w.r.t. a two-additive capacity.

Example 4 (Example 2 cont.). After eliciting the tactical operator preferences, the aggregation functions are given by:

Node 7: There is suspicion of illegal activity whenever either drug or human smuggling is detected. Hence there is redundancy between criteria 2 and 3. As human smuggling (crit. 3) is slightly more important than criterion 2, we obtain $v_7^U(x) = 0.8 \, v_3^U(x) + v_3^U(x) - 0.8 \, v_2^U(x) \times v_3^U(x)$;

Node 8: $v_8^U(x) = (v_4^U(x) + v_5^U(x))/2$;

Node 9: Nodes 6 and 8 are redundant, since there is a high risk that the ship escapes interception when it is either close to the shore (crit. 6) or very fast (node 8). Hence $v_9^U(x) = 0.8 \, v_6^U(x) + 0.8 \, v_8^U(x) - 0.6 \, v_6^U(x) \times v_8^U(x)$,

Node 10: Nodes 1 and 7 are redundant since there is a suspicion on the ship when the score is high on either node 1 or 7. Nodes 7 and 9 are complementary as the risk is not so high for a suspicious ship (high value at node 7) that is easy to intercept (low value at node 9), or for a ship that is difficult to intercept but that is not suspicious. We have the same behavior between nodes 1 and 9. Hence $v_{10}^U(x) = \left(v_1^U(x) + v_7^U(x) - v_1^U(x) \times v_7^U(x) + v_1^U(x) \times v_9^U(x) + v_7^U(x) \times v_9^U(x) \right)/3$.

For $x = (+, -, -, +, +, +)$, we obtain $u_2(x) = u_3(x) = 0$, $u_i(x) = 1$ for $i \in \{1, 4, 5, 6\}$, $v_7^U(x) = 0$, $v_8^U(x) = v_9^U(x) = 1$ and $U(x) = v_{10}^U(x) = \frac{2}{3}$. ∎

5.2   Expression of the Influence Index for the Multi-linear Model

We consider the case where all aggregations functions are multi-linear models.

We now give the main result of this paper.

Theorem 2. *Assume that the aggregation function at node r_l (for $l \in \{0, \ldots, t-1\}$) is done with a multi-linear extension w.r.t. Möbius coefficients m_{r_l}. Then*

$$I_i(x, y; U, T_{[J]}) = (u_i(y_i) - u_i(x_i)) \times \prod_{l=0}^{t-1} \Phi_l, \tag{9}$$

where

$$\Phi_l = \sum_{T \subseteq V_{l+1}', \, T \cup \{r_{l+1}\} \in \mathcal{S}_{l+1}} m_{r_l}(T \cup \{r_{l+1}\}) \times \sum_{S' \subseteq T} \prod_{m \in T \cap S'} y_m \times \prod_{m \in T \setminus S'} x_m \times$$

$$\left[\sum_{s''=0}^{|V_{l+1}| - |T| - 1} \frac{(|V_{l+1}| - |T| - 1)!}{s''!(|V_{l+1}| - |T| - 1 - s'')!} \frac{(|S'| + s'')!(|V_{l+1}| - |S'| - s'' - 1)!}{|V_{l+1}|!} \right].$$

In the generic expression of the influence index (see (5)), the complexity of the computation of I_i grows exponentially with the number t of layers (see Lemma 3).

Thanks to the previous result, one readily sees that the computation of the influence only grows linearly with the depth of the tree for the multi-linear model. In (9), the influence index takes the form of a product of an influence computed for each layer, where Φ_l is the local influence at aggregation node r_l. Hence the computation of (9) becomes very fast, whatever the depth of the tree and the number of aggregation functions, as the number of children at each aggregation node is small in practice (in general between 2 and 6). We note that there are strong similarities with the case of a weighted sum – see (7). The weighted is a particular case of a multi-linear model where all Möbius coefficients for the subsets of two or more elements are zero. In this case, Φ_l subsumes to $m_{r_l}(\{r_{l+1}\})$, which is equal to the weight $w_{r_l}(r_{l+1})$ of node r_{l+1} at aggregation node r_l in a weighted sum. Hence (9) subsumes to (7) for a weighted sum.

Lemma 4. *The number of terms in (5) is of order*

$$C_{\mathrm{MultiLin}}(T) := 1 + \sum_{l=1}^{t} |V_l'| \times \sum_{T \subseteq V_l', \, T \cup \{r_l\} \in \mathcal{S}_l} 2^{|T|} \leq 1 + \sum_{l=1}^{t} |V_l'| 3^{|V_l'|}. \quad (10)$$

If all multi-linear models are two-additive, the complexity becomes

$$C_{\mathrm{MultiLin}}(T) = 1 + \sum_{l=1}^{t} |V_l'| \left[1 + 2|V_l'| \right].$$

We now illustrate Theorem 2 on the running example.

Example 5. (Example 4 cont.). We consider the two options $x = (+, -, -, -, -, -)$ and $y = (+, +, +, +, +, +)$. We have

$$u_1(x) = 1, \; u_2(x) = u_3(x) = u_4(x) = u_5(x) = u_6(x) = 0,$$
$$u_1(y) = u_2(y) = u_3(y) = u_4(y) = u_5(y) = u_6(y) = 1.$$

Then the influence of node say 4 is equal to

$$I_4(x, y; U, T_{[J]}) = (u_4(y) - u_4(x)) \times \Phi_0 \times \Phi_1 \times \Phi_2,$$

where Φ_l is the contribution at aggregation node r_l to the influence. We have

$$\Phi_0 = m_{10}(\{9\}) + m_{10}(\{1, 9\}) \frac{u_1(x) + u_1(y)}{2} + m_{10}(\{7, 9\}) \frac{u_7(x) + u_7(y)}{2}$$
$$\text{(as } m_{10}(\{1, 7, 9\}) = 0),$$
$$\Phi_1 = m_9(\{8\}) + m_9(\{6, 9\}) \frac{u_6(x) + u_6(y)}{2},$$
$$\Phi_2 = m_8(\{4\}) \quad \text{(as } m_8(\{4, 5\}) = 0).$$

Hence $\Phi_0 = \frac{1}{2}, \Phi_1 = \frac{1}{2}, \Phi_2 = \frac{1}{2}$ and $I_4(x, y; U, T_{[J]}) = 0.125$. ∎

5.3 Computation of the Influence Index with Imprecise Values for the Multi-linear Model

As in Sect. 5.2, we now assume that all aggregation functions are multi-linear models. From Theorem 2,

$$I_i(x, y; U, T_{[J]}) = (y_i - x_i) \times \prod_{l=0}^{t-1} \Phi_l(x, y),$$

where

$$\Phi_l(x, y) = \sum_{S \subseteq V_l'} \frac{|S|! \times (|V_l| - |S| - 1)!}{|V_l|!} \times \sum_{T \subseteq V_l'} m_l(T \cup \{r_l\}) \times \prod_{j \in T \cap S} y_j \times \prod_{j \in T \setminus S} x_j.$$

Let us start with the computation of the lower bound of the influence of criterion i:

$$\underline{I_i}(\hat{x}, \hat{y}, T, U) = \min_{x_{-i} \in \hat{x}_{-i}} \min_{y_{-i} \in \hat{y}_{-i}} I_i\left((\overline{x}_i, x_{-i}), (\underline{y}_i, y_{-i}); T, U\right)$$

$$= (\underline{y}_i - \overline{x}_i) \times \prod_{l=1}^{t} \underline{\Phi}_l,$$

where $\underline{\Phi}_l = \min_{x_{-i} \in \hat{x}_{-i}} \min_{y_{-i} \in \hat{y}_{-i}} \Phi_l(x, y)$. Let $k \in V_l'$. Let us analyse the monotonicity of variables x_k and y_k on $\Phi_l(x, y)$:

$$\Phi_l(x, y) = \sum_{S \subseteq V_l' \setminus \{k\}} \sum_{T \subseteq V_l' \setminus \{k\}} \left[\frac{|S|! \times (|V_l| - |S| - 1)!}{|V_l|!} m_l(T \cup \{r_l\}) \right. \tag{11}$$

$$+ \frac{(|S| + 1)! \times (|V_l| - |S| - 2)!}{|V_l|!} m_l(T \cup \{r_l\})$$

$$+ \frac{|S|! \times (|V_l| - |S| - 1)!}{|V_l|!} m_l(T \cup \{r_l, k\}) x_k$$

$$+ \left. \frac{(|S| + 1)! \times (|V_l| - |S| - 2)!}{|V_l|!} m_l(T \cup \{r_l, k\}) y_k \right] \times \prod_{j \in T \cap S} y_j \times \prod_{j \in T \setminus S} x_j.$$

The first two terms in the bracket are constant w.r.t. x_k and y_k. Hence Φ_l is linear in x_k and in y_k. This implies that the minimum value in $\Phi_l(x, y)$ is attained at an extreme point of the intervals. As this holds for every k, we obtain

$$\underline{\Phi}_l = \min_{x_{-i} \in \prod_{j \neq i} \{\underline{x}_j, \overline{x}_j\}} \min_{y_{-i} \in \prod_{j \neq i} \{\underline{y}_j, \overline{y}_j\}} \Phi_l(x, y).$$

The optimal value can be obtained by enumerating the extreme values. This is not so time consuming as the number of elements in V_l' is not large. A similar approach can be performed to compute $\overline{I}_i(\hat{x}, \hat{y}, T, U)$.

A more efficient approach can be derived to compute $\underline{I}_i(\hat{x}, \hat{y}, T, U)$ and $\overline{I}_i(\hat{x}, \hat{y}, T, U)$ under assumptions on m_l. By (11), if $m_l(T \cup \{r_l, k\}) \geq 0$ (resp. ≤ 0)

for all $T \subseteq V_l' \setminus \{k\}$, then Φ_l is monotonically increasing (resp. decreasing) w.r.t. x_k and y_k. Hence the minimum $\underline{\Phi}_l$ is attained at $x_k = \underline{x}_k$ (resp. at $x_k = \overline{x}_k$). This is in particular the case when the Möbius coefficients are 2-additive. Indeed, for a 2-additive capacity, $m(T \cup \{r_l, k\})$ can be non-zero only for $T = \emptyset$.

6 Conclusion and Perspectives

The problem of generating explanations is of particular importance in many applications. It is also very challenging. We have considered the problem of explaining a hierarchical multi-criteria decision aiding problem using influence indices extending the Shapley value. The main drawback of this approach is that its computation complexity grows exponentially with the depth of the tree. We have shown that this complexity remains linear when the aggregation functions are multi-linear models. Secondly, we considered in the case where the values of the alternatives are imprecise on the criteria. The influence indices become thus imprecise. An efficient computation approach is proposed for the multi-linear model.

The work can be extended in several directions. In applications where a multi-linear model is not suitable, it is crucial to obtain efficient algorithms for other classes of aggregation models, such as the Choquet integral. One can also check the validity of the explanations on real users.

References

1. Bacchus, F., Grove, A.: Graphical models for preference and utility. In: Conference on Uncertainty in Artificial Intelligence (UAI), Montreal, Canada, pp. 3–10, July 1995
2. Beliakov, G., Pradera, A., Calvo, T.: Aggregation Functions: A Guide for Practitioners. Studies in Fuzziness and Soft Computing, vol. 221. Springer, Heidelberg (2007). https://doi.org/10.1007/978-3-540-73721-6
3. Choquet, G.: Theory of capacities. Annales de l'Institut Fourier **5**, 131–295 (1953)
4. Datta, A., Sen, S., Zick, Y.: Algorithmic transparency via quantitative input influence. In: IEEE Symposium on Security and Privacy, San Jose, CA, USA, May 2016
5. Diestel, R.: Graph Theory. Springer, New York (2005)
6. Fishburn, P.: Interdependence and additivity in multivariate, unidimensional expected utility theory. Int. Econ. Rev. **8**, 335–342 (1967)
7. Grabisch, M., Marichal, J., Mesiar, R., Pap, E.: Aggregation Functions. Cambridge University Press, Cambridge (2009)
8. Junker, U.: QUICKXPLAIN: preferred explanations and relaxations for over-constrained problems. In: Proceedings of the 19th National Conference on Artificial Intelligence (AAAI 2004), San Jose, California, pp. 167–172, July 2004
9. Labreuche, C., Fossier, S.: Explaining multi-criteria decision aiding models with an extended Shapley value. In: Proceedings of the Twenty-Seventh International Joint Conference on Artificial Intelligence (IJCAI 2018), Stockholm, Sweden, pp. 331–339, July 2018
10. Lundberg, S., Lee, S.: A unified approach to interpreting model predictions. In: Guyon, I., et al. (eds.) 31st Conference on Neural Information Processing Systems (NIPS 2017), Long Beach, CA, USA, pp. 4768–4777 (2017)
11. Owen, G.: Multilinear extensions of games. Management Sci. **18**, 64–79 (1972)

12. Ribeiro, M., Singh, S., Guestrin, C.: "Why Should I Trust You?": explaining the predictions of any classifier. In: KDD 2016 Proceedings of the 22nd ACM SIGKDD International Conference on Knowledge Discovery and Data Mining, San Francisco, California, USA, pp. 1135–1144 (2016)

13. Rota, G.: On the foundations of combinatorial theory I. Theory of Möbius functions. Zeitschrift für Wahrscheinlichkeitstheorie und Verwandte Gebiete **2**, 340–368 (1964)

14. Shapley, L.S.: A value for n-person games. In: Kuhn, H.W., Tucker, A.W. (eds.) Contributions to the Theory of Games, Vol. II. Annals of Mathematics Studies, no. 28, pp. 307–317. Princeton University Press, Princeton (1953)

15. Sugeno, M.: Fuzzy measures and fuzzy integrals. Trans. S.I.C.E. **8**(2), 218–226 (1972)

16. Wachter, S., Mittelstadt, B., Russell, C.: Counterfactual explanations without opening the black box: automated decisions and the GDPR. Harvard J. Law Technol. **31**(2), 841–887 (2018)

Assertional Removed Sets Merging of DL-Lite Knowledge Bases

Salem Benferhat[1], Zied Bouraoui[1], Odile Papini[2], and Eric Würbel[2(✉)]

[1] CRIL-CNRS UMR 8188, Univ Artois, Arras, France
{benferhat,bouraoui}@cril.univ-artois.fr
[2] LIS-CNRS UMR 7020, Aix Marseille Univ, Université de Toulon, Marseille, France
{papini,wurbel}@univ-amu.fr

Abstract. *DL-Lite* is a tractable family of Description Logics that underlies the *OWL-QL* profile of the ontology web language, which is specifically tailored for query answering. In this paper, we consider the setting where the queried data are provided by several and potentially conflicting sources. We propose a merging approach, called "Assertional Removed Sets Fusion" (ARSF) for merging *DL-Lite* assertional bases. This approach stems from the inconsistency minimization principle and consists in determining the minimal subsets of assertions, called assertional removed sets, that need to be dropped from the original assertional bases in order to resolve conflicts between them. We give several merging strategies based on different definitions of minimality criteria, and we characterize the behaviour of these strategies with respect to rational properties. The last part of the paper shows how to use the notion of hitting sets for computing the assertional removed sets, and the merging outcome.

1   Introduction

In the last years, there has been an increasing use of ontologies in many application areas including query answering, Semantic Web and information retrieval. Description Logics (DLs) have been recognized as powerful formalisms for both representing and reasoning about ontologies. A DL knowledge base is built upon two distinct components: a terminological base (called *TBox*), representing generic knowledge about an application domain, and an assertional base (called *ABox*), containing assertional facts that instantiate terminological knowledge. Among Description Logics, a lot of attention was given to *DL-Lite* [12], a lightweight family of DLs specifically tailored for applications that use huge volumes of data for which query answering is the most important reasoning task. *DL-Lite* guarantees a low computational complexity of the reasoning process.

In many practical situations, data are provided by several and potentially conflicting sources, where getting meaningful answers to queries is challenging. While the available sources are individually consistent, gathering them together may lead to inconsistency. Dealing with inconsistency in query answering has received a lot of attention in recent years. For example, a general framework for inconsistency-tolerant semantics

© Springer Nature Switzerland AG 2019
N. Ben Amor et al. (Eds.): SUM 2019, LNAI 11940, pp. 207–220, 2019.
https://doi.org/10.1007/978-3-030-35514-2_16

was proposed in [4,5]. This framework considers two key notions: modifiers and inference strategies. Inconsistency tolerant query answering is seen as made out of a modifier, which transforms the original ABox into a set of repairs, i.e. subsets of the original ABox which are consistent w.r.t. the TBox, and an inference strategy, which evaluates queries from these repairs. Interestingly enough, such setting covers the main existing works on inconsistency-tolerant query answering (see *e.g.* [2,9,22]). Pulling together the data provided by available sources and then applying inconsistency-tolerant query answering semantics provides a solution to deal with inconsistency. However, in this case valuable information about the sources will be lost. This information is indeed important when trying to find better strategies to deal with inconsistency during merging process.

This paper addresses query answering by merging data sources. Merging consists in achieving a synthesis between pieces of information provided by different sources. The aim of merging is to provide a consistent set of information, making maximum use of the information provided by the sources while not favoring any of them. Merging is an important issue in many fields of Artificial Intelligence [10]. Within the classical logic setting belief merging has been studied according different standpoints. One can distinguish model-based approaches that perform selection among the interpretations which are the closest to original belief bases. Postulates characterizing the rational behaviour of such merging operators, known as IC postulates, which have been proposed by Revesz [25] and improved by Konieczny and Pérez [21] in the same spirit as the seminal AGM [1] postulates for revision. Several concrete merging operators have been proposed [11,20,21,23,26]. In contrast to model-based approaches, the formula-based approaches perform selection on the set of formulas that are explicitly encoded in the initial belief bases. Some of these approaches have been adapted in the context of DL-Lite [13]. Falappa et al. [14] proposed a set of postulates to characterize the behaviour of belief bases merging operators and concrete merging operators have been proposed [6,8,14,17,19,24]. Among these formula-based merging approaches, Removed Sets Fusion approach has been proposed in [17,18] for merging propositional belief bases. This approach stems from removing a minimal subset of formulae, called removed set, to restore consistency. The minimality in Removed Sets Fusion stems from the operator used to perform merging, which can be the sum (Σ), the cardinality ($Card$), the maximum (Max), the lexicographic ordering ($GMax$). This approach has shown interesting properties: it is not too cautious and satisfies most rational IC postulates when extended to belief sets revision.

This paper studies *DL-Lite* Assertional Removed Sets Fusion (ARSF). The main motivation in considering ARSF is to take advantage of the tractability of *DL-Lite* for the merging process and the rational properties satisfied by ARSF operators. We consider in particular $DL\text{-}Lite_R$ as member of the *DL-Lite* family, which offers a good compromise between expressive power and computational complexity and underlies the *OWL2-QL* profile. We propose several merging strategies based on different definitions of minimality criterion, and we give a characterization of these merging strategies. The last section contains algorithms based on the notion hitting sets for computing the merging outcome.

2   Background

In this paper, we only consider *DL-Lite$_R$*, denoted by $\mathcal{L}$, which underlies *OWL2-QL*. However, results of this work can be easily generalized for several members of the *DL-Lite* family (see [3] for more details about the *DL-Lite* family).

Syntax. A *DL-Lite* knowledge base $\mathcal{K} = \langle \mathcal{T}, \mathcal{A} \rangle$ is built upon a set of atomic concepts (i.e. unary predicates), a set of atomic roles (i.e. binary predicates) and a set of individuals (i.e. constants). Complex concepts and roles are formed as follows:

$$B \longrightarrow A|\exists R, C \longrightarrow B|\neg B, R \longrightarrow P|P^-, E \longrightarrow R|\neg R,$$

where A (*resp.* P) is an atomic concept (*resp.* role). B (*resp.* C) are called basic (*resp.* complex) concepts and roles R (*resp.* E) are called basic (*resp.* complex) roles. The TBox $\mathcal{T}$ consists of a finite set of *inclusion axioms between concepts* of the form: $B \sqsubseteq C$ and *inclusion axioms between roles* of the form: $R \sqsubseteq E$. The ABox $\mathcal{A}$ consists of a finite set of *membership assertions* on atomic concepts and on atomic roles of the form: $A(a_i), P(a_i, a_j)$, where a_i and a_j are individuals. For the sake of simplicity, in the rest of this paper, when there is no ambiguity we simply use *DL-Lite* instead of *DL-Lite$_R$*.

Semantics. The *DL-Lite* semantics is given by an interpretation $\mathcal{I} = (\Delta^{\mathcal{I}}, .^{\mathcal{I}})$ which consists of a nonempty domain $\Delta^{\mathcal{I}}$ and an interpretation function $.^{\mathcal{I}}$. The function $.^{\mathcal{I}}$ assigns to each individual a an element $a^{\mathcal{I}} \in \Delta^{\mathcal{I}}$, to each concept C a subset $C^{\mathcal{I}} \subseteq \Delta^{\mathcal{I}}$ and to each role R a binary relation $R^{\mathcal{I}} \subseteq \Delta^{\mathcal{I}} \times \Delta^{\mathcal{I}}$ over $\Delta^{\mathcal{I}}$. Moreover, the interpretation function $.^{\mathcal{I}}$ is extended for all constructs of *DL-Lite$_R$*. For instance: $(\neg B)^{\mathcal{I}} = \Delta^{\mathcal{I}} \backslash B^{\mathcal{I}}$, $(\exists R)^{\mathcal{I}} = \{x \in \Delta^{\mathcal{I}} | \exists y \in \Delta^{\mathcal{I}} \, such\, that\, (x,y) \in R^{\mathcal{I}}\}$ and $(P^-)^{\mathcal{I}} = \{(y,x) \in \Delta^{\mathcal{I}} \times \Delta^{\mathcal{I}} | (x,y) \in P^{\mathcal{I}}\}$. Concerning the TBox, we say that $\mathcal{I}$ satisfies a concept (*resp.* role) inclusion axiom, denoted by $\mathcal{I} \models B \sqsubseteq C$ (*resp.* $\mathcal{I} \models R \sqsubseteq E$), iff $B^{\mathcal{I}} \subseteq C^{\mathcal{I}}$ (*resp.* $R^{\mathcal{I}} \subseteq E^{\mathcal{I}}$). Concerning the ABox, we say that $\mathcal{I}$ satisfies a concept (*resp.* role) membership assertion, denoted by $\mathcal{I} \models A(a_i)$ (*resp.* $\mathcal{I} \models P(a_i, a_j)$), iff $a_i^{\mathcal{I}} \in A^{\mathcal{I}}$ (*resp.* $(a_i^{\mathcal{I}}, a_j^{\mathcal{I}}) \in P^{\mathcal{I}}$). Finally, an interpretation $\mathcal{I}$ is said to satisfy $\mathcal{K} = \langle \mathcal{T}, \mathcal{A} \rangle$ iff $\mathcal{I}$ satisfies every axiom in $\mathcal{T}$ and every assertion in $\mathcal{A}$. Such interpretation is said to be a model of $\mathcal{K}$.

Incoherence and Inconsistency. Two kinds of inconsistency can be distinguished in DL setting: incoherence and inconsistency [7]. A knowledge base is said to be inconsistent iff it does not admit any model and it is said to be incoherent if there exists at least a non-satisfiable concept, namely for each interpretation $\mathcal{I}$ which is a model of $\mathcal{T}$, we have $C^{\mathcal{I}} = \emptyset$. In *DL-Lite* setting a TBox $\mathcal{T} = \{\text{PIs, NIs}\}$ can be viewed as composed of positive inclusion axioms, denoted by (PIs), and negative inclusion axioms, denoted by (NIs). PIs are of the form $B_1 \sqsubseteq B_2$ or $R_1 \sqsubseteq R_2$ and NIs are of the form $B_1 \sqsubseteq \neg B_2$ or $R_1 \sqsubseteq \neg R_2$. The negative closure of $\mathcal{T}$, denoted by $cln(\mathcal{T})$, represents the propagation of the NIs using both PIs and NIs in the TBox (see [12] for more details). Important properties have been established in [12] for consistency checking in *DL-Lite*: $\mathcal{K}$ is consistent if and only if $\langle cln(\mathcal{T}), \mathcal{A} \rangle$ is consistent. Moreover, every *DL-Lite* knowledge base with only PIs in its TBox is always satisfiable. However when $\mathcal{T}$ contains NI axioms then the *DL-Lite* knowledge base may be inconsistent and in an assertional-based approach only elements of ABoxes are removed to restore consistency [13].

3   Assertional Removed Sets Fusion

In this section, we study removed sets fusion to merge a set $\{\mathcal{A}_1, \cdots, \mathcal{A}_n\}$ of n assertional bases, representing different sources of information, linked to a *DL-lite* ontology $\mathcal{T}$. As representation formalism, we consider $\mathcal{M}_\mathcal{K} = \langle \mathcal{T}, \mathcal{M}_\mathcal{A} \rangle$, an MBox knowledge base where $\mathcal{M}_\mathcal{A} = \{\mathcal{A}_1, \ldots, \mathcal{A}_n\}$ is called an MBox. An MBox is simply a multi-set of membership assertions, where each $\mathcal{A}_i$ is an assertional base linked to $\mathcal{T}$. We assume that $\mathcal{M}_\mathcal{K}$ is coherent, i.e. $\mathcal{T}$ is coherent and for each $\mathcal{A}_i$, $1 \leq i \leq n$, $\langle \mathcal{T}, \mathcal{A}_i \rangle$ is consistent. However, the MBox $\mathcal{M}_\mathcal{K}$ may be inconsistent since the assertional bases $\mathcal{A}_i$ may be conflicting w.r.t. $\mathcal{T}$. We define the notion of conflict as a minimal inconsistent subset of $\mathcal{A}_1 \cup \ldots \cup \mathcal{A}_n$, more formally:

Definition 1. *Let $\mathcal{M}_\mathcal{K} = \langle \mathcal{T}, \mathcal{M}_\mathcal{A} \rangle$ be an inconsistent MBox DL-Lite knowledge base. A conflict C is a set of membership assertions such that (i) $C \subseteq \mathcal{A}_1 \cup \cdots \cup \mathcal{A}_n$, (ii) $\langle \mathcal{T}, C \rangle$ is inconsistent, (iii) $\forall C'$, if $C' \subset C$ then $\langle \mathcal{T}, C' \rangle$ is consistent.*

We denote by $\mathcal{C}(\mathcal{M}_\mathcal{K})$ the collection of conflicts in $\mathcal{M}_\mathcal{K}$. Since $\mathcal{M}_\mathcal{K}$ is assumed to be finite, if $\mathcal{M}_\mathcal{K}$ is inconsistent then $\mathcal{C}(\mathcal{M}_\mathcal{K}) \neq \emptyset$ is also finite.

Within the *DL-Lite* framework, in order to restore consistency, the following definition introduces the notion of potential assertional removed set.

Definition 2. *Let $\mathcal{M}_\mathcal{K} = \langle \mathcal{T}, \mathcal{M}_\mathcal{A} \rangle$ be a MBox DL-Lite knowledge base. A potential assertional removed set, denoted by X, is a set of membership assertions such that (i) $X \subseteq \mathcal{A}_1 \cup \cdots \cup \mathcal{A}_n$, (ii) $\langle \mathcal{T}, (\mathcal{A}_1 \cup \cdots \cup \mathcal{A}_n) \backslash X \rangle$ is consistent, (iii) $\forall X'$, if $X' \subset X \subseteq \mathcal{A}_1 \cup \cdots \cup \mathcal{A}_n$ then $\langle \mathcal{T}, (\mathcal{A}_1 \cup \cdots \cup \mathcal{A}_n) \backslash X' \rangle$ is inconsistent.*

We denote by $\mathcal{PR}(\mathcal{M}_\mathcal{K})$ the set of potential assertional removed sets of $\mathcal{M}_\mathcal{K}$. If $\mathcal{M}_\mathcal{K}$ is consistent then $\mathcal{PR}(\mathcal{M}_\mathcal{K}) = \{\emptyset\}$. The concept of potential assertional removed sets is to some extent dual to the concept of repairs (maximally consistent subbase). Namely, if X is a potential assertional removed set then $(\mathcal{A}_1 \cup \cdots \cup \mathcal{A}_n) \backslash X$ is a repair, and conversely.

Example 1. Let $\mathcal{M}_\mathcal{K} = \langle \mathcal{T}, \mathcal{M}_\mathcal{A} \rangle$ be an inconsistent MBox *DL-Lite* knowledge base such that $\mathcal{T} = \{A \sqsubseteq \neg B, C \sqsubseteq \neg D\}$ and $\mathcal{M}_\mathcal{A} = \{\mathcal{A}_1, \mathcal{A}_2, \mathcal{A}_3\}$ where $\mathcal{A}_1 = \{A(a), C(a)\}$ $\mathcal{A}_2 = \{A(a), A(b)\}$ and $\mathcal{A}_3 = \{B(a), D(a), C(b)\}$. By Definition 1, $\mathcal{C}(\mathcal{M}_\mathcal{K}) = \{\{A(a), B(a)\}, \{C(a), D(a)\}\}$. Hence, by Definition 2, $\mathcal{PR}(\mathcal{M}_\mathcal{K}) = \{\{A(a), C(a)\}, \{A(a), D(a)\}, \{B(a), C(a)\}, \{B(a), D(a)\}\}$.

In order to cope with conflicting sources, merging aims at exploiting the complementarity between the sources providing the ABoxes, so merging strategies are necessary. These merging strategies are captured by total pre-orders on potential assertional removed sets. Let X and Y be two potential assertional removed sets, for each strategy P a total pre-order $\leq_P$ over the potential assertional removed sets is defined. $X \leq_P Y$ means that X is preferred to Y according to the strategy P. We define $<_P$ as the strict total pre-order associated to $\leq_P$ (i.e. $X <_P Y$ if and only if $X \leq_P Y$ and $Y \not\leq_P X$).

Definition 3. *Let $\mathcal{M}_\mathcal{K} = \langle \mathcal{T}, \mathcal{M}_\mathcal{A} \rangle$ be a MBox DL-Lite knowledge base. An assertional removed set according to the strategy P, denoted by X, is a set of membership assertions such that (i) X is a potential assertional removed set of $\mathcal{M}_\mathcal{K}$; (ii) there does not exist any Y such that Y is a potential assertional removed set of $\mathcal{M}_\mathcal{K}$ and $Y <_P X$.*

We denote by $\mathcal{R}_P(\mathcal{M}_\mathcal{K})$ the set of assertional removed sets according to the strategy P of $\mathcal{M}_\mathcal{K}$. If $\mathcal{M}_\mathcal{K}$ is consistent then $\mathcal{R}_P(\mathcal{M}_\mathcal{K}) = \{\emptyset\}$. The usual merging strategies sum-based (Σ), cardinality-based ($Card$), maximum-based (Max) and lexicographic ordering ($GMax$) are captured by the following total pre-orders. We denote by $s(\mathcal{M}_\mathcal{A})$ the ABox obtained from $\mathcal{M}_\mathcal{K}$ where every assertion expressed more than once is reduced to a singleton.

(Σ): $X \leq_\Sigma Y$ if $\sum_{1 \leq i \leq n} | X \cap \mathcal{A}_i | \leq \sum_{1 \leq i \leq n} | Y \cap \mathcal{A}_i |$.

$(Card)$: $X \leq_{Card} Y$ if $|X \cap s(\mathcal{M}_\mathcal{A})| \leq |Y \cap s(\mathcal{M}_\mathcal{A})|$.

(Max): $X \leq_{Max} Y$ if $\max_{1 \leq i \leq n} | X \cap \mathcal{A}_i | \leq \max_{1 \leq i \leq n} | Y \cap \mathcal{A}_i |$.

$(GMax)$: For every potential assertional removed set X and every ABox $\mathcal{A}_i$, we define $p_X^{\mathcal{A}_i} = | X \cap \mathcal{A}_i |$. Let $L_X^{\mathcal{M}_\mathcal{A}}$ be the sequence $(p_X^{\mathcal{A}_1}, \ldots, p_X^{\mathcal{A}_n})$ sorted by decreasing order. Let X and Y be two potential assertional removed sets of $\mathcal{M}_\mathcal{K}$, $X \leq_{GMax} Y$ if $L_X^{\mathcal{M}_\mathcal{A}} \leq_{lex} L_Y^{\mathcal{M}_\mathcal{A}}$[1].

The Σ strategy minimizes the number of assertions to remove from $\mathcal{M}_\mathcal{A}$. The $Card$ strategy attempts, similarly to Σ, to minimize the number of removed assertions. But it does not take into account assertions which are expressed several times. Note that the Σ and $Card$ strategies only differ if there are redundant assertions. The Max strategy tries to distribute to the best the assertions to be removed among to ABoxes. It tries to do so by removing the less possible assertions in the most hit ABox. The $GMax$ strategy is a lexicographic refinement of the Max strategy. Note that when there is only one source, all strategies become equivalent.

We now present assertional-based $DL\text{-}Lite_R$ merging operators. A merging operator is a function that maps an MBox $DL\text{-}Lite_R$ $\mathcal{M}_\mathcal{K} = \langle \mathcal{T}, \mathcal{M}_\mathcal{A} \rangle$ to a knowledge base $\Delta(\mathcal{M}_\mathcal{K}) = \langle \mathcal{T}, \Delta(\mathcal{M}_\mathcal{A}) \rangle$, where the function Δ defined from $\mathcal{L} \times \ldots \times \mathcal{L}$ to $\mathcal{L}$, merges according to a strategy a multiset of assertions $\mathcal{M}_\mathcal{A}$ into a set of assertions denoted by $\Delta(\mathcal{M}_\mathcal{A})$. In the $DL\text{-}Lite$ language, it is not possible to find a set of assertions which represents the disjunction of such possible merged sets of assertions. If we want to keep the result of merging in $DL\text{-}Lite$, several options are possible. The first one is to consider the intersection of all possible merged set of assertions however this option may be too cautious since it could remove too many assertions and contradicts in some sense the minimal change principle. Another option is to define a selection function which allows us to define the family of ARSF operators. In this paper we consider the family of selection functions that select exactly one assertional removed set as follows.

Definition 4. *A selection function f is a mapping from $\mathcal{R}_P(\mathcal{M}_\mathcal{K})$ to $\mathcal{A}_1 \cup \ldots \cup \mathcal{A}_n$ such that (i) $f(\mathcal{R}_P(\mathcal{M}_\mathcal{K})) = X$ with $X \in \mathcal{R}_P(\mathcal{M}_\mathcal{K})$, (ii) $f(\{\emptyset\}) = \emptyset$.*

Definition 5. *Let $\mathcal{M}_\mathcal{K} = \langle \mathcal{T}, \mathcal{M}_\mathcal{A} \rangle$ be a MBox DL-Lite knowledge base, f be a selection function, and P be a strategy, the merged DL-Lite knowledge base, denoted by $\Delta_P^{arsf}(\mathcal{M}_\mathcal{K})$, is such that $\Delta_P^{arsf}(\mathcal{M}_\mathcal{K}) = \left\langle \mathcal{T}, \Delta_P^{arsf}(\mathcal{M}_\mathcal{A}) \right\rangle$ where $\Delta_P^{arsf}(\mathcal{M}_\mathcal{A}) = (\mathcal{A}_1 \cup \ldots \cup \mathcal{A}_n) \backslash f(\mathcal{R}_P(\mathcal{M}_\mathcal{K}))$.*

Let $\mathcal{M}_\mathcal{K} = \langle \mathcal{T}, \mathcal{M}_\mathcal{A} \rangle$ be a MBox *DL-Lite* knowledge base, and $q(x)$ a query. Querying multiple data sources is performed by querying merged data sources and $\langle \mathcal{T}, \mathcal{M}_\mathcal{A} \rangle \models q(x)$ amounts to $\left\langle \mathcal{T}, \Delta_P^{arsf}(\mathcal{M}_\mathcal{A}) \right\rangle \models q(x)$.

---

[1] $(X_1, \cdots, X_n) \leq_{lex} (Y_1, \cdots, Y_n)$ if $\exists i$, $1 \leq i \leq n$, (i) $X_i \leq Y_i$, (ii) $\forall j$, $1 \leq j < i$ $X_i = Y_i$.

Example 2. Let $\mathcal{M}_{\mathcal{K}} = \langle \mathcal{T}, \mathcal{M}_{\mathcal{A}} \rangle$ be the MBox of Example 1. The potential assertional removed sets are $X_1 = \{A(a), C(a)\}$, $X_2 = \{A(a), D(a)\}$, $X_3 = \{B(a), C(a)\}$ and $X_4 = \{B(a), D(a)\}$. As illustrated in the table below[2], we have $\mathcal{R}_{\Sigma}(\mathcal{M}_{\mathcal{K}}) = \{X_3, X_4\}$. Suppose the selection function f is such that $f(\mathcal{R}_{\Sigma}(\mathcal{M}_{\mathcal{K}})) = X_4$ we have $\Delta_{\Sigma}^{arsf}(\mathcal{M}_{\mathcal{A}}) = \{A(a), C(a), A(b), C(b)\}$. We have $\mathcal{R}_{Card}(\mathcal{M}_{\mathcal{K}}) = \{X_1, X_2, X_3, X_4\}$. Suppose the selection function f is such that $f(\mathcal{R}_{Card}(\mathcal{M}_{\mathcal{K}})) = X_1$ we have $\Delta_{Card}^{arsf}(\mathcal{M}_{\mathcal{A}}) = \{A(b), B(a), D(a), C(b)\}$. We have $\mathcal{R}_{Max}(\mathcal{M}_{\mathcal{K}}) = \{X_2, X_3\}$. Suppose the selection function f is such that $f(\mathcal{R}_{Card}(\mathcal{M}_{\mathcal{K}})) = X_2$ we have $\Delta_{Max}^{arsf}(\mathcal{M}_{\mathcal{A}}) = \{C(a), A(b), B(a), C(b)\}$. We have $\mathcal{R}_{GMax}(\mathcal{M}_{\mathcal{K}}) = \{X_3\}$ and $\Delta_{GMax}^{arsf}(\mathcal{M}_{\mathcal{A}}) = \{A(a), D(a), A(b), C(b)\}$.

X_i	$|X_i \cap \mathcal{A}_1|$	$|X_i \cap \mathcal{A}_2|$	$|X_i \cap \mathcal{A}_3|$	Σ	$Card$	Max	$GMax$
X_1	2	1	0	3	**2**	2	210
X_2	1	1	1	3	**2**	1	111
X_3	1	0	1	**2**	**2**	1	**110**
X_4	0	0	2	**2**	**2**	2	200

4   Logical Properties

Within the context of propositional logic, postulates have been proposed in order to classify reasonable belief bases merging operators [14–16][3]. In order to give logical properties of ARSF operators, we first rephrase these postulates within the DL-Lite framework, and then analyse to which extent the proposed operators satisfy these postulates for any selection function.

Let $\mathcal{M}_{\mathcal{K}} = \langle \mathcal{T}, \mathcal{M}_{A} \rangle$ and $\mathcal{M}_{\mathcal{K}'} = \langle \mathcal{T}, \mathcal{M}'_{A} \rangle$ be two MBox *DL-Lite* knowledge bases, let Δ be an assertional-based merging operator and $\langle \mathcal{T}, \Delta(\mathcal{M}_{A}) \rangle$ be the *DL-Lite* knowledge base resulting from merging, where $\Delta(\mathcal{M}_{A})$ is a set of assertions. Let σ be a permutation over $\{1, \ldots n\}$, and $\mathcal{M}_{A} = \{\mathcal{A}_1, \ldots, \mathcal{A}_n\}$ be a multiset of assertions, $\overline{\sigma}(\mathcal{M}_{A})$ denotes the set $\{\mathcal{A}_{\sigma(1)}, \ldots, \mathcal{A}_{\sigma(n)}\}$. We rephrase the postulates as follows:

Inclusion $\Delta(\mathcal{M}_{A}) \subseteq \mathcal{A}_1 \cup \ldots \cup \mathcal{A}_n$.
Symmetry For any permutation σ over $\{1, \ldots n\}$, $\Delta(\overline{\sigma}(\mathcal{M}_{A})) = \Delta(\mathcal{M}_{A})$.
Consistency $\langle \mathcal{T}, \Delta(\mathcal{M}_{A}) \rangle$ is consistent.
Congruence If $\mathcal{A}_1 \cup \ldots \cup \mathcal{A}_n = \mathcal{A}'_1 \cup \ldots \cup \mathcal{A}'_n$ then $\Delta(\mathcal{M}_{A}) = \Delta(\mathcal{M}_{A'})$.
Vacuity If $\langle \mathcal{T}, \mathcal{M}_{A} \rangle$ is consistent then $\Delta(\mathcal{M}_{A}) = \mathcal{A}_1 \cup \ldots \cup \mathcal{A}_n$.
Reversion If $\langle \mathcal{T}, \mathcal{M}_{A} \rangle$ and $\langle \mathcal{T}, \mathcal{M}_{A'} \rangle$ have the same minimal inconsistent subsets then $(\mathcal{A}_1 \cup \ldots \cup \mathcal{A}_n) \backslash \Delta(\mathcal{M}_{A}) = (\mathcal{A}'_1 \cup \ldots \cup \mathcal{A}'_n) \backslash \Delta(\mathcal{M}_{A'})$.

---

[2] On each column the assertional removed sets are in bold.
[3] We do not consider the IC postulates [21] since they apply to belief sets and not to belief bases.

Core-retainment If $\alpha \in \mathcal{A}_1 \cup \ldots \cup \mathcal{A}_n$ and $\alpha \notin \Delta(\mathcal{M}_A)$ then there exists $\mathcal{A}'$ s. t. $\mathcal{A}' \subseteq \mathcal{A}_1 \cup \ldots \cup \mathcal{A}_n$, $\mathcal{A}'$ is consistent but $\mathcal{A}' \cup \{\alpha\}$ is inconsistent.

Relevance If $\alpha \in \mathcal{A}_1 \cup \ldots \cup \mathcal{A}_n$ and $\alpha \notin \Delta(\mathcal{M}_A)$ then there exists $\mathcal{A}'$ s. t. $\Delta(\mathcal{M}_A) \subseteq \mathcal{A}' \subseteq \mathcal{A}_1 \cup \ldots \cup \mathcal{A}_n$, $\mathcal{A}'$ is consistent but $\mathcal{A}' \cup \{\alpha\}$ is inconsistent.

Inclusion states that the union of the initial ABoxes is the upper bound of any merging operation. *Symmetry* establishes that all ABoxes are considered of equal importance. *Consistency* requires the consistency of the result of merging. *Congruence* requires that the result of merging should not depend on syntactic properties of the ABoxes. *Vacuity* says that if the union of the ABoxes is consistent w.r.t. $\mathcal{T}$ then the result of merging equals this union. *Reversion* says that if ABoxes have the same minimal inconsistent subsets w.r.t. $\mathcal{T}$ then the assertions erased in the respective ABoxes are the same. *Core-retainment* and *Relevance* express the intuition that nothing is removed from the original ABoxes unless its removal in some way contribute to make the result consistent.

Proposition 1. *Let $\mathcal{M}_\mathcal{K} = \langle \mathcal{T}, \mathcal{M}_A \rangle$ be a MBox DL-Lite knowledge base. For any selection function, $\forall P \in \{\Sigma, Card, Max, GMax\}$, Δ_P^{arsf} satisfies the* Inclusion, Symmetry, Consistency, Vacuity, Core-retainment *and* Relevance. *Δ_{Card}^{arsf} satisfies* Congruence *and* Reversion, *but $\forall P \in \{\Sigma, Max, GMax\}$, Δ_P^{arsf} does not satisfy* Congruence *nor* Reversion.

(sketch of the proof) For any selection function, by Definitions 4 and 5, $\forall P \in \{\Sigma, Card, Max, GMax\}$, Δ_P^{arsf} satisfies *Inclusion, Symmetry, Consistency, Vacuity* and *Core-retainment*.

Relevance: By Definition 5, for any selection function f, $\forall P \in \{\Sigma, Card, GMax\}$, if $\alpha \in \mathcal{A}_1 \cup \ldots \cup \mathcal{A}_n$ and $\alpha \notin \Delta_P^{arsf}(\mathcal{M}_A)$ then $\alpha \in f(\mathcal{R}_P(\mathcal{M}_\mathcal{K}))$. Let $\mathcal{A}' = \Delta_P^{arsf}(\mathcal{M}_A)$, $\mathcal{A}'$ is consistent and $A' \cup \{\alpha\}$ is inconsistent since $\alpha \in f(\mathcal{R}_P(\mathcal{M}_\mathcal{K}))$ and $f(\mathcal{R}_P(\mathcal{M}_\mathcal{K}))$ is an assertional removed set. By Definition 5, Δ_{Card}^{arsf} satisfies *Congruence* and *Reversion* since every assertion expressed more than once is reduced to a singleton. We provide a counter-example for Δ_P^{arsf}, $\forall P \in \{\Sigma, Max, GMax\}$. Let $\mathcal{M}_\mathcal{K} = \langle \mathcal{T}, \mathcal{M}_A \rangle$ be an inconsistent MBox *DL-Lite* knowledge base such that $\mathcal{T} = \{A \sqsubseteq \neg B\}$ and $A_1 = \{A(a)\}$, $A_2 = \{A(b), B(a)\}$, $A_3 = \{B(a), A(b)\}$. The potential assertional removed sets are $\mathcal{PR}(\mathcal{M}_\mathcal{K}) = \{X_1, X_2, X_3, X_4\}$ with $X_1 = \{A(a), A(b)\}$, $X_2 = \{A(a), B(b)\}$, $X_3 = \{B(a), A(b)\}$, $X_4 = \{B(a), B(b)\}$ and the sets of assertional removed sets are $\mathcal{R}_\Sigma(\mathcal{M}_\mathcal{K}) = \{X_1, X_2\}$, $\mathcal{R}_{Max}(\mathcal{M}_\mathcal{K}) = \{X_1, X_2\}$ and $\mathcal{R}_{GMax}(\mathcal{M}_\mathcal{K}) = \{X_1, X_2\}$.

X_i	$\|X_i \cap \mathcal{A}_1\|$	$\|X_i \cap \mathcal{A}_2\|$	$\|X_i \cap \mathcal{A}_3\|$	Σ	Max	$GMax$
X_1	1	1	0	2	1	110
X_2	1	0	1	2	1	110
X_3	0	2	1	3	2	210
X_4	0	1	2	3	2	210

Besides, let $\mathcal{M}_{\mathcal{K}'} = \langle \mathcal{T}, \mathcal{M}'_A \rangle$ be an inconsistent MBox *DL-Lite* knowledge base such that $\mathcal{T} = \{A \sqsubseteq \neg B\}$ and $A'_1 = \{A(a), B(b)\}$, $A'_2 = \{B(a)\}$, $A'_3 = \{A(a), A(a)\}$. We have $(A_1 \cup A_2 \cup A_3) = (A'_1 \cup A'_2 \cup A'_3)$ and $\mathcal{PR}(\mathcal{M}_{\mathcal{K}}) = \mathcal{PR}(\mathcal{M}_{\mathcal{K}'})$, and the sets of assertional removed sets are $\mathcal{R}_{\Sigma}(\mathcal{M}_{\mathcal{K}'}) = \{X_3, X_4\}$, $\mathcal{R}_{Max}(\mathcal{M}_{\mathcal{K}'}) = \{X_3, X_4\}$ and $\mathcal{R}_{GMax}(\mathcal{M}_{\mathcal{K}'}) = \{X_3, X_4\}$.

X_i	$|X_i \cap A'_1|$	$|X_i \cap A'_2|$	$|X_i \cap A'_3|$	Σ	Max	$GMax$
X_1	1	0	2	3	2	210
X_2	2	0	1	3	2	110
X_3	0	1	1	2	1	**110**
X_4	1	1	0	2	1	**110**

$\forall P \in \{\Sigma, Max, GMax\}$ we have $\mathcal{R}_P(\mathcal{M}_{\mathcal{K}}) \neq \mathcal{R}_P(\mathcal{M}_{\mathcal{K}'})$, and there is no selection function such that $f(\mathcal{R}_P(\mathcal{M}_{\mathcal{K}})) \in \mathcal{R}_P(\mathcal{M}_{\mathcal{K}'})$ therefore $\Delta_P^{arsf}(\mathcal{M}_A) \neq \Delta_P^{arsf}(\mathcal{M}_{A'})$.

5   Computing ARSF Merging Outcome

We first show the one to one correspondence between potential assertional removed sets and minimal hitting sets w.r.t. set inclusion [28]. We recall that a set H is a *hitting set* of a collection of sets $\mathcal{C}$ iff $\forall C \in \mathcal{C}, C \cap H \neq \emptyset$.

Proposition 2. *Let X be such that $X \subseteq \cup_{1 \leq i \leq n} A_i$. X is an potential assertional removed set of $\mathcal{M}_{\mathcal{K}}$ if and only if X is minimal hitting set w.r.t. set inclusion of $\mathcal{C}(\mathcal{M}_{\mathcal{K}})$.*

The proof is straightforward following Definition 2. Notice that the algorithm for the computation of the set of conflicts $\mathcal{C}(\mathcal{M}_{\mathcal{K}})$ is done in polynomial w.r.t. the size of $\mathcal{M}_{\mathcal{K}}$. This can be found *e.g.* in [7]. In the following, we provide a single algorithm to compute the potential assertional removed sets and the assertional removed sets according to the strategies $Card$, Σ, Max and $Gmax$. We give explanations on the different use cases of this algorithm hereafter. For a given assertional base $\mathcal{M}_{\mathcal{K}}$, the outcome of Algorithm 1 depends on the value of the parameter P: if $P \in \{Card, \Sigma, Max, Gmax\}$, then the result is $\mathcal{R}_P(\mathcal{M}_{\mathcal{K}})$. Otherwise the result is $\mathcal{PR}(\mathcal{M}_{\mathcal{K}})$.

Let us first focus on the computation of $\mathcal{PR}(\mathcal{M}_{\mathcal{K}})$. The algorithm is an adaptation of the algorithm for the computation of the minimal hitting sets w.r.t. set inclusion of a collection of sets described in [28]. It relies on the breadth-first construction of a directed acyclic graph called an *HS-dag*. An HS-dag T is a dag with labeled nodes and edges such that: (i) The root is labeled with $\emptyset$ if $\mathcal{C}(\mathcal{M}_{\mathcal{K}})$ is empty, otherwise it is labeled with an arbitrary element of $\mathcal{C}(\mathcal{M}_{\mathcal{K}})$; (ii) for each node n of T, we denote by $H(n)$ the set of edge labels on the path from n to the root of T; (iii) The label of a node n is any set $C \in \mathcal{C}(\mathcal{M}_{\mathcal{K}})$ such that $C \cap H(n) = \emptyset$ if such a set exists. Otherwise n is labeled with $\emptyset$. Nodes labeled with $\emptyset$ are called *terminal nodes*; (iv) If n is labeled by a set C, then for each $\alpha \in C$, n has a successor n_α, joined to n by an edge labeled by α.

Algorithm 1. Computes the elements of $\mathcal{R}_P(\mathcal{M_K})$ or the elements of $\mathcal{PR}(\mathcal{M_K})$ depending on the P parameter value.

```
1: function COMPUTE-ASSERTIONAL-RS(M_K, P)
                                                                    ▷ P: strategy
2:     M_K = ⟨T, M_A⟩, M_A = {A_1, · · · , A_n}
3:     level ← 0
4:     label(root) ← an element C ∈ C(M_K)                          ▷ root is the root node
5:     PrevQ ← {root}                                               ▷ Queue of nodes in the previous level
6:     if P ∈ {Σ, Max, Gmax} then
7:         MinNodes ← ∅                                             ▷ set of optimal nodes
8:         MinCost ← ∞                                     ▷ ∞ for Σ and Max, (∞, . . . , ∞) for GMax
                                                                          ⏟ n times
9:     mincard ← false                                             ▷ used by Card strategy
10:    while PrevQ ≠ ∅ and not mincard do
11:        level ← level + 1
12:        CurQ ← ∅
13:        for all no ∈ PrevQ do
14:            if label(no) ≠ ∅ and label(no) ≠ ⊠ then
15:                label(no) = {α, β}
16:                label(left_branch(no)) ← α
17:                label(right_branch(no)) ← β
18:                left_child(no) ←PROCESSCHILD(α, no, CurQ, M_K, MinCost, MinNodes, P)
19:                right_child(no) ←PROCESSCHILD(β, no, CurQ, M_K, MinCost, MinNodes, P)
20:                if label(left_child(no)) = ∅ or label(right_child(no)) = ∅ and P = Card then
21:                    mincard ← true
22:        PrevQ ← CurQ
23:    if P ∉ {Σ, Max, Gmax} then
24:        MinNodes ← all nodes labelled with ∅
25:    return MinNodes
```

Algorithm 2. Process a child branch of a node. Return a node (new or recycled).

```
1: function PROCESSCHILD(b_label, pa, CurQ, M_K, MinCost, MinNodes, P)
                                                    ▷ b_label: label of the branch to the new node
                                                    ▷ pa: the parent node
                                    ▷ CurQ: queue of nodes already processed at the current level (input/output parameter)
                                    ▷ MinCost: current minimum cost (input/output parameter)
                                    ▷ MinNodes: set of current minimum cost nodes (input/output parameter)
                                                    ▷ P: strategy
2:     M_K = ⟨T, M_A⟩
3:     M_A = {A_1, · · · , A_n}
4:     if ∃n′ ∈ CurQ such that H(n′) = H(pa) ∪ {b_label} then
5:         child_node ← n′                                          ▷ no new node creation
6:     else if ∃n′ ∈ T such that H(n′) ⊂ H(pa) ∪ {b_label} and label(n′) = ∅ then
7:         child_node ← a new node
8:         label(child_node) ← ⊠                                    ▷ this is a closed node
9:     else if P ∈ {Σ, Max, Gmax} and COST(P, H(pa) ∪ {b_label}) > MinCost then
10:        child_node ← a new node
11:        label(child_node) ← ⊠                                    ▷ this is a closed node
12:    else
13:        child_node ← a new node
14:        label(child_node) ← an element C ∈ C(M_K) such that C ∩ (H(pa) ∪ {b_label}) = ∅
15:        CurQ ← CurQ ∪ {child_node}
16:        if P ∈ {Σ, Max, Gmax} and label(child_node) = ∅ then
17:            if COST(P, H(pa) ∪ {b_label}) < MinCost then
                                                    ▷ Close current level nodes which are no more optimal
18:                for all nopt ∈ MinNodes do
19:                    label(nopt) ← ⊠
20:                MinNodes ← ∅
21:                MinCost ←COST(P, H(pa) ∪ {b_label})
22:            MinNodes ← MinNodes ∪ {child_node}
23:    return child_node
```

In our case, the elements of $C \in \mathcal{C}(\mathcal{M_K})$ are such that $|C| = 2$ (see [12]), so the HS-dag is binary. Algorithm 1 computes the potential assertional removed sets by computing the minimal hitting sets w.r.t. set inclusion of $\mathcal{C}(\mathcal{M_K})$. It builds a *pruned HS-dag* in a breadth-first order, using some pruning rules to avoid a complete development of the branches. We move the processing of the left and right children nodes in a separate function (described in Algorithm 2), as it first permits to keep the algorithm short and simple, and second facilitates the extension of this algorithm to the computation of the assertional removed sets according to the different strategies.

$PrevQ$ and $CurQ$ are sets containing respectively the nodes of the previous and the current level. $label(n)$ denotes the label of a node n. In a similar way, if b is a branch, $label(b)$ represents the label of b. $left\_branch(n)$ (*resp.* $right\_branch(n)$) denotes the left (*resp.* right) branch under the node n. $left\_child(n)$ (*resp.* $right\_child(n)$) represent the left (*resp.* right) child node of the node n. The algorithm iterates the nodes of a level and tries to develop the branches under each of these nodes. The central property is that the conflict C labeling a node n is such that $C \cap H(n) = \emptyset$.

Pruning rules are applied when trying to develop the left and right branches of some parent node pa (lines 4–22 in function PROCESSCHILD, Algorithm 2). Let us briefly describe them: (i) if there exists a node n' on the same level as the currently developed child branch such that $H(n') = H(pa) \cup \{b\_label\}$ (b_label being the label of the currently developed child branch), we connect the child branch to n', and there is no node creation (line 4); (ii) if there exists a node n' in the HS-dag such that $H(n') \subset H(pa) \cup \{b\_label\}$ and n' is a terminal node, then the node connected to the child branch is a *closed node* (which is marked with ⊠) (line 6); (iii) otherwise the node connected to the child branch is labelled by a conflict C such that $H(pa) \cup \{b\_label\} \cap C = \emptyset$. This new node is added to the current level queue.

Now we explain the aspects of the computation of the assertional removed sets according to each strategy P. *Card strategy.* The $Card$ strategy is the simplest one to implement. First, observe that the level of a node n in the HS-dag is equal to the cardinality of $H(n)$. This means that if n is an end node (a node labeled with $\emptyset$), the cardinality of the corresponding minimal hitting set is $H(n)$. Thus, there is no need to continue the construction of the HS-dag, as we are only interested in hitting sets which are minimal w.r.t. cardinality. In the light of the preceding observation, The only modification of the algorithm is the use of a boolean flag $mincard$ which halts the computation at the end of the level where the first potential assertional removed set has been detected. Σ, Max and $GMax$ strategies. As regards these strategies, we have no guarantee that the assertional removed sets reside in the same level of the tree, as illustrated by the following example for the Σ strategy.

Example 3. Let $\mathcal{M_K} = \langle \mathcal{T}, \mathcal{M_A} \rangle$ be an inconsistent MBox $DL\text{-}Lite$ knowledge base such that $\mathcal{T} = \{A \sqsubseteq \neg B, C \sqsubseteq \neg B\}$, and $\mathcal{A}_1 = \{A(a)\}$, $\mathcal{A}_2 = \{C(a)\}$, $\mathcal{A}_3 = \{B(a)\}$, $\mathcal{A}_4 = \{B(a)\}$, $\mathcal{A}_5 = \{B(a)\}$. We have $\mathcal{PR}(\mathcal{M_K}) = \{\{A(a), C(a)\}, \{B(a)\}\}$ and $\mathcal{R}_\Sigma(\mathcal{M_K}) = \{\{A(a), C(a)\}\}$. Thus the only assertional removed set is found at level 2, while the first potential assertional removed set is found at level 1.

Similar examples can be exhibited for the Max and $GMax$ strategies. The search strategy and associated pruning techniques for Σ, Max and $Gmax$ are located in lines 9 and 16 of Algorithm 2. They rely on a cost function which takes as parameters a strategy and a set S of ABox assertions. The different cost functions are defined according to the strategies, that is, given an MBox $\mathcal{M_A} = \{\mathcal{A}_1 \cup \ldots \cup \mathcal{A}_n\}$: For the Σ strategy COST(Σ, S) computes $|S \cap \mathcal{A}_1| + \ldots + |S \cap \mathcal{A}_n|$. For the Max strategy COST(Max, S) computes $\max(|S \cap \mathcal{A}_1|, \ldots, |S \cap \mathcal{A}_n|)$, For the $GMax$ strategy, using $p_X^{\mathcal{A}_i} = |X \cap \mathcal{A}_i|$, COST($GMax, S$) computes $L_X^{\mathcal{M_A}}$, which is the sequence $(p_X^{\mathcal{A}_1}, \ldots, p_X^{\mathcal{A}_n})$ sorted by decreasing lexicographic order.

The variable $MinCost$ maintains the current minimal cost. In line 9 of Algorithm 2, if the cost of the current node is greater than $MinCost$, then the node is closed, as is cannot be optimal. Otherwise we create a new node, labelled with a conflict which does not intersect $H(pa) \cup \{b\_label\}$. If such a label cannot be found (line 16), i.e. the current node is a terminal node then, at this point: (i) we are assured that $\mathrm{COST}(P, H(pa) \cup \{b\_label\}) \leq MinCost$, so we add the new node to the set of currently optimal nodes (line 22); (ii) if the cost of the current node is strictly less than $MinCost$, then we close all nodes currently believed to be optimal, empty the set containing them, and update $MinCost$ (lines 18–21).

Example 4. We illustrate the operation of the algorithm with the computation of the assertional removed sets of Example 2. Figure 1 depicts the HS-dag built by Algorithm 1. Circled numbers shows the ordering of nodes (apart from root which is obviously the first node).

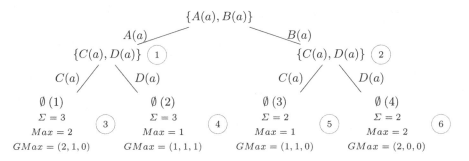

Fig. 1. Computing the removed sets of Example 2.

In order to facilitate the description, we denote by $MinNodes_P$ the variable $MinNode$ when considering strategy P. The same applies for $MinCost$. At the end of the execution of the processing of a node (PROCESSCHILD function), a state of these variables is given.

root The root is labelled with a conflict.

level 1
- Left and right branches of root node are labelled respectively with $A(a)$ and $B(a)$, the members of the root label (lines 16–17 of Algorithm 1).
- PROCESSCHILD$(\alpha, no, CurQ, \mathcal{M}_K, MinCost, MinNodes, P)$ is called. None of the pruning conditions in lines 4, 6 and 9 apply, so node ① is created, and labelled with a conflict not intersecting $H(①) = A(a)$, namely $\{C(a), D(a)\}$. The same processing leads to the creation of node ②.
  State: $MinNodes = \emptyset$, $MinCost = \infty$ for any strategy

level 2
- Left and right branches of node ① are labelled respectively with $C(a)$ and $D(a)$, the members of the label (lines 16–17 of Algorithm 1).
- PROCESSCHILD$(\alpha, no, CurQ, \mathcal{M}_K, MinCost, MinNodes, P)$ (left branch of node ①) is called. None of the pruning conditions in lines 4, 6 and 9 apply, so node ③ is created. As there is no conflict C such that $C \cap H(③) = \emptyset$,

the new node is labelled with $\emptyset$. Whatever the strategy is, its cost is necessarily less than $MinCost$ which has been initialized to ∞. Thus $MinCost$ is updated to the cost of node ③ depending on the strategy and node ③ is added to the $MinNodes$ set.

State: $MinNodes = \{③\}$, $MinCost_\Sigma = 3$, $MinCost_{Max} = 2$, $MinCost_{GMax} = (2, 1, 0)$.

- PROCESSCHILD($\beta, no, CurQ, \mathcal{M}_K, MinCost, MinNodes, P$) (right branch of node ①) is called. None of the pruning conditions in lines 4, 6 and 9 apply, so node ④ is created. As there is no conflict C such that $C \cap H(④) = \emptyset$, the new node is labelled with $\emptyset$. For strategy Σ, the cost of node ④ is equal to $MinCost$, thus node ④ is added to the $MinNodes$ set. For strategies Max and $GMax$, the cost of node ④ is less than $MinCost$: node ③ is closed (line 18), set $MinNodes$ is emptied, and $MinCost$ is updated.

  State: $MinNodes_\Sigma = \{③, ④\}$, $MinNodes_{Max} = \{④\}$, $MinNodes_{GMax} = \{④\}$, $MinCost_\Sigma = 3$, $MinCost_{Max} = 1$, $MinCost_{GMax} = (1, 1, 1)$.

- Left and right branches of node ② are labelled respectively with $C(a)$ and $D(a)$, the members of the label (lines 16–17 of Algorithm 1).

- PROCESSCHILD($\alpha, no, CurQ, \mathcal{M}_K, MinCost, MinNodes, P$) (left branch of node ②) is called. None of the pruning conditions in lines 4, 6 and 9 apply, so node ⑤ is created. As there is no conflict C such that $C \cap H(⑤) = \emptyset$, the new node is labelled with $\emptyset$. For strategy Σ, The cost of node ⑤ (2) is less than $MinCost$. The same applies for $GMax$

  State: $MinNodes_\Sigma = \{⑤\}$, $MinNodes_{Max} = \{④, ⑤\}$, $MinNodes_{GMax} = \{⑤\}$, $MinCost_\Sigma = 2$, $MinCost_{Max} = 1$, $MinCost_{GMax} = (1, 1, 0)$.

- PROCESSCHILD($\beta, no, CurQ, \mathcal{M}_K, MinCost, MinNodes, P$) (right branch of node ②) is called. None of the pruning conditions apply, so node ⑥ is created. As there is no conflict C such that $C \cap H(⑥) = \emptyset$, the new node is labelled with $\emptyset$. For strategy Σ, The cost of node ⑥ (2) is equal to $MinCost$.

  State: $MinNodes_\Sigma = \{⑤, ⑥\}$, $MinNodes_{Max} = \{④, ⑤\}$, $MinNodes_{GMax} = \{⑤\}$, $MinCost_\Sigma = 2$, $MinCost_{Max} = 1$, $MinCost_{GMax} = (1, 1, 0)$.

6   Conclusion

In this paper, we proposed new family of assertional-based merging operators, called Assertional Removed Sets Fusion (ARSF) operators, following several merging strategies (Σ, $Card$, Max, $GMax$). We studied the behaviour of ARSF operators with respect to a set of logical postulates (initially stated for propositional formula-based merging), which we rephrased within the DL-Lite framework. From a computational point of view, we proposed algorithms, stemming from the notion of hitting set, for computing the potential assertional removed sets as well as the assertional removed sets according to the different used strategies.

Belief change has been investigated within the framework of DL-Lite. Calvanese et al. [13] adapted formula-based and model-based approaches of ABox and Tbox belief revision and update, however they did not consider belief merging. Wang et al. [27] addressed the problem of TBox DL-Lite KB merging by adapting classical model-based belief merging to DL-Lite. This approach differs from the one we propose since we extend formula-based merging to DL lite.

In a future work we plan to conduct a complexity analysis of the proposed algorithm for the different used merging strategies. Moreover, we also want to focus on the implementation of ARSF operators and on an experimental study on real world applications, in particular 3D surveys within the context of underwater archaeology and handling conflicts in dances' videos. Furthermore, the ARSF operators stem from a selection function that selects one assertional removed set, we also plan to investigate operators stemming from other selection functions as well as other strategies and other approaches than ARSF for performing assertional-based merging.

Acknowledgements. This work is partially supported by the European project H2020-MSCA-RISE: AniAge (High Dimensional Heterogeneous Data based Animation Techniques for Southeast Asian Intangible Cultural Heritage). Zied Bouraoui was supported by CNRS PEPS INS2I MODERN.

References

1. Alchourrón, C., Gärdenfors, P., Makinson, D.: On the logic of theory change: partial meet contraction and revision functions. J. Symb. Log. **50**(2), 510–530 (1985)
2. Arenas, M., Bertossi, L.E., Chomicki, J.: Consistent query answers in inconsistent databases. In: Proceedings of the Eighteenth ACM SIGACT-SIGMOD-SIGART Symposium on Principles of Database Systems, Philadelphia, Pennsylvania, USA, pp. 68–79 (1999)
3. Artale, A., Calvanese, D., Kontchakov, R., Zakharyaschev, M.: The DL-Lite family and relations. J. Artif. Intell. Res. (JAIR) **36**, 1–69 (2009)
4. Baget, J.F., et al.: A general modifier-based framework for inconsistency-tolerant query answering. In: Principles of Knowledge Representation and Reasoning: Proceedings of the Fifteenth International Conference, KR 2016, Cape Town, South Africa, 25–29 April 2016, pp. 513–516 (2016)
5. Baget, J.F., et al.: Inconsistency-tolerant query answering: rationality properties and computational complexity analysis. In: Michael, L., Kakas, A. (eds.) JELIA 2016. LNCS (LNAI), vol. 10021, pp. 64–80. Springer, Cham (2016). https://doi.org/10.1007/978-3-319-48758-8_5
6. Baral, C., Kraus, S., Minker, J., Subrahmanian, V.S.: Combining knowledge bases consisting of first order theories. Comp. Intell. **8**(1), 45–71 (1992)
7. Benferhat, S., Bouraoui, Z., Papini, O., Würbel, E.: Assertional-based removed sets revision of DL-Lite$_R$ knowledge bases. In: ISAIM (2014)
8. Benferhat, S., Dubois, D., Kaci, S., Prade, H.: Possibilistic merging and distance-based fusion of propositional information. Stud. Logica. **58**(1), 17–45 (1997)
9. Bienvenu, M.: On the complexity of consistent query answering in the presence of simple ontologies. In: Proceedings of the Twenty-Sixth AAAI Conference on Artificial Intelligence (2012)
10. Bloch, I., Hunter, A., et al.: Fusion: general concepts and characteristics. Int. J. Intell. Syst. **16**(10), 1107–1134 (2001)
11. Bloch, I., Lang, J.: Towards mathematical morpho-logics. In: Bouchon-Meunier, B., Gutiérrez-Ríos, J., Magdalena, L., Yager, R.R. (eds.) Technologies for Constructing Intelligent Systems 2. STUDFUZZ, vol. 90, pp. 367–380. Physica, Heidelberg (2002). https://doi.org/10.1007/978-3-7908-1796-6_29

12. Calvanese, D., Giacomo, G.D., Lembo, D., Lenzerini, M., Rosati, R.: Tractable reasoning and efficient query answering in description logics: the DL-Lite family. J. Autom. Reasoning **39**(3), 385–429 (2007)

13. Calvanese, D., Kharlamov, E., Nutt, W., Zheleznyakov, D.: Evolution of *DL - Lite* knowledge bases. In: Patel-Schneider, P.F., et al. (eds.) ISWC 2010. LNCS, vol. 6496, pp. 112–128. Springer, Heidelberg (2010). https://doi.org/10.1007/978-3-642-17746-0_8

14. Falappa, M.A., Kern-Isberner, G., Reis, M.D.L., Simari, G.R.: Prioritized and non-prioritized multiple change on belief bases. J. Philos. Log. **41**, 77–113 (2012)

15. Falappa, M.A., Kern-Isberner, G., Simari, G.R.: Explanations, belief revision and defeasible reasoning. Artif. Intell. **141**(1/2), 1–28 (2002)

16. Fuhrmann, A.: An Essay on Contraction. CSLI Publications, Stanford (1997)

17. Hue, J., Papini, O., Würbel, E.: Syntactic propositional belief bases fusion with removed sets. In: Mellouli, K. (ed.) ECSQARU 2007. LNCS (LNAI), vol. 4724, pp. 66–77. Springer, Heidelberg (2007). https://doi.org/10.1007/978-3-540-75256-1_9

18. Hué, J., Würbel, E., Papini, O.: Removed sets fusion: performing off the shelf. In: Proceedings of ECAI 2008 (FIAI 178), pp. 94–98 (2008)

19. Konieczny, S.: On the difference between merging knowledge bases and combining them. In: Proceedings of KR 2000, pp. 135–144 (2000)

20. Konieczny, S., Lang, J., Marquis, P.: DA2 merging operators. Artif. Intell. **157**, 49–79 (2004)

21. Konieczny, S., Pérez, R.P.: Merging information under constraints. J. Log. Comput. **12**(5), 773–808 (2002)

22. Lembo, D., Lenzerini, M., Rosati, R., Ruzzi, M., Savo, D.F.: Inconsistency-tolerant query answering in ontology-based data access. J. Web Sem. **33**, 3–29 (2015)

23. Lin, J., Mendelzon, A.: Knowledge base merging by majority. In: Pareschi, R., Fronhoefer, B. (eds.) In Dynamic Worlds: From the Frame Problem to Knowledge Management. Kluwer, Dordrecht (1999)

24. Meyer, T., Ghose, A., Chopra, S.: Syntactic representations of semantic merging operations. In: Ishizuka, M., Sattar, A. (eds.) PRICAI 2002. LNCS (LNAI), vol. 2417, p. 620. Springer, Heidelberg (2002). https://doi.org/10.1007/3-540-45683-X_88

25. Revesz, P.Z.: On the semantics of theory change: arbitration between old and new information. In: 12th ACM SIGACT-SGMIT-SIGART Symposium on Principes of Databases, pp. 71–92 (1993)

26. Revesz, P.Z.: On the semantics of arbitration. J. Algebra Comput. **7**, 133–160 (1997)

27. Wang, Z., Wang, K., Jin, Y., Qi, G.: Ontomerge a system for merging DL-Lite ontologies. In: CEUR Workshop Proceedings, vol. 969, pp. 16–27 (2014)

28. Wilkerson, R.W., Greiner, R., Smith, B.A.: A correction to the algorithm in Reiter's theory of diagnosis. Artif. Intell. **41**, 79–88 (1989)

An Interactive Polyhedral Approach for Multi-objective Combinatorial Optimization with Incomplete Preference Information

Nawal Benabbou and Thibaut Lust[(✉)]

Sorbonne Université, CNRS, Laboratoire d'Informatique de Paris 6, LIP6,
75005 Paris, France
{nawal.benabbou,thibaut.lust}@lip6.fr

Abstract. In this paper, we develop a general interactive polyhedral approach to solve multi-objective combinatorial optimization problems with incomplete preference information. Assuming that preferences can be represented by a parameterized scalarizing function, we iteratively ask preferences queries to the decision maker in order to reduce the imprecision over the preference parameters until being able to determine her preferred solution. To produce informative preference queries at each step, we generate promising solutions using the extreme points of the polyhedron representing the admissible preference parameters and then we ask the decision maker to compare two of these solutions (we propose different selection strategies). These extreme points are also used to provide a stopping criterion guaranteeing that the returned solution is optimal (or near-optimal) according to the decision maker's preferences. We provide numerical results for the multi-objective spanning tree and traveling salesman problems with preferences represented by a weighted sum to demonstrate the practical efficiency of our approach. We compare our results to a recent approach based on minimax regret, where preference queries are generated during the construction of an optimal solution. We show that better results are achieved by our method both in terms of running time and number of questions.

Keywords: Multi-objective combinatorial optimization · Minimum spanning tree problem · Traveling salesman problem · Incremental preference elicitation · Minimax regret

1 Introduction

The increasing complexity of applications encountered in Computer Science significantly complicates the task of decision makers who need to find the best solution among a very large number of options. Multi-objective optimization is concerned with optimization problems involving several (conflicting) objectives/criteria to be optimized simultaneously (e.g., minimizing costs while maximizing profits). Without preference information, we only know that the best

© Springer Nature Switzerland AG 2019
N. Ben Amor et al. (Eds.): SUM 2019, LNAI 11940, pp. 221–235, 2019.
https://doi.org/10.1007/978-3-030-35514-2_17

solution for the decision maker (DM) is among the Pareto-optimal solutions (a solution is called Pareto-optimal if there exists no other solution that is better on all objectives while being strictly better on at least one of them). The main problem with this kind of approach is that the number of Pareto-optimal solutions can be intractable, that is exponential in the size of the problem (e.g. [13] for the multicriteria spanning tree problem). One way to address this issue is to restrict the size of the Pareto set in order to obtain a "well-represented" Pareto set; this approach is often based on a division of the objective space into different regions (e.g., [15]) or on ϵ-dominance (e.g., [18]). However, whenever the DM needs to identify the best solution, it seems more appropriate to refine the Pareto dominance relation with preferences to determine a single solution satisfying the subjective preferences of the DM. Of course, this implies the participation of the DM who has to give us some insights and share her preferences.

In this work, we assume that the DM's preferences can be represented by a parameterized scalarizing function (e.g., a weighted sum), allowing some trade-off between the objectives, but the corresponding preference parameters (e.g., the weights) are initially not known; hence, we have to consider the set of all parameters compatible with the collected preference information. An interesting approach to deal with preference imprecision has been recently developed [19, 21, 30] and consists in determining the *possibly* optimal solutions, that is the solutions that are optimal for at least one instance of the preference parameters. The main drawback of this approach, though, is that the number of possibly optimal solutions may still be very large compared to the number of Pareto-optimal solutions; therefore there is a need for elicitation methods aiming to specify the preference model by asking preference queries to the DM.

In this paper, we study the potential of incremental preference elicitation (e.g., [23, 27]) in the framework of multi-objective combinatorial optimization. Preference elicitation on combinatorial domains is an active topic that has been recently studied in various contexts, e.g. in multi-agents systems [1, 3, 6], in stable matching problems [9], in constraint satisfaction problems [7], in Markov Decision Processes [11, 24, 28] and in multi-objective optimization problems [4, 14, 16]. Our aim here is to propose a general interactive approach for multi-objective optimization with imprecise preference parameters. Our approach identifies informative preference queries by exploiting the extreme points of the polyhedron representing the admissible preference parameters. Moreover, these extreme points are also used to provide a stopping criterion which guarantees the determination of the (near-)optimal solution. Our approach is general in the sense that it can be applied to any multi-objective optimization problem, providing that the scalarizing function is linear in its preference parameters (e.g., weighted sums, Choquet integrals [8, 12]) and that there exists an efficient algorithm to solve the problem when preferences are precisely known (e.g., [17, 22] for the minimum spanning tree problem with a weighted sum).

The paper is organized as follows: We first give general notations and recall the basic principles of regret-based incremental elicitation. We then propose a new interactive method based on the minimax regret decision criterion and extreme points generation. Finally, to show the efficiency of our method, we provide numerical results for two well-known problems, namely the multicriteria traveling salesman and multicriteria spanning tree problems; for the latter, we compare our results with those obtained by the state-of-the-art method.

2 Multi-objective Combinatorial Optimization

In this paper, we consider a general multi-objective combinatorial optimization (MOCO) problem with n objective functions $y_i, i \in \{1, \ldots, n\}$, to be minimized. This problem can be defined as follows:

$$\underset{x \in \mathcal{X}}{\text{minimize}}\big(y_1(x), \ldots, y_n(x)\big)$$

In this definition, $\mathcal{X}$ is the feasible set in the decision space, typically defined by some constraint functions (e.g., for the multicriteria spanning tree problem, $\mathcal{X}$ is the set of all spanning trees of the graph). In this problem, any solution $x \in \mathcal{X}$ is associated with a cost vector $y(x) = (y_1(x), \ldots, y_n(x)) \in \mathbb{R}^n$ where $y_i(x)$ is the evaluation of x on the i-th criterion/objective. Thus the image of the feasible set in the objective space is defined by $\{y(x) : x \in \mathcal{X}\} \subset \mathbb{R}^n$.

Solutions are usually compared through their images in the objective space (also called points) using the *Pareto dominance* relation: we say that point $u = (u_1, \ldots, u_n) \in \mathbb{R}^n$ *Pareto dominates* point $v = (v_1, \ldots, v_n) \in \mathbb{R}^n$ (denoted by $u \prec_P v$) if and only if $u_i \leq v_i$ for all $i \in \{1, \ldots, n\}$, with at least one strict inequality. Solution $x^* \in \mathcal{X}$ is called *efficient* if there does not exist any other feasible solution $x \in \mathcal{X}$ such that $y(x) \prec_P y(x^*)$; its image in objective space is then called a non-dominated point.

3 Minimax Regret Criterion

We assume here that the DM's preferences over solutions can be represented by a parameterized scalarizing function f_ω that is linear in its parameters ω. Solution $x \in \mathcal{X}$ is preferred to solution $x' \in \mathcal{X}$ if and only if $f_\omega(y(x)) \leq f_\omega(y(x'))$. To give a few examples, function f_ω can be a weighted sum (i.e. $f_\omega(y(x)) = \sum_{i=1}^n \omega_i y_i(x)$) or a Choquet integral with capacity ω [8,12]. We also assume that parameters ω are not known initially. Instead, we consider a (possibly empty) set Θ of pairs $(u, v) \in \mathbb{R}^n \times \mathbb{R}^n$ such that u is known to be preferred to v; this set can be obtained by asking preference queries to the DM. Let Ω_Θ be the set of all parameters ω that are compatible with Θ, i.e. all parameters ω that satisfy the constraints $f_\omega(u) \leq f_\omega(v)$ for all $(u, v) \in \Theta$. Thus, since f_ω is linear in ω, we can assume that Ω_Θ is a convex polyhedron throughout the paper. The problem is now to determine the most promising solution under the preference imprecision (defined by Ω_Θ). To do so, we use the minimax regret approach (e.g., [7]) which is based on the following definitions:

Definition 1 (Pairwise Max Regret). *The Pairwise Max Regret (PMR) of solution $x \in \mathcal{X}$ with respect to solution $x' \in \mathcal{X}$ is:*

$$PMR(x, x', \Omega_\Theta) = \max_{\omega \in \Omega_\Theta} \{f_\omega(y(x)) - f_\omega(y(x'))\}$$

In other words, $PMR(x, x', \Omega_\Theta)$ is the worst-case loss when choosing solution x instead of solution x'.

Definition 2 (Max Regret). *The Max Regret (MR) of solution $x \in \mathcal{X}$ is:*

$$MR(x, \mathcal{X}, \Omega_\Theta) = \max_{x' \in \mathcal{X}} PMR(x, x', \Omega_\Theta)$$

Thus $MR(x, \mathcal{X}, \Omega_\Theta)$ is the worst-case loss when selecting solution x instead of any other feasible solution $x' \in \mathcal{X}$. We can now define the minimax regret:

Definition 3 (Minimax Regret). *The MiniMax Regret (MMR) is:*

$$MMR(\mathcal{X}, \Omega_\Theta) = \min_{x \in \mathcal{X}} MR(x, \mathcal{X}, \Omega_\Theta)$$

According to the minimax regret criterion, an optimal solution is a solution that achieves the minimax regret (i.e., any solution in $\arg\min_{x \in \mathcal{X}} MR(x, \mathcal{X}, \Omega_\Theta)$), allowing to minimize the worst-case loss. Note that if $MMR(\mathcal{X}, \Omega_\Theta) = 0$, then any optimal solution for the minimax regret criterion is necessarily optimal according to the DM's preferences.

4 An Interactive Polyhedral Method

Our aim is to produce an efficient regret-based interactive method for the determination of a (near-)optimal solution according to the DM's preferences. Note that the value $MMR(\mathcal{X}, \Omega_\Theta)$ can only decrease when inserting new preference information in Θ, as observed in previous works (see e.g., [5]). Therefore, the general idea of regret-based incremental elicitation is to ask preference queries to the DM in an iterative way, until the value $MMR(\mathcal{X}, \Omega_\Theta)$ drops below a given threshold $\delta \geq 0$ representing the maximum allowable gap to optimality; one can simply set $\delta = 0$ to obtain the preferred solution (i.e., the optimal solution according to the DM's preferences).

At each iteration step, the minimax regret $MMR(\mathcal{X}, \Omega_\Theta)$ could be obtained by computing the pairwise max regrets $PMR(x, x', \Omega_\Theta)$ for all pairs (x, x') of distinct solutions in $\mathcal{X}$ (see Definitions 2 and 3). However, this would not be very efficient in practice due to the large size of $\mathcal{X}$ (recall that $\mathcal{X}$ is the feasible set of a MOCO problem). This observation has led a group of researchers to propose a new approach consisting in combining preference elicitation and search by asking preference queries during the construction of the (near-)optimal solution (e.g., [2]). In this work, we propose to combine incremental elicitation and search in a different way: at each iteration step, we generate a set of promising solutions using the extreme points of Ω_Θ (the set of admissible parameters), we ask the

DM to compare two of these solutions, we update Ω_Θ according to her answer and we stop the process whenever a (near-)optimal solution is detected (i.e. a solution $x \in \mathcal{X}$ such that $MR(x, \mathcal{X}, \Omega_\Theta) \leq \delta$ holds). More precisely, taking as input a MOCO problem P, a tolerance threshold $\delta \geq 0$, a scalarizing function f_ω with unknown parameters ω and an initial set of preference statements Θ, our algorithm iterates as follows:

1. First, the set of all extreme points of polyhedron Ω_Θ are generated. This set is denoted by EP_Θ and its kth element is denoted by ω^k.
2. Then, for every point $\omega^k \in EP_\Theta$, P is solved considering the *precise* scalarizing function f_{ω^k} (the corresponding optimal solution is denoted by x^k).
3. Finally $MMR(X_\Theta, \Omega_\Theta)$ is computed, where $X_\Theta = \{x^k : k \in \{1, \ldots, |EP_\Theta|\}\}$. If this value is strictly larger than δ, then the DM is asked to compare two solutions $x, x' \in X_\Theta$ and Ω_Θ is updated by imposing the linear constraint $f_\omega(x) \leq f_\omega(x')$ (or $f_\omega(x) \geq f_\omega(x')$ depending on her answer); the algorithm stops otherwise.

Our algorithm, called IEEP (for Incremental Elicitation based on Extreme Points), is summarized in Algorithm 1. The implementation details of Select, Optimizing and ExtremePoints procedures are given in the numerical section. Note however that Optimizing is a procedure that depends on the optimization problem (e.g., Prim algorithm could be used for the spanning tree problem). The following proposition establishes the validity of our interactive method:

Proposition 1. *For any positive tolerance threshold δ, algorithm IEEP returns a solution $x^* \in \mathcal{X}$ such that the inequality $MR(x^*, \mathcal{X}, \Omega_\Theta) \leq \delta$ holds.*

Proof. Let x^* be the returned solution and let K be the number of extreme points of Ω_Θ at the end of the execution. For all $k \in \{1, \ldots, K\}$, let ω^k be the kth extreme point of Ω_Θ and let x^k be a solution minimizing function f_{ω^k}. Let $X_\Theta = \{x^k : k \in \{1, \ldots, K\}\}$. We know that $MR(x^*, X_\Theta, \Omega_\Theta) \leq \delta$ holds at the end of the while loop (see the loop condition); hence we have $f_\omega(x^*) - f_\omega(x^k) \leq \delta$ for all solutions $x^k \in X_\Theta$ and all parameters $\omega \in \Omega_\Theta$ (see Definition 2).

We want to prove that $MR(x^*, \mathcal{X}, \Omega_\Theta) \leq \delta$ holds at the end of execution. To do so, it is sufficient to prove that $f_\omega(x^*) - f_\omega(x) \leq \delta$ holds for all $x \in \mathcal{X}$ and all $\omega \in \Omega_\Theta$. Since Ω_Θ is a convex polyhedron, for any $\omega \in \Omega_\Theta$, there exists a vector $\lambda = (\lambda_1, \ldots, \lambda_K) \in [0,1]^K$ such that $\sum_{k=1}^{K} \lambda^k = 1$ and $\omega = \sum_{k=1}^{K} \lambda_k \omega^k$. Therefore, for all solutions $x \in \mathcal{X}$ and for all parameters $\omega \in \Omega_\Theta$, we have:

$$f_\omega(x^*) - f_\omega(x) = \sum_{k=1}^{K} \left[\lambda^k (f_{\omega^k}(x^*) - f_{\omega^k}(x)) \right] \text{ by linearity}$$

$$\leq \sum_{k=1}^{K} \left[\lambda^k (f_{\omega^k}(x^*) - f_{\omega^k}(x^k)) \right] \text{ since } x^k \text{ is } f_{\omega^k}\text{-optimal}$$

$$\leq \sum_{k=1}^{K} \left[\lambda^k \times \delta \right] \text{ since } f_\omega(x^*) - f_\omega(x^k) \leq \delta$$

$$= \delta \times \sum_{k=1}^{K} \lambda^k$$

$$= \delta. \qquad \qquad \square$$

For illustration proposes, we now present the execution of our algorithm on a small instance of the multicriteria spanning tree problem.

Algorithm 1. IEEP

IN ↓ P: a MOCO problem; δ: a threshold; f_ω: a scalarizing function with unknown parameters ω; Θ: a set of preference statements.
OUT ↑: a solution x^* with a max regret smaller than δ.

--| Initialization of the convex polyhedron:
$\Omega_\Theta \leftarrow \{\omega : \forall(u,v) \in \Theta, f_\omega(u) \leq f_\omega(v)\}$
--| Generation of the extreme points of the polyhedron:
$EP_\Theta \leftarrow$ ExtremePoints(Ω_Θ)
--| Generation of the optimal solutions attached to EP_Θ:
$X_\Theta \leftarrow$ Optimizing(P, EP_Θ)
while $MMR(X_\Theta, \Omega_\Theta) > \delta$ **do**
   --| Selection of two solutions to compare:
   $(x, x') \leftarrow$ Select(X_Θ)
   --| Question:
   query(x, x')
   --| Update preference information:
   if x is preferred to x' **then**
      $\Theta \leftarrow \Theta \cup \{(y(x), y(x'))\}$
   else
      $\Theta \leftarrow \Theta \cup \{(y(x'), y(x))\}$
   end
   $\Omega_\Theta \leftarrow \{\omega : \forall(u,v) \in \Theta, f_\omega(u) \leq f_\omega(v)\}$
   --| Generation of the extreme points of the polyhedron:
   $EP_\Theta \leftarrow$ ExtremePoints(Ω_Θ)
   --| Generation of the optimal solutions attached to EP_Θ:
   $X_\Theta \leftarrow$ Optimizing(P, EP_Θ)
end
return a solution $x^* \in X_\Theta$ minimizing $MR(x, X_\Theta, \Omega_\Theta)$

Example 1. Consider the multicriteria spanning tree problem with 5 nodes and 7 edges given in Fig. 1. Each edge is evaluated with respect to 3 criteria. Assume that the DM's preferences can be represented by a weighted sum f_ω with unknown parameters ω. Our goal is to determine an optimal spanning tree for the DM ($\delta = 0$), i.e. a connected acyclic sub-graph with 5 nodes that is f_ω-optimal.

We now apply algorithm IEEP on this instance, starting with an empty set of preference statements (i.e. $\Theta = \emptyset$).

Initialization: As $\Theta = \emptyset$, Ω_Θ is initialized to the set of all weighting vectors $\omega = (\omega_1, \omega_2, \omega_3) \in [0,1]^3$ such that $\omega_1 + \omega_2 + \omega_3 = 1$. In Fig. 2, Ω_Θ is represented by the triangle ABC in the space (ω_1, ω_2); value ω_3 is implicitly defined by $\omega_3 = 1 - \omega_1 - \omega_2$. Hence the initial extreme points are the vectors of the natural basis of the Euclidean space, corresponding to Pareto dominance [29]; in other words, we have $EP_\Theta = \{\omega^1, \omega^2, \omega^3\}$ with $\omega^1 = (1,0,0)$, $\omega^2 = (0,1,0)$ and $\omega^3 = (0,0,1)$. We then optimize according to all weighting vectors in EP_Θ using Prim algorithm [22], and we obtain the following three solutions: for ω^1, we have a spanning tree x^1 evaluated by $y(x^1) = (15, 17, 14)$; for ω^2, we obtain a spanning tree x^2 with $y(x^2) = (23, 8, 16)$; for ω^3, we find a spanning tree x^3 such that $y(x^3) = (17, 16, 11)$. Hence we have $X_\Theta = \{x^1, x^2, x^3\}$.

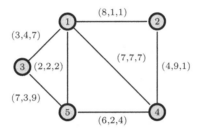

Fig. 1. A three-criteria minimum spanning tree problem.

Iteration Step 1: Since $MMR(X_\Theta, \Omega_\Theta) = 8 > \delta = 0$, we ask the DM to compare two solutions in X_Θ, say x^1 and x^2. Assume that the DM prefers x^2. In that case, we perform the following updates: $\Theta = \{((23, 8, 16), (15, 17, 14))\}$ and $\Omega_\Theta = \{\omega : f_\omega(23, 8, 16) \le f_\omega(15, 17, 14)\}$; in Fig. 3, Ω_Θ is represented by triangle BFE. We then compute the set EP_Θ of its extreme points (by applying the algorithm in [10] for example) and we obtain $EP_\Theta = \{\omega^1, \omega^2, \omega^3\}$ with $\omega^1 = (0.53, 0.47, 0)$, $\omega^2 = (0, 0.18, 0.82)$ and $\omega^3 = (0, 1, 0)$. We optimize according to these weights and we obtain three spanning trees: $X_\Theta = \{x^1, x^2, x^3\}$ with $y(x^1) = (23, 8, 16)$, $y(x^2) = (17, 16, 11)$ and $y(x^3) = (19, 9, 14)$.

Iteration Step 2: Here $MMR(X_\Theta, \Omega_\Theta) = 1.18 > \delta = 0$. Therefore, we ask the DM to compare two solutions in X_Θ, say x^1 and x^2. Assume she prefers x^2. We then obtain $\Theta = \{((23, 8, 16), (15, 17, 14)), ((17, 16, 11), (23, 8, 16))\}$ and we set $\Omega_\Theta = \{\omega : f_\omega(23, 8, 16) \le f_\omega(15, 17, 14) \wedge f_\omega(17, 16, 11) \le f_\omega(23, 8, 16)\}$. We compute the corresponding extreme points which are given by $EP_\Theta = \{(0.43, 0.42, 0.15), (0, 0.18, 0.82), (0, 0.38, 0.62)\}$ (see triangle HGE in Fig. 4); finally we have $X_\Theta = \{x^1, x^2\}$ with $y(x^1) = (17, 16, 11)$ and $y(x^2) = (19, 9, 14)$.

Iteration Step 3: Now $MMR(X_\Theta, \Omega_\Theta) = 1.18 > \delta = 0$. Therefore we ask the DM to compare x^1 and x^2. Assuming that she prefers x^2, we update Θ by inserting the preference statement $((19, 9, 14), (17, 16, 11))$ and we update Ω_Θ by imposing the following additional constraint: $f_\omega(19, 9, 14) \leq f_\omega(17, 16, 11)$ (see Fig. 5); the corresponding extreme points are given by $EP_\Theta = \{(0.18, 0.28, 0.54), (0, 0.3, 0.7), (0, 0.38, 0.62), (0.43, 0.42, 0.15)\}$. Now the set X_Θ only includes one spanning tree x^1 and $y(x^1) = (19, 9, 14)$. Finally, the algorithm stops (since we have $MMR(X_\Theta, \Omega_\Theta) = 0 \leq \delta = 0$) and it returns solution x^1 (which is guaranteed to be the optimal solution for the DM).

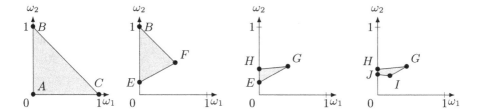

Fig. 2. Initial set.   **Fig. 3.** After step 1.   **Fig. 4.** After step 2.   **Fig. 5.** After step 3.

5   Experimental Results

We now provide numerical results aiming to evaluate the performance of our interactive approach. At each iteration step of our procedure, the DM is asked to compare two solutions selected from the set X_Θ until $MR(X_\Theta, \Omega_\Theta) \leq \delta$. Therefore, we need to estimate the impact of procedure Select on the performances of our algorithm. Here we consider the following query generation strategies:

- **Random:** The two solutions are randomly chosen in X_Θ.
- **Max-Dist:** We compute the Euclidean distance between all solutions in the objective space and we choose a pair of solutions maximizing the distance.
- **CSS:** The Current Solution Strategy (CSS) consists in selecting a solution that minimizes the max regret and one of its adversary's choice [7][1].

These strategies are compared using the following indicators:

- **time:** The running time given in seconds.
- **eval:** The number of evaluations, i.e. the number of problems with known preferences that are solved during the execution; recall that we solve one optimization problem per extreme point at each iteration step (see Optimizing).
- **queries:** The number of preference queries generated during the execution.
- **qOpt:** The number of preference queries generated until the determination of the preferred solution (but not yet proved optimal).

---

[1] Note that these three strategies are equivalent when only considering two objectives since the number of extreme points is always equal to two in this particular case.

We assume here that the DM's preferences can be represented by a weighted sum f_ω but the weights $\omega = (\omega_1, \ldots, \omega_n)$ are not known initially. More precisely, we start the execution with an empty set of preference statements (i.e. $\Theta = \emptyset$ and $\Omega_\Theta = \{\omega \in \mathbb{R}_+^n : \sum_{i=1}^n \omega_i = 1\}$) and then any new preference statement $(u, v) \in \mathbb{R}^2$ obtained from the DM induces the following linear constraint over the weights: $\sum_{i=1}^n \omega_i u_i \leq \sum_{i=1}^n \omega_i v_i$. Hence Ω_Θ is a convex polyhedron. In our experiments, the answers to queries are simulated using a weighting vector ω randomly generated before running the algorithm, using the procedure presented in [25], to guarantee a uniform distribution of the weights.

Implementation Details. Numerical tests were performed on a Intel Core i7-7700, at 3.60 GHz, with a program written in C. At each iteration step of our algorithm, the extreme points associated to the convex polyhedron Ω_Θ are generated using the `polymake` library[2]. Moreover, at each step, we do not compute PMR values using a linear programming solver. Instead, we only compute score differences since the maximum value is always obtained for an extreme point of the convex polyhedron. Furthermore, to reduce the number of PMR computations, we use Pareto dominance tests between the extreme points to eliminate dominated solutions, as proposed in [20].

5.1   Multicriteria Spanning Tree

In these experiments, we consider instances of the multicriteria spanning tree (MST) problem, which is defined by a connected graph $G = (V, E)$ where each edge $e \in E$ is valued by a cost vector giving its cost with respect to different criteria/objectives (every criterion is assumed to be additive over the edges). A spanning tree of G is a connected sub-graph of G which includes every vertex $v \in V$ while containing no cycle. In this problem, $\mathcal{X}$ is the set of all spanning trees of G. We generate instances of $G = (V, E)$ with a number of vertices $|V|$ varying between 50 and 100 and a number of objectives n ranging from 2 to 6. The edge costs are drawn within $\{1, \ldots, 1000\}^n$ uniformly at random. For the MST problem, procedure $\texttt{Optimizing}(P, EP_\Theta)$ proceeds as follows: First, for all extreme points $\omega^k \in EP_\Theta$, an instance of the spanning tree problem with a single objective is created by simply aggregating the edge costs of G using weights ω^k. Then, Prim algorithm is applied on the resulting graphs. The results obtained by averaging over 30 runs are given in Table 1 for $\delta = 0$.

---

[2] https://polymake.org.

Table 1. MST: comparison of the different query strategies (best values in bold).

n	|V|	IEEP - Random				IEEP - Max-Dist				IEEP - CSS			
		time(s)	queries	eval	qOpt	time(s)	queries	eval	qOpt	time(s)	queries	eval	qOpt
2	50	8.6	**7.4**	**9.4**	**4.6**	8.0	**7.4**	**9.4**	**4.6**	7.7	**7.4**	**9.4**	**4.6**
3	50	16.9	16.2	34.9	10.9	**16.5**	**15.2**	**33.1**	**10.2**	17.9	16.9	35.9	12.0
4	50	27.5	25.7	117.3	19.7	**26.4**	**24.6**	**112.3**	**17.2**	30.7	28.9	130.8	20.1
5	50	37.7	35.0	363.2	27.2	**36.2**	**34.3**	**358.4**	**23.3**	42.3	39.8	404.7	30.6
6	50	46.1	43.3	**1056.3**	35.3	**45.5**	**42.7**	1075.2	**32.8**	62.6	57.6	1537.9	43.6
2	100	10.0	**8.6**	**10.6**	**5.7**	8.9	**8.6**	**10.6**	**5.7**	9.2	**8.6**	**10.6**	**5.7**
3	100	**18.7**	17.6	37.8	14.0	19.0	**17.4**	**37.2**	13.9	19.0	17.7	37.7	**13.0**
4	100	32.0	29.9	134.0	23.3	**30.1**	**28.4**	**129.9**	**22.3**	34.8	32.5	147.0	24.1
5	100	**41.8**	39.8	**404.4**	31.3	42.1	**39.2**	411.5	**31.0**	55.9	51.7	564.8	40.6
6	100	55.9	51.5	1306.1	40.0	**52.3**	**49.1**	1259.3	**38.7**	84.0	75.7	2329.6	62.1

Running Time and Number of Evaluations. We observe that Random and Max-Dist strategies are much faster than CSS strategy; for instance, for $n = 6$ and $|V| = 100$, Random and Max-Dist strategies end before one minute whereas CSS needs almost a minute and a half. Note that time is mostly consumed by the generation of extreme points, given that the evaluations are performed by Prim algorithm which is very efficient. Since the number of evaluations with CSS drastically increases with the size of the problem, we may expect the performance gap between CSS and the two other strategies to be much larger for MOCO problems with a less efficient solving method.

Number of Generated Preference Queries. We can see that Max-Dist is the best strategy for minimizing the number of generated preference queries. More precisely, for all instances, the preferred solution is detected with less than 40 queries and the optimality is established after at most 50 queries. In fact, we can reduce even further the number of preference queries by considering a strictly positive tolerance threshold; to give an example, if we set $\delta = 0.1$ (i.e. 10% of the "maximum" error computed using the ideal point and the worst objective vector), then our algorithm combined with Max-Dist strategy generates at most 20 queries in all considered instances. In Table 1, we also observe that CSS strategy generates many more queries than Random, which is quite surprising since CSS strategy is intensively used in incremental elicitation (e.g., [4,7]). To better understand this result, we have plotted the evolution of minimax regret with respect to the number of queries for the bigger instance of our set ($|V| = 100$, $n = 6$). We have divided the figure in two parts: the first part is when the number of queries is between 1 and 20 and the other part is when the number of queries is between 20 and 50 (see Fig. 6). In the first figure, we observe that there is almost no difference between the three strategies, and the minimax regret is already close to 0 after only 20 questions (showing that we are very close to the optimum relatively quickly). However, there is a significant difference between the three strategies in the second figure: the minimax regret with CSS starts to reduce

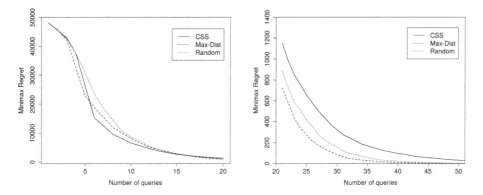

Fig. 6. MST problem with $n = 6$ and $|V| = 100$: evolution of the minimax regret between 1 and 20 queries (left) and between 21 and 50 queries (right).

less quickly after 30 queries, remaining strictly positive after 50 queries, whereas the optimal solution is found after about 40 queries with the other strategies. Thus, queries generated with CSS gradually becomes less and less informative than those generated by the two other strategies. This can be explained by the following: CSS always selects the minimax regret optimal solution and one of its worst adversary. Therefore, when the minimax regret optimal solution does not change after asking a query, the same solution is used for the next preference query. This can be less informative than asking the DM to compare two solutions for which we have no preference information at all; Random and Max-Dist strategies select the two solutions to compare in a more diverse way.

Comparison with the State-of-the-Art Method. In this subsection, we compare our interactive method with the state-of-the-art method proposed in [2]. The latter consists essentially in integrating incremental elicitation into Prim algorithm [22]; therefore, this method will be called IE-Prim hereafter. The main difference between IE-Prim and IEEP is that IE-Prim is constructive: queries are not asked on complete solutions but on partial solutions (edges of the graph). We have implemented ourselves IE-Prim, using the same programming language and data structures than IEEP, in order to allow a fair comparison between these methods. Although IE-Prim was only proposed and tested with CSS in [2], we have integrated the two other strategies (i.e., Max-Dist and Random) in IE-Prim.

In Table 2, we compare IEEP with Max-Dist and IE-Prim in terms of running times and number of queries[3]. We see that IEEP outperforms IE-Prim in all settings, allowing the running time and the number of queries to be divided by three in our biggest instances. Note that Max-Dist and Random strategies improve the performances of IE-Prim (compared to CSS), but it is still not enough to achieve results comparable to IEEP. This shows that asking queries during the

---

[3] Note that we cannot compute qOpt and eval for IE-Prim since it is constructive and makes no evaluation.

Table 2. MST: comparison between IEEP and IE-Prim (best values in bold).

n	$\|V\|$	IEEP - Max-Dist		IE-Prim - Random		IE-Prim - Max-Dist		IE-Prim - CSS	
		time(s)	queries	time(s)	queries	time(s)	queries	time(s)	queries
2	50	**8.0**	**7.4**	13.3	12.3	12.1	11.2	13.0	12.3
3	50	**16.5**	**15.2**	28.6	26.7	26.1	24.5	31.9	29.6
4	50	**26.4**	**24.6**	45.0	42.1	42.5	39.7	55.6	50.8
5	50	**36.2**	**34.3**	59.7	55.5	56.9	53.2	80.4	73.4
6	50	**45.5**	**42.7**	78.7	73.4	79.4	73.5	117.8	108.1
2	100	**8.9**	**8.6**	15.9	15.1	14.6	13.6	16.1	15.0
3	100	**19.0**	**17.4**	34.6	32.4	33.6	31.1	36.9	35.3
4	100	**30.1**	**28.4**	55.6	51.6	54.7	51.2	66.6	61.6
5	100	**42.1**	**39.2**	75.4	70.7	76.4	71.7	103.7	95.3
6	100	**52.3**	**49.1**	103.7	96.0	100.3	93.5	162.3	146.2

construction of the solutions is less informative than asking queries using the extreme points of the polyhedron representing the preference uncertainty.

Now we want to estimate the performances of our algorithm seen as an anytime algorithm (see Fig. 7). For each iteration step i, we compute the error obtained when deciding to return the solution that is optimal for the minimax regret criterion at step i (i.e., after i queries); this error is here expressed in terms of percentage from the optimal solution. For the sake of comparison, we also include the results obtained with IE-Prim. However IE-Prim cannot be seen as an anytime algorithm since it is constructive. Therefore, to vary the number of queries, we used different tolerance thresholds: $\delta = 0.3, 0.2, 0.1, 0.05$ and 0.01.

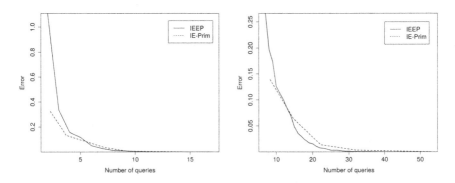

Fig. 7. MST problem with $|V| = 100$: Comparison of the errors with respect to the number of queries for $n = 3$ (left) and for $n = 6$ (right).

In Fig. 7, we observe that the error drops relatively quickly for both procedures. Note however that the error obtained with IE-Prim is smaller than with

IEEP when the number of queries is very low. This may suggest to favor IE-Prim over IEEP whenever the interactions are very limited and time is not an issue.

5.2 Multicriteria Traveling Salesman Problem

We now provide numerical results for the multicriteria traveling salesman problem (MTSP). In our tests, we consider existing Euclidean instances of the MTSP with 50 and 100 cities, and $n = 2$ to 6 objectives[4]. Moreover, we use the exact solver Concorde[5] to perform the optimization part of IEEP algorithm (see procedure Optimizing). Contrary to the MST, there exist no interactive constructive algorithms to solve the MTSP. Therefore, we only provide the results obtained by our algorithm IEEP with the three proposed query generation strategies (namely Random, Max-Dist and CSS). The results obtained by averaging over 30 runs are given in Table 3 for $\delta = 0$.

In this table, we see that Max-Dist remains the best strategy for minimizing the number of generated preference queries. Note that the running times are much higher for the MTSP than for the MST (see Table 1), as the traveling salesman problem is much more difficult to solve exactly with known preferences.

Table 3. MTSP: comparison of the different query strategies (best values in bold)

n	$|V|$	IEEP - Random				IEEP - Max-Dist				IEEP - CSS			
		time(s)	queries	eval	qOpt	time(s)	queries	eval	qOpt	time(s)	queries	eval	qOpt
2	50	**8.0**	**6.3**	**8.3**	**3.7**	8.8	**6.3**	**8.3**	**3.7**	10.0	**6.3**	**8.3**	**3.7**
3	50	**21.2**	14.3	31.3	10.0	23.5	**13.3**	29.5	**9.5**	24.5	14.9	32.4	10.6
4	50	**38.7**	22.6	101.5	**16.0**	50.2	**20.7**	93.6	16.2	67.7	24.2	109.1	16.9
5	50	210.9	31.2	331.1	22.7	**95.1**	28.6	304.8	**19.2**	137.1	38.5	387.7	23.9
6	50	390.8	41.0	1044.5	26.2	**238.8**	37.3	949.3	**24.4**	584.9	58.4	1531.0	28.9
2	100	12.2	**7.6**	**9.6**	4.3	**11.3**	**7.6**	**9.6**	4.3	19.1	**7.6**	**9.6**	4.3
3	100	28.3	15.9	34.7	12.4	**27.3**	15.4	**33.7**	12.1	42.2	16.5	35.6	**11.7**
4	100	73.1	26.7	121.1	20.0	**69.9**	25.4	115.8	**18.1**	94.8	28.4	124.9	19.6
5	100	241.9	36.4	**380.6**	27.3	**237.0**	35.5	383.0	**24.4**	361.8	44.7	481.3	31.2
6	100	981.2	45.0	1106.8	32.8	**586.3**	41.7	1014.5	**30.2**	1618.3	68.8	1865.3	39.2

6   Conclusion and Perspectives

In this paper, we have proposed a general method for solving multi-objective combinatorial optimization problems with unknown preference parameters. The method is based on a sharp combination of (1) regret-based incremental preference elicitation and (2) the generation of promising solutions using the extreme

---

[4] https://eden.dei.uc.pt/~paquete/tsp/.
[5] http://www.math.uwaterloo.ca/tsp/concorde.

points of the polyhedron representing the admissible preference parameters; several query generation strategies have been proposed in order to improve its performances. We have shown that our method returns the optimal solution according to the DM's preferences. Our method has been tested on the multicriteria spanning tree and multicriteria traveling salesman problems until 6 criteria and 100 vertices. We have provided numerical results showing that our method achieves better results than IE-Prim (the state-of-the-art method for the MST problem) both in terms of number of preference queries and running times.

Thus, in practice, our algorithm outperforms IE-Prim which is an algorithm that runs in polynomial time and generates no more than a polynomial number of queries. However, our algorithm does not have these performance guarantees. More precisely, the performances of our interactive method strongly depend on the number of extreme points at each iteration step, which can be exponential in the number of criteria (see e.g., [26]). Therefore, the next step could be to identify an approximate representation of the polyhedron which guarantees that the number of extreme points is always polynomial, while still being able to determine a (near-)optimal solution according to the DM's preferences.

References

1. Benabbou, N., Di Sabatino Di Diodoro, S., Perny, P., Viappiani, P.: Incremental preference elicitation in multi-attribute domains for choice and ranking with the Borda count. In: Schockaert, S., Senellart, P. (eds.) SUM 2016. LNCS (LNAI), vol. 9858, pp. 81–95. Springer, Cham (2016). https://doi.org/10.1007/978-3-319-45856-4_6
2. Benabbou, N., Perny, P.: On possibly optimal tradeoffs in multicriteria spanning tree problems. In: Walsh, T. (ed.) ADT 2015. LNCS (LNAI), vol. 9346, pp. 322–337. Springer, Cham (2015). https://doi.org/10.1007/978-3-319-23114-3_20
3. Benabbou, N., Perny, P.: Solving multi-agent knapsack problems using incremental approval voting. In: Proceedings of ECAI 2016, pp. 1318–1326 (2016)
4. Benabbou, N., Perny, P.: Interactive resolution of multiobjective combinatorial optimization problems by incremental elicitation of criteria weights. EURO J. Decis. Process. **6**(3–4), 283–319 (2018)
5. Benabbou, N., Perny, P., Viappiani, P.: Incremental elicitation of Choquet capacities for multicriteria choice, ranking and sorting problems. Artif. Intell. **246**, 152–180 (2017)
6. Bourdache, N., Perny, P.: Active preference elicitation based on generalized Gini functions: application to the multiagent knapsack problem. In: Proceedings of AAAI 2019 (2019)
7. Boutilier, C., Patrascu, R., Poupart, P., Schuurmans, D.: Constraint-based optimization and utility elicitation using the minimax decision criterion. Artif. Intell. **170**(8–9), 686–713 (2006)
8. Choquet, G.: Theory of capacities. Annales de l'Institut Fourier **5**, 31–295 (1953)
9. Drummond, J., Boutilier, C.: Preference elicitation and interview minimization in stable matchings. In: Proceedings of AAAI 2014, pp. 645–653 (2014)
10. Dyer, M., Proll, L.: An algorithm for determining all extreme points of a convex polytope. Math. Program. 12–81 (1977)

11. Gilbert, H., Spanjaard, O., Viappiani, P., Weng, P.: Reducing the number of queries in interactive value iteration. In: Walsh, T. (ed.) ADT 2015. LNCS (LNAI), vol. 9346, pp. 139–152. Springer, Cham (2015). https://doi.org/10.1007/978-3-319-23114-3_9

12. Grabisch, M., Labreuche, C.: A decade of application of the Choquet and Sugeno integrals in multi-criteria decision aid. Ann. Oper. Res. **175**(1), 247–286 (2010)

13. Hamacher, H., Ruhe, G.: On spanning tree problems with multiple objectives. Ann. Oper. Res. **52**, 209–230 (1994)

14. Kaddani, S., Vanderpooten, D., Vanpeperstraete, J.M., Aissi, H.: Weighted sum model with partial preference information: application to multi-objective optimization. Eur. J. Oper. Res. **260**, 665–679 (2017)

15. Karasakal, E., Köksalan, M.: Generating a representative subset of the nondominated frontier in multiple criteria. Oper. Res. **57**(1), 187–199 (2009)

16. Korhonen, P.: Interactive methods. In: Figueira, J., Greco, S., Ehrogott, M. (eds.) Multiple Criteria Decision Analysis: State of the Art Surveys. ISOR, vol. 78, pp. 641–661. Springer, New York (2005). https://doi.org/10.1007/0-387-23081-5_16

17. Kruskal, J.B.: On the shortest spanning subtree of a graph and the traveling salesman problem. Proc. Am. Math. Soc. **7**, 48–50 (1956)

18. Laumanns, M., Thiele, L., Deb, K., Zitzler, E.: Combining convergence and diversity in evolutionary multiobjective optimization. Evol. Comput. **10**(3), 263–282 (2002)

19. Lust, T., Rolland, A.: Choquet optimal set in biobjective combinatorial optimization. Comput. OR **40**(10), 2260–2269 (2013)

20. Marinescu, R., Razak, A., Wilson, N.: Multi-objective constraint optimization with tradeoffs. In: Schulte, C. (ed.) CP 2013. LNCS, vol. 8124, pp. 497–512. Springer, Heidelberg (2013). https://doi.org/10.1007/978-3-642-40627-0_38

21. Marinescu, R., Razak, A., Wilson, N.: Multi-objective influence diagrams with possibly optimal policies. In: Proceedings of AAAI 2017, pp. 3783–3789 (2017)

22. Prim, R.C.: Shortest connection networks and some generalizations. Bell Syst. Tech. J. **36**, 1389–1401 (1957)

23. White III, C.C., Sage, A.P., Dozono, S.: A model of multiattribute decisionmaking and trade-off weight determination under uncertainty. IEEE Trans. Syst. Man Cybern. **14**(2), 223–229 (1984)

24. Regan, K., Boutilier, C.: Eliciting additive reward functions for Markov decision processes. In: Proceedings of IJCAI 2011, pp. 2159–2164 (2011)

25. Rubinstein, R.: Generating random vectors uniformly distributed inside and on the surface of different regions. Eur. J. Oper. Res. **10**(2), 205–209 (1982)

26. Schrijver, A.: Combinatorial Optimization - Polyhedra and Efficiency. Springer, Heidelberg (2003)

27. Wang, T., Boutilier, C.: Incremental utility elicitation with the minimax regret decision criterion, pp. 309–316 (2003)

28. Weng, P., Zanuttini, B.: Interactive value iteration for Markov decision processes with unknown rewards. In: Proceedings of IJCAI 2013, pp. 2415–2421 (2013)

29. Wiecek, M.M.: Advances in cone-based preference modeling for decision making with multiple criteria. Decis. Making Manuf. Serv. **1**(1–2), 153–173 (2007)

30. Wilson, N., Razak, A., Marinescu, R.: Computing possibly optimal solutions for multi-objective constraint optimisation with tradeoffs. In: Proceedings of IJCAI 2015, pp. 815–822 (2015)

Open-Mindedness of Gradual Argumentation Semantics

Nico Potyka[(✉)]

Institute of Cognitive Science, University of Osnabrück, Osnabrück, Germany
npotyka@uos.de

Abstract. Gradual argumentation frameworks allow modeling arguments and their relationships and have been applied to problems like decision support and social media analysis. Semantics assign strength values to arguments based on an initial belief and their relationships. The final assignment should usually satisfy some common-sense properties. One property that may currently be missing in the literature is *Open-Mindedness*. Intuitively, *Open-Mindedness* is the ability to move away from the initial belief in an argument if sufficient evidence against this belief is given by other arguments. We generalize and refine a previously introduced notion of *Open-Mindedness* and use this definition to analyze nine gradual argumentation approaches from the literature.

Keywords: Gradual argumentation · Weighted argumentation · Semantical properties

1 Introduction

The basic idea of abstract argumentation is to study the acceptability of arguments abstracted from their content, just based on their relationships [13]. While arguments can only be accepted or rejected under classical semantics, gradual argumentation semantics consider a more fine-grained scale between these two extremes [3,6–8,10,16,20,22]. Arguments may have a base score that reflects a degree of belief that the argument is accepted when considered independent of all the other arguments. Semantics then assign strength values to all arguments based on their relationships and the base score if provided.

Of course, strength values should not be assigned in an arbitrary manner, but should satisfy some common-sense properties. Baroni, Rago and Toni recently showed that 29 properties from the literature can be reduced to basically two fundamental properties called *Balance* and *Monotonicity* [8] that we will discuss later. *Balance* and *Monotonicity* already capture a great deal of what we should expect from strength values of arguments, but they do not (and do not attempt to) capture everything. One desiderata that may be missing in many applications is *Open-Mindedness*. To illustrate the idea, suppose that we evaluate arguments by strength values between 0 and 1, where 0 means that we fully reject and 1 means that we fully accept an argument. Then, as we increase the number of

© Springer Nature Switzerland AG 2019
N. Ben Amor et al. (Eds.): SUM 2019, LNAI 11940, pp. 236–249, 2019.
https://doi.org/10.1007/978-3-030-35514-2_18

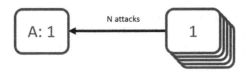

Fig. 1. Argument attacked by N other arguments.

supporters of an argument while keeping everything else equal, we should expect that its strength steadily approaches 1. Symmetrically, as we increase the number of attackers of an argument, we should expect that its strength approaches 0. To illustrate this, consider the graph in Fig. 1 that shows an argument A that is initially accepted (base score 1), but has N attackers that are initially accepted as well. For example, we could model a trial in law, where A corresponds to the argument that we should find the accused not guilty because we do not want to convict an innocent person. The N attackers correspond to pieces of evidence without reasonable doubt. Then, as N grows, we should expect that the strength of A goes to 0. Similar, in medical diagnosis, it is reasonable to initially accept that a patient with an infectious disease has a common cold because this is usually the case. However, as the number of symptoms for a more serious disease grows, we should be able to reject our default diagnosis at some point. Of course, we should expect a dual behaviour for support relations: if we initially reject A and have N supporters that are initially accepted, we should expect that the strength of A goes to 1 as N increases. A gradual argumentation approach that respects this idea is called open-minded. *Open-Mindedness* may not be necessary in every application, but it seems natural in many domains. Therefore, our goal here is to investigate which gradual argumentation semantics from the literature respect this property.

2   Compact QBAFs, Balance and Monotonicity

In our investigation, we consider quantitative bipolar argumentation frameworks (QBAFs) similar to [8]. However, for now, we will restrict to frameworks that assign values from a compact real interval to arguments in order to keep the formalism simple. At the end of this article, we will explain how the idea can be extended to more general QBAFs.

Definition 1 (Compact QBAF). *Let $\mathcal{D}$ be a compact real interval. A QBAF over $\mathcal{D}$ is a quadruple $(\mathcal{A}, \mathrm{Att}, \mathrm{Sup}, \beta)$ consisting of a set of arguments $\mathcal{A}$, two binary relations Att and Sup called attack and support and a function $\beta : \mathcal{A} \to \mathcal{D}$ that assigns a base score $\beta(a)$ to every argument $a \in \mathcal{A}$.*

Typical instantiations of the interval $\mathcal{D}$ are $[0,1]$ and $[-1,1]$. Sometimes non-compact intervals like open or unbounded intervals are considered as well, but we exclude these cases for now. We can consider different *subclasses of QBAFs* that use only some of the possible building blocks [8]. Among others, we will look at subclasses that contain QBAFs of the following restricted forms:

Attack-only: $(\mathcal{A}, \mathrm{Att}, \mathrm{Sup}, \beta)$ where $\mathrm{Sup} = \emptyset$,
Support-only: $(\mathcal{A}, \mathrm{Att}, \mathrm{Sup}, \beta)$ where $\mathrm{Att} = \emptyset$,
Bipolar without Base Score: $(\mathcal{A}, \mathrm{Att}, \mathrm{Sup}, \beta)$ where β is a constant function.

In order to interpret a given QBAF, we want to assign strength values to every argument. The strength values should be connected in a reasonable way to the base score of an argument and the strength of its attackers and supporters. Of course, this can be done in many different ways. However, eventually we want a function that assigns a strength value to every argument.

Definition 2 (QBAF interpretation). *Let $Q = (\mathcal{A}, \mathrm{Att}, \mathrm{Sup}, \beta)$ be a QBAF over a real interval $\mathcal{D}$. An interpretation of Q is a function $\sigma : \mathcal{A} \to \mathcal{D}$ and $\sigma(a)$ is called the strength of a for all $a \in \mathcal{A}$.*

Gradual argumentation semantics can define interpretations for the whole class of QBAFs or for a subclass only. One simple example is the *h-categorizer semantics* from [10] that interprets only acyclic attack-only QBAFs without base score. For all $a \in \mathcal{A}$, the *h-categorizer semantics* defines $\sigma(a) = \frac{1}{1+\sum_{(b,a)\in\mathrm{Att}} \sigma(b)}$. That is, unattacked arguments have strength 1, and the strength of all other arguments decreases monotonically based on the strength of their attackers. Since it only interprets acyclic QBAFs, the strength values can be evaluated in topological order, so that the strength values of all parents are known when interpreting the next argument.

Of course, we do not want to assign final strength values in an arbitrary way. Many desirable properties for different families of QBAFs have been proposed in the literature, see, e.g., [2–4,16,22]. Dependent on whether base scores, only attack, only support or both relations are considered, different properties have been proposed. However, as shown in [8], most properties can be reduced to basically two fundamental principles that are called *Balance* and *Monotonicity*. Roughly speaking, *Balance* says that the strength of an argument should be equal to its base score if its attackers and supporters are equally strong and that it should be smaller (greater) if the attackers are stronger (weaker) than the supporters. *Monotonicity* says, intuitively, that if the same impact (in terms of base score, attack and support) acts on two arguments a_1, a_2, then they should have the same strength, whereas if the impact on a_1 is more positive, it should have a larger strength than a_2. Several variants of *Balance* and *Monotonicity* have been discussed in [8]. For example, the stronger-than relationship between arguments can be formalized in a qualitative (focusing on the number of attackers and supporters) or quantitative manner (focusing on the strength of attackers and supporters). We refer to [8] for more details.

3   Open-Mindedness

Intuitively, it seems that *Balance* and *Monotonicity* could already imply *Open-Mindedness*. After all, they demand that adding attacks (supports) increases (decreases) the strength in a sense. However, this is not sufficient to guarantee that the strength can be moved arbitrarily close to the boundary values. To illustrate this, let us consider the Euler-based semantics that has been introduced for the whole class of QBAFs in [4]. Strength values are defined by

$$\sigma(a) = 1 - \frac{1 - \beta(a)^2}{1 + \beta(a) \cdot \exp(\sum_{(b,a) \in \mathrm{Sup}} \sigma(b) - \sum_{(b,a) \in \mathrm{Att}} \sigma(b))}$$

Note that if there are no attackers or supporters, the strength becomes just $1 - \frac{(1+\beta(a))(1-\beta(a))}{1+\beta(a)\cdot 1} = \beta(a)$. If the strength of a's attackers accumulates to a larger (smaller) value than the strength of a's supporters, the strength will be smaller (larger) than the base score. The Euler-based semantics satisfies the basic *Balance* and *Monotonicity* properties in most cases, see [4] for more details. However, it does not satisfy *Open-Mindedness* as has been noted in [21] already. There are two reasons for this. The first reason is somewhat weak and regards the boundary case $\beta(a) = 0$. In this case, the strength becomes $1 - \frac{1-0^2}{1+0} = 0$ independent of the supporters. In this boundary case, the Euler-based semantics does not satisfy *Balance* and *Monotonicity* either. The second reason is more profound and corresponds to the fact that the exponential function always yields positive values. Therefore, $1 + \beta(a) \cdot \exp(x) \geq 1$ and $\sigma(a) \geq 1 - \frac{1-\beta(a)^2}{1} = \beta(a)^2$ independent of the attackers. Hence, the strength value can never be smaller than the base score squared. The reason that the Euler-based semantics can still satisfy *Balance* and *Monotonicity* is that the limit $\beta(a)^2$ can never actually be taken, but is only approximated as the number of attackers goes to infinity.

Hence, *Open-Mindedness* is indeed a property that is currently not captured by *Balance* and *Monotonicity*. To begin with, we give a formal definition for a restricted case. We assume that larger values in $\mathcal{D}$ are stronger to avoid tedious case differentiations. This assumption is satisfied by the first eight semantics that we consider. We will give a more general definition later that also makes sense when this assumption is not satisfied. *Open-Mindedness* includes two dual conditions, one for attack- and one for support-relations. Intuitively, we want that in every QBAF, the strength of every argument with arbitrary base score can be moved arbitrarily close to $\min(\mathcal{D})$ ($\max(\mathcal{D})$) if we only add a sufficient number of strong attackers (supporters). In the following definition, ϵ captures the closeness and N the sufficiently large number.

Definition 3 (Open-Mindedness). *Consider a semantics that defines an interpretation $\sigma : \mathcal{A} \to \mathcal{D}$ for every QBAF from a particular class $\mathcal{F}$ of QBAFs over a compact interval $\mathcal{D}$. We call the semantics open-minded if for every QBAF $(\mathcal{A}, \mathrm{Att}, \mathrm{Sup}, \beta)$ in $\mathcal{F}$, for every argument $a \in \mathcal{A}$ and for every $\epsilon > 0$, the following condition is satisfied: there is an $N \in \mathbb{N}$ such that when adding N new arguments $A_N = \{a_1, \ldots, a_N\}$, $\mathcal{A} \cap A_N = \emptyset$, with maximum base score, then*

1. *if $\mathcal{F}$ allows attacks, then for $(\mathcal{A} \cup A_N, \mathrm{Att} \cup \{(a_i, a) \mid 1 \leq i \leq N\}, \mathrm{Sup}, \beta')$, we have $|\sigma(a) - \min(\mathcal{D})| < \epsilon$ and*
2. *if $\mathcal{F}$ allows supports, then for $(\mathcal{A} \cup A_N, \mathrm{Att}, \mathrm{Sup} \cup \{(a_i, a) \mid 1 \leq i \leq N\}, \beta')$, we have $|\sigma(a) - \max(\mathcal{D})| < \epsilon$,*

where $\beta'(b) = \beta(b)$ for all $b \in \mathcal{A}$ and $\beta'(a_i) = \max(\mathcal{D})$ for $i = 1, \ldots, n$.

Some explanations are in order. Note that we do not make any assumptions about the base score of a in Definition 3. Hence, we demand that the strength of a must become arbitrary small (large) within the domain $\mathcal{D}$, no matter what its base score is. One may consider a weaker notion of *Open-Mindedness* that excludes the boundary base scores for a. However, this distinction does not make a difference for our investigation and so we will not consider it here. Note also that we do not demand that the strength of a ever takes the extreme value $\max(\mathcal{D})$ $(\min(\mathcal{D}))$, but only that it can become arbitrarily close. Finally note that item 1 in Definition 3 is trivially satisfied for support-only QBAFs, and item 2 for attack-only QBAFs.

3.1 Attack-Only QBAFs over $\mathcal{D} = [0, 1]$

In this section, we consider three semantics for attack-only QBAFs over $\mathcal{D} = [0, 1]$. Recall from Sect. 2 that the *h-categorizer semantics* from [10] interprets acyclic attack-only QBAFs without base score. The definition has been extended to arbitrary (including cycles) attack-only QBAFs and base scores from $\mathcal{D} = [0, 1]$ in [6]. The strength of an argument under the *weighted h-categorizer semantics* is then defined by

$$\sigma(a) = \frac{\beta(a)}{1 + \sum_{(b,a) \in \mathrm{Att}} \sigma(b)} \tag{1}$$

for all $a \in \mathcal{A}$. Note that the original definition of the *h-categorizer semantics* from [10] is obtained when all base scores are 1. The strength values in (cyclic) graphs can be computed by initializing the strength values with the base scores and applying formula (1) repeatedly to all arguments simultaneously until the strength values converge [6]. It is not difficult to see that the *weighted h-categorizer semantics* satisfies *Open-Mindedness*. However, in order to illustrate our definition, we give a detailed proof of the claim.

Proposition 1. *The* weighted h-categorizer semantics *is open-minded.*

Proof. In the subclass of attack-only QBAFs, it suffices to check the first condition of Definition 3. Consider an arbitrary attack-only QBAF $(\mathcal{A}, \mathrm{Att}, \emptyset, \beta)$, an arbitrary argument $a \in \mathcal{A}$ and an arbitrary $\epsilon > 0$. Let $N = \lceil \frac{1}{\epsilon} \rceil + 1$ and consider the QBAF $(\mathcal{A} \cup \{a_1, \ldots, a_N\}, \mathrm{Att} \cup \{(a_i, a) \mid 1 \leq i \leq N\}, \mathrm{Sup}, \beta')$ as defined in Definition 3. Recall that the N new attackers $\{a_1, \ldots, a_N\}$ have base score 1 and do not have any attackers. Therefore, $\sigma(a_i) = \frac{\beta(a_i)}{1} = 1$ for $i = 1, \ldots, n$ and $\sum_{(b,a) \in \mathrm{Att}} \sigma(a) \geq \sum_{i=1}^{N} \sigma(a_i) = N$. Furthermore, we have $\beta(a) \leq 1$ because $\mathcal{D} = [0, 1]$. Hence, $|\sigma(a) - 0| = \frac{\beta(a)}{1 + \sum_{(b,a) \in \mathrm{Att}} \sigma(a)} < \frac{1}{N} < \epsilon$. $\qquad\square$

The *weighted max-based semantics* from [6] can be seen as a variant of the *h-categorizer semantics* that aggregates the strength of attackers by means of the maximum instead of the sum. The strength of arguments is defined by

$$\sigma(a) = \frac{\beta(a)}{1 + \max_{(b,a) \in \mathrm{Att}} \sigma(b)}. \tag{2}$$

If there are no attackers, the maximum yields 0 by convention. The motivation for using the maximum is to satisfy a property called *Quality Precedence*, which guarantees that when arguments a_1 and a_2 have the same base score, but a_1 has an attacker that is stronger than all attackers of a_2, then the strength of a_1 must be smaller than the strength of a_2. The strength values under the *weighted max-based semantics* can again be computed iteratively [6]. Since all strength values are in $[0, 1]$ and the maximum is used for aggregating the strength values, we can immediately see that $\sigma(a) \geq \frac{\beta(a)}{2}$. Therefore, the *weighted max-based semantics* is clearly not open-minded. For example, if $\beta(a) = 1$, the final strength cannot be smaller than $\frac{1}{2}$, no matter how many attackers there are.

Proposition 2. *The* weighted max-based semantics *is not open-minded.*

One may wonder if *Quality Precedence* and *Open-Mindedness* are incompatible. This is actually not the case. For example, when defining strength values by

$$\sigma(a) = \beta(a) \cdot \left(1 - \max_{(b,a) \in \mathrm{Att}} \sigma(b)\right)$$

both *Quality Precedence* and *Open-Mindedness* are satisfied. In particular, the strength now decreases linearly from $\beta(a)$ to 0 with respect to the strongest attacker, which makes this perhaps a more natural way to satisfy *Quality Precedence* when it is desired.

The *weighted card-based semantics* from [6] is another variant of the *h-categorizer semantics*. Instead of putting extra emphasis on the strength of attackers, it now puts extra emphasis on the number of attackers. Let $\mathrm{Att}^+ = \{(a, b) \in \mathrm{Att} \mid \beta(a) > 0\}$. Then the strength of arguments is defined by

$$\sigma(a) = \frac{\beta(a)}{1 + |\mathrm{Att}^+| + \frac{\sum_{(b,a) \in \mathrm{Att}^+} \sigma(b)}{|\mathrm{Att}^+|}}. \tag{3}$$

When reordering terms in the denominator, we can see that the only difference to the *h-categorizer semantics* is that every attacker b with non-zero strength adds $1 + \sigma(b)$ instead of just $\sigma(b)$ in the sum in the denominator (attacker with strength 0 do not add anything anyway). This enforces a property called *Cardinality Precedence*, which basically means that when arguments a_1 and a_2 have the same base score and a_1 has a larger number of non-rejected attackers ($\sigma(b) > 0$) than a_2, then the strength of a_1 must be smaller than the strength of a_2. The strength values under the *weighted card-based semantics* can again be computed iteratively [6]. Analogously to the *weighted h-categorizer semantics*, it can be checked that the *weighted card-based semantics* satisfies *Open-Mindedness*.

Proposition 3. *The* weighted card-based semantics *is open-minded.*

3.2   Support-Only QBAFs over $\mathcal{D} = [0, 1]$

We now consider three semantics for support-only QBAFs over $\mathcal{D} = [0, 1]$. For all semantics, the strength of arguments is defined by equations of the form

$$\sigma(a) = \beta(a) + (1 - \beta(a)) \cdot S(a),$$

where $S(a)$ is an aggregate of the strength of a's supporters. Therefore, the question whether *Open-Mindedness* is satisfied boils down to the question whether $S(a)$ converges to 1 as we keep adding supporters.

The *top-based semantics* from [3] defines the strength of arguments by

$$\sigma(a) = \beta(a) + (1 - \beta(a)) \max_{(b,a)\in\text{Sup}} \sigma(b). \tag{4}$$

If there are no supporters, the maximum again yields 0 by convention. Similar to the semantics in the previous section, the strength values can be computed iteratively by setting the initial strength values to the base score and applying formula (4) repeatedly until the values converge [3]. It is easy to check that the *top-based semantics* is open-minded. In fact, a single supporter with strength 1 is sufficient to move the strength all the way to 1 independently of the base score.

Proposition 4. *The* top-based semantics *is open-minded.*

The *aggregation-based semantics* from [3] defines the strength of arguments by the formula

$$\sigma(a) = \beta(a) + (1 - \beta(a))\frac{\sum_{(b,a)\in\text{Sup}} \sigma(b)}{1 + \sum_{(b,a)\in\text{Sup}} \sigma(b)}. \tag{5}$$

The strength values can again be computed iteratively [6]. It is easy to check that the *aggregation-based semantics* is open-minded. Just note that the fraction in (5) has the form $\frac{N}{1+N}$ and therefore approaches 1 as $N \to \infty$. Therefore, the strength of an argument will go to 1 as we keep adding supporters under the *aggregation-based semantics*.

Proposition 5. *The* aggregation-based semantics *is open-minded.*

The *reward-based semantics* from [3] is based on the idea of *founded arguments*. An argument a is called founded if there exists a sequence of arguments $(a_0, \ldots, a_n)$ such that $a_n = a$, $(a_{i-1}, a_i) \in \text{Sup}$ for $i = 1, \ldots, n$ and $\beta(a_0) > 0$. That is, a has non-zero base score or is supported by a sequence of supporters such that the first argument in the sequence has a non-zero base score. Intuitively, this implies that a must have non-zero strength. We let $\text{Sup}^+ = \{(a,b) \in \text{Sup} \mid a \text{ is founded}\}$ denote the founded supports. For every $a \in \mathcal{A}$, we let $N(a) = |\text{Sup}^+|$ denote the number of founded supporters of a and $M(a) = \frac{\sum_{(b,a)\in\text{Sup}^+} \sigma(b)}{N(A)}$ the mean strength of the founded supporters. Then the strength of a is defined as

$$\sigma(a) = \beta(a) + (1 - \beta(a))\left(\sum_{i=1}^{N(a)-1} \frac{1}{2^i} + \frac{M(a)}{2^{N(a)}}\right). \tag{6}$$

The strength values can again be computed iteratively [6]. As we show next, the reward-based semantics also satisfies *Open-Mindedness*.

Proposition 6. *The* reward-based semantics *is open-minded.*

Proof. In the subclass of support-only QBAFs, it suffices to check the second condition of Definition 3. Let us first note that $\sum_{i=1}^{N(a)-1} \frac{1}{2^i}$ is a geometric sum without the first term and therefore evaluates to

$$\frac{1 - \frac{1}{2^{N(a)}}}{1 - \frac{1}{2}} - 1 = 1 - \frac{1}{2^{N(a)-1}}$$

Note that this term already goes to 1 as the number of founded supporters $N(a)$ increases. We additionally add the non-negative term $\frac{M(a)}{2^{N(a)}} = \frac{\sum_{(b,a)\in\mathrm{Sup}^+} \sigma(b)}{N(A)\cdot 2^{N(a)}}$ which is bounded from above by $\frac{1}{2^{N(a)}}$. Therefore, the factor $\left(\sum_{i=1}^{N(a)-1} \frac{1}{2^i} + \frac{M(a)}{2^{N(a)}}\right)$ is always between 0 and 1 and approaches 1 as $|N(A)| \to \infty$.

To complete the proof, consider any support-only QBAF $(\mathcal{A}, \emptyset, \mathrm{Sup}, \beta)$, any argument $a \in \mathcal{A}$, any $\epsilon > 0$ and let $(\mathcal{A} \cup \{a_1, \ldots, a_N\}, \mathrm{Att}, \mathrm{Sup} \cup \{(a_i, a) \mid 1 \leq i \leq N\}, \beta')$ be the QBAF defined in Definition 3 for some $N \in \mathbb{N}$. Note that every argument in $\{a_1, \ldots, a_N\}$ is a founded supporter of a. Therefore, $N(A) \geq N$ and $\sigma(a) \to \beta(a) + (1 - \beta(a)) = 1$ as $N \to \infty$. This then implies that there exists an $N_0 \in \mathbb{N}$ such that $|\sigma(a) - 1| < \epsilon$. $\square$

3.3   Bipolar QBAFs Without Base Score over $\mathcal{D} = [-1, 1]$

In this section, we consider two semantics for bipolar QBAFs without base score over $\mathcal{D} = [-1, 1]$ that have been introduced in [7]. It has not been explained how the strength values are computed in [7]. However, given an acyclic graph, the strength values can again be computed in topological order because the strength of every argument depends only on the strength of its parents. For cyclic graphs, one may consider an iterative procedure as before, but convergence may be an issue. In our investigation, we will just assume that the strength values are well-defined.

Following [8], we call the first semantics from [7], the *loc-max semantics*. It defines strength values by the formula

$$\sigma(a) = \frac{\max_{(b,a)\in\mathrm{Sup}} \sigma(b) - \max_{(b,a)\in\mathrm{Att}} \sigma(b)}{2} \tag{7}$$

By convention, the maximum now yields -1 if there are no supporters/attackers (this is consistent with the previous conventions in that -1 is now the minimum of the domain, whereas the minimum was 0 before). If a has neither attackers nor supporters, then $\sigma(a) = \frac{-1-(-1)}{2} = 0$. As we keep adding supporters (attackers), the first (second) term in the numerator will take the maximum strength value. From this we can see that the *loc-sum semantics* is open-minded for attack-only QBAFs without base score and for support-only QBAFs without base score. However, it is not open-minded for bipolar QBAFs without base score. For example, suppose that a has a single supporter b', which has a single supporter b'' and no attackers. Further assume that b'' has neither attackers nor supporters, so that $\sigma(b'') = 0$, $\sigma(b') = \frac{0-(-1)}{2} = \frac{1}{2}$ and $\sigma(a) \geq \frac{\frac{1}{2}-\max_{(b,a)\in\text{Att}}\sigma(b)}{2}$. Since the maximum of the attackers can never become larger than 1, we have $\sigma(a) \geq \frac{\frac{1}{2}-1}{2} \geq -\frac{1}{4}$, no matter how many attackers we add. Thus, the first condition of *Open-Mindedness* is violated. Using a symmetrical example, we can show that the second condition can be violated as well.

Proposition 7. *The* loc-max *semantics is not open-minded. It is open-minded when restricting to attack-only QBAFs without base score or to support-only QBAFs without base score.*

Following [8], we call the second semantics from [7], the *loc-sum semantics*. It defines strength values by the formula

$$\sigma(a) = \frac{1}{1+\sum_{(b,a)\in\text{Att}}\frac{\sigma(b)+1}{2}} - \frac{1}{1+\sum_{(b,a)\in\text{Sup}}\frac{\sigma(b)+1}{2}} \tag{8}$$

Note that if there are neither attackers nor supporters, then both fractions are 1 such that their difference is just 0. As we keep adding attackers (supporters), the first (second) fraction goes to 0. It follows again that the *loc-sum semantics* is open-minded for attack-only QBAFs without base score and for support-only QBAFs without base score. However, it is again not open-minded for bipolar QBAFs without base score. For example, if a has a single supporter b' that has neither attackers nor supporters, then $\sigma(b') = 0$ and the second fraction evaluates to $\frac{1}{1+\frac{1}{2}} = \frac{2}{3}$. As we keep adding attackers, the first fraction will to 0 so that the strength of a will converge to $-\frac{2}{3}$ rather than to -1 as the first condition of *Open-Mindedness* demands. It is again easy to construct a symmetrical example to show that the second condition of *Open-Mindedness* can be violated as well.

Proposition 8. *The* loc-sum *semantics is not open-minded. It is open-minded when restricting to attack-only QBAFs without base score or to support-only QBAFs without base score.*

4   General QBAFs and Open-Mindedness

We now consider the general form of QBAFs as introduced in [8]. The domain $\mathcal{D} = (S, \preceq)$ is now an arbitrary set along with a preorder $\preceq$, that is, a reflexive and transitive relation over S. We further assume that there is an infimum $\inf(S)$ and a supremum $\sup(S)$ that may or may not be contained in S. For example, the open interval $(0, \infty)$, contains neither its infimum 0 nor its supremum ∞, whereas the half-open interval $[0, \infty)$ contains its infimum, but not its supremum.

Definition 4 (QBAF). *A QBAF over $\mathcal{D}$ = $(S, \preceq)$ is a quadruple $(\mathcal{A}, \text{Att}, \text{Sup}, \beta)$ consisting of a set of arguments $\mathcal{A}$, a binary attack relation Att, a binary support relation Sup and a function $\beta : \mathcal{A} \rightarrow \mathcal{D}$ that assigns a base score $\beta(a)$ to every argument $a \in \mathcal{A}$.*

We now define a generalized form of *Open-Mindedness* for general QBAFs. We have to take account of the fact that there may no longer exist a minimum or maximum of the set. So instead we ask that strength values can be made smaller/larger than every element from $S \setminus \{\inf(S), \sup(S)\}$ by adding a sufficient number of attackers/supporters. Intuitively, we want to add strong supporters. In Definition 3, we just assumed that the maximum corresponds to the strongest value, but there are semantics that regard smaller values as stronger and, again, S may neither contain a maximal nor a minimal element. Therefore, we will just demand that there is some base score s^*, such that adding attackers/supporters with base score s^* has the desired consequence.

Definition 5 (Open-Mindedness (General Form)). *Consider a semantics that defines an interpretation $\sigma : \mathcal{A} \to \mathcal{D}$ for every QBAF from a particular class $\mathcal{F}$ of QBAFs over $\mathcal{D} = (S, \preceq)$. We call the semantics open-minded if for every QBAF $(\mathcal{A}, \text{Att}, \text{Sup}, \beta)$ in $\mathcal{F}$, for every argument $a \in \mathcal{A}$ and for every $s \in S \setminus \{\inf(S), \sup(S)\}$, the following condition is satisfied: there is an $N \in \mathbb{N}$ and an $s^* \in S$ such that when adding N new arguments $A_N = \{a_1, \ldots, a_N\}$, $\mathcal{A} \cap A_N = \emptyset$, then*

1. *if $\mathcal{F}$ allows attacks, then for $(\mathcal{A} \cup A_N, \text{Att} \cup \{(a_i, a) \mid 1 \le i \le N\}, \text{Sup}, \beta')$, we have $\sigma(a) \preceq s$ and*
2. *if $\mathcal{F}$ allows supports, then for $(\mathcal{A} \cup A_N, \text{Att}, \text{Sup} \cup \{(a_i, a) \mid 1 \le i \le N\}, \beta')$, we have $s \preceq \sigma(a)$,*

where $\beta'(b) = \beta(b)$ for all $b \in \mathcal{A}$ and $\beta'(a_i) = s^*$ for $i = 1, \ldots, n$.

Note that if S is a compact real interval, $s \in S \setminus \{\inf(S), \sup(S)\}$ can be chosen arbitrarily close to $\sup(S) = \max(S)$ or $\inf(S) = \min(S)$, so that s in Definition 5 plays the role of ϵ in Definition 3. In particular, if Definition 3 is satisfied, Definition 5 can be satisfied as well for an arbitrary s by choosing base score $s^* = \max(S)$ and choosing N with respect to $\epsilon = \frac{\max(S) - s}{2}$ or $\epsilon = \frac{s - \min(S)}{2}$. Definitions 5 and 3 are actually equivalent for compact real intervals provided that $\max(S)$ corresponds to the strongest initialization of the base score under the given semantics, which is indeed the case in all previous examples.

As an example, for more general QBAFs, let us now consider the α-*burden-semantics* from [5]. It defines strength values for attack-only QBAFs without base score over the half-open interval $[1, \infty)$. As opposed to our previous examples, the minimum 1 now corresponds to the strongest value and increasing values correspond to less plausibility. The α-*burden-semantics* defines strength values via the formula

$$\sigma(a) = 1 + \Big(\sum_{(b,a) \in \text{Att}} \frac{1}{(\sigma(b))^\alpha} \Big)^{\frac{1}{\alpha}}. \tag{9}$$

α is called the *burden-parameter* and can be used to modify the semantics, see [5] for more details about the influence of α. For $\alpha \in [1, \infty) \cup \{\infty\}$, (9) is equivalent to arranging the reciprocals of strength values of all attackers in a vector v and to take the p-norm $\|v\|_p = \big(\sum_i v_i^p \big)^{\frac{1}{p}}$ of this vector with respect to $p = \alpha$

and adding 1. Popular examples of p-norms are the Manhattan-, Euclidean- and Maximum-norm that are obtained for $p = 1$, $p = 2$ and the limit-case $p = \infty$, respectively. An unattacked argument has just strength 1 under the α-*burden-semantics*. Hence, when adding N new attackers to a, we have $\sigma(a) \geq 1 + N^{\frac{1}{\alpha}}$ for $\alpha \in [1, \infty)$. Hence, the α-*burden-semantics* is clearly open-minded in this case, even though it becomes more conservative as α increases. In particular, for the limit case $\alpha = \infty$, it is not open-minded. This can be seen from the observation, that the second term in (9) now corresponds to the maximum norm. Since the strength of each attacker is in $[1, \infty)$, their reciprocals are in $(0, 1]$. Therefore, $\sigma(a) \leq 2$ independent of the number of attackers of a.

Proposition 9. *The α-burden-semantics is open-minded for $\alpha \in [1, \infty)$, but is not open-minded for $\alpha = \infty$.*

5 Related Work

Gradual argumentation has become a very active research area and found applications in areas like information retrieval [24], decision support [9,22] and social media analysis [1,12,16]. Our selection of semantics followed the selection in [8]. One difference is that we did not consider social abstract argumentation [16] here. The reason is that social abstract argumentation has been formulated in a very abstract form, which makes it difficult to formulate interesting conditions under which *Open-Mindedness* is guaranteed. Instead, we added the α-*burden-semantics* from [5] because it gives a nice example for a more general semantics that neither uses strength values from a compact interval nor regards larger values as stronger.

The authors in [8] also view ranking-based semantics [11] as gradual argumentation frameworks. In their most general form, ranking-based semantics just order arguments qualitatively, so that our notion of *Open-Mindedness* is not very meaningful. A variant may be interesting, however, that demands, that in every argumentation graph, every argument can become first or last in the order if only a sufficient number of supporters or attackers is added to this argument. However, in many cases, this notion of *Open-Mindedness* may be entailed by other properties already. For example, *Cardinality Precedence* [11] states that if argument a_1 has more attackers than a_2, then a_1 must be weaker than a_2. In finite argumentation graphs, this already implies that a_1 will be last in the order if we add a sufficient number of attackers.

There are other quantitative argumentation frameworks like probabilistic argumentation frameworks [14,15,17,19,23]. In this area, *Open-Mindedness* would simply state that the probability of an argument must go to 0 (1) as we keep adding attackers (supporters). It may be interesting to perform a similar analysis for probabilistic argumentation frameworks.

An operational definition of *Open-Mindedness* for the class of modular semantics [18] for weighted bipolar argumentation frameworks has been given in [21]. The *Df-QuAD semantics* [22] and the *Quadratic-energy Semantics* [20] satisfy this notion of open-mindedness [21]. However, in case of DF-QuAD and some other semantics, this is actually counterintuitive because they cannot move the strength of an argument towards 0 if there is a supporter with non-zero strength. Indeed, DF-QuAD does not satisfy Open-Mindedness as defined here (every QBAF with a non-zero strength supporter provides a counterexample). However, the quadratic energy model from [21] still satisfies the more restrictive definition of *Open-Mindedness* that we considered here.

Another interesting property for bipolar QBAFs that is not captured by *Balance* and *Monotonicity* is *Duality* [20]. Duality basically states that attack and support should behave in a symmetrical manner. Roughly speaking, when we convert an attack relation into a support relation or vice versa, the effect of the relation should just be inverted. *Duality* is satisfied by the *Df-QuAD semantics* [22] and the *Quadratic-energy Semantics* [20], but not by the *Euler-based semantics* [4]. A formal analysis can be found in [20,21].

6   Conclusions

We investigated 9 gradual argumentation semantics from the literature. 5 of them satisfy *Open-Mindedness* unconditionally. This includes the *weighted h-categorizer semantics* and the *weighted card-based semantics* for attack-only QBAFs from [6] and all three semantics for support-only QBAFs from [3]. The α-*burden-semantics* for attack-only QBAFs without base score from [5] is open-minded for $\alpha \in [1, \infty)$, but not for the limit case $\alpha = \infty$. The *loc-max semantics* and the *loc-sum semantics* for bipolar QBAFs without base score from [7] are only open-minded when restricted to either attack-only or to support-only QBAFs. Finally, the *weighted max-based semantics* for attack-only QBAFs from [6] is not open-minded. However, as we saw, it can easily be adapted to satisfy both *Open-Mindedness* and *Quality Precedence*.

In future work, it may be interesting to complement *Open-Mindedness* with a *Conservativeness* property that demands that the original base scores are not given up too easily. For the class of modular semantics [18] that iteratively compute strength values by repeatedly aggregating strength values and combining them with the base score, *Conservativeness* can actually be quantified analytically [21]. Intuitively, this can be done by analyzing the maximal local growth of the aggregation and influence functions. There is actually an interesting relationship between *Conservativeness* and *Well-Definedness* of strength values. For general QBAFs, procedures that compute strength values iteratively, can actually diverge [18] so that some strength values remain undefined. However, the mechanics that make semantics more conservative, simultaneously improve convergence guarantees [21]. In other words, convergence guarantees can often be improved by giving up *Open-Mindedness*. The extreme case would be the naive semantics that just assigns the base score as final strength to every argument independent of the attackers and supporters. This semantics is clearly most conservative and always well-defined, but does not make much sense.

My personal impression is indeed that gradual argumentation semantics for general QBAFs with strong convergence guarantees are too conservative at the moment. Some well-defined semantics for general QBAFs have been presented recently in [18], but they are not open-minded. I am indeed unaware of any semantics for general QBAFs that is generally well-defined and open-minded. It is actually possible to define for every $k \in \mathbb{N}$, an open-minded semantics that is well-defined for all QBAFs where arguments have at most k parents. One example is the 1-*max(k) semantics*, see Corollary 3.5 in [21]. However, as k grows, these semantics become more and more conservative even though they remain open-minded. More precisely, every single argument can change the strength value of another argument by at most $\frac{1}{k}$, so that at least k arguments are required to move the strength all the way from 0 to 1 and vice versa. A better way to improve convergence guarantees may be to define strength values not by discrete iterative procedures, but to replace them with continuous procedures that maintain the strength values in the limit, but improve convergence guarantees [20,21]. However, while I find this approach promising, I admit that it requires further analysis.

In conclusion, I think that *Open-Mindedness* is an interesting property that is important for many applications. It is indeed satisfied by many semantics from the literature. For others, like the *weighted max-based semantics*, we may be able to adapt the definition. One interesting open question is whether we can define semantics for general QBAFs that are generally well-defined and open-minded.

References

1. Alsinet, T., Argelich, J., Béjar, R., Fernández, C., Mateu, C., Planes, J.: Weighted argumentation for analysis of discussions in Twitter. Int. J. Approximate Reasoning **85**, 21–35 (2017)
2. Amgoud, L., Ben-Naim, J.: Axiomatic foundations of acceptability semantics. In: International Conference on Principles of Knowledge Representation and Reasoning (KR), pp. 2–11 (2016)
3. Amgoud, L., Ben-Naim, J.: Evaluation of arguments from support relations: axioms and semantics. In: International Joint Conferences on Artificial Intelligence (IJCAI), p. 900 (2016)
4. Amgoud, L., Ben-Naim, J.: Evaluation of arguments in weighted bipolar graphs. In: Antonucci, A., Cholvy, L., Papini, O. (eds.) ECSQARU 2017. LNCS (LNAI), vol. 10369, pp. 25–35. Springer, Cham (2017). https://doi.org/10.1007/978-3-319-61581-3_3
5. Amgoud, L., Ben-Naim, J., Doder, D., Vesic, S.: Ranking arguments with compensation-based semantics. In: International Conference on Principles of Knowledge Representation and Reasoning (KR) (2016)
6. Amgoud, L., Ben-Naim, J., Doder, D., Vesic, S.: Acceptability semantics for weighted argumentation frameworks. In: IJCAI, vol. 2017, pp. 56–62 (2017)
7. Amgoud, L., Cayrol, C., Lagasquie-Schiex, M.C., Livet, P.: On bipolarity in argumentation frameworks. Int. J. Intell. Syst. **23**(10), 1062–1093 (2008)
8. Baroni, P., Rago, A., Toni, F.: How many properties do we need for gradual argumentation? In: AAAI Conference on Artificial Intelligence (AAAI), pp. 1736–1743. AAAI (2018)

9. Baroni, P., Romano, M., Toni, F., Aurisicchio, M., Bertanza, G.: An argumentation-based approach for automatic evaluation of design debates. In: Leite, J., Son, T.C., Torroni, P., van der Torre, L., Woltran, S. (eds.) CLIMA 2013. LNCS (LNAI), vol. 8143, pp. 340–356. Springer, Heidelberg (2013). https://doi.org/10.1007/978-3-642-40624-9_21

10. Besnard, P., Hunter, A.: A logic-based theory of deductive arguments. Artif. Intell. **128**(1–2), 203–235 (2001)

11. Bonzon, E., Delobelle, J., Konieczny, S., Maudet, N.: A comparative study of ranking-based semantics for abstract argumentation. In: AAAI Conference on Artificial Intelligence (AAAI), pp. 914–920 (2016)

12. Cocarascu, O., Rago, A., Toni, F.: Extracting dialogical explanations for review aggregations with argumentative dialogical agents. In: International Conference on Autonomous Agents and MultiAgent Systems (AAMAS), pp. 1261–1269. International Foundation for Autonomous Agents and Multiagent Systems (2019)

13. Dung, P.M.: On the acceptability of arguments and its fundamental role in nonmonotonic reasoning, logic programming and n-person games. Artif. Intell. **77**(2), 321–357 (1995)

14. Hunter, A., Polberg, S., Potyka, N.: Updating belief in arguments in epistemic graphs. In: International Conference on Principles of Knowledge Representation and Reasoning (KR), pp. 138–147 (2018)

15. Hunter, A., Thimm, M.: Probabilistic reasoning with abstract argumentation frameworks. J. Artif. Intell. Res. **59**, 565–611 (2017)

16. Leite, J., Martins, J.: Social abstract argumentation. In: International Joint Conferences on Artificial Intelligence (IJCAI), vol. 11, pp. 2287–2292 (2011)

17. Li, H., Oren, N., Norman, T.J.: Probabilistic argumentation frameworks. In: Modgil, S., Oren, N., Toni, F. (eds.) TAFA 2011. LNCS (LNAI), vol. 7132, pp. 1–16. Springer, Heidelberg (2012). https://doi.org/10.1007/978-3-642-29184-5_1

18. Mossakowski, T., Neuhaus, F.: Modular semantics and characteristics for bipolar weighted argumentation graphs. arXiv preprint arXiv:1807.06685 (2018)

19. Polberg, S., Doder, D.: Probabilistic abstract dialectical frameworks. In: Fermé, E., Leite, J. (eds.) JELIA 2014. LNCS (LNAI), vol. 8761, pp. 591–599. Springer, Cham (2014). https://doi.org/10.1007/978-3-319-11558-0_42

20. Potyka, N.: Continuous dynamical systems for weighted bipolar argumentation. In: International Conference on Principles of Knowledge Representation and Reasoning (KR), pp. 148–157 (2018)

21. Potyka, N.: Extending modular semantics for bipolar weighted argumentation. In: International Conference on Autonomous Agents and MultiAgent Systems (AAMAS), pp. 1722–1730. International Foundation for Autonomous Agents and Multiagent Systems (2019)

22. Rago, A., Toni, F., Aurisicchio, M., Baroni, P.: Discontinuity-free decision support with quantitative argumentation debates. In: International Conference on Principles of Knowledge Representation and Reasoning (KR), pp. 63–73 (2016)

23. Rienstra, T., Thimm, M., Liao, B., van der Torre, L.: Probabilistic abstract argumentation based on SCC decomposability. In: International Conference on Principles of Knowledge Representation and Reasoning (KR), pp. 168–177 (2018)

24. Thiel, M., Ludwig, P., Mossakowski, T., Neuhaus, F., Nürnberger, A.: Web-retrieval supported argument space exploration. In: ACM SIGIR Conference on Human Information Interaction and Retrieval (CHIIR), pp. 309–312. ACM (2017)

Approximate Querying on Property Graphs

Stefania Dumbrava[1]([✉]), Angela Bonifati[2], Amaia Nazabal Ruiz Diaz[2],
and Romain Vuillemot[3]

[1] ENSIIE Évry & CNRS Samovar, Évry, France
`stefania.dumbrava@ensiie.fr`
[2] University of Lyon 1 & CNRS LIRIS, Lyon, France
`{angela.bonifati,amaia.nazabal-ruiz-diaz}@univ-lyon1.fr`
[3] École Centrale Lyon & CNRS LIRIS, Lyon, France
`romain.vuillemot@ec-lyon.fr`

Abstract. Property graphs are becoming widespread when modeling data with complex structural characteristics and enhancing edges and nodes with a list of properties. In this paper, we focus on the approximate evaluation of counting queries involving recursive paths on property graphs. As such queries are already difficult to evaluate over pure RDF graphs, they require an ad-hoc graph summary for their approximate evaluation on property graphs. We prove the intractability of the optimal graph summarization problem, under our algorithm's conditions. We design and implement a novel property graph summary suitable for the above queries, along with an approximate query evaluation module. Finally, we show the compactness of the obtained summaries as well as the accuracy of answering counting recursive queries on them.

1 Introduction

A tremendous amount of information stored in the LOD can be inspected, by leveraging the already mature query capabilities of SPARQL, relational, and graph databases [14]. However, arbitrarily complex queries [2,3,7], entailing rather intricate, possibly recursive, graph patterns prove difficult to evaluate, even on small-sized graph datasets [4,5]. On the other hand, the usage of these queries has radically increased in real-world query logs, as shown by recent empirical studies on SPARQL queries from large-scale Wikidata and DBPedia corpuses [8,17]. As a tangible example of this growth, the percentage of SPARQL property paths has increased from 15% to 40%, from 2017 to beginning 2018 [17], for user-specified Wikidata queries. In this paper, we focus on regular path queries (RPQs) that identify paths labeled with regular expressions and aim to offer an approximate query evaluation solution. In particular, we consider counting queries with regular paths, which are a notable fragment of graph analytical queries. The exact evaluation of counting queries on graphs is $\#P-$complete [21] and is based on another result on enumeration of simple graph paths.

© Springer Nature Switzerland AG 2019
N. Ben Amor et al. (Eds.): SUM 2019, LNAI 11940, pp. 250–265, 2019.
https://doi.org/10.1007/978-3-030-35514-2_19

Due to this intractability, an *efficient and highly-accurate approximation* of these queries is desirable, which we address in this paper.

Approximate query processing on relational data and the related sampling methods are not applicable to graphs, since the adopted techniques are based on the linearity assumption [15], i.e., the existence of a linear relationship between the sample size and execution time, typical of relational query processing. As such, we design a novel query-driven graph summarization approach tailored for *property graphs*. These significantly differ from RDF and relational data models, as they attach data values to property lists on both nodes and edges [7].

To the best of our knowledge, ours is the first work on approximate property graph analytics addressing counting estimation on top of navigational graph queries. We illustrate our query fragment with the running example below.

Example 1 (Social Network Advertising). Let $\mathcal{G}_{SN}$ (see Fig. 1) be a property graph (see Sect. 2) encoding a social network, whose schema is inspired by the LDBC benchmark [12][1]. Entities are people (type Person, P_i) that *know* (l_0) and/or *follow* (l_1) either each other or certain forums (type Forum, F_i). These are *moderated* (l_2) by specific persons and can *contain* (l_3) messages/ads (type Message, M_i), to which persons can *author* (l_4) other messages in *reply* (l_5).

We focus on a RPQ [3,23] dialect with counting, capturing following query types ($Q_1 - Q_7$) (see Fig. 2): *(1) Simple/Optional Label.* The number of pairs satisfying Q_1, i.e., $() \rightarrow l_5()$, counts the ad *reactions*, while that for Q_2, i.e., $() \rightarrow l_2?()$, indicates the number of *potential moderators. (2) Kleene Plus/Kleene Star.* The number of the *connected/potentially connected acquaintances* is the count of node pairs satisfying Q_3, i.e., $() \leftarrow l_0^+()$, respectively, Q_4, i.e., $() \leftarrow l_0^*()$. *(3) Disjunction.* The number of the *targeted subscribers* is the sum of counting all node pairs satisfying Q_5, i.e., $() \xleftarrow{l_4} ()$ or $() \xleftarrow{l_1} ()$. *(4) Conjunction.* The *direct reach* of a company via its page ads is the count of node pairs satisfying Q_6, i.e., $() \xleftarrow{l_4} () \rightarrow l_5()$. *(5) Conjunction with Property Filters.* Recommendation systems can further refine the Q_6 estimates. Thus, one can compute the *direct demographic reach* and target people within an age group, e.g., 18–24, by counting all node pairs that satisfy Q_7, i.e. $(x) \xleftarrow{l_4} () \rightarrow l_5()$, s.t $x.age \geq 18$ and $x.age \leq 24$.

Contributions. Our paper provides the following main contributions:

- We design a property graph summarization algorithm for approximately evaluating counting regular path queries (Sect. 3).
- We prove the intractability of the optimal graph summarization problem under the conditions of our summarization algorithm (Sect. 3).
- We define a query translation module, ensuring that queries on the initial and summary property graphs are expressible in the same fragment (Sect. 4).
- Based on this, we experimentally exhibit the small relative errors of various workloads, in the expressive query fragment from Example 1. We measure the relative response time between estimating counting recursive queries on

---

[1] One of the few benchmarks currently available for generating property graphs.

summaries and on the original graphs. For non-recursive queries, we compare with SumRDF [19], a baseline graph summary for RDF datasets (Sect. 5).

In Sect. 2, we revisit the property graph model and query language. We present related work in Sect. 6 and conclude the paper in Sect. 7.

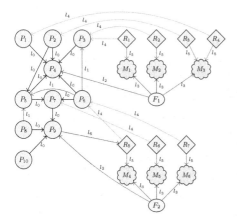

Fig. 1. Example social graph $\mathcal{G}_{SN}$

$Q_1(l_5)$	$Ans(count(\_)) \leftarrow l_5(\_,\_)$
$Q_2(l_2)$	$Ans(count(\_)) \leftarrow l_2?(\_,\_)$
$Q_3(l_0)$	$Ans(count(\_)) \leftarrow l_0^+(\_,\_)$
$Q_4(l_0)$	$Ans(count(\_)) \leftarrow l_0^*(\_,\_)$
$Q_5(l_4, l_1)$	$Ans(count(\_)) \leftarrow l_4 + l_1(\_,\_)$
$Q_6(l_4, l_5)$	$Ans(count(\_)) \leftarrow l_4^- \cdot l_5(\_,\_)$
$Q_7(l_4, l_5)$	$Ans(count(x)) \leftarrow l_4^- \cdot l_5(x,\_), \geq (x.age, 18),$ $\leq (x.age, 24)$

○ Forum   ○ Message   ◇ Reply   ○ Person

⁃ knows (l_0) – follows (l_1) – moderates (l_2) – contains (l_3)

⁃ authors (l_4) – replies (l_5) – reshares (l_6)

Fig. 2. Targeted advertising queries

2   Preliminaries

Graph Model. We take the *property graph model* (PGM) [7] as our foundation. Graph instances are multi-edge digraphs; its objects are represented by typed, data vertices and their relationships, by typed, labeled edges. Vertices and edges can have any number of *properties* (key/value pairs). Let L_V and L_E be disjoint sets of vertex (edge) labels and $\mathcal{G} = (V, E)$, with $E \subseteq V \times L_E \times V$, a *graph instance*. Vertices $v \in V$ have an id label, l_v, and a set of property labels (attributes, l_i), each with a (potentially undefined) term value. For $e \in E$, we use the binary notation $e = l_e(v_1, v_2)$ and abbreviate v_1, as $e.1$, and v_2, as $e.2$. We denote the number of occurrences of l_e, as $\#l_e$, and the set of all edge labels in $\mathcal{G}$, as $\Lambda(\mathcal{G})$. Other key notations henceforth used are given in Table 1.

Clauses	C	$::=$	$A \leftarrow A_1, \ldots, A_n \mid Q \leftarrow A_1, \ldots, A_n$
Queries	Q	$::=$	$Ans(count(\_)) \mid Ans(count(l_v)) \mid Ans(count(l_{v_1}, l_{v_2}))$
Atoms	A	$::=$	$\pi(l_{v_1}, l_{v_2}) \mid op(l_{v_1}.l_i, l_{v_2}.l_j) \mid op(l_{v_1}.l_i, k), op \in \{<, \leq, >, \geq\}, k \in \mathbb{R}$
Paths	π	$::=$	$\epsilon \mid l_e \mid l_e? \mid l_e^{-1} \mid l_e^* \mid l_{e_1} \cdot l_{e_2} \mid \pi + \pi$

Fig. 3. Graph query language

Graph Query Language. To query the above property graph model, we rely on an RPQ [10,11] fragment with aggregate operators (see Fig. 3). RPQs correspond to SPARQL 1.1 property paths and are a well-studied query class tailored to express *graph patterns* of one or more *label-constrained reachability paths*. For labels l_e^i and vertices v_i, the *labeled path* π, corresponding to $v_1 \rightarrow l_e^1 v_2 \ldots v_{k-1} \rightarrow l_e^k v_k$, is the concatenation $l_e^1 \cdot \ldots \cdot l_e^k$. In their full generality, RPQs allow one to select vertices connected via such labeled paths in a *regular language* over L_E. We restrict RPQs to handle *atomic paths* – bidirectional, optional, single-labeled (l_e, $l_e?$, and l_e^-) and transitive single-labeled (l_e^*) – and *composite paths* – conjunctive and disjunctive composition of atomic paths ($l_e \cdot l_e$ and $\pi + \pi$). While not as general as SPARQL, our fragment already captures more than 60% of the property paths found in practice in SPARQL query logs [8]. Moreover, it captures property path queries, as found in the large Wikidata corpus studied in [9]. Indeed, almost all the property paths in the considered logs contain Kleene-star expressions over *single* labels. In our work, we enrich the above query classes with the *count* operator and support basic graph reachability estimates.

3 Graph Summarization

We introduce a novel algorithm that summarizes any property graph into one tailored for approximately counting reachability queries. The key idea is that, as nodes and edges are compressed, informative properties are iteratively added to the corresponding newly formed structures, to enable accurate estimations.

The **grouping phase** (Sect. 3.1) computes Φ, a label-driven $\mathcal{G}$-partitioning into *subgroupings*, following the connectivity on the most frequent labels in $\mathcal{G}$. A first summarization collapses the vertices and inner-edges of each subgrouping into *s-nodes* and the edges connecting s-nodes, into *s-edges*. The **merge phase** (Sect. 3.2), based on further label-reachability conditions, specified by a **heuristic mode** m, collapses s-nodes into *h-nodes* and s-edges into *h-edges*.

Table 1. Notation table

$\mathcal{G}, \Phi, v, V, e, E$	≜	Graph, graph partitioning, vertex (set), edge (set)
$\mathcal{G}^*, v^*, V^*, e^*, E^*$	≜	S-graph, s-node (set), s-edge (set)
$\hat{\mathcal{G}}, \hat{v}, \hat{V}, \hat{e}, \hat{E}$	≜	H-graph, h-node (set), h-edge (set)
$\lambda(\mathcal{G})$	≜	label on which a graph $\mathcal{G}$ is maximally l-connected
$\Lambda_d(v^*), d \in \{1, 2\}$	≜	set of edge labels with direction d w.r.t v^* (1-incoming, 2-outgoing)

3.1   Grouping Phase

For each frequently occurring label l in $\mathcal{G}$, in descending order, we iteratively partition $\mathcal{G}$ into Φ, containing components that are connected on l, as below.

Definition 1 (Maximal L-Connectivity). *A $\mathcal{G}$-subgraph[2], $\mathcal{G}' = (V', E')$, is maximally l-connected, i.e., $\lambda(\mathcal{G}') = l$, iff (1) $\mathcal{G}'$ is weakly-connected, (2) removing any l-labeled edge from E', there exists a V' node pair not connected by a l^+-labeled undirected path, (3) no l-labeled edge connects a V' node to $V \setminus V'$.*

Example 2. In Fig. 1, $\mathcal{G}_1$ is maximally l_0-connected, since it is weakly-connected, not connected by an l_0-labeled edge to the rest of $\mathcal{G}$, and such that, by removing $P_8 \to l_0 P_9$, no undirected, l_0^+-labeled path unites P_8 and P_9.

We call each such component a *subgrouping*. The procedure (see Algorithm 1) computes, as the first *grouping*, all the subgroupings for the most frequent label, l_1, and then identifies those corresponding to the rest of the graph and to l_2. At the end, all remaining nodes are collected into a final subgrouping. We illustrate this in Fig. 4, on the running example below.

Example 3 (Grouping). In Fig. 1, $\#l_0 = 11$, $\#l_1 = 3$, $\#l_2 = 2$, $\#l_3 = 6$, $\#l_4 = \#l_5 = 7$, $\#l_6 = 1$, and $\overrightarrow{\Lambda(\mathcal{G})} = [l_0, l_5, l_4, l_3, l_1, l_2, l_6]$, as $\#l_4 = \#l_5$ allows arbitrary ordering. We add the maximal l_0-connected subgraph, $\mathcal{G}_1$, to Φ. Hence, $V = \{R_{i \in \overline{1,7}}, M_{i \in \overline{1,6}}, F_1, F_2\}$. Next, we add $\mathcal{G}_2$, regrouping the maximal l_5-connected subgraph. Hence, $V = \{F_1, F_2\}$; we add $\mathcal{G}_3$ and output $\Phi = \{\mathcal{G}_1, \mathcal{G}_2, \mathcal{G}_3\}$.

Algorithm 1. GROUPING($\mathcal{G}$)

Input: $\mathcal{G}$ – a graph; **Output:** Φ – a graph partitioning

1: $n \leftarrow |\Lambda(\mathcal{G})|$, $\overrightarrow{\Lambda(\mathcal{G})} \leftarrow [l_1, \ldots, l_n]$, $\Phi \leftarrow \emptyset$, $i \leftarrow 1$ ▷*Descending frequency label list $\overrightarrow{\Lambda(\mathcal{G})}$*
2: **for all** $l_i \in \overrightarrow{\Lambda(\mathcal{G})}$ **do**   ▷*Label-driven partitioning computation*
3:    $\Phi \leftarrow \Phi \cup \{\mathcal{G}_k^* = (V_k^*, E_k^*) \subseteq \mathcal{G} \mid \lambda(\mathcal{G}_k^*) = l_i\}$ ▷*Maximally l_i-Connected Subgraphs*
4:    $V \leftarrow V \setminus \{v \in V_k^* \mid k \in \mathbb{N}\}$ ▷*Discard Already Considered Nodes*
5:    $i \leftarrow i + 1$
6: $\Phi \leftarrow \Phi \cup \{\mathcal{G}_i = (V_i^*, E_i^*) \subseteq \mathcal{G} \mid V_i^* = V \setminus V^*\}$ ▷*Collect Remains in Final Subgroup*
7: **return** Φ

A $\mathcal{G}$-partitioning Φ (see Fig. 4a) is transformed into a *s-graph* $\mathcal{G}^* = (V^*, E^*)$ (see Fig. 4b). As such, each s-node gathers all the nodes and inner edges of a Φ-subgrouping, $\mathcal{G}_j^*$, and each *s-edge*, all same-labeled *cross-edges* (edges between pairwise distinct s-nodes). During this phase, we compute analytics concerning the regrouped entities. We leverage PGM's expressivity to internalize these as properties, e.g., Fig. 5 (right)[3]. Hence, to every s-edge, e^*, we attach *EWeight*,

---

[2] $\mathcal{G}'$ is a $\mathcal{G}$-subgraph iff $V' \subseteq V$ and $E' \subseteq E$ and is *weakly connected* iff there exists an undirected path between any pair of vertices.
[3] All corresponding formulas are provided in the additional material.

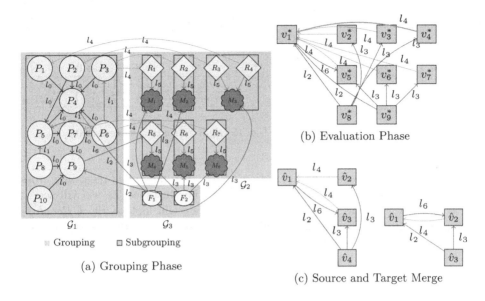

(b) Evaluation Phase

(a) Grouping Phase

(c) Source and Target Merge

Fig. 4. Summarization phases for $\mathcal{G}_{SN}$

its number of compressed edges, e.g., in Fig. 4b, all s-edges have weight 1, except $e^*(v_4^*, v_1^*)$, with weight 2. To every s-node, v^*, we attach properties concerning: *(1) Compression.* $VWeight$ and $EWeight$ store its number of *inner vertices/edges.* *(2) Inner-Connectivity.* The percentage of its l-labeled inner edges is $LPercent$ and the number of its vertex pairs, connected with an l-labeled edge, is $LReach$. These first two types of properties will be useful in Sect. 4, for estimating Kleene paths, as the labels of inner-edges in s-nodes are not unique, e.g., both l_0 and l_1 appear in v_1^*. *(3) Outer-Connectivity.* For pairs of labels and direction indices with respect to v^* ($d = 1$, for incoming edges, and $d = 2$, for outgoing ones), we compute *cross-connectivity,* $CReach$, as the number of binary cross-edge paths that start/end in v^*. Analogously, we record that of binary *traversal paths,* i.e., formed of an inner v^* edge and of a cross-edge, as $TReach$. Also, for a label l and given direction, we store, as V_F, the number of *frontier vertices* on l, i.e., that of v^* nodes at either endpoint of a l-labeled s-edge.

We can thus record *traversal connectivity* information, $LPart$, dividing the number of traversal paths by that of the frontier vertices on the cross-edge label. Intuitively, this is due to the fact that, traversal connectivity, as opposed to cross connectivity, also needs to account for the "dispersion" of the inner-edge label of the path, within the s-node it belongs to. For example, for a traversal path $l_c \cdot l_i$, formed of a cross-edge, l_c, and an inner one, l_i, not all frontier nodes l_c are endpoints of l_i labeled inner-edges, as we will see in the example below.

Example 4 (Outer-Connectivity). Figure 5 (left) depicts a stand-alone example, such that circles denote s-nodes, labeled arrows denote the s-edges relating them, and crosses represent nameless vertices, as we only label relevant ones,

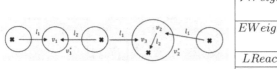

$\mathcal{V}Weight$	$v_1^* \mapsto 10, v_{\{2,3,5,6,7\}}^* \mapsto 2,$
	$v_4^* \mapsto 3, v_{\{8,9\}}^* \mapsto 1$
$EWeight$	$v_1^* \mapsto 14, v_{\{2,3,5,6,7\}}^* \mapsto 1,$
	$v_4^* \mapsto 3, v_{\{8,9\}}^* \mapsto 0$
$LReach$	$(v_1^*, l_0) \mapsto 11, (v_1^*, l_1) \mapsto 3$
$LPercent$	$(v_1^*, l_0) \to 79, (v_1^*, l_1) \to 21$

Fig. 5. Selected properties for Fig. 4b (right); Frontier vertices (left)

for simplicity. We use this configuration to illustrate analytics regarding cross and traversal connectivity on labels l_1 and l_2. For instance, as we will see in Sect. 4, when counting $l_1 \cdot l_2^-$ *cross-edge paths*, we will look at the *CReach* s-node properties mentioning these labels and note that there is a single such one, i.e., that corresponding to l_1 and l_2 appearing on edges incoming v_1^*, i.e., $CReach(v_1^*, l_1, l_2, 1, 1) = 1$. When counting $l_1 \cdot l_2$ *traversal paths*, for the case when l_1 appears on the cross-edge, we will look at the properties of s-nodes containing l_2 inner-edges. Hence, for v_2^*, we note that there is a single such path, formed by an outgoing l_2 edge and incoming l_1 edge, as $TReach(v_2^*, l_1, l_2, 1, 1) = 1$. To estimate the *traversal connectivity* we will divide this by the number of frontier vertices on incoming l_1 edges. As, $V_F(v_2^*, l_1, 1) = \{v_2, v_3\}$, we have that $LPart(v_2^*, l_1, l_2, 1, 1) = 0.5$.

3.2 Merge Phase

We take as input the graph computed by Algorithm 1, and a label set and output a compressed graph, $\hat{\mathcal{G}} = (\hat{V}, \hat{E})$. During this phase, sets of *h-nodes*, $\hat{V}$, and *h-edges*, $\hat{E}$, are created. At each step, as previously, $\hat{\mathcal{G}}$ is enriched with approximation-relevant precomputed properties (see Sect. 4).

Each *h-node*, $\hat{v}$, merges all s-nodes, $v_i^*, v_j^* \in V^*$, that are maximally label connected on the same label, i.e., $\lambda(v_i^*) = \lambda(v_j^*)$, and that have either the same set of incoming (*source-merge*) or outgoing (*target-merge*) edge labels, i.e., $\Lambda_d(v_i^*) = \Lambda_d(v_j^*)$, $d \in \{1, 2\}$ (see Algorithm 2). Each *h-edge*, $\hat{e}$, merges all s-edges in E^* with the same label and orientation, i.e., $e_i^*.d = e_j^*.d$, for $d \in \{1, 2\}$.

Algorithm 2. MERGE(V^*, Λ, m)

Input: V^* – s-nodes; Λ – labels; m – heuristic mode; **Output:** $\hat{V}$ – h-nodes
1: **for all** $v^* \in V^*$ **do**
2:    $\Lambda_d(v^*) \leftarrow \{l \in \Lambda \mid \exists e^* = l(\_, \_) \in E^* \wedge e.d = v^*\}$ ▷ *Labels Incoming/Outgoing* v^*
3: **for all** $v_1^*, v_2^* \in V^*$ **do**   ▷ *Pair-wise S-node Inspection*
4:    $b_\lambda \leftarrow \lambda(v_1^*) \stackrel{?}{=} \lambda(v_2^*)$, $b_d \leftarrow \Lambda_d(v_1^*) \stackrel{?}{=} \Lambda_d(v_2^*)$, $d \in \{1, 2\}$ ▷ *Boolean Conditions*
5:    **if** $m = $ **true then** $\hat{v} \leftarrow \{v_1^*, v_2^* \mid b_\lambda \wedge b_1 = $ **true**$\}$ ▷ *Target-Merge*
6:    **else** $\hat{v} \leftarrow \{v_1^*, v_2^* \mid b_\lambda \wedge b_2 = $ **true**$\}$ ▷ *Source-Merge*
7: $\hat{V} \leftarrow \{\hat{v}_k \mid k \in [1, |V^*|]\}$ ▷ *H-node Computation*
8: **return** $\hat{V}$

To each h-node, we attach properties, whose values, except *LPercent*, are the sum of those corresponding to each of its *s-nodes*. For the label percentage, these values record the weighted percentage mean. Next, we merge *s-edges* into *h-edges*, if they have the same label and endpoints, and attach to each h-edge, its number of compressed s-edges, *EWeight*. We also record the avg. s-node weight, $\mathcal{V}^*Weight$, to estimate how many nodes a h-node compresses.

To formally characterize the graph transformation corresponding to our summarization technique, we first define the following function.

Definition 2 (Valid Summarization). *For $\mathcal{G} = (\mathcal{V}, E)$, a valid summarization function $\chi_\Lambda : \mathcal{V} \to \mathbb{N}$ assigns vertex identifiers, s.t., any vertices with the same identifier are either in the same maximally l-connected $\mathcal{G}$-subgraph, or in different ones, not connected by an l-labeled edge.*

A *valid summary* is thus obtained from $\mathcal{G}$, by collapsing vertices with the same χ_Λ into h-nodes and edges with the same (depending on the heuristic, ingoing/outgoing) label into h-edges. We illustrate this below.

Example 5 (Graph Compression). The graphs in Fig. 4c are obtained from $\mathcal{G}^* = (V^*, E^*)$, after the **merge phase**. Each h-node contains the s-nodes (see Fig. 4b) collapsed via the target-merge (left) and source-merge (right) heuristics.

We study our summarization's *optimality*, i.e., the size of the obtained compressed graph, to graphs its tractability. Specifically, we investigate the following *MinSummary* problem, to establish whether one can always *minimize* the number of nodes of an input graph, when constructing its *valid* summary.

Problem 1 (Minimal Summary). Let *MinSummary* be the problem that, for a graph $\mathcal{G}$ and an integer $k' \geq 2$, decides if there exists a label-driven partitioning Φ of $\mathcal{G}$, $|\Phi| \leq k'$, such that χ_Λ is a *valid summarization*.

Each MinSummary h-node is thus intended to regroup as many nodes from the original graph as possible, while ensuring these are connected by frequently occurring labels. This condition (see Definition 2) reflects the central idea of our framework, namely that the connectivity of such prominent labels can serve to both compress a graph and to approximately evaluate label-constrained reachability queries. Next, we establish the difficulty of solving MinSummary.

Theorem 1 (MinSummary NP-completeness). *Even for undirected graphs, $|\Lambda(\mathcal{G})| \leq 2$, and $k' = 2$, MinSummary is NP-complete*[4].

The intractability of constructing an optimal summary thus justifies our search for heuristics with good performance in practice.

---

[4] Proof given at: http://web4.ensiie.fr/~stefania.dumbrava/SUM19_appx.pdf.

4   Approximate Query Evaluation

Query Translation. For $\mathcal{G}$ and a counting reachability query Q, we approximate $[\![Q]\!]_{\mathcal{G}}$, the evaluation of Q over $\mathcal{G}$. We translate Q into a query Q^T, evaluated over the summarization $\hat{\mathcal{G}}$ of $\mathcal{G}$, s.t $[\![Q^T]\!]_{\hat{\mathcal{G}}} \approx [\![Q]\!]_{\mathcal{G}}$. The translations by input query type are given in Fig. 6, with PGQL as concrete syntax. *(1) Simple and Optional Label Queries.* A label l occurs in $\hat{\mathcal{G}}$ either within a h-node or on a cross-edge. Thus, we either cumulate the number of l-labeled h-node inner-edges or the l-labeled cross-edge weights. To account for the potential absence of l, we also estimate, in the optional-label queries, the number of nodes in $\hat{\mathcal{G}}$, by cumulating those in each h-node. *(2) Kleene Plus and Kleene Star Queries.* To estimate l^+, we cumulate the counts within h-nodes containing l-labeled inner-edges and the weights on l-labeled cross-edges. For the former, we distinguish whether the l_+ reachability is due to: (1) inner-connectivity – we use the property counting the inner l-paths; (2) incoming cross-edges – we cumulate the l-labeled in-degrees of h-nodes; or (3) outgoing cross-edges – we cumulate the number of outgoing l-paths. To handle the ϵ-label in l^*, we also estimate the number of nodes in $\hat{\mathcal{G}}$. *(3) Disjunction.* We treat each possible configuration, on both labels. Hence, we either cumulate the number of h-node inner-edges or that of cross-edge weights, with either label. *(4) Binary Conjunction.* We distinguish whether the label pair appears on an inner h-node path, on a cross-edge path, or on a traversal one.

Example 6. We illustrate the approximate evaluation of these query types on Fig. 4. To evaluate the number of single-label atomic paths, e.g., $Q_L^T(l_5)$, as l_5 only occurs inside h-node $\hat{v}_2$, $[\![l_5]\!]_{\hat{\mathcal{G}}}$ is the amount of l_5-labeled inner edges in $\hat{v}_2$, i.e., $EWeight(\hat{v}_2, l_5) * LPercent(\hat{v}_2, l_5) = 7$. To estimate the number of optional label atomic paths, e.g., $Q_O^T(\ )$, we add to $Q_L^T(\ )$ the total number of graph vertices, $\sum_{\hat{v} \in \hat{v}} \mathcal{V}^*Weight(\hat{v}) * \mathcal{V}Weight(\hat{v})$ (empty case). As $\ $ only appears on a h-edge of weight 2 and there are 25 initial vertices, $[\![\ ]\!]_{\hat{\mathcal{G}}}$ is 27. To estimate Kleene-plus queries, e.g., $Q_P^T(l_0)$, as no h-edge has label l_0, we return $LReach(\hat{v}_1, l_0)$, i.e., the number of l_0-connected vertex pairs. Thus, $[\![l_0^+]\!]_{\hat{\mathcal{G}}}$ is 15. For Kleene-star, we add to this, the previously computed total number of vertices and obtain that $[\![l_0^*]\!]_{\hat{\mathcal{G}}}$ is 40. For disjunction queries, e.g., $[\![l_4 + l_1]\!]_{\hat{\mathcal{G}}}$, we cumulate the single-labeled atomic paths on each label, yielding 14. For binary conjunctions, e.g., $[\![l_4 \cdot l_5]\!]_{\hat{\mathcal{G}}}$, we rely on the traversal connectivity, $LPart(v^*, l_4, l_5, 2, 2)$, as l_4 appears on a h-edge and, l_5, inside h-nodes; we thus count 7 node pairs.

$Q_L(l)$	`SELECT COUNT(*) MATCH () -[:l]-> ()`
$Q_L^T(l)$	`SELECT SUM(x.LPERCENT_L * x.EWEIGHT) MATCH (x)` `SELECT SUM(e.EWEIGHT) MATCH () -[e:l]-> ()`
$Q_O(l)$	`SELECT COUNT(*) MATCH () -[:l?]-> ()`

Q_O^T	`SELECT SUM(x.LPERCENT_L * x.EWEIGHT) MATCH (x)`
	`SELECT SUM(e.EWEIGHT) MATCH () -[e:l]-> ()`
	`SELECT SUM(x.AVG_SN_VWEIGHT * x.VWEIGHT) MATCH (x)`

$Q_P(l)$	`SELECT COUNT(*) MATCH () -/:l+/-> ()`
$Q_P^T(l)$	`SELECT SUM(x.LREACH_L) MATCH (x) WHERE x.LREACH_L > 0` `SELECT SUM(e.EWEIGHT) MATCH () -[e:l]-> ()`
$Q_S(l)$	`SELECT COUNT(*) MATCH () -/:l*/-> ()`

$Q_S^T(l)$	`SELECT SUM(x.LREACH_L) MATCH (x) WHERE x.LREACH_L > 0`
	`SELECT SUM(e.EWEIGHT) MATCH () -[e:l]-> ()`
	`SELECT SUM(x.AVG_SN_VWEIGHT * x.VWEIGHT) MATCH (x)`

$Q_D(l_1, l_2)$	`SELECT COUNT(*) MATCH () -[:l1|l2]-> ()`
$Q_D^T(l_1, l_2)$	`SELECT SUM(x.LPERCENT_L1 * x.EWEIGHT + x.LPERCENT_L2 * x.EWEIGHT) MATCH (x)` `SELECT SUM(e.EWEIGHT) MATCH () -[e:l1|l2]-> ()`
$Q_C(l_1, l_2, 1, 1)$	`SELECT COUNT(*) MATCH () -[:l1]-> () <-[:l2]- ()`
$Q_C(l_1, l_2, 1, 2)$	`SELECT COUNT(*) MATCH () -[:l1]-> () -[:l2]-> ()`
$Q_C(l_1, l_2, 2, 1)$	`SELECT COUNT(*) MATCH () <-[:l1]- () <-[:l2]- ()`
$Q_C(l_1, l_2, 2, 2)$	`SELECT COUNT(*) MATCH () <-[:l1]- () -[:l2]-> ()`

$Q_C^T(l_1, l_2, d_1, d_2)$	`SELECT SUM((x.LPART_L2_L1_D2_D1 * e.EWEIGHT)/(x.LPERCENT_L1 * x.VWEIGHT))`
	`MATCH (x) -[e:l2] -> () WHERE x.LPERCENT_L1 > 0`
	`SELECT SUM((y.LPART_L1_L2_D1_D2 * e.EWEIGHT)/(y.LPERCENT_L2 * y.VWEIGHT))`
	`MATCH () -[e:l1] -> (y) WHERE y.LPERCENT_L2 >0`
	`SELECT SUM(x.CREACH_L1_L2_D1_D2) MATCH (x)`
	`SELECT SUM(x.EWEIGHT * min(x.LPERCENT_L1, x.LPERCENT_L2)) MATCH (x)`

Fig. 6. Query translations onto the graph summary.

5   Experimental Analysis

In this section, we present an empirical evaluation of our graph summarization, recording (1) the succinctness of our summaries and the efficiency of the underlying algorithm and (2) the suitability of our summaries for approximate evaluation of counting label-constrained reachability queries.

Setup, Datasets and Implementation. The summarization and approximation modules are implemented in Java using OpenJDK 1.8[5]. As the underlying graph database backend, we have used Oracle Labs PGX 3.1, which is the only property graph engine allowing for the evaluation of complex RPQs.

To implement the intermediate graph analysis operations (e.g., weakly connected components), we used the Green-Marl domain-specific language and modified the methods to fit the construction of node properties required by our summarization algorithm. We base our analysis on the graph datasets in Fig. 7, encoding: a Bibliographic network (*bib*), the LDBC social network schema [12] (*social*), Uniprot knowledge graphs (*uniprot*), and the WatDiv schema [1] (*shop*).

We obtained these datasets using gMark [5], a synthetic graph instance and query workload generator. As gMark tries to construct the instance that best fits

---

[5] Available at: https://github.com/grasp-algorithm/label-driven-summarization.

Dataset	$|L_V|$	$|L_E|$	~1K		~5K		~25K		~50K		~100K		~200K	
			$|V|$	$|E|$	$|V|$	$|E|$	$|V|$	$|E|$	$|V|$	$|E|$	$|V|$	$|E|$	$|V|$	$|E|$
bib	5	4	916	1304	4565	6140	22780	3159	44658	60300	88879	119575	179356	240052
social	15	27	897	2127	4434	10896	22252	55760	44390	110665	88715	223376	177301	450087
uniprot	5	7	2170	3898	6837	18899	25800	97059	47874	192574	91600	386810	177799	773082
shop	24	82	3136	4318	6605	10811	17893	34052	31181	56443	57131	93780	109205	168934

Fig. 7. Datasets: no. of vertices $|V|$, edges $|E|$, vertex $|L_V|$ and edge labels $|L_E|$.

the size parameter and schema constraints, the resulting sizes vary (especially for the very dense graphs *social* and *shop*). Next, on the same datasets, we generated workloads of varying sizes, for each type in Sect. 2. These datasets and related query workloads have been chosen since they provide the most recent benchmarks for recursive graph queries and also to ensure a comparison with SumRDF [19] (as shown next) on a subset of those supported by the latter. Studies [8,17] have shown that practical graph pattern queries formulated by users in online query endpoints are often small: 56.5% of real-life SPARQL queries consist of a single edge (RDF triple), whereas 90.8% use 6 edges at most. Hence, we select small-sized template queries with frequently occurring topologies, such as chains [8], and formulate them on our datasets, for workloads of ∼600 queries.

Experiments ran on a cloud VM with Intel Xeon E312xx, 4 cores, 1.80 GHz CPU, 128 GB RAM, and Ubuntu 16.04.4 64-bit. Each data point corresponds to repeating an experiment 6 times, removing the first value from the average.

Summary Compression Ratios. First, we evaluate the effect that using the source-merge and target-merge heuristics has on the *summary construction time* (SCT). We also assess the *compression ratio* (CR) on the original graph's vertices and edges, by measuring $(1 - |\hat{\mathcal{V}}|/|\mathcal{V}|) * 100$ and, respectively, $(1 - |\hat{\mathcal{E}}|/|\mathcal{E}|) * 100$.

Next, we compare the results for source and target merge. In Fig. 8(a-d), the most homogeneous datasets, *bib* and *uniprot*, achieve very high CR (close to 100%) and steadily maintain it with varying graph sizes. As far as heterogeneity significantly grows for *shop* and *social*, the CR becomes eagerly sensitive to the dataset size, starting with low values, for smaller graphs, and stabilizing between **85%** and **90%**, for larger ones. Notice also that the most heterogeneous datasets, *shop* and *social*, although similar, display a symmetric behavior for the vertex and edge CRs: the former better compresses vertices, while the latter, edges. Concerning the SCT runtime in Fig. 8(e-f), all datasets keep a reasonable performance for larger sizes, even the most heterogeneous one *shop*. The runtime is, in fact, not affected by heterogeneity, but is rather sensitive, for larger sizes, to $|E|$ variations (up to $450K$ and $773K$, for *uniprot* and *social*). Also, while the source and target merge SCT runtimes are similar, the latter achieves better CRs for *social*. Overall, the dataset with the worst CR for the two heuristics is *shop*, with the lowest CR for smaller sizes. This is also due to the high number of labels in the initial *shop* instances, and, hence, to the high number of properties its summary needs: on average, for all considered sizes, 62.33 properties, against 17.67, for *social* graph, 10.0, for *bib*, and 14.0, for *uniprot*. These experiments

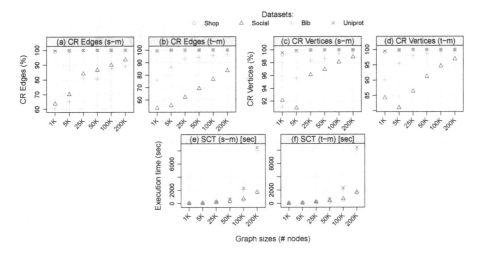

Fig. 8. CRs for vertices and edges, along with SCT runtime for various dataset sizes, for both source-merge (a-c-e), and target-merge (b-d-f).

show that, despite its high complexity, our summarization provides high CRs and low SCT runtimes, even for large, heterogeneous graphs.

Approximate Evaluation Accuracy. We assess the *accuracy* and *efficiency* of our engine with the *relative error* and *time gain* measures, respectively. The relative error (per query Q_i) is $1 - min(Q_i(\mathcal{G}), Q_i^T(\hat{\mathcal{G}}))/ max(Q_i(\mathcal{G}), Q_i^T(\hat{\mathcal{G}}))$ (in %), where $Q_i(\mathcal{G})$ computes (with PGX) the counting query Q_i, on the original graph, and $Q_i^T(\hat{\mathcal{G}})$ computes (with our engine) the translated query Q_i^T, on the summary. The time gain is: $t_{\mathcal{G}} - t_{\hat{\mathcal{G}}}/max(t_{\mathcal{G}}, t_{\hat{\mathcal{G}}})$ (in %), where $t_{\mathcal{G}}$ and $t_{\hat{\mathcal{G}}}$ are the query evaluation times of Q_i on the original graph and on the summary.

For the Disjunction, Kleene-plus, Kleene-star, Optional and Single Label query types, we have generated workloads of different sizes, bound by the number of labels in each dataset. For the concatenation workloads, we considered binary conjunctive queries (CQs) without disjunction, recursion, or optionality. Note that, currently, our summaries do not support compositionality.

Figure 9(a) and (b) show the relative error and average time gain for the Disjunction, Kleene-plus, Kleene-star, Optional and Single Label workloads. In Fig. 9(a), we note that the avg. relative error is kept low in all cases and is bound by 5.5%, for the Kleene-plus and Kleene-star workloads of the *social* dataset. In all the other cases, including the Kleene-plus and Kleene-star workloads of the *shop* dataset, the error is relatively small (near 0%). This confirms the effectiveness of our graph summaries for approximate evaluation of graph queries. In Fig. 9(b), we studied the efficiency of approximate evaluation on our summaries by reporting the time gain (in %) compared with the query evaluation on the original graphs for the four datasets. We notice a positive time gain ($\geq$**75%**) in most cases, but for disjunction. While the relative approximation error is still advantageous for disjunction, disjunctive queries are time-consuming for

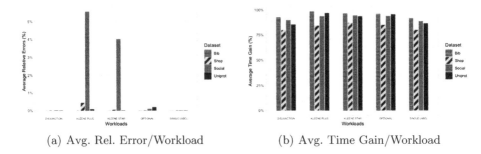

(a) Avg. Rel. Error/Workload   (b) Avg. Time Gain/Workload

Fig. 9. Rel. Error (a), Time Gain (b) per Workload, per Dataset, 200K nodes.

ID	Query Body	Approx. Answer		Rel. Error (%)		Runtime (ms)	
		SumRDF	APP	SumRDF	APP	SumRDF	APP
Q_1	(x0)-[:producer]->()<-[:paymentAccepted]-(x1)	75	76	1.32	0.00	136.30	38.2
Q_2	(x0)-[:totalVotes]->()<-[:price]-(x1)	42.4	44	3.64	0.00	50.99	17
Q_3	(x0)-[:jobTitle]->()<-[:keywords]-(x1)	226.7	221	2.51	0.18	463.85	12.8
Q_4	(x0)<-[:title]-()-[:performedIn]->(x1)	19.5	20	2.50	0.00	831.72	8.8
Q_5	(x0)-[:artist]->()<-[:employee]-(x1)	143.3	133	7.19	0.37	196.77	10.6
Q_6	(x0)-[:follows]->()<-[:editor]-(x1)	524	528	0.38	0.48	1295.83	19

Fig. 10. Performance Comparison: SumRDF vs. APP (our approach): approx. eval. of binary CQs, **SELECT COUNT(*) MATCH** Q_i, on the summaries of a shop graph instance (31K nodes, 56K edges); comparing estimated *cardinality* (no. of computed answers), *rel. error* w.r.t the original graph results, and *query runtime*.

approximate evaluation on our summaries, especially for extremely heterogeneous datasets, such as shop (having the most labels). This is due to the overhead introduced by considering all possible connectivity combinations on the disjunctive labels. The problem of scaling our method, without prohibitive accuracy loss, to queries involving multiple labels and further compositionality, e.g., Kleene-star over disjunctions [22], is challenging and falls under the scope of future work.

Baseline for Approximate Query Evaluation Performance. The closest system to ours is SumRDF [19] (see Sect. 6), which, however, operates on a *simpler edge-labeled model rather than on property graphs and is tailored for estimating the results of conjunctive queries only*. As a performance baseline, we considered the shop dataset in gMark [5], simulating the WatDiv benchmark [1] (also a benchmark in [19]). From this dataset with 31K nodes and 56K edges, we generated the corresponding SumRDF and our summaries. We obtained a better CR than SumRDF, with **2737** nodes vs. **3480** resources and **17430** edges vs. **29621** triples. This comparison is, however, tentative, as our approach compresses vertices independently of the edges, while SumRDF returns triples. We then considered the same CQ types as in Fig. 10. Comparing our approach vs.

SumRDF (see Fig. 10), we recorded an *average relative error* of estimation of only **0.15%.** vs. **2.5%** and an *average query runtime* of only **27.55 ms** vs. **427.53 ms.** As SumRDF does not support disjunctions, Kleene-star/plus queries and optional queries, further comparisons were not possible.

6 Related Work

Preliminary work on approximate graph analytics in a distributed setting has recently been pursued in [15]. They rather focus on a graph sparsification technique and small samples, in order to approximate the results of specific graph algorithms, such as PageRank and triangle counting on undirected graphs. In contrast, our approach operates in a centralized setting and relies on query-driven graph summarization for graph navigational queries with aggregates.

RDF graph summarization for cardinality estimation has been tackled in [19], albeit for a less expressive data model than ours (plain RDF vs. property graphs). They focus on Basic Graph Patterns (BGP), hence their considered query fragment has limited overlap with ours. As shown in Sect. 5, our approximate evaluation is faster and more accurate on a common set of (non recursive) queries.

An algorithm for answering graph reachability queries, using graph simulation based pattern matching, is given in [13], to construct query preserving summaries. However, it does not consider property graphs or aggregates.

Aggregation-based graph summarization [16] is at the heart of previous approaches, the most notable of which is SNAP [20]. This method is mainly devoted to discovery-driven graph summarization of heterogeneous networks and is unsuitable for approximate query evaluation.

More recently, Rudolf et al. [18] have introduced a graph summary suitable for property graphs based on a set of input summarization rules. However, it does not support the label-constrained reachability queries in this paper. Graph summaries for answering subgraphs returned by keyword queries on large networks are studied in [24]. Our query classes significantly differ from theirs.

7 Conclusion

Our paper focuses on a novel graph summarization method that is suitable for property graph querying. As the underlying MinSummary decision problem is NP-complete, this technique builds on an heuristic that compresses label frequency information in the nodes of the graph summary. We show the practical effectiveness of our approach, in terms of compression ratios, error rates and query evaluation time. As future work, we plan to investigate the feasibility of our graph summary for other query classes, such as those described in [22]. Also, we aim to apply formal methods, as described in [6], to ascertain the correctness of our approximation algorithm, with provably tight error bounds.

References

1. Aluç, G., Hartig, O., Özsu, M.T., Daudjee, K.: Diversified stress testing of RDF data management systems. In: Mika, P., et al. (eds.) ISWC 2014. LNCS, vol. 8796, pp. 197–212. Springer, Cham (2014). https://doi.org/10.1007/978-3-319-11964-9_13

2. Angles, R., et al.: G-CORE: a core for future graph query languages. In: SIGMOD, pp. 1421–1432 (2018)

3. Angles, R., Arenas, M., Barceló, P., Hogan, A., Reutter, J.L., Vrgoc, D.: Foundations of modern query languages for graph databases. ACM Comput. Surv. **50**(5), 68:1–68:40 (2017)

4. Arenas, M., Conca, S., Pérez, J.: Counting beyond a Yottabyte, or how SPARQL 1.1 property paths will prevent adoption of the standard. In: WWW, pp. 629–638 (2012)

5. Bagan, G., Bonifati, A., Ciucanu, R., Fletcher, G.H.L., Lemay, A., Advokaat, N.: gMark: schema-driven generation of graphs and queries. IEEE Trans. Knowl. Data Eng. **29**(4), 856–869 (2017)

6. Bonifati, A., Dumbrava, S., Arias, E.J.G.: Certified graph view maintenance with regular datalog. TPLP **18**(3–4), 372–389 (2018)

7. Bonifati, A., Fletcher, G., Voigt, H., Yakovets, N.: Querying Graphs. Synthesis Lectures on Data Management. Morgan & Claypool Publishers (2018)

8. Bonifati, A., Martens, W., Timm, T.: An analytical study of large SPARQL query logs. PVLDB **11**(2), 149–161 (2017)

9. Bonifati, A., Martens, W., Timm, T.: Navigating the maze of Wikidata query logs. In: WWW, pp. 127–138 (2019)

10. Calvanese, D., De Giacomo, G., Lenzerini, M., Vardi, M.Y.: Rewriting of regular expressions and regular path queries. J. Comput. Syst. Sci. **64**(3), 443–465 (2002)

11. Cruz, I.F., Mendelzon, A.O., Wood, P.T.: A graphical query language supporting recursion. In: SIGMOD, pp. 323–330 (1987)

12. Erling, O., et al.: The LDBC social network benchmark: interactive workload. In: SIGMOD, pp. 619–630 (2015)

13. Fan, W., Li, J., Wang, X., Wu, Y.: Query preserving graph compression. In: SIGMOD, pp. 157–168 (2012)

14. Hernández, D., Hogan, A., Riveros, C., Rojas, C., Zerega, E.: Querying Wikidata: comparing SPARQL, relational and graph databases. In: Groth, P., et al. (eds.) ISWC 2016. LNCS, vol. 9982, pp. 88–103. Springer, Cham (2016). https://doi.org/10.1007/978-3-319-46547-0_10

15. Iyer, A.P., et al.: Bridging the GAP: towards approximate graph analytics. In: GRADES, pp. 10:1–10:5 (2018)

16. Khan, A., Bhowmick, S.S., Bonchi, F.: Summarizing static and dynamic big graphs. PVLDB **10**(12), 1981–1984 (2017)

17. Malyshev, S., Krötzsch, M., González, L., Gonsior, J., Bielefeldt, A.: Getting the most out of Wikidata: semantic technology usage in Wikipedia's knowledge graph. In: Vrandečić, D., et al. (eds.) ISWC 2018. LNCS, vol. 11137, pp. 376–394. Springer, Cham (2018). https://doi.org/10.1007/978-3-030-00668-6_23

18. Rudolf, M., Voigt, H., Bornhövd, C., Lehner, W.: SynopSys: foundations for multidimensional graph analytics. In: Castellanos, M., Dayal, U., Pedersen, T.B., Tatbul, N. (eds.) BIRTE 2013-2014. LNBIP, vol. 206, pp. 159–166. Springer, Heidelberg (2015). https://doi.org/10.1007/978-3-662-46839-5_11

19. Stefanoni, G., Motik, B., Kostylev, E.V.: Estimating the cardinality of conjunctive queries over RDF data using graph summarisation. In: WWW, pp. 1043–1052 (2018)
20. Tian, Y., Hankins, R.A., Patel, J.M.: Efficient aggregation for graph summarization. In: SIGMOD, pp. 567–580. ACM (2008)
21. Valiant, L.G.: The complexity of enumeration and reliability problems. SIAM J. Comput. **8**(3), 410–421 (1979)
22. Valstar, L.D.J., Fletcher, G.H.L., Yoshida, Y.: Landmark indexing for evaluation of label-constrained reachability queries. In: SIGMOD, pp. 345–358 (2017)
23. Wood, P.T.: Query languages for graph databases. SIGMOD Rec. **41**(1), 50–60 (2012)
24. Wu, Y., Yang, S., Srivatsa, M., Iyengar, A., Yan, X.: Summarizing answer graphs induced by keyword queries. PVLDB **6**(14), 1774–1785 (2013)

Learning from Imprecise Data: Adjustments of Optimistic and Pessimistic Variants

Eyke Hüllermeier[1], Sébastien Destercke[2(✉)], and Ines Couso[3]

[1] Heinz Nixdorf Institute and Department of Computer Science, Intelligent Systems
and Machine Learning Group, Paderborn University, Paderborn, Germany
eyke@upb.de

[2] UMR CNRS 7253 Heudiasyc, Sorbonne Universités, Université de Technologie de
Compiègne, Compiègne, France
sebastien.destercke@hds.utc.fr

[3] Department of Statistics and Operations Research,
University of Oviedo, Oviedo, Spain
couso@uniovi.es

Abstract. The problem of learning from imprecise data has recently
attracted increasing attention, and various methods to tackle this problem have been proposed. In this paper, we discuss and compare two quite
opposite approaches, an "optimistic" one that interprets imprecise data
in a way that is most favourable for a candidate model, and a "pessimistic" one in which model choice is guided by the most unfavourable
interpretation. To avoid an overly extreme behaviour, a modified version
of the latter has recently been proposed, which we complement by an
adjusted version of the optimistic approach. By presenting the various
methods within a common (loss minimization) framework and discussing
illustrative examples, we hope to provide some insight into important
properties and differences, thereby paving the way for a more formal
analysis.

1 Introduction

Superset learning is a specific type of learning from weak supervision, in which
the outcome (response) associated with a training instance is only characterized
in terms of a set of possible candidates. There are numerous applications in which
supervision is partial in that sense [9]. Correspondingly, the superset learning
problem has received increasing attention in recent years, and has been studied
under various names, such as *learning from ambiguously labelled examples* or
learning from partial labels [2,10]. The contributions so far also differ with regard
to their assumptions on the incomplete information being provided, and how it
has been produced. In this paper, we only assume the actual outcome to be
covered by the subset—hence the name *superset* learning.

In spite of the ambiguous, set-valued training data, the goal that is commonly
considered in superset learning is to induce a *unique* model, or a set of models

© Springer Nature Switzerland AG 2019
N. Ben Amor et al. (Eds.): SUM 2019, LNAI 11940, pp. 266–279, 2019.
https://doi.org/10.1007/978-3-030-35514-2_20

that are all deemed optimal (in the sense of fitting the observed data equally well) and not differentiated any further. This differs from approaches that allow for a set of incomparable, undominated models, resulting for instance from the interval order induced by set-valued loss functions [3], or by the application of conservative, imprecise Bayesian updating rules [11].

In this paper, we reconsider the principle of generalized loss minimization based on the so-called *optimistic superset loss* (OSL) as introduced in [7]. To better understand its nature and possible deficiencies, we contrast the latter with another, in a sense diametral approach based on a "pessimistic" inference principle. Moreover, to compensate for a bias that might be caused by an overly optimistic attitude, we propose an adjustment of the OSL, which can be seen as a counterpart of a corresponding modification of the pessimistic approach [6]. Presenting the various methods within a common framework of loss minimization in supervised learning allows us to highlight some important properties and differences through illustrative examples.

2 Preliminaries

2.1 Setting and Notation

The OSL was introduced in a standard setting of supervised learning with an input (instance) space $\mathcal{X}$ and an output space $\mathcal{Y}$. The goal is to learn a mapping from $\mathcal{X}$ to $\mathcal{Y}$ that captures, in one way or the other, the dependence of outputs (responses) on inputs (predictors). The learning problem essentially consists of choosing an optimal model (hypothesis) h^* from a given model space (hypothesis space) $\mathcal{H}$, based on a set of training data

$$\mathcal{D} = \left\{ (\boldsymbol{x}_n, y_n) \right\}_{n=1}^{N} \in (\mathcal{X} \times \mathcal{Y})^N \ . \tag{1}$$

More specifically, optimality typically refers to optimal prediction accuracy, i.e., a model is sought whose expected prediction loss or *risk*

$$\mathcal{R}(h) = \int L\big(y, h(\boldsymbol{x})\big) \, d\mathbf{P}(\boldsymbol{x}, y) \tag{2}$$

is minimal; here, $L : \mathcal{Y} \times \mathcal{Y} \longrightarrow \mathbb{R}$ is a loss function, and $\mathbf{P}$ is an (unknown) probability measure on $\mathcal{X} \times \mathcal{Y}$ modeling the underlying data generating process.

In the following, we assume hypotheses to be uniquely defined in terms of a parameter θ from an underlying parameter space Θ: $\mathcal{H} = \{h_\theta \,|\, \theta \in \Theta\}$, where h_θ is the hypothesis associated with θ. Selecting an optimal hypothesis $h^* \in \mathcal{H}$ thus reduces to estimating an optimal parameter $\theta^* \in \Theta$.

We are interested in the case where parts of the data are not observed precisely. More specifically, focusing on the output values[1] $y_n \in \mathcal{Y}$, we assume that only supersets $Y_n \subseteq \mathcal{Y}$ are observed. Thus, the learning algorithm does not

---

[1] The principles of optimistic (and likewise pessimistic) loss minimization also extend to the case of imprecision in the instance features.

have direct access to the (precise) data (1), but only to the (imprecise, coarse, ambiguous) observations

$$\mathcal{O} = \left\{(\boldsymbol{x}_n, Y_n)\right\}_{n=1}^{N} \in (\mathcal{X} \times 2^{\mathcal{Y}})^N \ . \tag{3}$$

In the following, we denote by $\mathbf{Y} = Y_1 \times Y_2 \times \cdots \times Y_N$ the (Cartesian) product of the supersets observed for $\boldsymbol{x}_1, \ldots, \boldsymbol{x}_N$. Moreover, each $\boldsymbol{y} = (y_1, \ldots, y_N) \in \mathbf{Y}$ is called an *instantiation* of the imprecisely observed data. More generally, we call a sample $\mathcal{D}$ in (1) an instantiation of $\mathcal{O}$ if the instances $\boldsymbol{x}_n$ coincide and $y_n \in Y_n$ for all $n \in [N] := \{1, \ldots, N\}$.

2.2   Optimistic and Pessimistic Learning

According to [7], a candidate $\theta \in \Theta$ is evaluated optimistically in terms of

$$\mathcal{R}_{emp}^{OPT}(\theta) := \min_{\boldsymbol{y} \in \mathbf{Y}} \frac{1}{N} \sum_{n=1}^{N} L\left(y_n, h_\theta(\boldsymbol{x}_n)\right) \ , \tag{4}$$

i.e., in terms of the empirical risk of h_θ in the case of a most favourable selection of the outcomes y_n. Moreover, given a loss L that is decomposable (over examples), the "optimism" can be moved into the loss:

$$\theta^* := \underset{\theta \in \Theta}{\operatorname{argmin}} \, \mathcal{R}_{emp}^{OPT}(\theta) = \underset{\theta \in \Theta}{\operatorname{argmin}} \, \frac{1}{N} \sum_{n=1}^{N} L_O\left(Y_n, h_\theta(\boldsymbol{x}_n)\right) \ , \tag{5}$$

with the *optimistic superset loss* (OSL)

$$L_O(Y, \hat{y}) = \min\left\{L(y, \hat{y}) \,|\, y \in Y\right\} \ , \tag{6}$$

which compares (precise) predictions with set-valued observations. A key motivation of the OSL is the idea of *data disambiguation*, i.e., the idea of simultaneously inducing the true model (parameter θ) and reconstructing the values of the underlying precise data.

A completely opposite principle is to replace the optimistic minimum in (4) by a pessimistic maximum [5]. More specifically, this principle was introduced in the realm of statistical inference (instead of supervised learning) with L the logistic loss, i.e., in the setting of maximum likelihood inference. The idea is to evaluate each candidate θ in terms of the worst likelihood it can achieve over all instantiations $\boldsymbol{y} \in \mathbf{Y}$, and to pick the best among these pessimistic evaluations. Expressed in terms of generic loss functions (possibly but not necessarily the logistic loss), this principle would amount to considering

$$\mathcal{R}_{emp}^{PESS}(\theta) := \max_{\boldsymbol{y} \in \mathbf{Y}} \frac{1}{N} \sum_{n=1}^{N} L\left(y_n, h_\theta(\boldsymbol{x}_n)\right) \ , \tag{7}$$

and (again assuming the loss to be decomposable) choosing

$$\theta_* := \operatorname*{argmin}_{\theta \in \Theta} \mathcal{R}_{emp}^{PESS}(\theta) = \operatorname*{argmin}_{\theta \in \Theta} \frac{1}{N} \sum_{n=1}^{N} L_P\left(Y_n, h_\theta(\boldsymbol{x}_n)\right) \tag{8}$$

as a presumably best model, with the *pessimistic superset loss* (PSL)

$$L_P(Y, \hat{y}) = \max\left\{L(y, \hat{y}) \,|\, y \in Y\right\} . \tag{9}$$

3 Illustrative Examples

Which of the two approaches to superset learning is more reasonable, the optimistic or the pessimistic one? This question is difficult (or actually impossible) to answer without further assumptions on the coarsening process, i.e., the process that turns precise data into imprecise observations. In the following, to get a better idea of the nature of the two approaches, we illustrate them by some simple examples. We shall refer to the optimistic approach (based on the OSL) as OPT and to the pessimistic one (based on the PSL) and PESS.

3.1 Linear Regression

In linear regression, $\mathcal{X} = \mathbb{R}^d$, $\mathcal{Y} = \mathbb{R}$, and the goal is to learn a linear predictor $h(\boldsymbol{x}) = \boldsymbol{x}^\top \theta = \langle \boldsymbol{x}, \theta \rangle$. Training data is typically assumed to be noisy observations $y_n = \boldsymbol{x}^\top \theta_0 + \epsilon$, where θ_0 is the *ground-truth* parameter and ϵ a noise term (with zero expectation). Correspondingly, in the setting of superset learning, we assume observations $Y_n \ni y_n$. Note, therefore, that Y_n does not necessarily cover the ideal outcome (e.g., the expected value $\mathbb{E}(y \,|\, \boldsymbol{x}_n) = \boldsymbol{x}^\top \theta_0$); instead, just like the precise observation y_n itself, it might be shifted by the noise.

To evaluate predictions $\hat{y} = h(\boldsymbol{x})$, the loss function most commonly used in linear regression is the squared error loss. For the case of interval-valued data $Y = [y_{min}, y_{max}]$, the OSL (6) is then given as follows (cf. Fig. 1):

$$L_O\left([y_{min}, y_{max}], \hat{y}\right) = \begin{cases} (y_{min} - \hat{y})^2 & \text{if } \hat{y} < y_{min} \\ 0 & \text{if } y_{min} \leq \hat{y} \leq y_{max} \\ (\hat{y} - y_{max})^2 & \text{if } y_{max} < \hat{y} \end{cases} \tag{10}$$

Thus, the loss is 0 if the prediction is inside the interval, i.e., if the regression function intersects with the interval, and grows quadratically with the distance from the interval outside. A small one-dimensional example of a set of interval-valued data together with a regression line minimizing (5) is shown in Fig. 2 (left).

The PSL version (9) of the squared error loss is given as follows (cf. Fig. 1):

$$L_P\left([y_{min}, y_{max}], \hat{y}\right) = \begin{cases} (y_{max} - \hat{y})^2 & \text{if } \hat{y} < \frac{1}{2}(y_{min} + y_{max}) \\ (\hat{y} - y_{min})^2 & \text{if } \hat{y} \geq \frac{1}{2}(y_{min} + y_{max}) \end{cases} \tag{11}$$

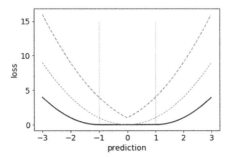

Fig. 1. The OSL (solid line in blue) and PSL (dashed line in red) as extensions of the squared error loss (gray line) in the case of an interval-valued observation (here the interval $[-1, 1]$, indicated by the vertical lines). (Color figure online)

As can be seen in Fig. 1, the PSL targets the midpoint of the interval as an optimal "compromise value"; this point minimizes the maximal prediction error possible, and hence the loss function. Moreover, the larger the interval, the stronger the loss function increases. Therefore, PESS is very similar to *weighted linear regression*, where the weight of an example increases with the width of the corresponding interval. The OSL behaves in a quite different way: the larger the interval, the smaller the loss function. Moreover, OSL does not prefer any values inside the interval (e.g., the midpoint) to any other values. Note that, if the data is completely coherent with a (noise-free) linear model, i.e., if there is a regression function intersecting all intervals, then any such function will be optimal for OPT, while this is not necessarily the case for PESS, as PESS may prefer a function not intersecting all intervals (see Fig. 2 (right) for an illustration). Obviously, since the OSL is no longer strictly convex (in contrast with PSL), the optimisation problem solved by OPT may no longer have a unique solution.

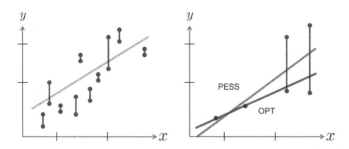

Fig. 2. Left: Linear regression with interval-valued data. Right: Comparison between PESS and OPT for linear regression.

We can also compare OPT and PESS from the point of view of *model updating* or *revision* in the case where new data is observed. Imagine, for example, that a new data point $(\boldsymbol{x}_{N+1}, Y_{N+1})$ is added to the data seen so far. OPT will check for how compatible its current model is with the interval Y_{N+1} and make adjustments only if necessary. In particular, if $\hat{y}_{N+1} = h_\theta(\boldsymbol{x}_{N+1}) \in Y_{N+1}$, i.e., the interval includes the current prediction, the model will not be changed at all, as it is considered fully coherent with the new observation. This also implies that an extremely wide interval will be ignored as being completely uninformative. PESS, on the other side, will always change its current estimate θ, unless $\hat{y}_{N+1} = h_\theta(\boldsymbol{x}_{N+1})$ corresponds exactly to the midpoint of Y_{N+1}; this is because any deviation from this "perfect" prediction is considered as a mistake (or at least a suboptimal choice) that ought to be mitigated.

From the above comments, it is clear that the two strategies may behave quite differently on the same data. OPT assumes that Y_n is a set of candidate values, one of which corresponds to the true measurement. Therefore, fitting one of these candidates, namely the one that is maximally coherent with the model assumption and the rest of the data, is enough. As opposed to this, PESS seeks to fit all values $y_n \in Y_n$ simultaneously, i.e., to find a good compromise prediction $\hat{y}_n$ that is not in conflict with any of the candidates.

It appears that OPT proceeds from a *disjunctive* interpretation of the set Y_n, and considers that the true data will not be chosen so as to systematically put the assumed model in default. In contrast, PESS is more in line with a *conjunctive* interpretation, which makes sense if all the candidates are indeed guaranteed to be possible measurements. One could imagine, for example, that $\boldsymbol{x}_n$ actually characterizes a whole set of entities, and that Y_n is the collection of outputs associated with these entities. As an illustration, suppose we would like to learn a control rule that prescribes an autonomous car the strength of braking depending on its current speed x. Since the optimal strength will also depend on other factors (such as weather conditions), which are ignored (or "integrated out") here, training examples might be interval-valued. For example, depending on further unknown conditions, the optimal strength could be in-between y_{min} and y_{max} for a speed of x Km/h. Adopting a "cautious" model, which minimizes the worst mistake it can make, may look like a reasonable strategy then.

3.2 Logistic Regression

In logistic regression, the goal is to learn a probabilistic classifier

$$h_\theta(\boldsymbol{x}) = \frac{1}{1 + \exp(-\langle \theta, \boldsymbol{x} \rangle)}, \tag{12}$$

where $h_\theta(\boldsymbol{x})$ is an estimate of the (conditional) probability $\mathbf{p}(y = 1 \,|\, \boldsymbol{x})$ of the positive class. Inference is done on the basis of the maximum likelihood principle, which is equivalent to minimizing the log-loss on the training data:

$$\theta^* = \underset{\theta \in \Theta}{\operatorname{argmin}} \sum_{n=1}^{N} L(y_n, h_\theta(\boldsymbol{x}_n))$$

with

$$L(y,p) = -\log\big(py + (1-p)(1-y)\big) = \begin{cases} -\log(p) & \text{if } y = 1 \\ -\log(1-p) & \text{if } y = 0 \end{cases}$$

Using the representation (12) for the probability p, and the class encoding $\mathcal{Y} = \{-1, +1\}$ instead of $\mathcal{Y} = \{0, 1\}$, the loss can also be written as follows:

$$L(y,s) = \log\big(1 + \exp(-ys)\big) ,$$

where $s = \langle \theta, x \rangle$ is the predicted score and ys is the *margin*, i.e., the distance from the decision boundary (to the right side) (Fig. 3).

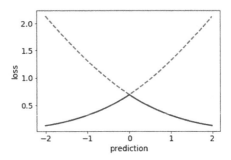

Fig. 3. OSL (blue, solid line) and PSL (red, dashed line) for the logistic loss function. (Color figure online)

Since $\mathcal{Y} = \{-1, +1\}$ contains only two elements, there is only one imprecise observation that can be made, namely $Y = \{-1, +1\} = \mathcal{Y}$, and the setting reduces to so-called semi-supervised learning (with a part of the data being precisely labeled, and another part without any supervision). Thus, the OSL is given by

$$L_O(Y,s) = \begin{cases} L(-1, s) & \text{if } Y = \{-1\} \\ L(+1, s) & \text{if } Y = \{+1\} \\ \min\{L(-1, s), L(+1, s)\} & \text{if } Y = \{-1, +1\} \end{cases},$$

and the pessimistic version L_P by the same expression with min in the third case replaced by max. As a consequence, if an imprecise observation is made, OPT will try to *disambiguate*, i.e., to choose θ such that $ys = y\langle\theta, x\rangle$ is large (and hence p is close to 0 or close to 1); this is in line with a large margin approach, i.e., the learner tries to move the decision boundary away from the data points. Indeed, the generalized loss L_O can be seen as the logistic version of the "hat loss" that is used in semi-supervised learning of support vector machines [1].

As opposed to this, PESS will try to choose θ such that $s \approx 0$ and hence $p \approx \frac{1}{2}$. Obviously, this may lead to drastically different solutions. An example is shown in Fig. 4, where a few labeled training examples are given (positive in blue and negative in red) and many unlabeled. OPT seeks to maximize the margin of the decision boundary, and hence puts it in-between the two clusters. This is in line with the goal of disambiguation: ideally, the unlabeled examples are far from the decision boundary, which means they are clearly identified as positive or negative. PESS is doing exactly the opposite and tries to have the unlabeled examples close to the decision boundary.

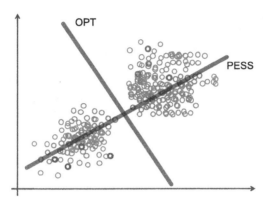

Fig. 4. Logistic regression in a semi-supervised setting: Solutions for OPT and PESS. (Color figure online)

This example suggests that PESS is not really appropriate for tackling discriminative learning tasks. To be fair, however, one has to acknowledge that PESS may produce more reasonable results in other scenarios. For example, if the unlabeled examples are not chosen arbitrarily but indeed correspond to those cases that are very close to the true decision boundary, i.e., for which the posterior probability is indeed close to $\frac{1}{2}$, and which could hence be hard to label, then PESS is just doing the right thing.

As another rather extreme example, suppose that the precise observations in Fig. 4 are just the "noisy" cases, whereas all "normal" cases are hidden (the blue class is actually in the upper right and the red class in the lower left). One can imagine, for example, an "adversarial" coarsening process that coarsens all normal cases and only reveals the noise in the data. In this scenario, it is clear that OPT will be completely misled and produce exactly the opposite of the right model. In such adversarial settings [8], PESS (and more generally minimax approaches) may indeed be considered a more reasonable strategy, as it may provide some guarantees in terms of protection with regard to the coarsening process. Anyway, what all these examples are showing is that the reasonableness of an approach strongly depends on which assumptions about the coarsening process can be considered as plausible.

3.3 Statistical Parameter Estimation

As already said, OPT and PESS have been introduced in different contexts. While generalized loss minimization with the OSL was mainly motivated by problems of supervised machine learning, PESS has mostly been considered in a setting of statistical parameter estimation, such as the estimation of the parameter θ of a Bernoulli distribution in coin tossing. In these cases, OPT may tend to produce rather extreme estimates. For example, consider a sample such as

$$1, 0, ?, 0, ?, 1, 1, 1, ?, ? \ ,$$

with p positive outcomes indicated by a 1 (e.g., a coin toss landing heads up), n negative outcomes indicated by a 0, and u unknowns indicated by a ?. One can check that, in the case where $p > n$, OPT will produce the estimate $\theta^* = {p+u}/{p+u+n}$, based on a corresponding disambiguation in which each unknown is replaced by a positive outcome. More generally, in a multinomial case, all unknowns are supposed to belong to the majority of the precise part of the data. This estimate maximizes the likelihood or, equivalently, minimizes the log-loss

$$L(\theta) = -\sum_{n=1}^{N} X_i \log(\theta) + (1 - X_i) \log(1 - \theta) \ .$$

Such an estimate may appear somewhat implausible. Why should all the unknowns be positive? Of course, one may not exclude that the coarsening process is such that only positives are hidden. In that case, OPT will exactly do the right thing. Still, the estimate remains rather extreme and hence arguable.

In contrast, PESS would try to maximize the entropy of the estimated distribution [4, Corollary 1], which is equivalent to having $\theta^* = 1/2$ in the example given above. While such an estimate may seem less extreme and more reasonable, there is again no compelling reason to consider it more (or less) legitimate than the one obtained by POSS, unless further assumptions are made about the coarsening process. Finally, note that neither POSS nor PESS can produce the estimate obtained by the classical coarsening-at-random (CAR) assumption, which would give $\theta^* = 2/3$.

As a first remark, let us repeat that generalized loss minimization based on OSL was actually not intended, or at least not motivated, by this sort of problem. To explain this point, let us compare the above (statistical estimation) example of coin tossing with the previous (machine learning) example of logistic regression. In fact, the former can be seen as a special case of the latter, with an instance space $\mathcal{X} = \{x_0\}$ consisting of a single instance, such that $\theta = \mathbf{p}(y = 1 \mid x_0)$. Correspondingly, since $\mathcal{X}$ has no structure, it is impossible to leverage any *structural assumptions* about the sought model $h : \mathcal{X} \longrightarrow \mathcal{Y}$, which is the basis of the idea of data disambiguation as performed by OPT.

In particular, in the case of coin flipping, each ? can be replaced by any (hypothetical) outcome, independently of all others and without violating any model assumptions. In other words, every instantiation of the coarse data is as plausible as any other. This is in sharp contrast with the case of logistic regression, where the assumption of a linear model, i.e., the assumption that the probability of success for an input x depends on the spatial position of that point, lets many disambiguations appear implausible. For example, in Fig. 5, the instantiation in

Fig. 5. Coarse data (left) together with two instantiations (middle and right).

the middle, where half of the unlabeled examples are disambiguated as positive and the other half as negative, is clearly more coherent with the assumption of (almost) linearly separable classes than the instantiation on the right, where all unknowns are assigned to the positive class.

In spite of this, examples like the one of coin tossing are indeed suggesting that OSL might be overly optimistic in certain cases. Even in discriminative learning, OSL makes the assumption that the chosen model class is the right one, which may lead to overly confident results should the model choice be wrong. This motivates a reconsideration of the optimistic inference principle and perhaps a suitable adjustment.

4   Adjustments of OSL and PSL

A noticeable property of the previous coin tossing example is a bias of the estimation (or learning) process, which is caused by the fact that a higher likelihood can principally be achieved with a more extreme θ. For example, with $\theta \in \{0, 1\}$, the probability of an "ideal" sample is 1, whereas for $\theta = 1/2$, the highest probability achievable on a sample of size N is $(1/2)^N$. Thus, it seems that, from the very beginning, the candidate estimate $\theta = 1/2$ is put at a systematic disadvantage.

This can also be seen as follows: Consider any sample produced by $\theta = 1$, i.e., a sequence of tosses with heads up. When coarsening the data by covering a subset of the sample, OPT will still produce $\theta = 1$ as an estimate. Roughly speaking, $\theta = 1$ is "robust" toward coarsening. As opposed to this, when coarsening a sample produced with $\theta = 1/2$, OPT will diverge and either produce a smaller or a larger estimate.

4.1   Regularized OSL

One way to counter a systematic bias in disfavour of certain parameters or hypotheses is to adopt a Bayesian approach. Instead of looking at the highest likelihood value $\max_{y \in \mathbf{Y}} \mathbf{p}(y \mid \theta)$ of θ across different instantiations of the imprecise data[2], one may start with a prior π on θ and look at the highest posterior[3]

$$\max_{y \in Y} \frac{\mathbf{p}(y \mid \theta)\, \pi(\theta)}{\mathbf{p}(y)} \, ,$$

or, equivalently,

$$\max_{y \in Y} \left\{ \log \mathbf{p}(y \mid \theta) - H(\theta, y) \right\} = \max_{y \in Y} \left\{ \sum_{i=1}^{N} \log \mathbf{p}(y_n \mid \theta) - H(\theta, y) \right\} \tag{13}$$

with

$$H(\theta, y) := \log \mathbf{p}(y) - \log \pi(\theta) \tag{14}$$

At the level of loss minimization, when ignoring the role of y in (14), this approach essentially comes down to adding a regularization term to the empirical risk, and hence to minimizing the *regularized* OSL

$$\mathcal{R}_{reg}^{OPT}(\theta) := \frac{1}{N} \sum_{n=1}^{N} L_O\big(Y_n, h_\theta(\boldsymbol{x}_n)\big) + F(h_\theta) \, , \tag{15}$$

where $F(h_\theta)$ is a suitable penalty term.

Coming back to our original motivation, namely that some parameters can principally achieve a higher likelihood than others, one instantiation of F one may think of is the maximal (log-)likelihood conceivable for θ (where the sample can be chosen freely and does not depend on the actual imprecise observations):

$$F(\theta) = - \max_{y \in \mathcal{Y}^N} \log \mathbf{p}(y \mid \theta) \tag{16}$$

In this case, $F(\theta)$ can again be moved inside the loss function L_O in (15):

$$\mathcal{R}_{reg}^{OPT}(\theta) := \frac{1}{N} \sum_{n=1}^{N} \mathcal{L}_O\big(Y_n, h_\theta(\boldsymbol{x}_n)\big) \tag{17}$$

with

$$\mathcal{L}_O(Y, \hat{y}) := \min_{y \in Y} L(y, \hat{y}) - \min_{y \in \mathcal{Y}} L(y, \hat{y}) \, . \tag{18}$$

For some losses, such as squared error loss in regression, the adjustment (18) has no effect, because $L(y, \hat{y}) = 0$ can always be achieved for at least one $y \in \mathcal{Y}$. For others, however, $\mathcal{L}_O$ may indeed differ from L_O. For the log-loss in binary

---

[2] We assume the $\boldsymbol{x}_n$ in the data $\{(\boldsymbol{x}_n, y_n)\}_{n=1}^N$ to be fixed.

[3] The obtained bound are similar to the upper expectation bound obtained by the updating rule discussed by Zaffalon and Miranda [11] in the case of a completely unknown coarsening process and precise prior information. However, Zaffalon and Miranda discussed generic robust updating schemes leading to sets of probabilities or sets of models, which is not the intent of the methods discussed in this paper.

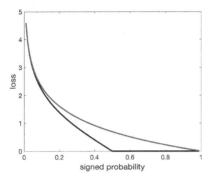

Fig. 6. The adjusted OSL version (19) of the logistic loss (black line) compared to the original version (red line). (Color figure online)

classification, for example, the normalizing term in (18) is $\min\{L(0,p), L(1,p)\}$, which means that

$$\mathcal{L}_O(Y,p) = \begin{cases} \log(1-p) - \log(p) & \text{if } Y = \{1\},\, p < 1/2 \\ \log(p) - \log(1-p) & \text{if } Y = \{0\},\, p > 1/2 \\ 0 & \text{otherwise} \end{cases} \quad . \tag{19}$$

A graphical representation of this loss function, which can be seen as a combination of the 0/1 loss (it is 0 for signed probabilities $\geq 1/2$) and the log-loss, is shown in Fig. 6.

4.2 Adjustment of PSL: Min-Max Regret

Interestingly, a similar adjustment, called *min-max regret* criterion, has recently been proposed for PESS [6]. The motivation of the latter, namely to assess a parameter θ in a *relative* rather than *absolute* way, is quite similar to ours. Adopting our notation, a candidate θ is evaluated in terms of

$$\max_{\boldsymbol{y} \in \mathbf{Y}} \left\{ \log \mathbf{p}(\boldsymbol{y} \,|\, \theta) - \max_{\hat{\theta}} \log \mathbf{p}(\boldsymbol{y} \,|\, \hat{\theta}) \right\} \quad . \tag{20}$$

That is, θ is assessed on a concrete instantiation $\boldsymbol{y} \in \mathbf{Y}$ by comparing it to the best estimation $\hat{\theta}_{\boldsymbol{y}}$ on that data, which defines the regret, and then the worst comparison over all possible instantiations (the maximum regret) is considered. Like in the case of OSL, this can again be seen as an approximation of (14) with

$$F(\boldsymbol{y}) = \max_{\hat{\theta}} \log \mathbf{p}(\boldsymbol{y} \,|\, \hat{\theta}) \,,$$

which now depends on $\boldsymbol{y}$ but not on θ (whereas the F in (15) depends on θ but not on $\boldsymbol{y}$). Obviously, the min-max regret principle is less pessimistic than the original PSL, and leads to an adjustment of PESS that is even somewhat comparable to OPT: The loss of a candidate θ on an instantiation $\boldsymbol{y}$ is corrected by the

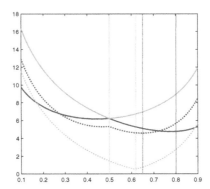

Fig. 7. Loss functions and optimal predictions of θ (minima of the losses indicated by vertical lines) in the case of coin tossing with observations 0, 0, 1, 1, 1, 1, ?, ?, ?: solid blue line for OSL, dashed blue for the regularized OSL version (14) with π the beta (5,5) distribution, solid red for PSL, and dashed red for the adjusted PSL (20). (Color figure online)

minimal loss $F(\boldsymbol{y})$ that can be achieved on this instantiation. Obviously, by doing so, the influence of instantiations that necessarily cause a high loss is reduced. But these instantiations are exactly those that are considered as "implausible" and down-weighted by OPT (cf. Sect. 3.3). See Fig. 7 for an illustrative comparison in the case of coin tossing as discussed in Sect. 3.3. Note that (20) does not permit an additive decomposition into losses on individual training examples, because the regret is defined on the entire set of data. Instead, a generalization of (20) to loss functions other than log-loss suggests evaluating each θ in terms of the maximal regret

$$\mathrm{MReg}(\theta) := \max_{\boldsymbol{y} \in \mathbf{Y}} \left(\mathcal{R}_{emp}(\theta, \boldsymbol{y}) - \min_{\hat{\theta}} \mathcal{R}_{emp}(\hat{\theta}, \boldsymbol{y}) \right), \tag{21}$$

where $\mathcal{R}_{emp}(\theta, \boldsymbol{y})$ denotes the empirical risk of θ on the data obtained for the instantiation $\boldsymbol{y}$. Computing the maximal regret (21), let alone finding the minimizer $\theta^* = \mathrm{argmin}_{\theta} \mathrm{MReg}(\theta)$, appears to be intractable except for trivial cases. In particular, the problem will be hard in cases like logistic regression, where the empirical risk minimizer $\min_{\hat{\theta}} \mathcal{R}_{emp}(\hat{\theta}, \boldsymbol{y})$ cannot be obtained analytically, because then even the evaluation of a single candidate θ on a single instantiation $\boldsymbol{y}$ requires the solution of a complete learning task—not to mention that the minimization over all instantiations $\boldsymbol{y}$ comes on top of this.

5   Concluding Remarks

The goal of our discussion was to provide some insight into the basic nature of the "optimistic" and the "pessimistic" approach to learning from imprecise data. To this end, we presented both of them in a unified framework and highlighted important properties and differences through illustrative examples.

As future work, we plan a more thorough comparison going beyond anecdotal evidence. Even if both approaches deliberately refrain from specific assumptions about the coarsening process, it would be interesting to characterize situations in which they are likely to produce accurate results, perhaps even with formal guarantees, and situations in which they may fail. In addition to a formal analysis of that kind, it would also be interesting to compare the approaches empirically. This is not an easy task, however, especially due to a lack of suitable (real) benchmark data. Synthetic data can of course be used as well, but as our examples have shown, it is always possible to create the data in favour of the one and in disfavour of the other approach.

References

1. Chapelle, O., Sindhwani, V., Keerthi, S.S.: Optimization techniques for semi-supervised support vector machines. J. Mach. Learn. Res. **9**, 203–233 (2008)
2. Cour, T., Sapp, B., Taskar, B.: Learning from partial labels. J. Mach. Learn. Res. **12**, 1501–1536 (2011)
3. Couso, I., Sánchez, L.: Machine learning models, epistemic set-valued data and generalized loss functions: an encompassing approach. Inf. Sci. **358**, 129–150 (2016)
4. Guillaume, R., Couso, I., Dubois, D.: Maximum likelihood with coarse data based on robust optimisation. In: Proceedings of the Tenth International Symposium on Imprecise Probability: Theories and Applications, pp. 169–180 (2017)
5. Guillaume, R., Dubois, D.: Robust parameter estimation of density functions under fuzzy interval observations. In: 9th International Symposium on Imprecise Probability: Theories and Applications (ISIPTA 2015), pp. 147–156 (2015)
6. Guillaume, R., Dubois, D.: A maximum likelihood approach to inference under coarse data based on minimax regret. In: Destercke, S., Denoeux, T., Gil, M.Á., Grzegorzewski, P., Hryniewicz, O. (eds.) SMPS 2018. AISC, vol. 832, pp. 99–106. Springer, Cham (2019). https://doi.org/10.1007/978-3-319-97547-4_14
7. Hüllermeier, E.: Learning from imprecise and fuzzy observations: data disambiguation through generalized loss minimization. Int. J. Approxim. Reasoning **55**(7), 1519–1534 (2014)
8. Laskov, P., Lippmann, R.: Machine learning in adversarial environments. Mach. Learn. **81**(2), 115–119 (2010)
9. Liu, L.P., Dietterich, T.G.: A conditional multinomial mixture model for superset label learning. In: Proceedings NIPS (2012)
10. Nguyen, N., Caruana, R.: Classification with partial labels. In: 14th International Conference on Knowledge Discovery and Data Mining Proceedings KDD, Las Vegas, USA, p. 2008 (2008)
11. Zaffalon, M., Miranda, E.: Conservative inference rule for uncertain reasoning under incompleteness. J. Artif. Intell. Res. **34**, 757–821 (2009)

On Cautiousness and Expressiveness in Interval-Valued Logic

Sébastien Destercke[(✉)] and Sylvain Lagrue

Université de Technologie de Compiègne, CNRS, UMR 7253 - Heudiasyc,
Centre de Recherche de Royallieu, Compiègne, France
{sebastien.destercke,sylvain.lagrue}@hds.utc.fr

Abstract. In this paper, we study how cautious conclusions should be taken when considering interval-valued propositional logic, that is logic where to each formula is associated a real-valued interval providing imprecise information about the penalty incurred for falsifying this formula. We work under the general assumption that the weights of falsified formulas are aggregated through a non-decreasing commutative function, and that an interpretation is all the more plausible as it is less penalized. We then formulate some dominance notions, as well as properties that such notions should follow if we want to draw conclusions that are at the same time informative and cautious. We then discuss the dominance notions in light of such properties.

Keywords: Logic · Imprecise weights · Skeptic inference · Robust inferences · Penalty logic

1 Introduction

Logical frameworks have always played an important role in artificial intelligence, and adding weights to logical formulas allow one to deal with a variety of problems with which classical logic struggles [3].

Usually, such weights are assumed to be precisely given, and associated to an aggregation function, such as the maximum in possibilistic logic [4] or the sum in penalty logic [5]. These approaches can typically find applications in non-monotonic reasoning [1] or preference handling [7].

However, as providing specific weights to each formula is likely to be a cognitively demanding tasks, many authors have considered extensions of these frameworks to interval-valued weights [2,6], where intervals are assumed to contain the true, ill-known weights. Such approaches can also be used, for instance, to check how robust conclusions obtained with precise weights are.

In this paper, we are interested in making cautious or robust inferences in such interval-valued frameworks. That is, we look for inference tools that will typically result in a partial order over the interpretations or world states, such that any preference statement made by this partial order is made in a skeptic way, i.e., it holds for any replacement of the weights by precise ones within the

© Springer Nature Switzerland AG 2019
N. Ben Amor et al. (Eds.): SUM 2019, LNAI 11940, pp. 280–288, 2019.
https://doi.org/10.1007/978-3-030-35514-2_21

intervals, and should not be reversed when gaining more information. We simply assume that the weights are positive and aggregated by a quite generic function, meaning that we include for instance possibilistic and penalty logics as special cases.

We provide the necessary notations and basic material in Sect. 2. In Sect. 3, we introduce different ways to obtain partial orders over interpretations, and discuss different properties that corresponding cautious inference tools could or should satisfy. Namely, that reducing the intervals will provide more informative and non-contradictory inferences, and that if an interpretation falsify a subset of formulas falsified by another one, then it should be at least as good as this latter one. Section 4 shows which of the introduced inference tools satisfy which property.

2   Preliminaries

We consider a finite propositional language $\mathcal{L}$. We denote by Ω the space of all interpretations of $\mathcal{L}$, and by ω an element of Ω. Given a formula ϕ, ω is a model of ϕ if it satisfies it, denoted $\omega \models \phi$.

A weighted formula is a tuple $\langle \phi, \alpha \rangle$ where α represents the importance of the rule, and the penalty incurred if it is not satisfied. This weight may be understood in various ways: as a degree of certainty, as a degree of importance of an individual preference, ... We assume that α take their values on an interval of $\mathbb{R}^+$, possibly extended to include ∞ (e.g., to represent formulas that cannot be falsified). In this paper, a formula with $\alpha = 0$ is understood as a totally unimportant formula that can be ignored, while a formula with maximal α is a formula that must be satisfied.

A (precisely) weighted knowledge base $K = \{\langle \phi_i, \alpha_i \rangle : i = 1, \ldots, n\}$ is a set of distinct weighted formulas. Since these formulas are weighted, an interpretation can (and sometimes must, if K without weights is inconsistent) falsify some of them, and still be considered as valid. In order to determine an ordering between different interpretations, we introduce two new notations:

- $F_K(\omega) = \{\phi_i : \omega \not\models \phi_i\}$, the set of formulas falsified by ω
- $F_K(\omega \setminus \omega') = \{\phi_i : \omega \not\models \phi_i \wedge \omega' \models \phi_i\}$

Let us furthermore consider an aggregation function $ag : \mathbb{R}^n \to \mathbb{R}$ that we assume to be non-decreasing, commutative, continuous and well-defined[1] for any finite number n.

We consider that $ag(\{\alpha_i : \phi_i \in F\})$ applied to a subset F of formulas of K measure the overall penalty corresponding to F, with $ag(\emptyset) = 0$. Given this, we also assume that if ag receives two vectors $\boldsymbol{a}$ and $\boldsymbol{b}$ of dimensions n and $n + m$ such that $\boldsymbol{b}$ has the same first n elements as $\boldsymbol{a}$, i.e., $\boldsymbol{b} = (\boldsymbol{a}, y_1, \ldots, y_m)$, then $ag(\boldsymbol{a}) \leq ag(\boldsymbol{b})$. The idea here is that adding (falsified) formulas to $\boldsymbol{a}$ can only increase the global penalty. Classical options correspond to possibilistic logic

---

[1] As we do not necessarily assume it to be associative.

(weights are in $[0, 1]$ and $ag = \max$) or penalty logic (weights are positive reals and $ag = \sum$). Based on this aggregation function, we define a given K the two following complete orderings between interpretations when weights are precise:

- $\omega \succeq_{All}^{K} \omega'$ iff $ag(\{\alpha_i : \phi_i \in F_K(\omega)\}) \leq ag(\{\alpha_i : \phi_i \in F_K(\omega')\})$.
- $\omega \succeq_{Diff}^{K} \omega'$ iff $ag(\{\alpha_i : \phi_i \in F_K(\omega \setminus \omega')\}) \leq ag(\{\alpha_i : \phi_i \in F_K(\omega' \setminus \omega)\})$.

Both orderings can be read as $\omega \succeq \omega'$ meaning that "ω is more plausible, or preferred to ω', given K".

When the weights are precise, it may be desirable for $\succeq_{All}$ and $\succeq_{Diff}$ to be consistent, that is not to have $\omega \succ_{All}^{K} \omega'$ and $\omega \prec_{Diff}^{K} \omega'$ for a given K. It may be hard to characterize the exact family of functions ag that will satisfy this, but we can show that adding associativity and strict increasigness[2] to the other mentioned properties ensure that results will be consistent.

Proposition 1. *If ag is continuous, commutative, strictly increasing and associative, then given a knowledge base K, we have that*

$$\omega \succeq_{All}^{K} \omega' \Leftrightarrow \omega \succeq_{Diff}^{K} \omega'$$

Proof. Let us denote the sets $\{\alpha_i : \phi_i \in F_K(\omega)\}$ and $\{\alpha_i : \phi_i \in F_K(\omega')\}$ as real-valued vectors $\boldsymbol{a} = (x_1, \ldots, x_n, y_{n+1}, \ldots, y_{n_a})$ and $\boldsymbol{b} = (x_1, \ldots, x_n, z_{n+1}, \ldots, z_{n_b})$, where $x_1, \ldots, x_n$ are the weights associated to the formulas that both interpretations falsify. Showing the equivalence of Proposition 1 then comes down to show

$$ag(\boldsymbol{a}) \geq ag(\boldsymbol{b}) \Leftrightarrow ag((y_{n+1}, \ldots, y_{n_a})) \geq ag((z_{n+1}, \ldots, z_{n_b})).$$

Let us first remark that, due to associativity,

$$ag(\boldsymbol{a}) = ag(ag((x_1, \ldots, x_n)), ag((y_{n+1}, \ldots, y_{n_a}))) := ag(A, B),$$

$$ag(\boldsymbol{b}) = ag(ag((x_1, \ldots, x_n)), ag((z_{n+1}, \ldots, z_{n_b}))) := ag(A, C).$$

Under these notations, we must show that $ag(A, B) \geq ag(A, C) \Leftrightarrow B \geq C$.

That $B \geq C \Rightarrow ag(A, B) \geq ag(A, C)$ is immediate, as ag is non-decreasing. To show that $B \geq C \Leftarrow ag(A, B) \geq ag(A, C)$, we can just see that if $B < C$, we have $ag(A, B) < ag(A, C)$ due to the strict increasingness of ag.

3   Interval-Valued Logic, Dominance Notions and Properties

In practice, it is a strong requirement to ask users to provide precise weights for each formula, and they may be more comfortable in providing imprecise ones. This is one of the reason why researchers proposed to extend weighted logics to interval-valued logics, where the knowledge base is assumed to have the form

---

[2] Which is also necessary, as $ag = \max$ will not always satisfy Property 1.

$K = \{\langle \phi_i, I_i \rangle : i = 1, \ldots, n\}$ with $I_i = [a_i, b_i]$ representing an interval of possible weights assigned to ϕ_i.

In practice, this means that the result of applying ag to a set of formulas F is no longer a precise value, but an interval $[\underline{ag}, \overline{ag}]$. As ag is a non-decreasing continuous function, computing this interval is quite easy as

$$\underline{ag} = \underline{ag}(\{a_i | \phi_i \in F\}),$$

$$\overline{ag} = \overline{ag}(\{b_i | \phi_i \in F\}),$$

which means that if the problem with precise weights is easy to solve, then solving it for interval-valued weights is equally easy, as it amounts to solve twice the problems for specific precise weights (i.e., the lower and upper bounds). A question is now to know how we should rank the various interpretations in a cautious way given these interval-valued formulas. In particular, this means that the resulting order between interpretations should be a partial order if we have no way to know whether one has a higher score than the other, given our imprecise information. But at the same time, we should try to not lose too much information by making things imprecise.

There are two classical ways to compare interval-valued scores that results in possible incomparabilities:

- Lattice ordering: $[a, b] \preceq_L [c, d]$ iff $a \leq c$ and $b \leq d$. We then have $[a, b] \prec_L [c, d]$ if one of the two inequalities is strict, and $[a, b] \simeq [c, d]$ iff $[a, b] = [c, d]$. Incomparability of $[a, b]$ and $[c, d]$ corresponds to one of the two set being strictly included in the other.
- Strict ordering: $[a, b] \preceq_S [c, d]$ iff $b \leq c$. We then have $[a, b] \prec_S [c, d]$ if $b < c$, and indifference will only happen when $a = b = c = d$ (hence never if intervals are non-degenerate). Incomparability of $[a, b]$ and $[c, d]$ corresponds to the two sets overlapping.

These two orderings can then be applied either to $\succeq_{All}$ or $\succeq_{Diff}$, resulting in four different extensions: $\succeq_{All,L}, \succeq_{All,S}, \succeq_{Diff,L}, \succeq_{Diff,S}$. A first remark is that strict comparisons are stronger than lattice ones, as the former imply the latter, that is if $[a, b] \preceq_S [c, d]$, then $[a, b] \preceq_L [c, d]$. In order to decide which of these orderings are the most adequate, let us first propose some properties they should follow when one wants to perform cautious inferences.

Property 1 (Informational monotonicity). *Assume that we have two knowledge bases* $K^1 = \{\langle \phi_i^1, I_i^1 \rangle : i = 1, \ldots, n\}$ *and* $K^2 = \{\langle \phi_i^2, I_i^2 \rangle : i = 1, \ldots, n\}$ *with* $\phi_i^1 = \phi_i^2$ *and* $I_i^1 \subseteq I_i^2$ *for all* i. *An aggregation method and the partial order* $\succ$ *it induces on interpretations is informational monotonic if*

$$\omega \succ^{K_2} \omega' \implies \omega \succ^{K_1} \omega'$$

That is, the more we gain information, the better we become at differentiating and ranking interpretations. If ω is strictly preferred to ω' before getting more precise assessments, it should remain so after the assessments become more precise[3]. A direct consequence of Property 1 is that we cannot have $w \succ^{K_2} w'$ and $w' \succ^{K_1} w$, meaning that $\succ^{K_1}$ will be a refinement of $\succ^{K_2}$. This makes sense if we aim for a cautious behaviour, as the conclusion we make in terms of preferred interpretations should be guaranteed, i.e., they should not be revised when we become more precise.

It should also be noted that violating this property means that the corresponding partial order is not skeptic in the sense advocated in the introduction, as a conclusion taken at an earlier step can be contradicted later on by gaining more information.

Property 2 (subset/implication monotonicity). *Assume that we have a knowledge base K. An aggregation method and the partial order $\succ$ it induces on interpretations follows subset monotonicity if*

$$F_K(w) \subseteq F_K(w') \implies w \succeq^K w' \text{ for any pair } w, w'$$

This principle is quite intuitive: if we are sure that w' falsifies the same formulas than w in addition to some others, then certainly w' should be less preferable/certain than w.

4   Discussing Dominance Notions

Let us now discuss the different partial orders in light of these properties, starting with the lattice orderings and then proceeding to interval orderings.

4.1   Lattice Orderings

Let us first show that $\succeq_{All,L}, \succeq_{Diff,L}$ do not satisfy Property 1 in general, by considering the following example:

Example 1. Consider the case where $a_i, b_i \in \mathbb{R}$ and $ag = \sum$, with the following knowledge base on the propositional variables $\{p, q\}$

$$\phi_1 = p, \quad \phi_2 = p \wedge q, \quad \phi_3 = \neg q$$

with the three following sets (respectively denoted K^1, K^2, K^3) of interval-valued scores

$$
\begin{array}{lll}
I_1^{K_1} = [2.5, 2.5], & I_2^{K_1} = [0, 4], & I_3^{K_1} = [1, 5], \\
I_1^{K_2} = [2.5, 2.5], & I_2^{K_2} = [4, 4], & I_3^{K_2} = [1, 5], \\
I_1^{K_3} = [2.5, 2.5], & I_2^{K_3} = [4, 4], & I_3^{K_3} = [1, 1],
\end{array}
$$

that are such that $I^{K_3} \subseteq I^{K_2} \subseteq I^{K_1}$ for all formulas. The resulting scores using the choice $\succeq_{All}$ following on the different interpretations are summarised in Table 1.

---

[3] Note that we consider the new assessments to be consistent with the previous ones, as $I_i{}^1 \subseteq I_i{}^2$.

Table 1. Interval-valued scores from Example 1

	p	q	ag^1	ag^2	ag^3
ω_0	0	0	[2.5,6.5]	[6.5,6.5]	[6.5,6.5]
ω_1	0	1	[3.5, 11.5]	[7.5,11.5]	[7.5,7.5]
ω_2	1	0	[0, 4]	[4, 4]	[4, 4]
ω_3	1	1	[1, 5]	[1, 5]	[1, 1]

Figure 1 shows the different partial orders between the interpretations, according to $\succeq_{All,L}$. We can see that ω_2 and ω_3 go from comparable to incomparable when going from $\succ_{All,L}^{K_1}$ to $\succ_{All,L}^{K_2}$, and that the preference or ranking between them is even reversed when going from $\succ_{All,L}^{K_1}$ to $\succ_{All,L}^{K_3}$.

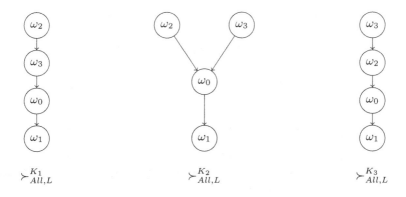

Fig. 1. Orderings $\succ_{All,L}$ of Example 1 on interpretations.

It should be noted that what happens to ω_2, ω_3 for $\succeq_{All,L}$ is also true for $\succeq_{Diff,L}$. Indeed, $F_K(\omega_2) = \{p \sqcap q\}$ and $F_K(\omega_3) = \{\neg q\}$, hence $F_K(\omega_2 \setminus \omega_3) = F_K(\omega_2)$ and $F_K(\omega_3 \setminus \omega_2) = \emptyset$. However, we can show that the two orderings based on lattice do satisfy subset monotonicity.

Proposition 2. *Given a knowledge base K, the two orderings $\succeq_{All,L}^K, \succeq_{Diff,L}^K$ satisfy subset monotonicity.*

Proof. For $\succeq_{Diff,L}$, it is sufficient to notice that if $F_K(\omega) \subseteq F_K(\omega')$, then $F_K(\omega \setminus \omega') = F_K(\emptyset)$. This means that $ag(\{\alpha_i : \phi_i \in F_K(\omega \setminus \omega')\}) = [0,0]$, hence we necessarily have $\omega \succeq_{Diff,L}^K \omega'$.

For $\succeq_{All,L}$, the fact that $F_K(\omega) \subseteq F_K(\omega')$ means that the vectors $\boldsymbol{a}$ and $\boldsymbol{a}'$ of lower values associated to $\{\alpha_i : \phi_i \in F_K(\omega)\}$ and $\{\alpha_i : \phi_i \in F_K(\omega')\}$ will be of the kind $\boldsymbol{a}' = (\boldsymbol{a}, a_1, \ldots, a_m)$, hence we will have $\underline{ag}(\boldsymbol{a}) \leq \underline{ag}(\boldsymbol{a}')$. The same reasoning applied to upper bounds means that we will also have $\overline{ag}(\boldsymbol{a}) \leq \overline{ag}(\boldsymbol{a}')$, meaning that $\omega \succeq_{All,L}^K \omega'$.

From this, we deduce that lattice orderings will tend to be too informative for our purpose[4], i.e., they will induce preferences between interpretations that should be absent if we want to make only those inferences that are guaranteed (i.e., hold whatever the value chosen within the intervals I_i).

4.2 Strict Orderings

In this section, we will study strict orderings, and will show in particular that while $\succeq_{All,S}$ provides orderings that are not informative enough for our purpose, $\succeq_{Diff,S}$ does satisfy our two properties.

As we did for lattice orderings, let us first focus on the notion of informational monotonicity, and show that both orderings satisfy it.

Proposition 3. *Given knowledge bases K^1, K^2 with $\phi_i^1 = \phi_i^2$ and $I_i^{\,1} \subseteq I_i^{\,2}$ for all $i \in \{1, \ldots, n\}$, the two orderings $\succeq_{All,S}, \succeq_{Diff,S}$ satisfy information monotonicity.*

Proof. Assume that $[a, b]$ and $[c, d]$ are the intervals obtained from K^2 respectively for ω and ω' after aggregation has been performed, with $b \leq c$, hence $\omega \succeq_{\ell,S}^{K_2} \omega'$ with $\ell \in \{All, Diff\}$.

Since ag is an increasing function, and as $I_i^{\,1} \subseteq I_i^{\,2}$, we will have that the intervals $[a', b']$ and $[c', d']$ obtained from K^1 for ω and ω' after aggregation will be such that $[a', b'] \subseteq [a, b]$ and $[c', d'] \subseteq [c, d]$, meaning that $b' \leq b \leq c \leq c'$, hence $\omega \succeq_{\ell,S}^{K_1} \omega'$, and this finishes the proof.

Let us now look at the property of subset monotonicity. From the knowledge base K^1 in Example 1, one can immediately see that $\succeq_{All,S}$ is not subset monotonic, as $F_K(\omega_3) \subseteq F_K(\omega_1)$ and $F_K(\omega_2) \subseteq F_K(\omega_0) \subseteq F_K(\omega_1)$, yet all intervals in Table 1 overlap, meaning that all interpretations are incomparable. Hence $\succeq_{All,S}$ will usually not be as informative as we would like a cautious ranking procedure to be. This is mainly due to the presence of redundant variables, or common formulas, in the comparison of interpretations. In contrast, $\succeq_{Diff,S}$ does not suffer from the same defect, as the next proposition shows.

Proposition 4. *Given a knowledge base K, the ordering $\succeq_{Diff,S}$ satisfies subset monotonicity.*

Proof. As for Proposition 2, it is sufficient to notice that if $F_K(\omega) \subseteq F_K(\omega')$, then $F_K(\omega \setminus \omega') = F_K(\emptyset)$. This means that $ag(\{\alpha_i : \phi_i \in F_K(\omega \setminus \omega')\}) = [0, 0]$, hence we necessarily have $\omega \succeq_{Diff,L}^K \omega'$.

Hence, the ordering $\succeq_{Diff,S}$ satisfies all properties we have considered desirable in our framework. It does not add unwanted comparisons, while not losing information that could be deduced without knowing the weights.

---

[4] Which does not prevent them to be suitable for other purposes.

Example 2. If we consider the knowledge base K^1 of Example 1, using $\succeq_{Diff,S}$ we could only deduce the rankings induced by the facts that $F_K(\omega_3) \subseteq F_K(\omega_1)$ and $F_K(\omega_2) \subseteq F_K(\omega_0) \subseteq F_K(\omega_1)$, as ω_3 does not falsify any of the formulas that ω_2 and ω_0 falsify, hence we can directly compare their intervals. The resulting ordering is pictured in Fig. 2.

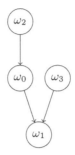

Fig. 2. Orderings $\succ_{Diff,S}$ of Example 2

5   Conclusions

In this paper, we have looked at the problem of making cautious inferences in weighted logics when weights are interval-valued, and have made first proposals to make such inferences. There is of course a lot that remains to be done, such as studying expressivity, representational or computational issues.

It should also be noted that our approach can easily be extended to cases where weights are given by other uncertainty models. If $\mathcal{I}_i$ is an uncertain quantity (modelled by a fuzzy set, a belief function, a probability, ...), we would then need to specify how to propagate them to obtain $ag(F)$, and how to compare these uncertain quantities.

References

1. Benferhat, S., Dubois, D., Prade, H.: Possibilistic and standard probabilistic semantics of conditional knowledge bases. J. Logic Comput. **9**(6), 873–895 (1999)
2. Benferhat, S., Hué, J., Lagrue, S., Rossit, J.: Interval-based possibilistic logic. In: Twenty-Second International Joint Conference on Artificial Intelligence (2011)
3. Dubois, D., Godo, L., Prade, H.: Weighted logics for artificial intelligence - an introductory discussion. Int. J. Approx. Reason. **9**(55), 1819–1829 (2014)
4. Dubois, D., Prade, H.: Possibilistic logic: a retrospective and prospective view. Fuzzy Sets Syst. **144**(1), 3–23 (2004)
5. Dupin De Saint-Cyr, F., Lang, J., Schiex, T.: Penalty logic and its link with Dempster-Shafer theory. In: Uncertainty Proceedings 1994, pp. 204–211. Elsevier (1994)

6. Gelain, M., Pini, M.S., Rossi, F., Venable, K.B., Wilson, N.: Interval-valued soft constraint problems. Ann. Math. Artif. Intell. **58**(3–4), 261–298 (2010)
7. Kaci, S., van der Torre, L.: Reasoning with various kinds of preferences: logic, non-monotonicity, and algorithms. Ann. Oper. Res. **163**(1), 89–114 (2008)

Preference Elicitation with Uncertainty: Extending Regret Based Methods with Belief Functions

Pierre-Louis Guillot and Sebastien Destercke[✉]

Heudiasyc laboratory, 60200 Compiègne, France
sebastien.destercke@hds.utc.fr

Abstract. Preference elicitation is a key element of any multi-criteria decision analysis (MCDA) problem, and more generally of individual user preference learning. Existing efficient elicitation procedures in the literature mostly use either robust or Bayesian approaches. In this paper, we are interested in extending the former ones by allowing the user to express uncertainty in addition of her preferential information and by modelling it through belief functions. We show that doing this, we preserve the strong guarantees of robust approaches, while overcoming some of their drawbacks. In particular, our approach allows the user to contradict herself, therefore allowing us to detect inconsistencies or ill-chosen model, something that is impossible with more classical robust methods.

Keywords: Belief functions · Preference elicitation · Multicriteria decision

1 Introduction

Preference elicitation, the process through which we collect preference from a user, is an important step whenever we want to model her preferences. It is a key element of domains such as *multi-criteria decision analysis* (MCDA) or preference learning [7], where one wants to build a ranking model on multivariate alternatives (characterised by criteria, features, ...). Our contribution is more specific to MCDA, as it focuses on getting preferences from a single user, and not a population of them.

Note that within this setting, preference modelling or learning can be associated with various decision problems. Such problems most commonly include the **ranking** problem that consists in ranking alternatives from best to worst, the **sorting** problem that consists in classifying alternatives into ordered classes, and finally the **choice** problem that consists in picking a single best candidate among available alternatives. This article only deals with the choice problem but can be extended towards the ranking problem in a quite straightforward manner – commonly known as the iterative choice procedure – by considering a ranking as a series of consecutive choices [1].

© Springer Nature Switzerland AG 2019
N. Ben Amor et al. (Eds.): SUM 2019, LNAI 11940, pp. 289–309, 2019.
https://doi.org/10.1007/978-3-030-35514-2_22

In order for the expert to make a recommendation in MCDA, she must first restrict her search to a set of plausible MCDA models. This is often done accordingly to *a priori* assumptions on the decision making process, possibly constrained by computational considerations.

In this paper, we will assume that alternatives are characterised by q real values, i.e. are represented by a vector in $\mathbb{R}^q$, and that preferences over them can be modelled by a value function $f : \mathbb{R}^q \rightarrow \mathbb{R}$ such that $a \succ b$ iff $f(a) > f(b)$. More specifically, we will look at weighted averages. The example below illustrate this setting. Our results can straightforwardly be extended to other evaluations functions (Choquet integrals, GAI, ...) in theory, but would face additional computational issues that would need to be solved.

Example 1 (choosing the best course). Consider a problem in which the DM is a student wanting to find the best possible course in a large set of courses, each of which has been previously associated a grade from 0 to 10–0 being the least preferred and 10 being the most preferred – according to 3 criteria: *usefulness*, *pedagogy* and *interest*. The expert makes the assumption that the DM evaluates each course according to a score computed by a weighted sum of its 3 grades. This is a strong assumption as it means for example that an increase of 0.5 in *usefulness* will have the same impact on the score regardless of the grades in *pedagogy* and *interest*. In such a set of models, a particular model is equivalent to a set of weights in $\mathbb{R}^3$. Assume that the DM preferences follow the model given by the weights $(0.1, 0.8, 0.1)$, meaning that she considers *pedagogy* to be eight time as important as *usefulness* and *interest* which are of equal importance. Given the grades reported in Table 1, she would prefer the *Optimization* course over the *Machine learning* course, as the former would have a 5.45 value, and the later a 3.2 value.

Table 1. Grades of courses

Machine learning:

usefulness	pedagogy	interest
8.5	1.5	10

Optimization:

usefulness	pedagogy	interest
3	5.5	2

Linear algebra:

usefulness	pedagogy	interest
7	5	5.5

Graph theory:

usefulness	pedagogy	interest
1	2	6

Beyond the choice of a model, the expert also needs to collect or elicit preferences that are specific to the DM, and that she could not have guessed according to *a priori* assumptions. Information regarding preferences that are specific to the DM can be collected by asking them to answer questions in several form such as the ranking of a subset of alternatives from best to worst or the choice of a preferred candidate among a subset of alternatives.

Example 2 (choosing the best course (continued)). In our example, directly asking for weights would make little sense (as our model may be wrong, and as the user cannot be expected to be an expert of the chosen model). A way to get this information from her would therefore be to ask her to pick her favorite out of two courses. Let's assume that when asked to choose between *Optimization* and *Graph theory*, she prefers the *Optimisation* course. The latter being better than the former in *pedagogy* and worse in *interest*, her answer is compatible with weights $(0.05, 0.9, 0.05)$ (strong preference for pedagogy over other criteria) but not with $(0.05, 0.05, 0.9)$ (strong preference for interest over other criteria). Her answer has therefore given the expert additional information on the preferential model underlying her decision. We will see later that this generates a linear constraint over the possible weights.

Provided we have made some preference model assumptions (our case here), it is possible to look for *efficient* elicitation methods, in the sense that they solve the decision problem we want to solve in a small enough, if not minimal number of questions. A lot of work has been specifically directed towards active elicitation methods, in which the set of questions to ask the DM is not given in advance but determined on the fly. In robust methods, this preferential information is assumed to be given with full certainty which leads to at least two issues. The first one is that elicitation methods thus do not account for the fact that the DM might doubt her own answers, and that they might not reflect her actual preferences. The second one, that is somehow implied by the first one, is that most robust active elicitation methods will never put the DM in a position where she could contradict either herself or assumptions made by the expert, as new questions will be built on the basis that previous answers are correct and hence should not be doubted. This is especially problematic when inaccurate preferences are given early on, or when the preference model is based on wrong assumptions.

This paper presents an extension of the Current Solution Strategy [3] that includes uncertainty in the answers of the DM by using the framework based on belief functions presented in [5]. Section 2 will present necessary preliminaries on both robust preference elicitation based on regret and uncertainty management based on belief functions. Section 3 will present our extension and some of the associated theoretical results and guarantees. Finally Sect. 4 will present some first numerical experiments that were made in order to test the method and its properties in simulations.

2 Preliminaries

2.1 Formalization

Alternatives and Models: We will denote $\mathcal{X}$ the space of possible alternatives, and $\mathbb{X} \subseteq \mathcal{X}$ the subset of **available** alternatives at the disposal of our DM and about which a recommendation needs to be made. In this paper we will consider alternatives summarised by q real values corresponding to criteria, hence $\mathcal{X} \subseteq \mathbb{R}^q$. For any $x \in \mathcal{X}$ and $1 \le i \le q$, we denote by $x^i \in \mathbb{R}$ the evaluation of alternative x according to criterion i. We also assume that for any $x, y \in \mathcal{X}$

such that $x^i > y^i$ for some $i \in \{1, \ldots, q\}$ and $x^l \geq y^l, \forall l \in \{1, \ldots, q\} \setminus \{i\}$, x will always be strictly preferred to y – meaning that preferences respect *ceteris paribus* monotonicity, and we assume that criteria utility scale is given.

$\mathbb{X}$ is a finite set of k alternative such that $\mathbb{X} = \{x_1, x_2, \ldots, x_k\}$ with x_j the j-th alternative of $\mathbb{X}$. Let $\mathcal{P}(\mathbb{X})$ be a preference relation over $\mathbb{X}$, and $x, y \in \mathbb{X}$ be two alternatives to compare. We will state that $x \succ_P y$ if and only if x is **strictly** preferred to y in the corresponding relation, $x \simeq_P y$ if and only if x and y are **equally** preferred in the corresponding relation, and $x \succeq_P y$ if and only if either x is strictly preferred to y or x and y are equally preferred.

Preference Modelling and Weighted Sums: In this work, we focus on the case where the hypothesis set Ω of preference models is the set of weighted sum models[1]. A singular model ω will be represented by its vector of weights in $\mathbb{R}^q$, and ω will be used to describe indifferently the decision model and the corresponding weight vector. Ω can therefore be described as:

$$\Omega = \left\{ \omega \in \mathbb{R}^q \; : \; \omega^i \geq 0 \text{ and } \sum_{i=1}^{q} \omega^i = 1 \right\}.$$

Each model ω is associated to the corresponding aggregating evaluation function

$$f_\omega(x) = \sum_{i=1}^{q} \omega^i x^i,$$

and any two potential alternatives x, y in $\mathcal{X}$ can then be compared by comparing their aggregated evaluation:

$$x \succeq_\omega y \iff f_\omega(x) \geq f_\omega(y) \tag{1}$$

which means that if the model ω is known, $\mathcal{P}_\omega(\mathbb{X})$ is a **total** preorder over $\mathbb{X}$, the set of existing alternatives. Note that $\mathcal{P}_\omega(\mathbb{X})$ can be determined using pairwise relations $\succeq_\omega$. Weighted averages are a key model of preference learning whose linearity usually allows the development of efficient methods, especially in regret-based elicitation [2]. It is therefore an ideal starting point to explore other more complex functions, such as those that are linear in their parameters once alternatives are known (i.e., Choquet integrals, Ordered weighted averages).

2.2   Robust Preference Elicitation

In theory, obtaining a unique true preference model requires both unlimited time and unbounded cognitive abilities. This means that in practice, the best we can do is to collect information identifying a subset Ω' of possible models, and act

---

[1] In principle, our methods apply to any value function with the same properties, but may have to solve computational issue that depends on the specific chosen hypothesis.

accordingly. Rather than choosing a unique model within Ω', robust methods usually look at the inferences that hold for every model in Ω'. Let Ω' be the subset of models compatible with all the given preferential information, then we can define $\mathcal{P}_{\Omega'}(\mathbb{X})$, a **partial** preorder of robust preferences over $\mathbb{X}$, as follows:

$$x \succeq_{\Omega'} y \iff \forall \omega \in \Omega' \ f_\omega(x) \geq f_\omega(y). \tag{2}$$

The research question we address here is to find elicitation strategies that reduce Ω' as quickly as possible, obtaining at the limit an order $\mathcal{P}_{\Omega'}(\mathbb{X})$ having only one maximal element[2]. In practice, one may have to stop collecting information before that point, explaining the need for heuristic indicators of the fitness of competing alternatives as potential choices.

Regret Based Elicitation: Regret is a common way to assess the potential loss of recommending a given alternative under incomplete knowledge. It can help both the problem of making a recommendation and finding an efficient question. Regret methods use various indicators, such as the **regret** $R_\omega(x, y)$ of choosing x over y according to model ω, defined as

$$R_\omega(x, y) = f_\omega(y) - f_\omega(x). \tag{3}$$

From this regret and a set Ω' of possible models, we can then define the **pairwise max regret** as

$$\mathrm{PMR}(x, y, \Omega') = \max_{\omega \in \Omega'} R_\omega(x, y) = \max_{\omega \in \Omega'} (f_\omega(y) - f_\omega(x)) \tag{4}$$

that corresponds to the maximum possible regret of choosing x over y for any model in Ω'. The **max regret** for an alternative x defined as

$$\mathrm{MR}(x, \Omega') = \max_{y \in \mathbb{X}} \mathrm{PMR}(x, y, \Omega') = \max_{y \in \mathbb{X}} \max_{\omega \in \Omega'} (f_\omega(y) - f_\omega(x)) \tag{5}$$

then corresponds to the worst possible regret one can have when choosing x. Finally the **min max regret** over a subset of models Ω' is

$$\mathrm{mMR}(\Omega') = \min_{x \in \mathbb{X}} \mathrm{MR}(x, \Omega') = \min_{x \in \mathbb{X}} \max_{y \in \mathbb{X}} \max_{\omega \in \Omega'} (f_\omega(y) - f_\omega(x)) \tag{6}$$

Picking as choice $x^* = \arg\min \mathrm{mMR}(\Omega')$ is then a robust choice, in the sense that it is the one giving the minimal regret in a worst-case scenario (the one leading to max regret).

Example 3 (choosing the best course (continued)). Let $\mathcal{X} = [0, 10]^3$ be the set of valid alternatives composed of 3 grades from 0 to 10 in respectively *pedagogy*, *usefulness* and *interest*. Let $\mathbb{X} = \{x_1, x_2, x_3, x_4\}$ be the set of available alternatives in which x_1 corresponds to the *Machine learning* course, x_2 corresponds

---

[2] Or in some cases a maximal set $\{x_1, \ldots, x_p\}$ of *equally preferred* elements s.t. $x_1 \simeq \ldots \simeq x_p$.

to the *Optimization* course, x_3 corresponds to the *Linear algebra* course and x_4 corresponds to the *Graph theory* course, as reported in Table 1. Let $x, y \in \mathbb{X}$ be two alternatives and Ω the set of weighted sum models, $\mathrm{PMR}(x, y, \Omega)$ can be computed by optimizing $\max_{\omega \in \Omega} \left[\omega^1(x^1 - y^1) + \omega^2(x^2 - y^2) + \omega^3(x^3 - y^3)\right]$. As this linear function of ω is optimized over a convex polytope Ω, it can easily be solved exactly using linear programming (LP). Results of $\mathrm{PMR}(x, y, \Omega)$ and $\mathrm{MR}(x, \Omega)$ are shown in Table 2. In this example, x_1 is the alternative with minimum max regret, and the most conservative candidate to answer the choice problem according to regret.

Table 2. Values of $\mathrm{PMR}(x, y, \Omega)$ (left) and $\mathrm{MR}(x, \Omega)$ (right)

x \\ y	x_1	x_2	x_3	x_4
x_1	0	4	3.5	0.5
x_2	8	0	4	4
x_3	4.5	0.5	0	0.5
x_4	7.5	3.5	6	0

x	MR
x_1	4
x_2	8
x_3	4.5
x_4	7.5

Regret indicators are also helpful for making the elicitation strategy *efficient* and helping the expert ask relevant questions to the DM. Let Ω' and Ω'' be two sets of models such that $\mathrm{mMR}(\Omega') < \mathrm{mMR}(\Omega'')$. In the worst case, we are certain that $x^*_{\Omega'}$ the optimal choice for Ω' is less regretted than $x^*_{\Omega''}$ the optimal choice for Ω'', which means that we would rather have Ω' be our set of models than Ω''. Let $\mathcal{I}, \mathcal{I}'$ be two pieces of preferential information and $\Omega^{\mathcal{I}}, \Omega^{\mathcal{I}'}$ the sets obtained by integrating this information. Finding which of the two is the most helpful statement in the progress towards a robust choice can therefore be done by comparing $\mathrm{mMR}(\Omega^{\mathcal{I}})$ and $\mathrm{mMR}(\Omega^{\mathcal{I}'})$. An optimal elicitation process (w.r.t. minimax regret) would then choose the question for which the **worst** possible answer gives us a restriction on Ω that is the **most** helpful in providing a robust choice. However, computing such a question can be difficult, and the heuristic we present next aims at picking a nearly optimal question in an efficient and tractable way.

The Current Solution Strategy: Let's assume that Ω' is the subset of decision models that is consistent with every information available so far to the expert. Let's restrict ourselves to questions that consist in comparing pairs x, y of alternatives in $\mathbb{X}$. The DM can only answer with $\mathcal{I}_1 = x \succeq y$ or $\mathcal{I}_2 = x \preceq y$. A pair helpful in finding a robust solution as fast as possible can be computed as a solution to the following optimization problem that consists in finding the pair minimizing the **worst-case min max regret**:

$$\min_{(x,y) \in \mathbb{X}^2} \mathrm{WmMR}(\{x, y\}) = \min_{(x,y) \in \mathbb{X}^2} \max \left\{ \mathrm{mMR}(\Omega' \cap \Omega^{x \succeq y}), \mathrm{mMR}(\Omega' \cap \Omega^{x \preceq y}) \right\}$$

$$(7)$$

The current solution strategy (referred to as CSS) is a heuristic answer to this problem that has proved to be efficient in practice [3]. It consists in asking the DM to compare $x^* \in \arg\mathrm{mMR}(\Omega')$ the least regretted alternative to $y^* = \arg\max_{y \in X} \mathrm{PMR}(x^*, y, \Omega')$ the one it could be the most regretted to (its "worst opponent"). CSS is efficient in the sense that it requires the computation of only one value of min max regret, instead of the $\mathcal{O}(q^2)$ required to solve (7).

Example 4 (Choosing the best course (continued).). Using the same example, according to Table 2, we have $\mathrm{mMR}(\Omega) = \mathrm{MR}(x_1, \Omega) = \mathrm{PMR}(x_1, x_2, \Omega)$, meaning that x_1 is the least regretted alternative in the worst case and x_2 is the one it is most regretted to. The CSS heuristic consists in asking the DM to compare x_1 and x_2, respectively the *Machine learning* course and the *Optimization* course.

2.3   Uncertain Preferential Information

Two key assumptions behind the methods we just described are that (1) the initial chosen set Ω of models can perfectly describe the DM's choices and (2) the DM is an oracle, in the sense that any answer she provides truly reflects her preferences, no matter how difficult the question. This certainly makes CSS an efficient strategy, but also an unrealistic one. This means in particular that if the DM makes a mistake, we will just pursue with this mistake all along the process and will never question what was said before, possibly ending up with sub-optimal recommendations.

Example 5 (choosing the best course (continued)). Let's assume similarly to Example 2 that the expert has not gathered any preference from the DM yet, and that this time she asks her to compare alternatives x_1 and x_2 – respectively the *Machine learning* course and the *Optimization* course. Let's also assume similarly to Example 1 that the DM makes decisions according to a weighted sum model with weights $\omega^* = (0.1, 0.8, 0.1)$. $f_{\omega^*}(x_2) = 5.45 > 3.2 = f_{\omega^*}(x_1)$, which means that she should prefer the *Optimization* course over the *Machine learning* course. However for some reason – such as her being unfocused or unsure about her preference – assume the DM's answer is inconsistent with ω^* and she states that $x_1 \succeq x_2$ rather than $x_2 \succeq x_1$.

Then Ω' the set of model consistent with available preferential information is such that $\Omega' = \Omega^{x_1 \succeq x_2} = \{\omega \in \Omega : \sum_{i=1}^{3} \omega^i (x_1^i - x_2^i) \geq 0\} = \{\omega \in \Omega : \omega^2 \leq \frac{2}{3} - \frac{5}{24}\omega^1\}$, as represented in Fig. 1. It is clear that $\omega^* \notin \Omega'$: subsequent questions will only ever restrict Ω' and the expert will never get quite close to modelling ω^*.

A similar point could be made if ω^*, the model according to which the DM makes her decision, does not even belong to Ω the set of weighted sums models that the expert chose.

As we shall see, one way to adapt min-max regret approaches to circumvent the two above difficulties can be to include a simple measure of how uncertain an answer is.

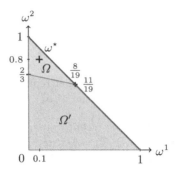

Fig. 1. Graphical representation of Ω, Ω' and ω^*

The Belief Function Framework. In classical CSS, the response to a query by the DM always implies a set of consistent models Ω' such that $\Omega' \subseteq \Omega$. Here, we allow the DM to give alongside her answer a confidence level $\alpha \in [0, 1]$, interpreted as how confident she is that this particular answer matches her preferences. In the framework developed in [5], such information is represented by a mass function on Ω', referred to as $m_\alpha^{\Omega'}$ and defined as:

$$m_\alpha^{\Omega'}(\Omega) = 1 - \alpha, \qquad m_\alpha^{\Omega'}(\Omega') = \alpha.$$

Such mass assignments are usually called simple support [13] and represent elementary pieces of uncertain information. A confidence level of 0 will correspond to a vacuous knowledge about the true model ω^*, and will in no way imply that the answer is wrong (as would have been the case in a purely probabilistic framework). A confidence level of 1 will correspond to the case of certainty putting a hard constraint on the subset of models to consider.

Remark 1. Note that values of α do not necessarily need to come from the DM, but can just be chosen by the analyst (in the simplest case as a constant) to weaken the assumptions of classical models. We will see in the experiments of Sect. 4 that such a strategy may indeed lead to interesting behaviours, without necessitating the DM to provide confidence degrees if she thinks the task is too difficult, or if the analyst thinks such self-assessed confidence is meaningless.

Dempster's Rule. Pieces of information corresponding to each answer will be combined through non-normalized Dempster's rule $+_\cap$. At step k, m_k the mass function capturing the current belief about the DM's decision model can thus be defined recursively as:

$$m_0 = m_1^\Omega \qquad \ldots \qquad m_k = m_{k-1} +_\cap m_{\alpha_k}^{\Omega_k}. \tag{8}$$

This rule, also known as TBM conjunctive rule, is meant to combine distinct pieces of information. It is central to the Transferable Belief Model, that intends to justify belief functions without using probabilistic arguments [14].

Note that an information fusion setting and the interpretation of the TBM fit our problem particularly well as it assumes the existence of a unique true model ω^* underlying the DM's decision process, that might or might not be in our predefined set of models Ω. Allowing for an open world is a key feature of the framework. Let us nevertheless recall that non-normalized Dempster's rule $+_\cap$ can also be justified without resorting to the TBM [8,9,11].

In our case this independence of sources associated with two mass assignments $m_{\alpha_i}^{\Omega_i}$ and $m_{\alpha_j}^{\Omega_j}$ means that even though both preferential information account for preferences of the same DM, the answer a DM gives to the ith question does not directly impact the answer she gives to the jth question: she would have answered the same thing had their ith answer been different for some reason. This seems reasonable, as we do not expect the DM to have a clear intuition about the consequences of her answers over the set of models, nor to even be aware that such a set – or axioms underlying it – exists. One must however be careful to not ask the exact same question twice in short time range.

Since combined masses are all possibility distributions, an alternative to assuming independence would be to assume complete dependence, simply using the minimum rule [6] which among other consequences would imply a loss of expressivity[3] but a gain in computation[4].

As said before, one of the key interest of using this rule (rather than its normalised version) is to allow $m(\emptyset) > 0$, notably to detect either mistakes in the DM's answer (considered as an unreliable source) or a bad choice of model (under an open world assumption). Determining where the conflict mainly comes from and acting upon it will be the topic of future works. Note that in the specific case of simple support functions, we have the following result:

Proposition 1. *If* $m_{\alpha_k}^{\Omega_k}$ *are simple support functions combined through Dempster's rule, then*

$$m(\emptyset) = 0 \Leftrightarrow \exists \omega, Pl(\{\omega\}) = 1$$

with $Pl(\{\omega\}) = \sum_{E \subseteq \Omega, \omega \in E} m(E)$ *the plausibility measure of model* ω.

Proof (Sketch). The $\Leftarrow$ part is obvious given the properties of Plausibility measure. The $\Rightarrow$ part follows from the fact that if $m(\emptyset) = 0$, then all focal elements are supersets of $\bigcap_{i \in \{1,\dots,k\}} \Omega_i$, hence all contains at least one common element. $\square$

This in particular shows that $m(\emptyset)$ can, in this specific case, be used as an estimate of the logical consistency of the provided information pieces.

Consistency with Robust, Set-Based Methods: when an information $\mathcal{I}$ is given with full certainty $\alpha = 1$, we retrieve a so-called categorical mass

---

[3] For instance, no new values of confidence would be created when using a finite set $\{\alpha_1, \dots, \alpha_M\}$ for elicitation.

[4] The number of focal sets increasing only linearly with the number of information pieces.

$m_k\left(\Omega^{\mathcal{I}}\right) = 1$. Combining a set $\mathcal{I}_1, ..., \mathcal{I}_k$ of such certain information will end up in the combined mass

$$m_k\left(\bigcap_{i \in \{1,...,k\}} \Omega_i\right) = 1$$

which is simply the intersection of all provided constraints, that may turn up either empty or non-empty, meaning that **inconsistency** will be a Boolean notion, i.e.,

$$m_k\left(\emptyset\right) = \begin{cases} 1 & \text{if } \bigcap_{i \in \{1,...,k\}} \Omega_i = \emptyset \\ 0 & \text{otherwise} \end{cases}$$

Recall that in the usual CSS or minimax regret strategies, such a situation can never happen.

3   Extending CSS Within the Belief Function Framework

We now present our proposed extension of the Current Solution Strategy integrating confidence degrees and uncertain answers. Note that in the two first Sects. 3.1 and 3.2, we assume that the mass on the empty set is null in order to parallel our approach with the usual one not including uncertainties. We will then consider the problem of conflict in Sect. 3.3.

3.1   Extending Regret Notions

Extending PMR: when uncertainty over possible models is defined through a mass function $2^\Omega \to [0, 1]$, subsets of Ω known as focal sets are associated to a value $m(\Omega')$ that correspond to the knowledge we have that ω belongs to Ω' and nothing more. The extension we propose averages the value of PMR on focal sets weighted by their corresponding mass:

$$\text{EPMR}(x, y, m) = \sum_{\Omega' \subseteq \Omega} m(\Omega').\text{PMR}(x, y, \Omega') \tag{9}$$

and we can easily see that in the case of certain answers ($\alpha = 1$), we do have

$$\text{EPMR}(x, y, m_k) = \text{PMR}\left(x, y, \left(\bigcap_{i \in \{1,...,k\}} \Omega_i\right)\right) \tag{10}$$

hence formally extending Eq. (4). When interpreting $m(\Omega')$ as the probability that ω belongs to Ω', EPMR could be seen as an expectation of PMR when randomly picking a set in 2^Ω.

Extending EMR: Similarly, we propose a weighted extension of maximum regret

$$\text{EMR}(x, m) = \sum_{\Omega' \subseteq \Omega} m(\Omega') . \text{MR}(x, \Omega') = \sum_{\Omega' \subseteq \Omega} m(\Omega') . \max_{y \in \mathbb{X}} \left\{ \text{PMR}(x, y, \Omega') \right\} .$$

(11)

EMR is the **expectation** of the **maximal** pairwise max regret taken each time between x and $y \in \mathbb{X}$ its worst adversary – as opposed to a **maximum** considering each $y \in \mathbb{X}$ of the **expected** pairwise max regret between x and the given y, described by $\text{MER}(x, m) = \max_y \text{EPMR}(x, y, m)$. Both approaches would be equivalent to MR in the certain case, meaning that if $\alpha_i = 1$ then

$$\text{EMR}(x, m_k) = \text{MER}(x, m_k) = \text{MR}\left(x, \left(\bigcap_{i \in \{1, \dots, k\}} \Omega_i \right) \right) .$$

(12)

However EMR seems to be a better option to assess the max regret of an alternative, as under the assumption that the true model ω^* is within the focal set Ω', it makes more sense to compare x to its worst opponent within Ω', which may well be different for two different focal sets. Indeed, if ω^* the true model does in fact belong to Ω', decision x is only as bad as how big the regret can get for **any** adversarial counterpart $y_{\Omega'} \in \mathbb{X}$.

Extending mMR: we propose to extend it as

$$\text{mEMR}(m) = \min_{x \in \mathbb{X}} \text{EMR}(x, m).$$

(13)

mEMR **minimizes** for each $x \in \mathbb{X}$ the **expectation** of max regret and is different from the **expectation** of the **minimal** max regret for whichever alternative x is optimal, described by $\text{EmMR}(m) = \sum_{\Omega'} m(\Omega') \min_{x \in \mathbb{X}} \text{MR}(x, \Omega')$. Again, these two options with certain answers boil down to mMR as we have

$$\text{mEMR}(m) = \text{EmMR}(m) = \text{mMR}\left(\bigcap_{i \in \{1, \dots, k\}} \Omega_i \right) .$$

(14)

The problem with EmMR is that it would allow for multiple possible best alternatives, leaving us with an unclear answer as to what is the best choice option, (arg min EmMR) not being defined. It indicates how robust in the sense of regret we expect **any** best answer to the choice problem to be, assuming there can be an optimal alternative for each focal set. In contrast, mEMR minimizes the max regret while restricting the optimal alternative x to be the same in all of them, hence providing a unique argument and allowing our recommendation system and elicitation strategy to give an optimal recommendation.

Extending CSS: our Evidential Current Solution Strategy (ECSS) then amounts, at step k with mass function m_k, to perform the following sequence of operations:

- Find $x^* = \arg \mathrm{mEMR}(m_k) = \arg \min_{x \in \mathbb{X}} \mathrm{EMR}(x, m_k)$;
- Find $y^* = \arg \max_{y \in \mathbb{X}} \mathrm{EPMR}(x^*, y, m_k)$;
- Ask the DM to compare x^*, y^* and provide α_k, obtaining $m_{\alpha_k}^{\Omega^k}$;
- Compute $m_{k+1} := m_k +_{\cap} m_{\alpha_k}^{\Omega^k}$
- Repeat until conflict is too high (red flag), budget of questions is exhausted, or $\mathrm{mEMR}(m_k)$ is sufficiently low

Finally, recommend $x^* = \arg \mathrm{mEMR}(m_k)$. Thanks to Eqs. (10), (12) and (14), it is easy to see that we retrieve CSS as the special case in which all answers are completely certain.

Example 6. Starting with intial mass function m_0 such that $m_0(\Omega) = 1$, the choice of CSS coincides with the choice of ECSS (all evidence we have is committed to $\omega \in \Omega$). With the values of PMR reported in Table 2 the alternatives the DM is asked to compare are x_1 the least regretted alternative and x_2 its most regretted counterpart. In accordance with her true preference model $\omega^* = (0.1, 0.8, 0.1)$, the DM states that $x_2 \succeq x_1$, i.e., she prefers the *Optimization* course over the *Machine learning* course, with confidence degree $\alpha = 0.7$. Let Ω_1 be the set of WS models in which x_2 can be preferred to x_1, which in symmetry with Example 5 can be defined as $\Omega_1 = \{\omega \in \Omega : \omega^2 \geq \frac{2}{3} - \frac{5}{24}\omega^1\}$, as represented in Fig. 2.

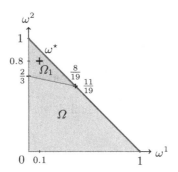

Fig. 2. Graphical representation of Ω, Ω_1 and ω^*

Available information on her decision model after step 1 is represented by mass function m_1 with

$$m_1(\ \Omega\ ) = 0.3$$
$$m_1(\ \Omega_1\ ) = 0.7$$

The values of PMR and MR on Ω_1 can be computed using LP as reported in Table 3. Values of PMR and MR and have been previously computed on Ω as reported in Table 2. Values of EPMR and EMR can be then be deduced by combining them according to Eqs. (9) and (11), as reported in Table 4. In this example x_3 minimizes both MR and EMR and our extension agrees with the robust version as to which recommendation is to be made. However the most regretted counterpart to which the DM has to compare x_3 in the next step differs, as ECSS would require that she compares x_3 and x_1 rather than x_3 and x_2 for CSS.

Table 3. Values of $\mathrm{PMR}(x, y, \Omega_1)$ (left) and $\mathrm{MR}(x, \Omega_1)$ (right)

x \ y	x_1	x_2	x_3	x_4
x_1	0	4	3.5	0.5
x_2	0	0	$\frac{53}{38} \simeq \mathbf{1.39}$	-1
x_3	$-\frac{5}{6} \simeq -0.83$	**0.5**	0	$-\frac{11}{6} \simeq -1.83$
x_4	$\frac{109}{38} \simeq 2.87$	3.5	$\frac{81}{19} \simeq \mathbf{4.26}$	0

x	MR
x_1	4
x_2	$\frac{53}{38} \simeq 1.39$
x_3	**0.5**
x_4	$\frac{81}{19} \simeq 4.26$

Table 4. Values of $\mathrm{EPMR}(x, y, m_1)$ (left) and $\mathrm{EMR}(x, m_1)$ (right)

x \ y	x_1	x_2	x_3	x_4
x_1	0	4	3.5	0.5
x_2	**2.4**	0	$\frac{827}{390} \simeq 2.18$	0.5
x_3	$\frac{23}{30} \simeq \mathbf{0.77}$	**0.5**	0	$-\frac{68}{60} \simeq -1.13$
x_4	$\frac{809}{190} \simeq 4.26$	3.5	$\frac{909}{190} \simeq \mathbf{4.78}$	0

x	MR
x_1	4
x_2	$\frac{1283}{380} \simeq 3.38$
x_3	$\frac{23}{30} \simeq \mathbf{0.7}$
x_4	$\frac{1389}{380} \simeq 5.23$

3.2   Preserving the Properties of CSS

This section discusses to what extent is ECSS consistent with three key properties of CSS:

1. CSS is monotonic, in the sense that the minmax regret mMR reduces at each iteration.
2. CSS provides strong guarantees, in the sense that the felt regret of the recommendation is ensured to be at least as bad as the computed mMR.
3. CSS produces questions that are non-conflicting (whatever the answer) with previous answers.

We would like to keep the first two properties at least in the absence of conflicting information, as they ensure respectively that the method will converge and will provide robust recommendations. However, we would like our strategy to raise questions possibly contradicting some previous answers, so as to raise the previously mentioned red flags in case of problems (unreliable DM or bad choice of model assumption). As shows the next property, our method also converges.

Proposition 2. *Let m_{k-1} and $m_{\alpha_k}^{\Omega^k}$ be two mass functions on Ω issued from ECSS such that $m_k(\varnothing) = \left[m_{k-1} +_\cap m_{\alpha_k}^{\Omega^k}\right](\varnothing) = 0$, then*

1. $EPMR(x, y, m_k) \leq EPMR(x, y, m_{k-1})$
2. $EMR(x, m_k) \leq EMR(x, m_{k-1})$
3. $mEMR(m_k) \leq mEMR(m_{k-1})$

Proof (sketch). The two first items are simply due to the combined facts that on one hand we know [15] that applying $+_\cap$ means that m_k is a specialisation of m_{k-1}, and on the other hand that for any $\Omega'' \subseteq \Omega'$ we have $f(x, y, \Omega'') \leq f(x, y, \Omega')$ for any $f \in \{\text{PMR}, \text{MR}\}$. The third item is implied by the second as it consists in taking a minimum over a set of values of EMR that are all smaller.

Note that the above argument applies to any combination rule producing a specialisation of the two combined masses, including possibilistic minimum rule [6], Denoeux's family of w-based rules [4], etc. We can also show that the evidential approach, if we provide it with questions computed through CSS, is actually more cautious than CSS:

Proposition 3. *Consider the subsets of models $\Omega_1, \ldots, \Omega_k$ issued from the answers of the CSS strategy, and some values $\alpha_1, \ldots, \alpha_k$ provided a posteriori by the DM. Let m_{k-1} and $m_{\alpha_k}^{\Omega^k}$ be two mass functions issued from ECSS on Ω such that $m_k(\emptyset) = 0$. Then we have*

1. $EPMR(x, y, m_k) \geq PMR\left(x, y, \left(\bigcap_{i \in \{1, \ldots, k\}} \Omega_i\right)\right)$
2. $EMR(x, m_k) \geq MR\left(x, \left(\bigcap_{i \in \{1, \ldots, k\}} \Omega_i\right)\right)$
3. $mEMR(m_k) \geq mMR\left(\left(\bigcap_{i \in \{1, \ldots, k\}} \Omega_i\right)\right)$

Proof (sketch). The first two items are due to the combined facts that on one hand all focal elements are supersets of $\left(\bigcap_{i \in \{1, \ldots, k\}} \Omega_i\right)$ and on the other hand that for any $\Omega'' \subseteq \Omega'$ we have $f(x, y, \Omega'') \leq f(x, y, \Omega')$ for any $f \in \{\text{PMR}, \text{MR}\}$. Any value of EPMR or EMR is a weighted average over terms all greater than their robust counterpart on $\left(\bigcap_{i \in \{1, \ldots, k\}} \Omega_i\right)$, and is therefore greater itself. The third item is implied by the second as the biggest value of EMR is thus necessarily bigger than all the values of MR.

This simply shows that, if anything, our method is even more cautious than CSS. It is in that sense probably slightly too cautious in an idealized scenario – especially as unlike robust indicators our evidential extensions will never reach 0 – but provides guarantees that are at least as strong.

While we find the two first properties appealing, one goal of including uncertainties in the DM answers is to relax the third property, whose underlying assumptions (perfectness of the DM and of the chosen model) are quite strong. In Sects. 3.3 and 4, we show that ECSS indeed satisfies this requirement, respectively on an example and in experiments.

3.3   Evidential CSS and Conflict

The following example simply demonstrates that, in practice, ECSS can lead to questions that are possibly conflicting with each others, a feature CSS does not have. This conflict is only a possibility: no conflict will appear should the DM provide answers completely consistent with the set of models and what she previously stated, and in that case at least one model will be fully plausible[5] (see Proposition 1).

Example 7 (Choosing the best course (continued)). At step 2 of our example the DM is asked to compare x_1 to x_3 in accordance with Table 4. Even though it conflicts with ω^* the model underlying her decision the DM **has the option** to state that $x_1 \succeq x_3$ with confidence degree $\alpha > 0$, putting weight on Ω_2 the set of consistent model defined by $\Omega_2 = \{\omega \in \Omega : \sum_{i=1}^{q} \omega^i \left(x_1^i - x_3^i\right) \geq 0\} = \{\omega \in \Omega : \omega^2 \leq \frac{9}{16} - \frac{3}{8}\omega^1\}$. However as represented in Fig. 3, $\Omega_1 \cap \Omega_2 = \varnothing$.

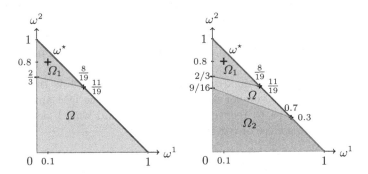

Fig. 3. Graphical representation of Ω, Ω_1, Ω_2 and ω^*

This means that $x_1 \succeq x_2$ and $x_3 \succeq x_1$ are not compatible preferences assuming the DM acts according to a weighted sum model. This can be either because she actually does not follow such a model, or because one of her answers did not reflect her actual preference (which would be the case here). Assuming she does state that $x_1 \succeq x_3$ with confidence $\alpha = 0.6$, information about the preference model at step 2 is captured through mass function m_2 defined as:

$$m_2 : \Omega \;\rightarrow 0.12 \quad \Omega_2 \rightarrow 0.18$$
$$\Omega_1 \rightarrow 0.28 \quad \varnothing \;\;\rightarrow 0.42$$

Meaning that ECSS detects a degree of inconsistency equal to $m_2(\varnothing) = 0.42$.

This illustrating example is of course not representative of real situations, having only four alternatives, but clearly shows that within ECSS, conflict may

---

[5] This contrasts with a Bayesian/probabilistic approach, where no model would receive full support in non-degenerate cases.

appear as an effect of the strategy. In other words, we do not have to modify it to detect consistency issues, it will automatically seek out for such cases if $\alpha_i < 1$. In our opinion, this is clearly a desirable property showing that we depart from the assumptions of CSS.

However, the inclusion of conflict as a focal element in the study raises new issues, the first one being how to extend the various indicators of regret to this situation. In other words, how can we compute $\mathrm{PMR}(x, y, \varnothing)$, $\mathrm{MR}(x, \varnothing)$ or $\mathrm{MMR}(\varnothing)$? This question does not have one answer and requires careful thinking, however the most straightforward extension is to propose a way to compute $\mathrm{PMR}(x, y, \varnothing)$, and then to plug in the different estimates of ECSS. Two possibilities immediately come to mind:

- $\mathrm{PMR}(x, y, \varnothing) = \max_{\omega \in \Omega} R_\omega(x, y)$, which takes the highest regret among all models, and would therefore be equivalent to consider conflict as ignorance. This amounts to consider Yager's rule [16] in the setting of belief functions. This rule would make the regret increases when conflict appears, therefore providing alerts, but this clearly means that the monotonicity of Proposition 3 would not longer hold. Yet, one could discuss whether such a property is desirable or not in case of conflicting opinions. It would also mean that elicitation methods are likely to try to avoid conflict, as it will induce a regret increase.
- $\mathrm{PMR}(x, y, \varnothing) = \min_{\omega \in \Omega} \max(0, R_\omega(x, y))$, considering $\emptyset$ as the limit intersection of smaller and smaller sets consistent with the DM answers. Such a choice would allow us to recover monotonicity, and would clearly privilege conflict as a good source of regret reduction.

Such distinctions expand to other indicators, but we leave such a discussion for future works.

3.4   On Computational Tractability

ECSS requires, in principle, to compute PMR values for every possible focal elements, which could lead to an exponential explosion of the computational burden. We can however show that in the case of weighted sums and more generally of linear constraints, where PMR has to be solved through a LP program, we can improve upon this worst-case bound. We introduce two simplifications that lead to more efficient methods providing exact answers:

Using the Polynomial Number of Elementary Subsets. The computational cost can be reduced by using the fact that if $P_j = \{\Omega_{i_1}, \ldots, \Omega_{i_k}\}$ is a partition of Ω_j, then:

$$\mathrm{PMR}(x, y, \Omega_j) = \max_{l \in \{i_1, \ldots, i_k\}} \mathrm{PMR}(x, y, \Omega_l)$$

Hence, computing PMR on the partition is sufficient to retrieve the global PMR through a simple max. Let us now show that, in our case, the size of this partition only increases polynomially. Let $\Omega_1, \ldots, \Omega_n$ be the set of models consistent with respectively the first to the nth answer, and $\Omega_i^C, \ldots, \Omega_n^C$ their respective complement in Ω.

Due to the nature of the conjunctive rule $+_\cap$, every focal set Ω' of $m_k = m_1^\Omega +_\cap m_{\alpha_1}^{\Omega_1} +_\cap \cdots +_\cap m_{\alpha_n}^{\Omega_n}$ is the union of elements of the partition $P_{\Omega'} = \{\tilde{\Omega}_1, \ldots, \tilde{\Omega}_s\}$, with:

$$\tilde{\Omega}_k = \Omega \bigcap_{i \in U_k} \Omega_i \bigcap_{i \in \{1, \ldots, n\} \setminus U_k} \Omega_i^C, U_k \subseteq \{1, \ldots, n\}$$

Which means that for each Ω''s PMR can be computed using the PMR of its corresponding partition. This still does not help much, as there is a total of 2^n possible value of $\tilde{\Omega}_k$. Yet, in the case of convex domains cut by linear constraints, which holds for the weighted sum, the following theorem shows that the total number of elementary subset in Ω only increases polynomially.

Theorem 1 *[12, P 39]. Let E be a convex bounded subset of F an euclidean space of dimension q, and $H = \{\eta_1, \ldots, \eta_n\}$ a set of n hyperplanes in F such that $\forall i \in \{1, \ldots, n\}$, η_i separates F into two subsets $F_0^{\eta_i}$ and $F_1^{\eta_i}$.*
To each of the 2^n possible $U \subseteq \{1, \ldots, n\}$ a subset $F_U^H = F \bigcap_{i \in U} F_1^{\eta_i}$
$\bigcap_{i \in \{1, \ldots, n\} \setminus U} F_0^{\eta_i}$ *can be associated.*
Let $\Theta_H = \{U \subseteq \{1, \ldots, n\} : F_U^H \cap E \neq \varnothing\}$ and $B_H = |\Theta_H|$, then

$$B_H \leq \Lambda_q^n = 1 + n + \binom{n}{2} + \cdots + \binom{n}{q}$$

Meaning that at most Λ_q^n of the F_U^H subsets have a non empty intersection with E.

In the above theorem (the proof of which can be found in [12], or in [10] for the specific case of $E \subset \mathbb{R}^3$), the subsets B_H are equivalent to $\tilde{\Omega}_k$, whose size only grow according to a polynomial whose power increases with q.

Using the Polynomial Number of Extreme Points in the Simplex Problem. Since we work with LP, we also know that optimal values will be obtained at extreme points. Optimization on focal sets can therefore ALL be done by maxing points at the intersection of q hyperplanes. This set of extreme point is

$$\mathcal{E} = \{\omega = \eta_i \cap \cdots \cap \eta_q : \{i_1, \ldots, i_q\} \in \{1, \ldots, n\}\} \tag{15}$$

with η_i the hyper-planes corresponding to the questions. We have $|\mathcal{E}| = \binom{n}{q} \in \mathcal{O}(n^q)$ which is reasonable whenever q is small enough (typically the case in MCDA). The computation of the coordinate of extreme points related to constraints of each subset can be done in advance for each $\omega \in \mathcal{E}$ and not once per subset and pair of alternatives, since $\mathcal{E}_{\Omega'}$ the set of extreme points of Ω' will always be such that $\mathcal{E}_{\Omega'} \subseteq \mathcal{E}$. The computation of the dot products necessary to compute $R_\omega(x,y)$ for all $\omega \in \mathcal{E}$, $x,y \in \mathbb{X}$ can also be done once for each $\omega \in \mathcal{E}$, and not be repeated in each subset Ω' s.t. $\omega \in \mathcal{E}_{\Omega'}$. Those results indicate us that when q (the model-space dimension) is reasonably low and questions correspond to cutting hyper-planes over a convex set, ECSS can be performed efficiently. This will be the case for several models such as OWA or k-additive Choquet integrals with low k, but not for others such as full Choquet integrals, whose dimension if we have k criteria is $2^k - 2$. In these cases, it seems inevitable that one would resort to approximations having a low numerical impact (e.g., merging or forgetting focal elements having a very low mass value).

4   Experiments

To test our strategy and its properties, we proceeded to simulated experiments, in which the confidence degree was always constant. Such experiments therefore also show what would happen if we did not ask confidence degrees to the DM, but nevertheless assumed that she could make mistakes with a very simple noise model.

The first experiment reported in Fig. 4 compares the extra cautiousness of EMR when compared to MR. To do so, simulations were made for several fixed degrees of confidence – including 1 in which case EMR coincides with MR – in which a virtual DM states her preferences with the given degree of confidence, and the value of EMR at each step is divided by the initial value so as to observe its evolution. Those EMR ratios were then averaged over 100 simulations for each degree. Results show that while high confidence degrees will have a limited impact, low confidence degrees (< 0.7) may greatly slow down the convergence.

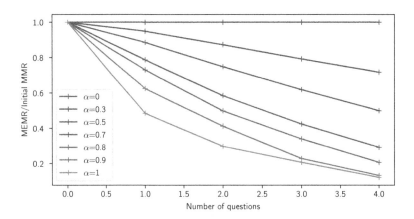

Fig. 4. Average evolution of min max regret with various degrees of confidence

The second experiment reported in Fig. 5 aims at finding if ECSS and CSS truly generate different question strategies. To do so, we monitored the two strategies for a given confidence degree, and identify the first step k for which the two questions are different. Those values were averaged over 300 simulations for several confidence degrees. Results show that even for a high confidence degree ($\alpha = 0.9$) it takes in average only 3 question to see a difference. This shows that the methods are truly different in practice.

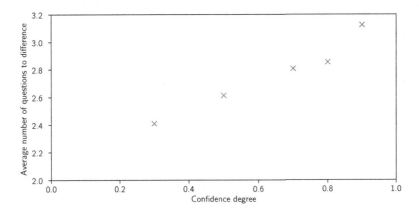

Fig. 5. Average position of the first different question in the elicitation process/degrees of confidence

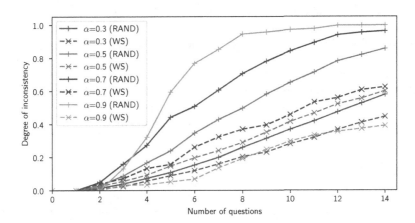

Fig. 6. Evolution of average inconsistency with a DM fitting the WS model and a randomly choosing DM

The third experiment reported in Fig. 6 is meant to observe how good $m(\varnothing)$ our measure of inconsistency is in practice as an indicator that something is wrong with the answers given by a DM. In order to do so simulations were made in which one of two virtual DMs answers with a fixed confidence degree and the value of $m(\varnothing)$ is recorded at each step. They were then averaged over 100 simulations for each confidence degree. The two virtual DMs behaved respectively completely randomly (RAND) and in accordance with a fixed weighted sum model (WS) with probability α and randomly with a probability $1 - \alpha$. So the first one is highly inconsistent with our model assumption, while the second is consistent with this assumptions but makes mistakes.

Results are quite encouraging: the inconsistency of the random DM with the model assumption is quickly identified, especially for high confidence degrees. For the DM that follows our model assumptions but makes mistakes, the results are similar, except for the fact that the conflict increase is not especially higher for lower confidence degrees. This can easily be explained that in case of low confidence degrees, we have more mistakes but those are assigned a lower weight, while in case of high confidence degrees the occasional mistake is quite impactful, as it has a high weight.

5   Conclusion

In this paper, we have proposed an evidential extension of the CSS strategy, used in robust elicitation of preferences.

We have studied its properties, notably comparing them to those of CSS, and have performed first experiments to demonstrate the utility of including confidence degrees in robust preference elicitation. Those latter experiments confirm the interest of our proposal, in the sense that it quickly identifies inconsistencies between the DM answer and model assumptions. It remains to check whether, in presence of mistakes from the DM, the real-regret (and not the computed one) obtained for ECSS is better than the one obtained for CSS.

As future works, we would like to work on the next step, i.e., identify the sources of inconsistency (whether it comes from bad model assumption or an unreliable DM) and propose correction strategies. We would also like to perform more experiments, and extend our approach to other decision models (Choquet integrals and OWA operators being the first candidates).

References

1. Benabbou, N., Gonzales, C., Perny, P., Viappiani, P.: Incremental elicitation of choquet capacities for multicriteria choice, ranking and sorting problems. Artif. Intell. **246**, 152–180 (2017)
2. Benabbou, N., Gonzales, C., Perny, P., Viappiani, P.: Minimax regret approaches for preference elicitation with rank-dependent aggregators. EURO J. Decis. Processes **3**(1–2), 29–64 (2015)
3. Boutilier, C., Patrascu, R., Poupart, P., Schuurmans, D.: Constraint-based optimization and utility elicitation using the minimax decision criterion. Artif. Intell. **170**(8–9), 686–713 (2006)
4. Denœux, T.: Conjunctive and disjunctive combination of belief functions induced by nondistinct bodies of evidence. Artif. Intell. **172**(2–3), 234–264 (2008)

5. Destercke, S.: A generic framework to include belief functions in preference handling and multi-criteria decision. Int. J. Approximate Reasoning **98**, 62–77 (2018)
6. Destercke, S., Dubois, D.: Idempotent conjunctive combination of belief functions: extending the minimum rule of possibility theory. Inf. Sci. **181**(18), 3925–3945 (2011)
7. Fürnkranz, J., Hüllermeier, E.: Preference Learning. Springer, Heidelberg (2010). https://doi.org/10.1007/978-3-642-14125-6
8. Klawonn, F., Schweke, E.: On the axiomatic justification of Dempster's rule of combination. Int. J. Intell. Syst. **7**(5), 469–478 (1992)
9. Klawonn, F., Smets, P.: The dynamic of belief in the transferable belief model and specialization-generalization matrices. In: Proceedings of the 8th Conference on Uncertainty in Artificial Intelligence (2013)
10. Orlik, P., Terao, H.: Arrangements of hyperplanes, vol. 300. Springer Science & Business Media, Heidelberg (2013)
11. Pichon, F., Denoeux, T.: The unnormalized Dempster's rule of combination: a new justification from the least commitment principle and some extensions. J. Autom. Reasoning **45**(1), 61–87 (2010)
12. Schläfli, L., Wild, H.: Theorie der vielfachen Kontinuität, vol. 38. Springer, Basel (1901). https://doi.org/10.1007/978-3-0348-5118-3
13. Shafer, G.: A Mathematical Theory of Evidence. Princeton University Press, Princeton (1976)
14. Smets, P.: The combination of evidence in the transferable belief model. IEEE Trans. Pattern Anal. Mach. Intell. **12**(5), 447–458 (1990)
15. Smets, P.: The application of the matrix calculus to belief functions. Int. J. Approxim. Reasoning **31**(1–2), 1–30 (2002)
16. Yager, R.R.: On the dempster-shafer framework and new combination rules. Inf. Sci. **41**(2), 93–137 (1987)

Evidence Propagation and Consensus Formation in Noisy Environments

Michael Crosscombe[1]([✉]) [iD], Jonathan Lawry[1] [iD], and Palina Bartashevich[2] [iD]

[1] University of Bristol, Bristol, UK
{m.crosscombe,j.lawry}@bristol.ac.uk
[2] Otto von Guericke University Magdeburg, Magdeburg, Germany
palina.bartashevich@ovgu.de

Abstract. We study the effectiveness of consensus formation in multi-agent systems where there is both belief updating based on direct evidence and also belief combination between agents. In particular, we consider the scenario in which a population of agents collaborate on the best-of-n problem where the aim is to reach a consensus about which is the best (alternatively, *true*) state from amongst a set of states, each with a different quality value (or level of evidence). Agents' beliefs are represented within Dempster-Shafer theory by mass functions and we investigate the macro-level properties of four well-known belief combination operators for this multi-agent consensus formation problem: Dempster's rule, Yager's rule, Dubois & Prade's operator and the averaging operator. The convergence properties of the operators are considered and simulation experiments are conducted for different evidence rates and noise levels. Results show that a combination of updating on direct evidence and belief combination between agents results in better consensus to the best state than does evidence updating alone. We also find that in this framework the operators are robust to noise. Broadly, Yager's rule is shown to be the better operator under various parameter values, i.e. convergence to the best state, robustness to noise, and scalability.

Keywords: Evidence propagation · Consensus formation ·
Dempster-Shafer theory · Distributed decision making · Multi-agent systems

1 Introduction

Agents operating in noisy and complex environments receive evidence from a variety of different sources, many of which will be at least partially inconsistent. In this paper we investigate the interaction between two broad categories of evidence: (i) direct evidence from the environment, and (ii) evidence received from other agents with whom an agent is interacting or collaborating to perform a task. For example, robots engaged in a search and rescue mission will receive data directly from sensors as well as information from other robots in the team.

© Springer Nature Switzerland AG 2019
N. Ben Amor et al. (Eds.): SUM 2019, LNAI 11940, pp. 310–323, 2019.
https://doi.org/10.1007/978-3-030-35514-2_23

Alternatively, software agents can have access to online data as well as sharing data with other agents.

The efficacy of combining these two types of evidence in multi-agent systems has been studied from a number of different perspectives. In social epistemology [6] has argued that agent-to-agent communications has an important role to play in propagating locally held information widely across a population. For example, interaction between scientists facilitates the sharing of experimental evidence. Simulation results are then presented which show that a combination of direct evidence and agent interaction, within the Hegselmann-Krause opinion dynamics model [10], results in faster convergence to the true state than updating based solely on direct evidence. A probabilistic model combining Bayesian updating and probability pooling of beliefs in an agent-based system has been proposed in [13]. In this context it is shown that combining updating and pooling leads to faster convergence and better consensus than Bayesian updating alone. An alternative methodology exploits three-valued logic to combine both types of evidence [2] and has been effectively applied to distributed decision-making in swarm robotics [3].

In this current study we exploit the capacity of Dempster-Shafer theory (DST) to fuse conflicting evidence in order to investigate how direct evidence can be combined with a process of iterative belief aggregation in the context of the best-of-n problem. The latter refers to a general class of problems in distributed decision-making [16,22] in which a population of agents must collectively identify which of n alternatives is the correct, or best, choice. These alternatives could correspond to physical locations as, for example, in a search and rescue scenario, different possible states of the world, or different decision-making or control strategies. Agents receive direct but limited feedback in the form of quality values associated with each choice, which then influence their beliefs when combined with those of other agents with whom they interact. It is not our intention to develop new operators in DST nor to study the axiomatic properties of particular operators at the local level (see [7] for an overview of such properties). Instead, our motivation is to study the macro-level convergence properties of several established operators when applied *iteratively* by a population of agents, over long timescales, and in conjunction with a process of evidential updating, i.e., updating beliefs based on evidence.

An outline of the remainder of the paper is as follows. In Sect. 2 we give a brief introduction to the relevant concepts from DST and summarise its previous application to dynamic belief revision in agent-based systems. Section 3 introduces a version of the best-of-n problem exploiting DST measures and combination operators. In Sect. 4 we then give the fixed point analysis of a dynamical system employing DST operators so as to provide insight into the convergence properties of such systems. In Sect. 5 we present the results from a number of agent-based simulation experiments carried out to investigate consensus formation in the best-of-n problem under varying rates of evidence and levels of noise. Finally, Sect. 6 concludes with some discussion.

2 An Overview of Dempster-Shafer Theory

In this section we introduce relevant concepts from Dempster-Shafer theory (DST) [5,19], including four well-known belief combination operators.

Definition 1. *Mass function (or agent's belief)*

  Given a set of states or frame of discernment $\mathbb{S} = \{s_1, ..., s_n\}$, let $2^{\mathbb{S}}$ denote the power set of $\mathbb{S}$. An agent's belief is then defined by a basic probability assignment or mass function $m : 2^{\mathbb{S}} \to [0, 1]$, where $m(\emptyset) = 0$ and $\sum_{A \subseteq \mathbb{S}} m(A) = 1$. The mass function then characterises a belief and a plausibility measure defined on $2^{\mathbb{S}}$ such that for $A \subseteq \mathbb{S}$:

$$Bel(A) = \sum_{B \subseteq A} m(B) \text{ and } Pl(A) = \sum_{B : B \cap A \neq \emptyset} m(B)$$

and hence where $Pl(A) = 1 - Bel(A^c)$.

  A number of operators have been proposed in DST for combining or fusing mass functions [20]. In this paper we will compare in a dynamic multi-agent setting the following operators: Dempster's rule of combination (**DR**) [19], Dubois & Prade's operator (**D&P**) [8], Yager's rule (**YR**) [25], and a simple averaging operator (**AVG**). The first three operators all make the assumption of independence between the sources of the evidence to be combined but then employ different techniques for dealing with the resulting inconsistency. DR uniformly reallocates the mass associated with non-intersecting pairs of sets to the overlapping pairs, D&P does not re-normalise in such cases but instead takes the union of the two sets, while YR reallocates all inconsistent mass values to the universal set $\mathbb{S}$. These four operators were chosen based on several factors: the operators are well established and have been well studied, they require no additional information about individual agents, and they are computationally efficient at scale (within the limits of DST).

Definition 2. *Combination operators*

  Let m_1 and m_2 be mass functions on $2^{\mathbb{S}}$. Then the combined mass function $m_1 \odot m_2$ is a function $m_1 \odot m_2 : 2^{\mathbb{S}} \to [0, 1]$ such that for $\emptyset \neq A, B, C \subseteq \mathbb{S}$:

$$(\mathbf{DR})\, m_1 \odot m_2(C) = \frac{1}{1 - K} \sum_{A \cap B = C \neq \emptyset} m_1(A) \cdot m_2(B),$$

$$(\mathbf{D\&P})\, m_1 \odot m_2(C) = \sum_{A \cap B = C \neq \emptyset} m_1(A) \cdot m_2(B) + \sum_{\substack{A \cap B = \emptyset, \\ A \cup B = C}} m_1(A) \cdot m_2(B),$$

$$(\mathbf{YR})\, m_1 \odot m_2(C) = \sum_{A \cap B = C \neq \emptyset} m_1(A) \cdot m_2(B) \text{ if } C \neq \mathbb{S}, \text{ and}$$

$$m_1 \odot m_2(\mathbb{S}) = m_1(\mathbb{S}) \cdot m_2(\mathbb{S}) + K,$$

$$(\mathbf{AVG})\, m_1 \odot m_2(C) = \frac{1}{2} \left(m_1(C) + m_2(C) \right),$$

where K is associated with conflict, i.e., $K = \sum_{A \cap B = \emptyset} m_1(A) \cdot m_2(B)$.

In the agent-based model of the best-of-n problem, proposed in Sect. 3, agents are required to make a choice as to which of n possible states they should investigate at any particular time. To this end we utilise the notion of *pignistic distribution* proposed by Smets and Kennes [21].

Definition 3. *Pignistic distribution*
  For a given mass function m, the corresponding pignistic distribution on $\mathbb{S}$ is a probability distribution obtained by reallocating the mass associated with each set $A \subseteq \mathbb{S}$ uniformly to the elements of that set, i.e., $s_i \in A$, as follows:

$$P(s_i|m) = \sum_{A:s_i \in A} \frac{m(A)}{|A|}.$$

DST has been applied to multi-agent dynamic belief revision in a number of ways. For example, [4] and [24] investigate belief revision where agents update their beliefs by taking a weighted combination of conditional belief values of other agents using Fagin-Halpern conditional belief measures. These measures are motivated by the probabilistic interpretation of DST according to which a belief and plausibility measure are characterised by a set of probability distributions on $\mathbb{S}$. Several studies [1,2,15] have applied a three-valued version of DST in multi-agent simulations. This corresponds to the case in which there are two states, $\mathbb{S} = \{s_1, s_2\}$, one of which is associated with the truth value *true* (e.g., s_1), one with *false* (s_2), and where the set $\{s_1, s_2\}$ is then taken as corresponding to a third truth state representing *uncertain* or *borderline*. One such approach based on subjective logic [1] employs the combination operator proposed in [11]. Another [15] uses Dempster's rule applied to combine an agent's beliefs with an aggregate of those of her neighbours. Similarly, [2] uses Dubois & Prade's operator for evidence propagation. Other relevant studies include [12] in which Dempster's rule is applied across a network of sparsely connected agents.

With the exception of [2], and only for *two* states, none of the above studies considers the interaction between direct evidential updating and belief combination. The main contribution of this paper is therefore to provide a detailed and general study of DST applied to dynamic multi-agent systems in which there is both direct evidence from the environment and belief combination between agents with partially conflicting beliefs. In particular, we will investigate and compare the consensus formation properties of the four combination operators (Definition 2) when applied to the best-of-n problem.

3   The Best-of-n Problem within DST

Here we present a formulation of the best-of-n problem within the DST framework. We take the n choices to be the states $\mathbb{S}$. Each state $s_i \in \mathbb{S}$ is assumed to have an associated quality value $q_i \in [0, 1]$ with 0 and 1 corresponding to minimal and maximal quality, respectively. Alternatively, we might interpret q_i as quantifying the level of available evidence that s_i corresponds to the true state of the world.

In the best-of-n problem agents explore their environment and interact with each other with the aim of identifying which is the highest quality (or true) state. Agents sample states and receive evidence in the form of the quality q_i, so that in the current context evidence E_i regarding state s_i takes the form of the following mass function;

$$m_{E_i} = \{s_i\} : q_i, \ \mathbb{S} : 1 - q_i.$$

Hence, q_i is taken as quantifying both the evidence directly in favour of s_i provided by E_i, and also the evidence directly against any other state s_j for $j \neq i$. Given evidence E_i an agent updates its belief by combining its current mass function m with m_{E_i} using a combination operator so as to obtain the new mass function given by $m \odot m_{E_i}$.

A summary of the process by which an agent might obtain direct evidence in this model is then as follows. Based on its current mass function m, an agent stochastically selects a state $s_i \in \mathbb{S}$ to investigate[1], according to the pignistic probability distribution for m as given in Definition 3. More specifically, it will update m to $m \odot m_{E_i}$ with probability $P(s_i|m) \times r$ for $i = 1, \ldots, n$ and leave its belief unchanged with probability $(1 - r)$, where $r \in [0, 1]$ is a fixed evidence rate quantifying the probability of finding evidence about the state that it is currently investigating. In addition, we also allow for the possibility of noise in the evidential updating process. This is modelled by a random variable $\epsilon \sim \mathcal{N}(0, \sigma^2)$ associated with each quality value. In other words, in the presence of noise the evidence E_i received by an agent has the form:

$$m_{E_i} = \{s_i\} : q_i + \epsilon, \mathbb{S} : 1 - q_i - \epsilon,$$

where if $q_i + \epsilon < 0$ then it is set to 0, and if $q_i + \epsilon > 1$ then it is set to 1. Overall, the process of updating from direct evidence is governed by the two parameters, r and σ, quantifying the availability of evidence and the level of associated noise, respectively.

In addition to receiving direct evidence we also include belief combination between agents in this model. This is conducted in a pairwise symmetric manner in which two agents are selected at random to combine their beliefs, with both agents then adopting this combination as their new belief, i.e., if the two agents have beliefs m_1 and m_2, respectively, then they both replace these with $m_1 \odot m_2$. However, in the case that agents are combining their beliefs under Dempster's rule and that their beliefs are completely inconsistent, i.e., when $K = 1$ (see Definition 2), then they do not form consensus and the process moves on to the next iteration.

In summary, during each iteration both processes of evidential updating and consensus formation take place[2]. However, while every agent in the population

---

[1] We utilise roulette wheel selection; a proportionate selection process.

[2] Due to the possibility of rounding errors occurring as a result of the multiplication of small numbers close to 0, we renormalise the mass function that results from each process.

has the potential to update its own belief, provided that it successfully receives a piece of evidence, the consensus formation is restricted to a *single pair* of agents for each iteration. That is, we assume that only two agents in the whole population are able to communicate and combine their beliefs during each iteration.

4   Fixed Point Analysis

In the following, we provide an insight into the convergence properties of the dynamical system described in Sect. 2. Consider an agent model in which at each time step t two agents are selected at random to combine their beliefs from a population of k agents $\mathcal{A} = \{a_1 \ldots, a_k\}$ with beliefs quantified by mass functions m_i^t : $i = 1, \ldots, k$. For any t the state of the system can be represented by a vector of mass functions $\langle m_1^t, \ldots, m_k^t \rangle$. Without loss of generality, we can assume that the updated state is then $\langle m_1^{t+1}, m_2^{t+1}, \ldots, m_k^{t+1} \rangle = \langle m_1^t \odot m_2^t, m_1^t \odot m_2^t, m_3^t, \ldots, m_k^t \rangle$. Hence, we have a dynamical system characterised by the following mapping:

$$\langle m_1^t, \ldots, m_k^t \rangle \rightarrow \langle m_1^t \odot m_2^t, m_1^t \odot m_2^t, m_3^t, \ldots, m_k^t \rangle.$$

The fixed points of this mapping are those for which $m_1^t = m_1^t \odot m_2^t$ and $m_2^t = m_1^t \odot m_2^t$. This requires that $m_1^t = m_2^t$ and hence the fixed point of the mapping are the fixed points of the operator, i.e., those mass functions m for which $m \odot m = m$. Let us analyse in detail the fixed points for the case in which there are 3 states $\mathbb{S} = \{s_1, s_2, s_3\}$. Let $m = \{s_1, s_2, s_3\} : x_7, \{s_1, s_2\} : x_4, \{s_1, s_3\} : x_5, \{s_2, s_3\} : x_6, \{s_1\} : x_1, \{s_2\} : x_2, \{s_3\} : x_3$ represent a general mass function defined on this state space and where without loss of generality we take $x_7 = 1 - x_1 - x_2 - x_3 - x_4 - x_5 - x_6$. For Dubois & Prade's operator the constraint that $m \odot m = m$ generates the following simultaneous equations.

$$x_1^2 + 2x_1x_4 + 2x_1x_5 + 2x_1x_7 + 2x_4x_5 = x_1$$
$$x_2^2 + 2x_2x_4 + 2x_2x_6 + 2x_2x_7 + 2x_4x_6 = x_2$$
$$x_3^2 + 2x_3x_5 + 2x_3x_6 + 2x_3x_7 + 2x_5x_6 = x_3$$
$$x_4^2 + 2x_1x_2 + 2x_4x_7 = x_4$$
$$x_5^2 + 2x_1x_3 + 2x_5x_7 = x_5$$
$$x_6^2 + 2x_2x_3 + 2x_6x_7 = x_6$$

The Jacobian for this set of equations is given by:

$$\mathbf{J} = \left(\frac{\partial}{\partial x_j} m \odot m(A_i) \right),$$

where $A_1 = \{s_1\}, A_2 = \{s_2\}, A_3 = \{s_3\}, A_4 = \{s_1, s_2\}, \ldots$ The stable fixed points are those solutions to the above equations for which the eigenvalues of the Jacobian evaluated at the fixed point lie within the unit circle on the complex plane. In this case the only stable fixed points are the mass functions $\{s_1\} : 1$,

$\{s_2\}$: 1 and $\{s_3\}$: 1. In other words, the only stable fixed points are those for which agents' beliefs are both certain and precise. That is where for some state $s_i \in \mathbb{S}$, $Bel(\{s_i\}) = Pl(\{s_i\}) = 1$. The stable fixed points for Dempster's rule and Yager's rule are also of this form. The averaging operator is idempotent and all mass functions are unstable fixed points.

The above analysis concerns agent-based systems applying a combination in order to reach consensus. However, we have yet to incorporate evidential updating into this model. As outlined in Sect. 3, it is proposed that each agent investigates a particular state s_i chosen according to its current beliefs using the pignistic distribution. With probability r this will result in an update to its beliefs from m to $m \odot m_{E_i}$. Hence, for convergence it is also required that agents only choose to investigate states for which $m \odot m_{E_i} = m$. Assuming $q_i > 0$, then there is only one such fixed point corresponding to $m = \{s_i\}$: 1. Hence, the consensus driven by belief combination as characterised by the above fixed point analysis will result in convergence of individual agent beliefs if we also incorporate evidential updating. That is, an agent with beliefs close to a fixed point of the operator, i.e., $m = \{s_i\}$: 1, will choose to investigate state s_i with very high probability and will therefore tend to be close to a fixed point of the evidential updating process.

5 Simulation Experiments

In this section we describe experiments conducted to understand the behaviour of the four belief combination operators in the context of the dynamic multi-agent best-of-n problem introduced in Sect. 3. We compare their performance under different evidence rates r, noise levels σ, and their scalability for different numbers of states n.

5.1 Parameter Settings

Unless otherwise stated, all experiments share the following parameter values. We consider a population $\mathcal{A}$ of $k = 100$ agents with beliefs initialised so that:

$$m_i^0 = \mathbb{S} : 1 \text{ for } i = 1, \ldots, 100.$$

In other words, at the beginning of each simulation every agent is in a state of complete ignorance as represented in DST by allocating all mass to the set of all states $\mathbb{S}$. Each experiment is run for a maximum of 5000 iterations, or until the population converges. Here, convergence requires that the beliefs of the population have not changed for 100 interactions, where an interaction may be the updating of beliefs based on evidence or the combination of beliefs between agents. For a given set of parameter values the simulation is run 100 times and results are then averaged across these runs.

Quality values are defined so that $q_i = \frac{i}{n+1}$ for $i = 1, \ldots, n$ and consequently s_n is the best state. In the following, $Bel(\{s_n\})$ provides a measure of convergence performance for the considered operators.

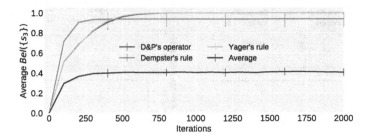

Fig. 1. Average $Bel(\{s_3\})$ plotted against iteration t with $r = 0.05$ and $\sigma = 0.1$. Comparison of all four operators with error bars displaying the standard deviation.

5.2 Convergence Results

Initially we consider the best-of-n problem where $n = 3$ with quality values $q_1 = 0.25$, $q_2 = 0.5$ and $q_3 = 0.75$. Figure 1 shows belief values for the best state s_3 averaged across agents and simulation runs for the evidence rate $r = 0.05$ and noise standard deviation $\sigma = 0.1$. For both Dubois & Prade's operator and Yager's rule there is complete convergence to $Bel(\{s_3\}) = 1$ while for Dempster's rule the average value of $Bel(\{s_3\})$ at steady state is approximately 0.9. The averaging operator does not converge to a steady state and instead maintains an average value of $Bel(\{s_3\})$ oscillating around 0.4. For all but the averaging operator, at steady state the average belief and plausibility values are equal. This is consistent with the fixed point analysis given for Dubois & Prade's operator in Sect. 4, showing that all agents converge to mass functions of the form $m = \{s_i\} : 1$ for some state $s_i \in \mathbb{S}$. Indeed, for both Dubois & Prade's operator and Yager's rule all agents converge to $m = \{s_3\} : 1$, while for Dempster's rule this happens in the large majority of cases. In other words, the combination of updating from direct evidence and belief combination results in agents reaching the certain and precise belief that s_3 is the true state of the world.

5.3 Varying Evidence Rates

In this section we investigate how the rate at which agents receive information from their environment affects their ability to reach a consensus about the true state of the world.

Figures 2a and b show average steady state values of $Bel(\{s_3\})$ for evidence rates in the lower range $r \in [0, 0.01]$ and across the whole range $r \in [0, 1]$, respectively. For each operator we compare the combination of evidential updating and belief combination (solid lines) with that of evidential updating alone (dashed lines). From Fig. 2a we see that for low values of $r \leq 0.02$ Dempster's rule converges to higher average values of $Bel(\{s_3\})$ than do the other operators. Indeed, for $0.001 \leq r \leq 0.006$ the average value of $Bel(\{s_3\})$ obtained using Dempster's rule is approximately 10% higher than is obtained using Dubois & Prade's operator and Yager's rule, and is significantly higher still than that of the averaging operator. However, the performance of Dempster's rule declines significantly for

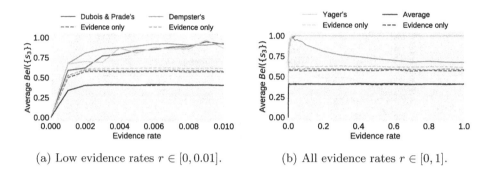

(a) Low evidence rates $r \in [0, 0.01]$.   (b) All evidence rates $r \in [0, 1]$.

Fig. 2. Average $Bel(\{s_3\})$ for evidence rates $r \in [0, 1]$. Comparison of all four operators both with and without belief combination between agents.

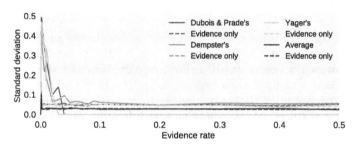

Fig. 3. Standard deviation for different evidence rates $r \in [0, 0.5]$. Comparison of all four operators both with and without belief combination between agents.

higher evidence rates and for $r > 0.3$ it converges to average values for $Bel(\{s_3\})$ of less than 0.8. At $r = 1$, when every agent is receiving evidence at each time step, there is failure to reach consensus when applying Dempster's rule. Indeed, there is polarisation with the population splitting into separate groups, each certain that a different state is the best. In contrast, both Dubois & Prade's operator and Yager's rule perform well for higher evidence rates and for all $r > 0.02$ there is convergence to an average value of $Bel(\{s_3\}) = 1$. Meanwhile the averaging operator appears to perform differently for increasing evidence rates and instead maintains similar levels of performance for $r > 0.1$. For all subsequent figures showing steady state results, we do not include error bars as this impacts negatively on readability. Instead, we show the standard deviation plotted separately against the evidence rate in Fig. 3. As expected, standard deviation is high for low evidence rates in which the sparsity of evidence results in different runs of the simulation converging to different states. This then declines rapidly with increasing evidence rates.

The dashed lines in Figs. 2a and b show the values of $Bel(\{s_3\})$ obtained at steady state when there is only updating based on direct evidence. In most cases the performance is broadly no better than, and indeed often worse than, the results which combine evidential updating with belief combination between agents. For low evidence rates where $r < 0.1$ the population does not tend to fully converge to a steady state since there is insufficient evidence available to allow convergence. For higher evidence rates under Dempster's rule, Dubois & Prade's operator and Yager's rule, the population eventually converges on a single state with complete certainty. However, since the average value of $Bel(\{s_3\})$ in both cases is approximately 0.6 for $r > 0.002$ then clearly convergence is often not to the best state. The averaging operator is not affected by the combined updating method and performs the same under evidential updating alone as it does in conjunction with consensus formation.

Overall, it is clear then that in this formulation of the best-of-n problem combining both updating from direct evidence and belief combination results in much better performance than obtained by using evidential updating alone for all considered operators except the averaging operator.

5.4   Noisy Evidence

Noise is ubiquitous in applications of multi-agent systems. In embodied agents such as robots this is often a result of sensor errors, but noise can also be a feature of an inherently variable environment. In this section we consider the effect of evidential noise on the best-of-n problem, as governed by the standard distribution σ of the noise.

Figure 4 shows the average value of $Bel(\{s_3\})$ at steady state plotted against $\sigma \in [0, 0.3]$ for different evidence rates $r \in \{0.01, 0.05, 0.1\}$. Figure 4 (left) shows that for an evidence rate $r = 0.01$, all operators except the averaging operator have very similar performance in the presence of noise. For example with no noise, i.e., $\sigma = 0$, Yager's rule converges to an average value of 0.97, Dubois & Prade's operator converges to an average of $Bel(\{s_3\}) = 0.96$, Dempster's rule to 0.95 on average, and the averaging operator to 0.4. Then, with $\sigma = 0.3$, Yager's rule converges to an average value of 0.8, Dubois & Prade's operator to an average value of $Bel(\{s_3\}) = 0.77$, Dempster's rule to 0.74, and the averaging operator converges to 0.29. Hence, all operators are affected by the noise to a similar extent given this low evidence rate.

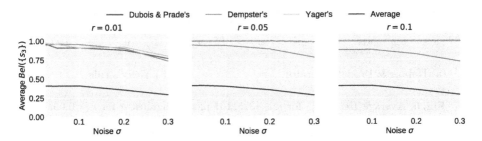

Fig. 4. Average $Bel(\{s_3\})$ for all four operators plotted against $\sigma \in [0, 0.3]$ for different evidence rates r. Left: $r = 0.01$. Centre: $r = 0.05$. Right: $r = 0.1$.

In contrast, for the evidence rates of $r = 0.05$ and $r = 0.1$, Fig. 4 (centre) and (right), respectively, we see that both Dubois & Prade's operator and Yager's rule are the most robust combination operators to increased noise. Specifically, for $r = 0.05$ and $\sigma = 0$, they both converge to an average value of $Bel(\{s_3\}) = 1$ and for $\sigma = 0.3$ they only decrease to 0.99. On the other hand, the presence of noise at this evidence rate has a much higher impact on the performance of Dempster's rule and the averaging operator. For $\sigma = 0$ Dempster's rule converges to an average value of $Bel(\{s_3\}) = 0.95$ but this decreases to 0.78 for $\sigma = 0.3$, and for the averaging operator the average value of $Bel(\{s_3\}) = 0.41$ and decreases to 0.29. The contrast between the performance of the operators in the presence of noise is even greater for the evidence rate $r = 0.1$ as seen in Fig. 4 (right). However, both Dubois & Prade's operator and Yager's rule differ in this context since, for both evidence rates $r = 0.05$ and $r = 0.1$, their average values of $Bel(\{s_3\})$ remain constant at approximately 1.

5.5 Scalability to Larger Numbers of States

In the swarm robotics literature most best-of-n studies are for $n = 2$ (see for example [17,23]). However, there is a growing interest in studying larger numbers of choices in this context [3,18]. Indeed, for many distributed decision-making applications the size of the state space, i.e., the value of n in the best-of-n problem, will be much larger. Hence, it is important to investigate the scalability of the proposed DST approach to larger values of n.

Having up to now focused on the $n = 3$ case, in this section we present additional simulation results for $n = 5$ and $n = 10$. As proposed in Sect. 5.1, the quality values are allocated so that $q_i = \frac{i}{n+1}$ for $i = 1, \ldots, n$. Here, we only consider Dubois & Prade's operator and Yager's rule due to their better performance when compared with the other two combination operators.

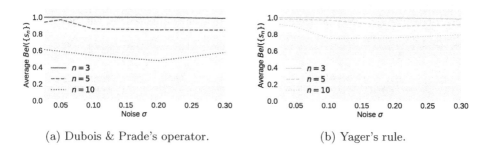

(a) Dubois & Prade's operator.   (b) Yager's rule.

Fig. 5. Average $Bel(\{s_n\})$ for $n \in \{3, 5, 10\}$ plotted against σ for $r = 0.05$.

Figure 5 shows the average values of $Bel(\{s_n\})$ at steady state plotted against noise $\sigma \in [0, 0.3]$ for evidence rate $r = 0.05$, where $Bel(\{s_n\})$ is the belief in the best state for $n = 3$, 5 and 10. For Dubois & Prade's operator, Fig. 5a shows the steady state values of $Bel(\{s_3\}) = 1$ independent of the noise level, followed closely by the values of $Bel(\{s_5\}) = 0.94$ at $\sigma = 0$ for the $n = 5$ case. However, for $n = 10$ the value of $Bel(\{s_{10}\})$ is 0.61 when $\sigma = 0$, corresponding to a significant decrease in performance. At the same time, from Fig. 5b, we can see that for Yager's rule performance declines much less rapidly with increasing n than for Dubois & Prade's operator. So at $\sigma = 0$ and $n = 5$ the average value at steady state for Yager's rule is almost the same as for $n = 3$, i.e. $Bel(\{s_5\}) = 0.98$, with a slight decrease in the performance $Bel(\{s_{10}\}) = 0.92$ for $n = 10$. As expected the performance of both operators decreases as σ increases, with Yager's rule being much more robust to noise than Dubois & Prade's operator for large values of n.

In this way, the results support only limited scalability for the DST approach to the best-of-n problem, at least as far as uniquely identifying the best state is concerned. Furthermore, as n increases so does sensitivity to noise. This reduced performance may in part be a feature of the way quality values have been allocated. Notice that as n increases, the difference between successive quality values $q_{i+1} - q_i = \frac{1}{n+1}$ decreases. This is likely to make it difficult for a population of agents to distinguish between the best state and those which have increasingly similar quality values. Furthermore, a given noise standard deviation σ results in an inaccurate ordering of the quality values the closer those values are to each other, making it difficult for a population of agents to distinguish between the best state and those which have increasingly similar quality values.

6   Conclusions and Future Work

In this paper we have introduced a model of consensus formation in the best-of-n problem which combines updating from direct evidence with belief combination between pairs of agents. We have utilised DST as a convenient framework for representing agents' beliefs, as well as the evidence that agents receive from the environment. In particular, we have studied and compared the macro-level convergence properties of several established operators applied iteratively in a dynamic multi-agent setting and through simulation we have identified several important properties of these operators within this context. Yager's rule and Dubois & Prade's operator are shown to be most effective at reducing polarisation and reaching a consensus for all except very low evidence rates, despite them not satisfying certain desirable properties, e.g., Dubois & Prade's operator is not associative while Yager's rule is only quasi-associative [7]. Both have also demonstrated robustness to different noise levels. However, Yager's rule is more robust to noise than Dubois & Prade's operator for large values of states $n > 3$. Although the performance of both operators decreases with an increase in the number of states n, Yager's rule is shown to be more scalable. We believe that underlying the difference in the performance of all but the averaging operator is the way in which they differ in their handling of inconsistent beliefs. Specifically, the manner in which they reallocate the mass associated with the inconsistent non-overlapping sets in the case of Dempster's rule, Dubois & Prade's operator and Yager's rule.

Further work will investigate the issue of scalability in more detail, including whether alternatives to the updating process may be applicable in a DST model, such as that of negative updating in swarm robotics [14]. We must also consider the increasing computational cost of DST as the size of the state space increases and investigate other representations such as possibility theory [9] as a means of avoiding exponential increases in the cost of storing and combining mass functions. Finally, we hope to adapt our method to be applied to a network, as opposed to a complete graph, so as to study the effects of limited or constrained communications on convergence.

Acknowledgments. This work was funded and delivered in partnership between Thales Group, University of Bristol and with the support of the UK Engineering and Physical Sciences Research Council, ref. EP/R004757/1 entitled "Thales-Bristol Partnership in Hybrid Autonomous Systems Engineering (T-B PHASE)".

References

1. Cho, J.H., Swami, A.: Dynamics of uncertain opinions in social networks. In: 2014 IEEE Military Communications Conference, pp. 1627–1632 (2014)
2. Crosscombe, M., Lawry, J.: A model of multi-agent consensus for vague and uncertain beliefs. Adapt. Behav. **24**(4), 249–260 (2016)
3. Crosscombe, M., Lawry, J., Hauert, S., Homer, M.: Robust distributed decision-making in robot swarms: exploiting a third truth state. In: 2017 IEEE/RSJ International Conference on Intelligent Robots and Systems (IROS), pp. 4326–4332. IEEE (September 2017). https://doi.org/10.1109/IROS.2017.8206297
4. Dabarera, R., Núñez, R., Premaratne, K., Murthi, M.N.: Dynamics of belief theoretic agent opinions under bounded confidence. In: 17th International Conference on Information Fusion (FUSION), pp. 1–8 (2014)
5. Dempster, A.P.: Upper and lower probabilities induced by a multivalued mapping. Ann. Math. Stat. **38**(2), 325–339 (1967)
6. Douven, I., Kelp, C.: Truth approximation, social epistemology, and opinion dynamics. Erkenntnis **75**, 271–283 (2011)
7. Dubois, D., Liu, W., Ma, J., Prade, H.: The basic principles of uncertain information fusion. An organised review of merging rules in different representation frameworks. Inf. Fus. **32**, 12–39 (2016). https://doi.org/10.1016/j.inffus.2016.02.006
8. Dubois, D., Prade, H.: Representation and combination of uncertainty with belief functions and possibility measures. Comput. Intell. **4**(3), 244–264 (1988). https://doi.org/10.1111/j.1467-8640.1988.tb00279.x
9. Dubois, D., Prade, H.: Possibility theory, probability theory and multiple-valued logics: a clarification. Ann. Math. Artif. Intell. **32**(1), 35–66 (2001). https://doi.org/10.1023/A:1016740830286
10. Hegselmann, R., Krause, U.: Opinion dynamics and bounded confidence: models, analysis and simulation. J. Artif. Soc. Soc. Simul. **5**, 2 (2002)
11. Jøsang, A.: The consensus operator for combining beliefs. Artif. Intell. **141**(1–2), 157–170 (2002). https://doi.org/10.1016/S0004-3702(02)00259-X
12. Kanjanatarakul, O., Denoux, T.: Distributed data fusion in the dempster-shafer framework. In: 2017 12th System of Systems Engineering Conference (SoSE), pp. 1–6. IEEE (June 2017). https://doi.org/10.1109/SYSOSE.2017.7994954

13. Lee, C., Lawry, J., Winfield, A.: Combining opinion pooling and evidential updating for multi-agent consensus. In: Proceedings of the Twenty-Seventh International Joint Conference on Artificial Intelligence (IJCAI-2018) Combining, pp. 347–353 (2018)

14. Lee, C., Lawry, J., Winfield, A.: Negative updating combined with opinion pooling in the best-of-n problem in swarm robotics. In: Dorigo, M., Birattari, M., Blum, C., Christensen, A.L., Reina, A., Trianni, V. (eds.) ANTS 2018. LNCS, vol. 11172, pp. 97–108. Springer, Cham (2018). https://doi.org/10.1007/978-3-030-00533-7_8

15. Lu, X., Mo, H., Deng, Y.: An evidential opinion dynamics model based on heterogeneous social influential power. Chaos Solitons Fractals **73**, 98–107 (2015). https://doi.org/10.1016/j.chaos.2015.01.007

16. Parker, C.A.C., Zhang, H.: Cooperative decision-making in decentralized multiple-robot systems: the best-of-n problem. IEEE/ASME Trans. Mechatron. **14**(2), 240–251 (2009). https://doi.org/10.1109/TMECH.2009.2014370

17. Reina, A., Bose, T., Trianni, V., Marshall, J.A.R.: Effects of spatiality on value-sensitive decisions made by robot swarms. In: Groß, R., et al. (eds.) Distributed Autonomous Robotic Systems. SPAR, vol. 6, pp. 461–473. Springer, Cham (2018). https://doi.org/10.1007/978-3-319-73008-0_32

18. Reina, A., Marshall, J.A.R., Trianni, V., Bose, T.: Model of the best-of-n nest-site selection process in honeybees. Phys. Rev. E **95**, 052411 (2017). https://doi.org/10.1103/PhysRevE.95.052411

19. Shafer, G.: A Mathematical Theory of Evidence. Princeton University Press, Princeton (1976)

20. Smets, P.: Analyzing the combination of conflicting belief functions. Inf. Fusion **8**(4), 387–412 (2007). https://doi.org/10.1016/j.inffus.2006.04.003

21. Smets, P., Kennes, R.: The transferable belief model. Artif. Intell. **66**, 387–412 (1994)

22. Valentini, G., Ferrante, E., Dorigo, M.: The best-of-n problem in robot swarms: formalization, state of the art, and novel perspectives. Front. Robot. AI **4**, 9 (2017). https://doi.org/10.3389/frobt.2017.00009

23. Valentini, G., Hamann, H., Dorigo, M.: Self-organized collective decision making: the weighted voter model. In: Proceedings of the 2014 International Conference on Autonomous Agents and Multi-agent Systems, pp. 45–52. AAMAS 2014. International Foundation for Autonomous Agents and Multiagent Systems, Richland (2014)

24. Wickramarathne, T.L., Premaratine, K., Murthi, M.N., Chawla, N.V.: Convergence analysis of iterated belief revision in complex fusion environments. IEEE J. Sel. Top. Signal Process. **8**(4), 598–612 (2014). https://doi.org/10.1109/JSTSP.2014.2314854

25. Yager, R.R.: On the specificity of a possibility distribution. Fuzzy Sets Syst. **50**(3), 279–292 (1992). https://doi.org/10.1016/0165-0114(92)90226-T

Order-Independent Structure Learning of Multivariate Regression Chain Graphs

Mohammad Ali Javidian, Marco Valtorta$^{(\boxtimes)}$, and Pooyan Jamshidi

University of South Carolina, Columbia, USA
javidian@email.sc.edu, {mgv,pjamshid}@cse.sc.edu

Abstract. This paper deals with multivariate regression chain graphs (MVR CGs), which were introduced by Cox and Wermuth in the nineties to represent linear causal models with correlated errors. We consider the PC-like algorithm for structure learning of MVR CGs, a constraint-based method proposed by Sonntag and Peña in 2012. We show that the PC-like algorithm is order-dependent, because the output can depend on the order in which the variables are given. This order-dependence is a minor issue in low-dimensional settings. However, it can be very pronounced in high-dimensional settings, where it can lead to highly variable results. We propose two modifications of the PC-like algorithm that remove part or all of this order-dependence. Simulations under a variety of settings demonstrate the competitive performance of our algorithms in comparison with the original PC-like algorithm in low-dimensional settings and improved performance in high-dimensional settings.

Keywords: Multivariate regression chain graph · Structural learning · Order independence · High-dimensional data · Scalable machine learning techniques

1 Introduction

Chain graphs were introduced by Lauritzen, Wermuth and Frydenberg [5,9] as a generalization of graphs based on undirected graphs and directed acyclic graphs (DAGs). Later Andersson, Madigan and Perlman introduced an alternative Markov property for chain graphs [1]. In 1993 [3], Cox and Wermuth introduced multivariate regression chain graphs (MVR CGs). The different interpretations of CGs have different merits, but none of the interpretations subsumes another interpretation [4].

Acyclic directed mixed graphs (ADMGs), also known as semi-Markov(ian) [12] models contain directed ($\rightarrow$) and bidirected ($\leftrightarrow$) edges subject to the restriction that there are no directed cycles [15]. An ADMG that has no partially directed cycle is called a *multivariate regression chain graph*. Cox and Wermuth represented these graphs using directed edges and dashed edges, but we follow Richardson [15] because bidirected edges allow the m-separation criterion

---

Supported by AFRL and DARPA (FA8750-16-2-0042).

N. Ben Amor et al. (Eds.): SUM 2019, LNAI 11940, pp. 324–338, 2019.
https://doi.org/10.1007/978-3-030-35514-2_24

(defined in Sect. 2) to be viewed more directly as an extension of d-separation than is possible with dashed edges [15].

Unlike in the other CG interpretations, the bidirected edge in MVR CGs has a strong intuitive meaning. It can be seen to represent one or more hidden common causes between the variables connected by it. In other words, in an MVR CG any bidirected edge $X \leftrightarrow Y$ can be replaced by $X \leftarrow H \rightarrow Y$ to obtain a Bayesian network representing the same independence model over the original variables, i.e. excluding the new variables H. These variables are called hidden, or latent, and have been marginalized away in the CG model. See [7,17,18] for details on the properties of MVR chain graphs.

Two *constraint-based* learning algorithms, that use a statistical analysis to test the presence of a conditional independency, exist for learning MVR CGs: (1) the PC-like algorithm [16], and (2) the answer set programming (ASP) algorithm [13]. The PC-like algorithm extends the original learning algorithm for Bayesian networks by **P**eter Spirtes and **C**lark Glymour [19]. It learns the structure of the underlying MVR chain graph in four steps: (a) determining the skeleton: the resulting undirected graph in this phase contains an undirected edge $u - v$ iff there is no set $S \subseteq V \setminus \{u, v\}$ such that $u \perp\!\!\!\perp v | S$; (b) determining the v-structures (unshielded colliders); (c) orienting some of the undirected/directed edges into directed/bidirected edges according to a set of rules applied iteratively; (d) transforming the resulting graph in the previous step into an MVR CG. The essential recovery algorithm obtained after step (c) contains all directed and bidirected edges that are present in every MVR CG of the same Markov equivalence class.

In this paper, we show that the PC-like algorithm is order-dependent, in the sense that the output can depend on the order in which the variables are given. We propose several modifications of the PC-like algorithm that remove part or all of this order-dependence, but do not change the result when perfect conditional independence information is used. When applied to data, the modified algorithms are partly or fully order-independent. Proofs, implementations in R, and details of experimental results can be found in the supplementary material at https://github.com/majavid/SUM2019.

2 Definitions and Concepts

Below we briefly list some of the most important concepts used in this paper.

If there is an arrow from a pointing towards b, a is said to be a parent of b. The set of parents of b is denoted as $pa(b)$. If there is a bidirected edge between a and b, a and b are said to be neighbors. The set of neighbors of a vertex a is denoted as $ne(a)$. The expressions $pa(A)$ and $ne(A)$ denote the collection of parents and neighbors of vertices in A that are not themselves elements of A. The boundary $bd(A)$ of a subset A of vertices is the set of vertices in $V \setminus A$ that are parents or neighbors to vertices in A.

A path of length n from a to b is a sequence $a = a_0, \ldots, a_n = b$ of distinct vertices such that $(a_i \rightarrow a_{i+1}) \in E$, for all $i = 1, \ldots, n$. A chain of length n from

a to b is a sequence $a = a_0, \ldots, a_n = b$ of distinct vertices such that $(a_i \rightarrow a_{i+1}) \in E$, or $(a_{i+1} \rightarrow a_i) \in E$, or $(a_{i+1} \leftrightarrow a_i) \in E$, for all $i = 1, \ldots, n$. We say that u is an ancestor of v and v is a descendant of u if there is a path from u to v in G. The set of ancestors of v is denoted as $an(v)$, and we define $An(v) = an(v) \cup v$. We apply this definition to sets: $an(X) = \{\alpha | \alpha$ is an ancestor of β for some $\beta \in X\}$. A partially directed cycle in a graph G is a sequence of n distinct vertices $v_1, \ldots, v_n (n \geq 3)$, and $v_{n+1} \equiv v_1$, such that $\forall i (1 \leq i \leq n)$ either $v_i \leftrightarrow v_{i+1}$ or $v_i \rightarrow v_{i+1}$, and $\exists j (1 \leq j \leq n)$ such that $v_i \rightarrow v_{i+1}$.

A graph with only undirected edges is called an undirected graph (UG). A graph with only directed edges and without directed cycles is called a directed acyclic graph (DAG). Acyclic directed mixed graphs, also known as semi-Markov(ian) [12] models contain directed ($\rightarrow$) and bidirected ($\leftrightarrow$) edges subject to the restriction that there are no directed cycles [15]. A graph that has no partially directed cycles is called a *chain graph*.

A non endpoint vertex ζ on a chain is a *collider* on the chain if the edges preceding and succeeding ζ on the chain have an arrowhead at ζ, that is, $\rightarrow \zeta \leftarrow$, or $\leftrightarrow \zeta \leftrightarrow$, or $\leftrightarrow \zeta \leftarrow$, or $\rightarrow \zeta \leftrightarrow$. A nonendpoint vertex ζ on a chain which is not a collider is a noncollider on the chain. A chain between vertices α and β in chain graph G is said to be m-connecting given a set Z (possibly empty), with $\alpha, \beta \notin Z$, if every noncollider on the path is not in Z, and every collider on the path is in $An_G(Z)$.

A chain that is not m-connecting given Z is said to be blocked given (or by) Z. If there is no chain m-connecting α and β given Z, then α and β are said to be m-*separated* given Z. Sets X and Y are m-separated given Z, if for every pair α, β, with $\alpha \in X$ and $\beta \in Y$, α and β are m-separated given Z (X, Y, and Z are disjoint sets; X, Y are nonempty). We denote the independence model resulting from applying the m-separation criterion to G, by $\Im_m(G)$. This is an extension of Pearl's d-separation criterion [11] to MVR chain graphs in that in a DAG D, a chain is d-connecting if and only if it is m-connecting.

We say that two MVR CGs G and H are Markov equivalent or that they are in the same Markov equivalence class iff $\Im_m(G) = \Im_m(H)$. If G and H have the same adjacencies and unshielded colliders, then $\Im_m(G) = \Im_m(H)$ [21].

Just like for many other probabilistic graphical models there might exist multiple MVR CGs that represent the same independence model. Sometimes it can however be desirable to have a unique graphical representation of the different representable independence models in the MVR CGs interpretation. A graph G^* is said to be the essential MVR CG of an MVR CG G if it has the same skeleton as G and contains all and only the arrowheads common to every MVR CG in the Markov equivalence class of G. One thing that can be noted here is that an essential MVR CG does not need to be a MVR CG. Instead these graphs can contain three types of edges, undirected, directed and bidirected [17].

3   Order-Dependent PC-Like Algorithm

In this section, we show that the PC-like algorithm proposed by Sonntag and Peña in [16] is order-dependent, in the sense that the output can depend on the order in which the variables are given. The PC-like algorithm for learning MVR CGs under the faithfulness assumption is formally described in Algorithm 1.

Algorithm 1. The order-dependent PC-like algorithm for learning MVR chain graphs [16]

Input: A set V of nodes and a probability distribution p faithful to an unknown MVR CG G and an ordering order(V) on the variables.

Output: An MVR CG G' s.t. G and G' are Markov equivalent and G' has exactly the minimum set of bidirected edges for its equivalence class.

1 Let H denote the complete undirected graph over V;

    /* Skeleton Recovery   */

2 **for** $i \leftarrow 0$ **to** $|V_H| - 2$ **do**

3   **while** *possible* **do**

4   Select any ordered pair of nodes u and v in H such that $u \in ad_H(v)$ and $|ad_H(u) \setminus v| \geq i$ using order(V);

                /* $ad_H(x) := \{y \in V | x \longrightarrow y, y \longrightarrow x,$ or $x \relbar\joinrel\relbar y\}$ */

5   **if** *there exists* $S \subseteq (ad_H(u) \setminus v)$ *s.t.* $|S| = i$ *and* $u \perp\!\!\!\perp_p v|S$ *(i.e., u is independent of v given S in the probability distribution p)* **then**

6   Set $S_{uv} = S_{vu} = S$;

7   Remove the edge $u \relbar\joinrel\relbar v$ from H;

8   **end**

9   **end**

10 **end**

    /* v-structure Recovery   */

11 **for** *each m-separator* S_{uv} **do**

12   **if** $u \,\circ\!\!\relbar\, w \relbar\!\!\circ\, v$ *appears in the skeleton and w is not in S_{uv}* **then**

        /* $u \,\circ\!\!\relbar\, w$ means $u \longleftarrow w$ or $u \relbar\joinrel\relbar w$. Also, $w \relbar\!\!\circ\, v$ means $w \longrightarrow v$ or $w \relbar\joinrel\relbar v$.   */

13   Determine a v-structure $u \,\circ\!\!\!\longrightarrow\, w \longleftarrow\!\!\circ\, v$;

14   **end**

15 **end**

16 Apply rules 1-3 in Figure 1 while possible;

    /* After this line, the learned graph is the *essential graph* of MVR CG G.   */

17 Let G'_u be the subgraph of G' containing only the nodes and the undirected edges in G';

18 Let T be the junction tree of G'_u;

    /* If G'_u is disconnected, the cliques belonging to different connected components can be linked with empty separators, as described in [6, Theorem 4.8]Golumbic.   */

19 Order the cliques $C_1, \cdots, C_n$ of G'_u s.t. C_1 is the root of T and if C_i is closer to the root than C_j in T then $C_i < C_j$;

20 Order the nodes such that if $A \in C_i$, $B \in C_j$, and $C_i < C_j$ then $A < B$;

21 Orient the undirected edges in G' according to the ordering obtained in line 21.

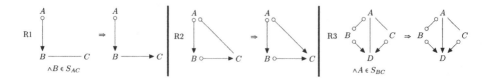

Fig. 1. The rules [16]

In applications, we do not have perfect conditional independence information. Instead, we assume that we have an i.i.d. sample of size n of variables $V = (X_1, \ldots, Xp)$. In the PC-like algorithm [16] all conditional independence queries are estimated by statistical conditional independence tests at some pre-specified significance level (p value) α. For example, if the distribution of V is multivariate Gaussian, one can test for zero partial correlation, see, e.g., [8]. For this purpose, we use the gaussCItest() function from the R package pcalg throughout this paper. Let order(V) denote an ordering on the variables in V. We now consider the role of order(V) in every step of the algorithm.

In the skeleton recovery phase of the PC-like algorithm [16], the order of variables affects the estimation of the skeleton and the separating sets. In particular, as noted for the special case of Bayesian networks in [2], for each level of i, the order of variables determines the order in which pairs of adjacent vertices and subsets S of their adjacency sets are considered (see lines 4 and 5 in Algorithm 1). The skeleton H is updated after each edge removal. Hence, the adjacency sets typically change within one level of i, and this affects which other conditional independencies are checked, since the algorithm only conditions on subsets of the adjacency sets. When we have perfect conditional independence information, all orderings on the variables lead to the same output. In the sample version, however, we typically make mistakes in keeping or removing edges, because conditional independence relationships have to be estimated from data. In such cases, the resulting changes in the adjacency sets can lead to different skeletons, as illustrated in Example 1.

Moreover, different variable orderings can lead to different separating sets in the skeleton recovery phase. When we have perfect conditional independence information, this is not important, because any valid separating set leads to the correct v-structure decision in the orientation phase. In the sample version, however, different separating sets in the skeleton recovery phase of the algorithm may yield different decisions about v-structures in the orientation phase. This is illustrated in Example 2.

Finally, we consider the role of order(V) on the orientation rules in the essential graph recovery phase of the sample version of the PC-like algorithm. Example 3 illustrates that different variable orderings can lead to different orientations, even if the skeleton and separating sets are order-independent.

Example 1 (Order-dependent skeleton of the PC-like algorithm). Suppose that the distribution of $V = \{a, b, c, d, e\}$ is faithful to the DAG in Fig. 2(a). This DAG encodes the following conditional independencies (using the notation defined in line 5 of Algorithm 1) with minimal separating sets: $a \perp\!\!\!\perp d|\{b, c\}$ and $a \perp\!\!\!\perp e|\{b, c\}$.

Suppose that we have an i.i.d. sample of (a, b, c, d, e), and that the following conditional independencies with minimal separating sets are judged to hold at some significance level α: $a \perp\!\!\!\perp d|\{b, c\}$, $a \perp\!\!\!\perp e|\{b, c, d\}$, and $c \perp\!\!\!\perp e|\{a, b, d\}$. Thus, the first two are correct, while the third is false.

We now apply the skeleton recovery phase of the PC-like algorithm with two different orderings: $\text{order}_1(V) = (d, e, a, c, b)$ and $\text{order}_2(V) = (d, c, e, a, b)$. The resulting skeletons are shown in Figs. 2(b) and (c), respectively.

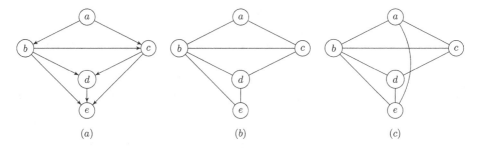

Fig. 2. (a) The DAG G, (b) the skeleton returned by Algorithm 1 with $\text{order}_1(V)$, (c) the skeleton returned by Algorithm 1 with $\text{order}_2(V)$.

We see that the skeletons are different, and that both are incorrect as the edge $c - e$ is missing. The skeleton for $\text{order}_2(V)$ contains an additional error, as there is an additional edge $a - e$. We now go through Algorithm 1 to see what happened. We start with a complete undirected graph on V. When $i = 0$, variables are tested for marginal independence, and the algorithm correctly does not remove any edge. Also, when $i = 1$, the algorithm correctly does not remove any edge. When $i = 2$, there is a pair of vertices that is thought to be conditionally independent given a subset of size two, and the algorithm correctly removes the edge between a and d. When $i = 3$, there are two pairs of vertices that are thought to be conditionally independent given a subset of size three. Table 1 shows the trace table of Algorithm 1 for $i = 3$ and $\text{order}_1(V) = (d, e, a, c, b)$.

Table 1. The trace table of Algorithm 1 for $i = 3$ and $\text{order}_1(V) = (d, e, a, c, b)$.

Ordered pair (u, v)	$ad_H(u)$	S_{uv}	Is $S_{uv} \subseteq ad_H(u) \setminus \{v\}$?	Is $u - v$ removed?
(e, a)	$\{a, b, c, d\}$	$\{b, c, d\}$	Yes	Yes
(e, c)	$\{b, c, d\}$	$\{a, b, d\}$	No	No
(c, e)	$\{a, b, d, e\}$	$\{a, b, d\}$	Yes	Yes

Table 2 shows the trace table of Algorithm 1 for $i = 3$ and $\text{order}_2(V) = (d, c, e, a, b)$.

Table 2. The trace table of Algorithm 1 for $i = 3$ and $\text{order}_2(V) = (d, c, e, a, b)$.

Ordered Pair (u, v)	$ad_H(u)$	S_{uv}	Is $S_{uv} \subseteq ad_H(u) \setminus \{v\}$?	Is $u \text{---} v$ removed?
(c, e)	$\{a, b, d, e\}$	$\{a, b, d\}$	Yes	Yes
(e, a)	$\{a, b, d\}$	$\{b, c, d\}$	No	No
(a, e)	$\{b, c, e\}$	$\{b, c, d\}$	No	No

Example 2 (Order-dependent separating sets and v-structures of the PC-like algorithm). Suppose that the distribution of $V = \{a, b, c, d, e\}$ is faithful to the DAG in Fig. 3(a). This DAG encodes the following conditional independencies with minimal separating sets: $a \perp\!\!\!\perp d|b, a \perp\!\!\!\perp e|\{b, c\}, a \perp\!\!\!\perp e|\{c, d\}, b \perp\!\!\!\perp c, b \perp\!\!\!\perp e|d$, and $c \perp\!\!\!\perp d$.

Suppose that we have an i.i.d. sample of (a, b, c, d, e). Assume that all true conditional independencies are judged to hold except $c \perp\!\!\!\perp d$. Suppose that $c \perp\!\!\!\perp d|b$ and $c \perp\!\!\!\perp d|e$ are thought to hold. Thus, the first is correct, while the second is false. We now apply the v-structure recovery phase of the PC-like algorithm with two different orderings: $\text{order}_1(V) = (d, c, b, a, e)$ and $\text{order}_3(V) = (c, d, e, a, b)$. The resulting CGs are shown in Figs. 3(b) and (c), respectively. Note that while the separating set for vertices c and d with $\text{order}_1(V)$ is $S_{dc} = S_{cd} = \{b\}$, the separating set for them with $\text{order}_2(V)$ is $S_{cd} = S_{dc} = \{e\}$.

This illustrates that order-dependent separating sets in the skeleton recovery phase of the sample version of the PC-algorithm can lead to order-dependent v-structures.

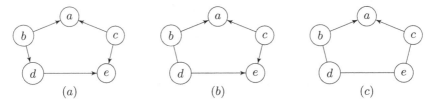

(a) (b) (c)

Fig. 3. (a) The DAG G, (b) the CG returned after the v-structure recovery phase of Algorithm 1 with $\text{order}_1(V)$, (c) the CG returned after the v-structure recovery phase of Algorithm 1 with $\text{order}_3(V)$.

Example 3 (Order-dependent orientation rules of the PC-like algorithm). Consider the graph in Fig. 4, and assume that this is the output of the sample version of the PC-like algorithm after v-structure recovery. Also, consider that $c \in S_{a,d}$ and $d \in S_{b,f}$. Thus, we have two v-structures, namely $a \longrightarrow c \longleftarrow e$ and $b \longrightarrow d \longleftarrow f$, and four unshielded triples, namely $(e, c, d), (c, d, f), (a, c, d)$, and (b, d, c). Thus, we then apply the orientation rules in the essential recovery phase of the algorithm, starting with rule R1. If one of the two unshielded triples (e, c, d) or (a, c, d) is considered first, we obtain $c \longrightarrow d$. On the other hand, if one of the unshielded triples (b, d, c) or (c, d, f) is considered first, then we obtain $c \longleftarrow d$. Note that we have no issues with overwriting of edges here, since as soon as the edge c —— d is oriented, all edges are oriented and no further orientation rules are applied. These examples illustrate that the essential graph recovery phase of the PC-like algorithm can be order-dependent regardless of the output of the previous steps.

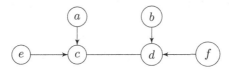

Fig. 4. Possible mixed graph after v-structure recovery phase of the sample version of the PC-like algorithm.

4 Order Independent Algorithms for Learning MVR CGs

We now propose several modifications of the original PC-like algorithm (and hence also of the related algorithms) that remove the order-dependence in the various stages of the algorithm, analogously to what Colombo and Maathuis [2] did for the original PC algorithm in the case of DAGs. For this purpose, we discuss the skeleton, v-structures, and the orientation rules, respectively.

4.1 Order-Independent Skeleton Recovery

We first consider estimation of the skeleton in the adjacency search of the PC-like algorithm. The pseudocode for our modification is given in Algorithm 2. The resulting PC-like algorithm in Algorithm 2 is called *stable PC-like*.

The main difference between Algorithms 1 and 2 is given by the for-loop on lines 3–5 in the latter one, which computes and stores the adjacency sets $a_H(v_i)$ of all variables after each new size i of the conditioning sets. These stored adjacency sets $a_H(v_i)$ are used whenever we search for conditioning sets of this given size i. Consequently, an edge deletion on line 10 no longer affects which conditional independencies are checked for other pairs of variables at this level of i.

In other words, at each level of i, Algorithm 2 records which edges should be removed, but for the purpose of the adjacency sets it removes these edges only when it goes to the next value of i. Besides resolving the order-dependence in the estimation of the skeleton, our algorithm has the advantage that it is easily parallelizable at each level of i. The stable PC-like algorithm is correct, i.e. it returns an MVR CG to which the given probability distribution is faithful (Theorem 1), and it yields order-independent skeletons in the sample version (Theorem 2). We illustrate the algorithm in Example 4.

Theorem 1. *Let the distribution of V be faithful to an MVR CG G, and assume that we are given perfect conditional independence information about all pairs of variables (u, v) in V given subsets $S \subseteq V \setminus \{u, v\}$. Then the output of the stable PC-like algorithm is an MVR CG that has exactly the minimum set of bidirected edges for its equivalence class.*

Theorem 2. *The skeleton resulting from the sample version of the stable PC-like algorithm is order-independent.*

Example 4 (Order-independent skeletons). We go back to Example 1, and consider the sample version of Algorithm 2. The algorithm now outputs the skeleton shown in Fig. 2(b) for both orderings $order_1(V)$ and $order_2(V)$. We again go through the algorithm step by step. We start with a complete undirected graph on V. No conditional independence found when $i = 0$. Also, when $i = 1$, the algorithm correctly does not remove any edge. When $i = 2$, the algorithm first computes the new adjacency sets: $a_H(v) = V \setminus \{v\}, \forall v \in V$. There is a pair of variables that is thought to be conditionally independent given a subset of size two, namely (a, d). Since the sets $a_H(v)$ are not updated after edge removals, it does not matter in which order we consider the ordered pair. Any ordering leads to the removal of edge between a and d. When $i = 3$, the algorithm first computes the new adjacency sets: $a_H(a) = a_H(d) = \{b, c, e\}$ and $a_H(v) = V \setminus \{v\}$, for $v = b, c, e$. There are two pairs of variables that are thought to be conditionally independent given a subset of size three, namely (a, e) and (c, e). Since the sets $a_H(v)$ are not updated after edge removals, it does not matter in which order we consider the ordered pair. Any ordering leads to the removal of both edges a —— e and c —— e.

Algorithm 2. The order-independent (stable) PC-like algorithm for learning MVR chain graphs.

Input: A set V of nodes and a probability distribution p faithful to an
unknown MVR CG G and an ordering $order(V)$ on the variables.
Output: An MVR CG G' s.t. G and G' are Markov equivalent and G' has
exactly the minimum set of bidirected edges for its equivalence class.

1   Let H denote the complete undirected graph over $V = \{v_1, \ldots, v_n\}$;
     /* Skeleton Recovery   */
2   **for** $i \leftarrow 0$ **to** $|V_H| - 2$ **do**
3   **for** $j \leftarrow 1$ **to** $|V_H|$ **do**
4 | Set $a_H(v_i) = ad_H(v_i)$;
5   **end**
6   **while** *possible* **do**
7   Select any ordered pair of nodes u and v in H such that $u \in a_H(v)$ and $|a_H(u) \setminus v| \geq i$ using $order(V)$;
8   **if** *there exists $S \subseteq (a_H(u) \setminus v)$ s.t. $|S| = i$ and $u \perp\!\!\!\perp_p v|S$ (i.e., u is independent of v given S in the probability distribution p)* **then**
9 | Set $S_{uv} = S_{vu} = S$;
10 | Remove the edge u —— v from H;
11   **end**
12   **end**
13 **end**
     /* v-structure Recovery and orientation rules   */
14 Follow the same procedures in Algorithm 1 (lines: 11–21).

4.2   Order-Independent v-structures Recovery

We propose two methods to resolve the order-dependence in the determination of the v-structures, using the conservative PC algorithm (CPC) of Ramsey et al. [14] and the majority rule PC-like algorithm (MPC) of Colombo & Maathuis [2].

The **Conservative PC-like algorithm (CPC-like algorithm)** works as follows. Let H be the undirected graph resulting from the skeleton recovery phase of the PC-like algorithm (Algorithm 1). For all unshielded triples (X_i, X_j, X_k) in H, determine all subsets S of $ad_H(X_i)$ and of $ad_H(X_k)$ that make X_i and X_k conditionally independent, i.e., that satisfy $X_i \perp\!\!\!\perp_p X_k | S$. We refer to such sets as separating sets. The triple (X_i, X_j, X_k) is labelled as *unambiguous* if at least one such separating set is found and either X_j is in all separating sets or in none of them; otherwise it is labelled as *ambiguous*. If the triple is unambiguous, it is oriented as v-structure if and only if X_j is in none of the separating sets. Moreover, in the v-structure recovery phase of the PC-like algorithm (Algorithm 1, lines 11–15), the orientation rules are adapted so that only unambiguous triples are oriented. The output of the CPC-like algorithm is a mixed graph in which ambiguous triples are marked. We refer to the combination of the stable PC-like and CPC-like algorithms as the *stable CPC-like algorithm*.

In the case of DAGs, Colombo and Maathuis [2] found that the CPC-algorithm can be very conservative, in the sense that very few unshielded triples are unambiguous in the sample version, where conditional independence relationships have to be estimated from data. They proposed a minor modification of the CPC approach, called *Majority rule PC algorithm (MPC)* to mitigate the (unnecessary) severity of CPC-like approach. We similarly propose the **Majority rule PC-like algorithm (MPC-like)** for MVR CGs. As in the CPC-like algorithm, we first determine all subsets S of $ad_H(X_i)$ and of $ad_H(X_k)$ that make X_i and X_k conditionally independent, i.e., that satisfy $X_i \perp\!\!\!\perp_p X_k | S$. The triple (X_i, X_j, X_k) is labelled as (α, β)-*unambiguous* if at least one such separating set is found or X_j is in no more than $\alpha\%$ or no less than $\beta\%$ of the separating sets, for $0 \le \alpha \le \beta \le 100$. Otherwise it is labelled as *ambiguous*. (As an example, consider $\alpha = 30$ and $\beta = 60$.) If a triple is unambiguous, it is oriented as a v-structure if and only if X_j is in less than $\alpha\%$ of the separating sets. As in the CPC-like algorithm, the orientation rules in the v-structure recovery phase of the PC-like algorithm (Algorithm 1, lines 11–15) are adapted so that only unambiguous triples are oriented, and the output is a mixed graph in which ambiguous triples are marked. Note that the CPC-like algorithm is the special case of the MPC-like algorithm with $\alpha = 0$ and $\beta = 100$. We refer to the combination of the stable PC-like and MPC-like algorithms as the *stable MPC-like algorithm*.

Theorem 3. *Let the distribution of V be faithful to an MVR CG G, and assume that we are given perfect conditional independence information about all pairs of variables (u, v) in V given subsets $S \subseteq V \setminus \{u, v\}$. Then the output of the (stable) CPC/MPC-like algorithm is an MVR CG that is Markov equivalent with G that has exactly the minimum set of bidirected edges for its equivalence class.*

Theorem 4. *The decisions about v-structures in the sample version of the stable CPC/MPC-like algorithm is order-independent.*

Example 5 (Order-independent decisions about v-structures). We consider the sample versions of the stable CPC/MPC-like algorithm, using the same input as in Example 2. In particular, we assume that all conditional independencies induced by the MVR CG in Fig. 3(a) are judged to hold except $c \perp\!\!\!\perp d$. Suppose that $c \perp\!\!\!\perp d|b$ and $c \perp\!\!\!\perp d|e$ are thought to hold. Let $\alpha = \beta = 50$.

Denote the skeleton after the skeleton recovery phase by H. We consider the unshielded triple (c, e, d). First, we compute $a_H(c) = \{a, d, e\}$ and $a_H(d) = \{a, b, c, e\}$, when $i = 1$. We now consider all subsets S of these adjacency sets, and check whether $c \perp\!\!\!\perp d|S$. The following separating sets are found: $\{b\}, \{e\}$, and $\{b, e\}$. Since e is in some but not all of these separating sets, the stable CPC-like algorithm determines that the triple is ambiguous, and no orientations are performed. Since e is in more than half of the separating sets, stable MPC-like determines that the triple is unambiguous and not a v-structure. The output of both algorithms is given in Fig. 3(c).

At this point it should be clear why the modified PC-like algorithm is labeled "conservative": it is more cautious than the (stable) PC-like algorithm in drawing unambiguous conclusions about orientations. As we showed in Example 5, the output of the (stable) CPC-like algorithm may not be collider equivalent with the true MVR CG G, if the resulting CG contains an ambiguous triple.

4.3   Order-Independent Orientation Rules

Even when the skeleton and the determination of the v-structures are order-independent, Example 3 showed that there might be some order-dependent steps left in the sample version. Regarding the orientation rules, we note that the PC-like algorithm does not suffer from conflicting v-structures (as shown in [2] for the PC-algorithm in the case of DAGs), because bi-directed edges are allowed. However, the three orientation rules still suffer from order-dependence issues (see Example 3 and Fig. 4). To solve this problem, we can use lists of candidate edges for each orientation rule as follows: we first generate a list of all edges that can be oriented by rule R1. We orient all these edges, creating bi-directed edges if there are conflicts. We do the same for rules R2 and R3, and iterate this procedure until no more edges can be oriented.

When using this procedure, we add the letter L (standing for lists), e.g., (stable) LCPC-like and (stable) LMPC-like. The (stable) LCPC-like and (stable) LMPC-like algorithms are fully order-independent in the sample versions. The procedure is illustrated in Example 6.

Theorem 5. *Let the distribution of V be faithful to an MVR CG G, and assume that we are given perfect conditional independence information about all pairs of variables (u, v) in V given subsets $S \subseteq V \setminus \{u, v\}$. Then the output of the (stable) LCPC/LMPC-like algorithm is an MVR CG that is Markov equivalent with G that has exactly the minimum set of bidirected edges for its equivalence class.*

Theorem 6. *The sample versions of stable CPC-like and stable MPC-like algorithms are fully order-independent.*

Example 6. Consider the structure shown in Fig. 4. As a first step, we construct a list containing all candidate structures eligible for orientation rule R1 in the phase of the essential graph recovery. The list contains the unshielded triples $(e, c, d), (c, d, f), (a, c, d)$, and (b, d, c). Now, we go through each element in the list and we orient the edges accordingly, allowing bi-directed edges. This yields the edge orientation $c \longleftrightarrow d$, regardless of the ordering of the variables.

5   Evaluation

In this section, we compare the performance of our algorithms (Table 3) with the original PC-like learning algorithm by running them on randomly generated MVR chain graphs in low-dimensional and high-dimensional data, respectively. We report on the Gaussian case only because of space limitations.

We evaluate the performance of the proposed algorithms in terms of the six measurements that are commonly used [2,8,10,20] for constraint-based learning algorithms: (a) the true positive rate (TPR) (also known as sensitivity, recall, and hit rate), (b) the false positive rate (FPR) (also known as fall-out), (c) the true discovery rate (TDR) (also known as precision or positive predictive value), (d) accuracy (ACC) for the skeleton, (e) the structural Hamming distance (SHD) (this is the metric described in [20] to compare the structure of the learned and the original graphs), and (f) run-time for the LCG recovery algorithms. In principle, large values of TPR, TDR, and ACC, and small values of FPR and SHD indicate good performance. All of these six measurements are computed on the essential graphs of the CGs, rather than the CGs directly, to avoid spurious differences due to random orientation of undirected edges.

Table 3. Order-dependence issues and corresponding modifications of the PC-like algorithm that remove the problem. "Yes" indicates that the corresponding aspect of the graph is estimated order-independently in the sample version.

	Skeleton	v-structures decisions	Edges orientations
PC-like	No	No	No
Stable PC-like	Yes	No	No
Stable CPC/MPC-like	Yes	Yes	No
Stable LCPC/LMPC-like	Yes	Yes	Yes

Figure 5 shows that: (a) as we expected [8,10], all algorithms work well on sparse graphs ($N = 2$), (b) for all algorithms, typically the TPR, TDR, and ACC increase with sample size, (c) for all algorithms, typically the SHD and FPR decrease with sample size, (d) a large significance level ($\alpha = 0.05$) typically yields

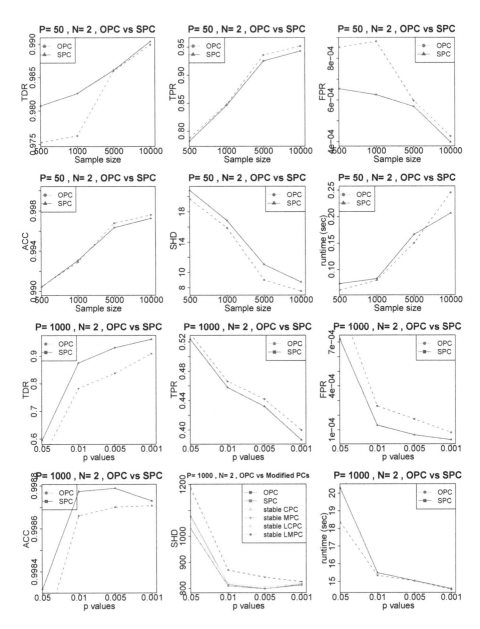

Fig. 5. The first two rows show the performance of the original (OPC) and stable PC-like (SPC) algorithms for randomly generated Gaussian chain graph models: average over 30 repetitions with 50 variables correspond to N = 2, and the significance level $\alpha = 0.001$. The last two rows show the performance of the original (OPC) and stable PC-like (SPC) algorithms for randomly generated Gaussian chain graph models: average over 30 repetitions with 1000 variables correspond to N = 2, sample size S = 50, and the significance level $\alpha = 0.05, 0.01, 0.005, 0.001$.

large TPR, FPR, and SHD, (e) while the stable PC-like algorithm has a better TDR and FPR in comparison with the original PC-like algorithm, the original PC-like algorithm has a better TPR (as observed in the case of DAGs [2]). This can be explained by the fact that the stable PC-like algorithm tends to perform more tests than the original PC-like algorithm, and (h) while the original PC-like algorithm has a (slightly) better SHD in comparison with the stable PC-like algorithm in low-dimensional data, the stable PC-like algorithm has a better SHD in high-dimensional data. Also, (very) small variances indicate that the order-independent versions of the PC-like algorithm in high-dimensional data are stable. When considering average running times versus sample sizes, as shown in Fig. 5, we observe that: (a) the average run time increases when sample size increases; (b) generally, the average run time for the original PC-like algorithm is (slightly) better than that for the stable PC-like algorithm in both low and high dimensional settings.

In summary, empirical simulations show that our algorithms achieve competitive results with the original PC-like learning algorithm; in particular, in the Gaussian case the order-independent algorithms achieve output of better quality than the original PC-like algorithm, especially in high-dimensional settings. Since we know of no score-based learning algorithms for MVR chain graphs (and, in fact, for any kind of chain graphs), we plan to investigate the feasibility of a scalable algorithm of this kind.

References

1. Andersson, S.A., Madigan, D., Perlman, M.D.: An alternative Markov property for chain graphs. In: Proceedings of UAI Conference, pp. 40–48 (1996)
2. Colombo, D., Maathuis, M.H.: Order-independent constraint-based causal structure learning. J. Mach. Learn. Res. **15**(1), 3741–3782 (2014)
3. Cox, D.R., Wermuth, N.: Linear dependencies represented by chain graphs. Stat. Sci. **8**(3), 204–218 (1993)
4. Drton, M.: Discrete chain graph models. Bernoulli **15**(3), 736–753 (2009)
5. Frydenberg, M.: The chain graph Markov property. Scand. J. Stat. **17**(4), 333–353 (1990)
6. Golumbic, M.: Algorithmic Graph Theory and Perfect Graphs. Academic Press, New York (1980)
7. Javidian, M.A., Valtorta, M.: On the properties of MVR chain graphs. In: Workshop Proceedings of PGM Conference, pp. 13–24 (2018)
8. Kalisch, M., Bühlmann, P.: Estimating high-dimensional directed acyclic graphs with the PC-algorithm. J. Mach. Learn. Res. **8**, 613–636 (2007)
9. Lauritzen, S., Wermuth, N.: Graphical models for associations between variables, some of which are qualitative and some quantitative. Ann. Stat. **17**(1), 31–57 (1989)
10. Ma, Z., Xie, X., Geng, Z.: Structural learning of chain graphs via decomposition. J. Mach. Learn. Res. **9**, 2847–2880 (2008)
11. Pearl, J.: Probabilistic Reasoning in Intelligent Systems: Networks of Plausible Inference. Morgan Kaufmann Publishers Inc., San Francisco, CA, USA (1988)
12. Pearl, J.: Causality: Models, Reasoning, and Inference. Cambridge University Press, Cambridge (2009)

13. Peña, J.M.: Alternative Markov and causal properties for acyclic directed mixed graphs. In: Proceedings of UAI Conference, pp. 577–586 (2016)

14. Ramsey, J., Spirtes, P., Zhang, J.: Adjacency-faithfulness and conservative causal inference. In: Proceedings of UAI Conference, pp. 401–408 (2006)

15. Richardson, T.S.: Markov properties for acyclic directed mixed graphs. Scand. J. Stat. **30**(1), 145–157 (2003)

16. Sonntag, D., Peña, J.M.: Learning multivariate regression chain graphs under faithfulness. In: Proceedings of PGM Workshop, pp. 299–306 (2012)

17. Sonntag, D., Peña, J.M.: Chain graph interpretations and their relations revisited. Int. J. Approx. Reason. **58**, 39–56 (2015)

18. Sonntag, D., Peña, J.M., Gómez-Olmedo, M.: Approximate counting of graphical models via MCMC revisited. Int. J. Intell. Syst. **30**(3), 384–420 (2015)

19. Spirtes, P., Glymour, C., Scheines, R.: Causation, Prediction and Search, 2nd edn. MIT Press, Cambridge (2000)

20. Tsamardinos, I., Brown, L.E., Aliferis, C.F.: The max-min hill-climbing Bayesian network structure learning algorithm. Mach. Learn. **65**(1), 31–78 (2006)

21. Wermuth, N., Sadeghi, K.: Sequences of regressions and their independences. Test **21**, 215–252 (2012)

Comparison of Analogy-Based Methods for Predicting Preferences

Myriam Bounhas[1,2(✉)], Marc Pirlot[3], Henri Prade[4], and Olivier Sobrie[3]

[1] Emirates College of Technology, Abu Dhabi, UAE
[2] LARODEC Lab, ISG de Tunis, Tunis, Tunisia
myriam_bounhas@yahoo.fr
[3] Faculté Polytechnique, Université de Mons, Mons, Belgium
marc.pirlot@umons.ac.be, olivier.sobrie@gmail.com
[4] IRIT, Université Paul Sabatier, 118 route de Narbonne,
31062 Toulouse cedex 09, France
prade@irit.fr

Abstract. Given a set of preferences between items taken by pairs and described in terms of nominal or numerical attribute values, the problem considered is to predict the preference between the items of a new pair. The paper proposes and compares two approaches based on analogical proportions, which are statements of the form "a is to b as c is to d". The first one uses triples of pairs of items for which preferences are known and which make analogical proportions, altogether with the new pair. These proportions express attribute by attribute that the change of values between the items of the first two pairs is the same as between the last two pairs. This provides a basis for predicting the preference associated with the fourth pair, also making sure that no contradictory trade-offs are created. Moreover, we also consider the option that one of the pairs in the triples is taken as a k-nearest neighbor of the new pair. The second approach exploits pairs of compared items one by one: for predicting the preference between two items, one looks for another pair of items for which the preference is known such that, attribute by attribute, the change between the elements of the first pair is the same as between the elements of the second pair. As discussed in the paper, the two approaches agree with the postulates underlying weighted averages and more general multiple criteria aggregation models. The paper proposes new algorithms for implementing these methods. The reported experiments, both on real data sets and on generated datasets suggest the effectiveness of the approaches. We also compare with predictions given by weighted sums compatible with the data, and obtained by linear programming.

1 Introduction

Predicting preferences has become a challenging topic in artificial intelligence, e.g., [9]. The idea of applying analogical proportion-based inference to this problem has been recently proposed [11] and different approaches have been successfully tested [1,7], following previous studies that obtained good results in

© Springer Nature Switzerland AG 2019
N. Ben Amor et al. (Eds.): SUM 2019, LNAI 11940, pp. 339–354, 2019.
https://doi.org/10.1007/978-3-030-35514-2_25

classification [3,10]. Analogical proportions are statements of the form "a is to b as c is to d", and express that the change (if any) between items a and b is the same as the one between c and d. Analogical inference relies on the idea that when analogical proportions hold, some related one may hold as well. Interestingly enough, analogical inference may work with rather small amounts of examples. There are two ways of making an analogical reading of preferences between items (taken by pairs), which lead to different prediction methods. In this paper, we explain the two analogical readings in Sect. 2 and how they give birth to new prediction algorithms in Sect. 3. They are compared in Sect. 4 on benchmarks with the two existing implementations, and with a linear programming method when applicable.

2   Analogy and Linear Utility

Analogical proportions are statements of the form "a is to b as c is to d", often denoted $a : b :: c : d$. As numerical proportions, they are quaternary relations that are supposed to satisfy the following postulates: (i) $a : b :: a : b$ holds (reflexivity); (ii) if $a : b :: c : d$ holds, $c : d :: a : b$ holds (symmetry); (iii) if $a : b :: c : d$ holds, $a : c :: b : d$ holds (central permutation). When $a : b :: c : d$ holds, it expresses that "a differs from b as c differs from d and b differs from a as d differs from c". This translates into a Boolean logical expression (see, e.g., [12–14]),

$$a : b :: c : d = ((a \wedge \neg b) \equiv (c \wedge \neg d)) \wedge ((b \wedge \neg a) \equiv (d \wedge \neg c)).$$

$a : b :: c : d$ is true only for the 6 following patterns $(0,0,0,0)$, $(1,1,1,1)$, $(1,0,1,0)$, $(0,1,0,1)$, $(1,1,0,0)$, and $(0,0,1,1)$ for (a,b,c,d). This can be generalized to nominal values; then $a : b :: c : d$ holds true if and only if $abcd$ is one of the following patterns $ssss$, $stst$, or $sstt$ where s and t are two possible distinct values of items a, b, c and d.

Analogical proportions extends to vectors describing items in terms of attribute values such as $\boldsymbol{a} = (a_1, ..., a_n)$, by stating $\boldsymbol{a} : \boldsymbol{b} :: \boldsymbol{c} : \boldsymbol{d}$ iff $\forall i \in [[1, n]]$, $a_i : b_i :: c_i : d_i$.

The basic analogical inference pattern applied to compared items is then, $\forall i \in [[1, n]]$,

$$a_i^1 : b_i^1 :: c_i^1 : d_i^1 \text{ and } a_i^2 : b_i^2 :: c_i^2 : d_i^2$$
$$\boldsymbol{a^1} \preceq \boldsymbol{a^2}$$
$$\boldsymbol{b^1} \preceq \boldsymbol{b^2}$$
$$\boldsymbol{c^1} \preceq \boldsymbol{c^2}$$
$$----------------$$
$$\boldsymbol{d^1} \preceq \boldsymbol{d^2}.$$

where $x \preceq y$ expresses that y is preferred to x (equivalently $y \succeq x$); other patterns (equivalent up to some rewriting) exist where, e.g., $\preceq$ is changed into $\succeq$ for (i) pairs (b^1, b^2) and (d^1, d^2), or in (ii) pairs (c^1, c^2) and (d^1, d^2) (since $a : b :: c : d$ is stable under central permutation) [1].

Following [11], the above pattern corresponds to a *vertical* reading, while another pattern corresponding to the *horizontal* reading can be stated as follows $\forall i \in [[1, n]]$,

$$a_i : b_i :: c_i : d_i$$
$$\boldsymbol{a} \preceq \boldsymbol{b},$$
$$------$$
$$\boldsymbol{c} \preceq \boldsymbol{d}.$$

The intuition behind the second pattern is simple: since $\boldsymbol{a}$ differs from $\boldsymbol{b}$ as $\boldsymbol{c}$ differs from $\boldsymbol{d}$ (and vice-versa), and $\boldsymbol{b}$ is preferred to $\boldsymbol{a}$, $\boldsymbol{d}$ should be preferred to $\boldsymbol{c}$ as well. The first pattern, which involves more items and more preferences, states that since the pair of items $(\boldsymbol{d^1}, \boldsymbol{d^2})$ makes an analogical proportion with the three other pairs $(\boldsymbol{a^1}, \boldsymbol{a^2}), (\boldsymbol{b^1}, \boldsymbol{b^2}), (\boldsymbol{c^1}, \boldsymbol{c^2})$, then the preference relation that holds for the 3 first pairs should hold as well for the fourth one.

Besides, the structure of the first pattern follows the axiomatics of additive utility functions, for which contradictory trade-offs are forbidden, namely: if $\forall i, j$,

$$\boldsymbol{a^1}_{-i}\alpha \preceq \boldsymbol{a^2}_{-i}\beta$$

$$\boldsymbol{a^1}_{-i}\gamma \succeq \boldsymbol{a^2}_{-i}\delta$$

$$\boldsymbol{c^1}_{-j}\alpha \succeq \boldsymbol{c^2}_{-j}\beta$$

one cannot have:

$$\boldsymbol{c^1}_{-j}\gamma \prec \boldsymbol{c^2}_{-j}\delta$$

where $\boldsymbol{x}_{-i}$ denotes the $n-1$-dimensional vector made of the evaluations of $\boldsymbol{x}$ on all criteria except the ith one for which the Greek letter denotes the substituted value. This property ensures that the differences of preference between α and β, on the one hand, and between γ and δ, on the other hand, can consistently be compared.

Thus, when applying the first pattern, one may also make sure that no contradictory trade-offs are introduced by the prediction mechanism. In the first pattern, analogical reasoning amounts here to finding triples of pairs of compared items $(\boldsymbol{a}, \boldsymbol{b}, \boldsymbol{c})$ appropriate for inferring the missing value(s) in $\boldsymbol{d}$. When there exist several suitable triples, possibly leading to different conclusions, one may use a majority vote for concluding.

Analogical proportions can be extended to numerical values, once the values are renormalized on scale $[0, 1]$, by a multiple-valued logic expression. The main option, which agrees with the Boolean case, and where truth is a matter of degree [6] is:

$A(a, b, c, d) = 1 - |(a - b) - (c - d)|$ if $a \geq b$ and $c \geq d$, or $a \leq b$ and $c \leq d$
$\qquad = 1 - max(|a - b|, |c - d|)$ otherwise.

Note that $A(a, b, c, d) = 1$ iff $a - b = c - d$.

We can then compute to what extent an analogical proportion holds between vectors:

$$A(\boldsymbol{a}, \boldsymbol{b}, \boldsymbol{c}, \boldsymbol{d}) = \frac{\Sigma_{i=1}^n A(a_i, b_i, c_i, d_i)}{n} \qquad (1)$$

Lastly, let us remark that the second, simpler, pattern agrees with the view that preferences wrt each criterion are represented by differences of evaluations. This includes the weighted sum, namely $\boldsymbol{b} \succeq \boldsymbol{a}$ iff $\sum_{i=1,n} w_i(b_i - a_i) \geq 0$, while analogy holds at degree 1 iff $\forall i \in [[1, n]], b_i - a_i = d_i - c_i$. This pattern does not agree with more general models of additive utility functions, while the first pattern is compatible with more general preference models.

3   Analogy-Based Preference Learning

In order to study the ability of analogical proportions to predict new preference relations from a given set of such relations, while avoiding the generation of con-tradictory trade-offs, we propose different "Analogy-based Preference Learning" algorithms (APL algorithms for short). The criteria are assumed to be evalu-ated on a scale $S = \{1, 2, ..., k\}$. Let $E = \{e^j : x^j \succeq y^j\}$ be a set of preference examples, where $\succeq$ is a preference relation telling us that choice x^j is preferred to choice y^j.

3.1   Methodology

Given a new pair of items $d = (d^1, d^2)$ for which preference is to be predicted, we present two types of algorithms for predicting preferences in the following, corre-sponding respectively to the "vertical reading" (first pattern) that exploits *triples* of pairs of items, and to the "horizontal reading" (second pattern) where pairs of items are taken one by one. This leads to algorithms APL_3 and APL_1 respectively.

APL_3: The basic principle of APL_3 is to find triples $t(a, b, c)$ of examples in E^3 that form with d either the non-contradictory trade-offs pattern (considered in first), or the analogical proportion-based inference pattern.

For each triple $t(a, b, c)$, we compute an analogical score $A_t(a, b, c, d)$ that estimates the extent to which it is in analogy with the item d using Formula 1. Then to guess the final preference of d, for each possible solution, we first cumulate these atomic scores provided by each of these triples in favor of this solution and finally we assign to d the solution with the highest score. In case of ties, a majority vote is applied.

The APL_3 can be described by this basic process:

– For a given d whose preference is to be predicted.
– Search for solvable triples $t \in E^3$ that make the analogical proportion, linking the 4 preference relations of the triple elements with d, valid (the preference relation between the 4 items satisfy one of the vertical pattern given in Sect. 2).
– For each triple t, compute the analogical score $A_t(a, b, c, d)$.
– Compute the sum of these scores for each possible solution for d and assign to d, the solution with the highest score.

APL_1: Applying the "horizontal reading" (second pattern), we consider *only* one item a at a time and apply a comparison with d in terms of pairs of vectors rather than comparing simultaneously 4 preferences, as with the first pattern. From a preference $a : a^1 \succeq a^2$ such that (a^1, a^2, d^1, d^2) is in analogical proportion, one extrapolates that the *same* preference still holds for $d : d^1 \succeq d^2$. A similar process is applied in [7] that they called *analogical transfer* of preferences. A comparison of APL_1 with the algorithm they recently proposed is presented in the subsection after the next one.

Following this logic, for each item a in the training set, an analogical score $A(a^1, a^2, d^1, d^2)$ is computed. As in case of the vertical reading, these atomic scores are accumulated for each possible solution for d (induced from items a). Finally, the solution with the highest score is assigned to d.

The APL_1 can be described by this basic process:

- For a given $d : d^1, d^2$ whose preference is to be predicted.
- For each item $a \in E$, compute the analogical score $A(a^1, a^2, d^1, d^2)$.
- Compute the sum of these scores for each possible solution for d and assign to d, the solution with the highest score.

3.2   Algorithms

Based on the above ideas, we propose two different algorithms for predicting preferences. Let E be a training set of examples whose preference is known. Algorithms 1 and 2 respectively describe the two previously introduced

---

Algorithm 1. $APL3$

Input: a training set E of examples with known preferences
a new item $d \notin E$ whose preference $P(d)$ is unknown.
SumA(p)=0 for each $p \in \{\preceq, \succeq\}$
$BestA_t$=0, $S = \emptyset$, $BestSol = \emptyset$
for each triple $t = (a, b, c)$ in E^3 **do**
    $S =$ FindCandidateTriples(t)
    for each candidate triple $ct = (a', b', c')$ in S **do**
        if $(P(a') : P(b') :: P(c') : x$ has solution $p)$ **then**
            $A_t = Min(A(a'_1, b'_1, c'_1, d_1), A(a'_2, b'_2, c'_2, d_2))$
            if $(A_t > BestA_t)$ **then**
                $BestA_t = A_t$
                $BestSol = Sol(ct)$
            end if
        end if
    end for
    $SumA(BestSol)+ = BestA_t$
end for
$maxi = max\{SumA\}$
if $(maxi \neq 0)$ **then**
    if $(unique(maxi, SumA))$ **then**
        $P(d) = argmax_p\{SumA\}$
    else
        Majority vote
    end if
else
    No Prediction
end if
return $P(d)$

---

procedures APL_3 and APL_1. Note that in Algorithm 1, to evaluate the analogical score $A_t(a, b, c, d)$ for each triple t, we choose to consider *all* the possible arrangements of items a, b and c, i.e., for each item x, both $x : x^1 \preceq x^2$ and $x' : x^2 \succeq x^1$ are to be evaluated. The function $FindCandidateTriples(t)$ helps to find such candidate triples. Since we are dealing with triples in this algorithm, 2^3 candidate triples are evaluated for each triple t. The final score for t is that corresponding to the *best* score among its candidate triples. In both Algorithms 1 and 2, $P(x)$ returns the preference sign of the preference relation for x. For APL_3, we also consider another alternative in order to drastically reduce the number of triples to be investigated. This alternative follows exactly the same process described by Algorithm 1 except one difference: instead of systematically surveying E^3, we restrict the search for solvable triples $t(a, b, c)$ by constraining c to be one of the k-nearest neighbors of d w.r.t. Manhattan distance (k is a parameter to be tuned). This option allows us to decrease the complexity of APL_3 that become quadratic instead of being cubic. A similar approach [3] showed good efficiency for classifying nominal or numerical data. We denote $APL_3(NN)$ the algorithm corresponding to this alternative.

Algorithm 2. $APL1$

    Input: a training set E of examples with known preferences
    a new item $d \notin E$ whose preference $P(d)$ is unknown.
    SumA(p)=0 for each $p \in \{\preceq, \succeq\}$
    $BestA=0, BestSol = \emptyset$
    for each a in E **do**
        $BestA = max(A(a_1, a_2, d_1, d_2), A(a_2, a_1, d_1, d_2))$
        if $(A(a_1, a_2, d_1, d_2) > A(a_2, a_1, d_1, d_2))$ **then**
            $BestSol = P(a)$
        else
            $BestSol = notP(a)$
        end if
        $SumA(BestSol)+ = BestA$
    end for
    $maxi = max\{SumA\}$
    if $(maxi \neq 0)$ **then**
        if $(unique(maxi, SumA))$ **then**
            $P(d) = argmax_p\{SumA\}$
        else
            Majority vote
        end if
    else
        No Prediction
    end if
    return $P(d)$

3.3   Related Work and Complexity

A panoply of research works has been developed to deal with preference learning problems, see, e.g., [8]. The goal of most of these works is to predict a total order function that agrees with a given preference relation. See, e.g., Cohen et al. [5] that developed a greedy ordering algorithm to build a total order on the input preferences given by an expert, and also suggest an approach to linearly combine a set of preferences functions. However, we are only intended to predict preferences relations in this paper and not a total order on preferences.

Even if the proposed approaches for predicting preferences may also look similar to the recommender systems models, these latter address problems that are somewhat different from ours. Observe that, in general, prediction recommender systems is based on examples where items are associated with absolute grades (e.g., from 1 to 5); namely examples are not made of comparisons between pairs of items as in our case.

To the best of our knowledge, the only approach also aiming at predicting preferences on an analogical proportion basis is the recent paper [7], which only investigates "the horizontal reading" of preference relations, leaving aside a preliminary version of "the vertical reading" [1]. The algorithms proposed in [1] only use the Boolean setting of analogical proportions, while the approach presented here deals with the multiple-valued setting (also applied in [7]). Moreover, Algorithm 2 in [1] used a set of preference examples completed by monotony in case no triples satisfying analogical proportion could be found, while in this work, this issue was simply solved by selecting triples satisfying the analogical proportions with some degree as suggested in the presentation of the multiple-valued extension in Sect. 2. That's why we compare deeply our proposed analogical proportions algorithms to this last work [7]. APL algorithms differ from this approach in two ways.

First, the focus of [7] is on learning to rank user preferences based on the evaluation of a loss function, while our focus is on predicting preferences (rather than getting a ranking) evaluated with the error rate of predictions.

Lastly, although Algorithm 1 in [7] may be useful for predicting preferences in the same way as our APL_1, the key difference between these two algorithms is that APL_1 exploits the summation of *all* valid analogical proportions for each possible solution for d to be predicted, while Algorithm 1 in [7] computes all valid analogical proportions, and then considers *only* the N most relevant ones for prediction, i.e., those having the largest analogical scores. To select such N *best* scores, a total order on valid proportions is required for each item d to be predicted which may seem computationally costly for a large number of items, as noted in [7], while no ordering on analogical proportions is required in the proposed APL_1.

To compare our APL algorithms to Algorithm 1 in [7], suitable for predicting preferences, we also re-implemented the latter as described in their paper (without considering their Algorithm 2). We also tuned the parameter N with the same input values as fixed by the authors [7].

In terms of complexity, due to the use of triples of pairs of items in APL_3, the algorithm has a cubic complexity while Algorithm $APL_3(NN)$ is quadratic. Both APL_1 and Algorithm 1 in [7] are linear, even if Algorithm 1 is slightly computationally more costly due to the ordering process.

4   Experimentations

To evaluate the proposed APL algorithms, we have developed a set of experiments that we describe in the following.

4.1   Datasets

The experimental study is based on five datasets, the two first ones are synthetic data generated from different functions: weighted average, Tversky's additive difference and Sugeno Integral described in the following. For each dataset, any possible combination of the feature values over the scale S is associated with the preference relation.

– **Datasets 1**: we consider only 3 criteria in each preference relation i.e., $n = 3$. We generate different type of datasets:
   1. Examples in this dataset are first generated using a weighted average function (denoted WA in Table 2) with $0.6, 0.3, 0.1$ weights respectively for criteria 1, 2 and 3.
   2. The second artificial dataset (denoted TV in Table 2) is generated using a Tversky's additive difference model [15], i.e. an alternative a is preferred over b if $\sum_{i=1}^{n} \Phi_i(a_i - b_i) \geq 0$, where Φ_i are increasing and odd real-valued functions. For generating this dataset, we used the piecewise linear functions given in appendix A.
   3. Then, we generate this dataset using weighted max and weighted min which are particular cases of Sugeno integrals, namely using the aggregation functions defined as follows:

$$S_{Max} = \max_{i=1}^{n}(\min(v_i, w_i)),$$

$$S_{Min} = \min_{i=1}^{n}(\max(v_i, 6 - w_i)),$$

   where v_i refers to the value of criterion i and w_i represents its weight. In this case, we tried two different sets of weights : $w_1 = 5, 4, 2$ and $w_2 = 5, 3, 3$, respectively for criteria 1, 2 and 3.
– **Datasets 2**: we expand each preference relation to support 5 criteria, i.e: $n = 5$. We apply the weights $0.4, 0.3, 0.1, 0.1, 0.1$ in case of weighted average function and $w_1 = 5, 4, 3, 2, 1$ and $w_2 = 5, 4, 4, 2, 2$ in case of Sugeno integral functions. For generating the second dataset (TV), we used the following piecewise linear functions given in Appendix A. For the two datasets, weights are fixed on a empirical basis, although other choices have been tested and have led to similar results.

We limit ourselves to 5 criteria since it is already a rather high number of criteria for the cognitive appraisal of an item by a human user in practice. For both datasets, each criterion is evaluated on a scale with 5 levels, i.e., $S = \{1, ..., 5\}$.

To check the applicability of APL algorithms, it is important to measure their efficiency on real data. For our experiments, data should be collected as *pairs* of choices/options for which a human is supposed to pick one of them. To the best of our knowledge, there is no such a dataset that is available in this format [4]. Note that in the following datasets, a user only provides an overall rating score for each choice. Instead, we first pre-process these datasets to generate the preferences into the needed format. For any two inputs with different ratings, we generate a preference relation. We use the three following datasets:

- The **Food dataset** (https://github.com/trungngv/gpfm) contains 4036 user preferences among 20 food menus picked by 212 users. Features represent 3 levels of user hunger; the study is restricted to 5 different foods.
- The **University dataset** (www.cwur.org) includes the top 100 universities from the world for 2017 with 9 numerical features such as national rank, quality of education, etc.
- The **Movie-Lens dataset** (https://grouplens.org) includes users responses in a survey on how serendipitous a particular movie was to them. It contains 2150 user preferences among different movies picked by different users.

Table 1 gives a summary of the datasets characteristics. In order to apply the multiple-valued definition of analogy, all numerical attributes are rescaled. Each numerical feature x is replaced by $\frac{x - x_{min}}{x_{max} - x_{min}}$, where x_{min} and x_{max} respectively represent the minimal and the maximal values for this feature computed using the training set only.

4.2   Validation Protocol

In terms of protocol, we apply a standard 10 fold cross-validation technique. To tune the parameter k of $APL_3(NN)$ as well as parameter N for Algorithm 1 in [7], for each fold, we only keep the corresponding training set and we perform again a 5-fold cross-validation with diverse values of the parameters. We consider $k \in \{10, 15, ..., 30\}$ and we keep the same optimization values for N fixed by the authors for a fair comparison with [7], i.e., $N \in \{10, 15, 20\}$. We then select the parameter values providing the best accuracy. These tuned parameters are then

Table 1. Datasets description

Dataset	Features	Ordinal	Binary	Numeric	Instances
Dataset1	3	3	–	–	200
Dataset2	5	5	–	–	200
Food	4	4	–	–	200
University	9	–	–	9	200
Movie-lens	17	9	8	–	200

used to perform the initial cross-validation. We run each algorithm (with the previous procedure) 10 times. Accuracies and parameters shown in Table 2 are the average values over the 10 different values (one for each run).

4.3　Results

Tables 2 and 3 provide prediction accuracies respectively for synthetic and real datasets for the three proposed APL algorithms as well as Algorithm 1 described in [7] (denoted here "FH18"). The best accuracies for each dataset size are highlighted in bold.

If we analyze results in Tables 2 and 3 we can conclude that:

– For synthetic data and in case of datasets generated from a weighted average, it is clear that APL_3 achieves the best performances for almost all dataset sizes. APL_1 is just after. Note that these two algorithms record *all* triples/items analogical scores for prediction. We may think that it is better to use *all* the training set for prediction to be compatible with weighted average examples.
– In case of datasets generated from a Sugeno integral, $APL1$ is significantly better than other algorithms for most datasets sizes and for the two weights W_1 and W_2.
– If we compare results of the three types of datasets: the one generated from a weighted average, from Tversky's additive difference or from a Sugeno integral, globally, we can see that the accuracy obtained for a Sugeno integral dataset is the best in case of datasets with 3 criteria (see for example $APL1$). For datasets with 5 criteria, results obtained on weighted average datasets are better than on the two others. While results obtained for Tversky's additive difference datasets seem less accurate in most cases.
– For real datasets, it appears that $APL_3(NN)$ is the best predictor for most tested datasets. To predict user preferences, rather than using *all* the training set for prediction, we can select a set of training examples, those where one of them is among the k-nearest neighbors.
– $APL_3(NN)$ seems less efficient in case of synthetic datasets. This is due to the fact that synthetic data is generated randomly and applying the NN-approach is less suitable in such cases.
– If we compare APL algorithms to Algorithm1 "FH18", we can see that APL_3 outperforms the latter in case of synthetic datasets. Moreover, $APL_3(NN)$ is better than "FH18" in case of real datasets.
– APL algorithms achieve the same accuracy as Algorithm2 in [1] with a very small dataset size (for the dataset with 5 criteria built from a weighted average, only 200 examples are used by APL algorithms instead of 1000 examples in [1] to achieve the best accuracy). The two algorithms have close results for the Food dataset.
– Comparing *vertical* and *horizontal* approaches, there is no clear superiority of one view (for the tested datasets).

To better investigate the difference between the two best algorithms noted in the previous study, we also develop a pairwise comparison at the instance level

Table 2. Prediction accuracies for Dataset 1 and Dataset 2

Data	Size		APL3	APL3(NN)	k^*	APL1	FH18	N^*
D1	50	WA	92.4 ± 11.94	89.0 ± 12.72	22	$\mathbf{92.9 \pm 11.03}$	90.6 ± 12.61	14
		TV	87.8 ± 12.86	88.6 ± 12.71	19	86.8 ± 14.28	$\mathbf{90.4 \pm 12.61}$	11
		SMax-W1	90.0 ± 12.24	88.8 ± 12.01	19	$\mathbf{90.2 \pm 11.66}$	88.0 ± 13.02	15
		SMax-W2	91.2 ± 12.91	88.4 ± 13.59	20	$\mathbf{95.8 \pm 7.92}$	90.2 ± 11.62	15
		SMin-W1	$\mathbf{90.0 \pm 13.24}$	89.4 ± 11.97	17	88.6 ± 13.18	86.4 ± 13.76	16
		SMin-W2	93.2 ± 10.37	89.6 ± 12.61	20	$\mathbf{94.2 \pm 9.86}$	92.4 ± 11.48	18
	100	WA	$\mathbf{94.75 \pm 6.79}$	91.55 ± 8.51	24	93.85 ± 7.51	93.5 ± 7.09	15
		TV	89.6 ± 9.43	86.5 ± 10.41	21	88.0 ± 9.6	$\mathbf{92.1 \pm 8.45}$	15
		SMax-W1	$\mathbf{93.4 \pm 6.5}$	91.2 ± 9.31	25	93.2 ± 6.98	91.2 ± 9.14	17
		SMax-W2	93.8 ± 7.17	88.8 ± 8.08	25	$\mathbf{95.3 \pm 5.66}$	92.9 ± 8.2	15
		SMin-W1	90.4 ± 9.66	89.6 ± 8.57	19	$\mathbf{90.6 \pm 9.43}$	89.4 ± 8.73	16
		SMin-W2	94.3 ± 6.56	90.1 ± 7.68	21	$\mathbf{96.3 \pm 5.09}$	94.6 ± 5.68	14
	200	WA	95.25 ± 4.94	92.25 ± 6.28	23	$\mathbf{95.4 \pm 4.33}$	94.55 ± 5.17	13
		TV	91.25 ± 5.48	91.7 ± 5.93	25	90.1 ± 5.18	$\mathbf{95.1 \pm 4.53}$	13
		SMax-W1	90.0 ± 5.98	89.3 ± 6.73	24	$\mathbf{90.2 \pm 6.24}$	89.4 ± 6.38	14
		SMax-W2	95.7 ± 3.28	92.6 ± 5.35	26	$\mathbf{97.2 \pm 2.71}$	95.6 ± 4.08	15
		SMin-W1	$\mathbf{92.0 \pm 7.17}$	91.1 ± 5.86	24	$\mathbf{92.0 \pm 5.82}$	90.3 ± 6.38	18
		SMin-W2	94.8 ± 4.88	91.55 ± 5.15	26	$\mathbf{97.4 \pm 3.35}$	95.0 ± 4.45	16
D2	50	WA	$\mathbf{88.3 \pm 12.41}$	86.2 ± 14.43	20	84.9 ± 15.17	83.5 ± 14.99	15
		TV	$\mathbf{89.4 \pm 15.08}$	86.0 ± 16.46	17	89.2 ± 14.85	87.6 ± 14.44	17
		SMax-W1	$\mathbf{86.6 \pm 13.77}$	83.8 ± 14.14	18	86.4 ± 12.18	83.2 ± 15.53	16
		SMax-W2	85.8 ± 15.77	82.4 ± 15.44	20	$\mathbf{87.0 \pm 14.58}$	81.2 ± 15.7	14
		SMin-W1	$\mathbf{86.6 \pm 14.34}$	86.2 ± 13.45	23	85.6 ± 14.61	84.0 ± 14.57	16
		SMin-W2	88.8 ± 11.5	86.2 ± 14.29	22	$\mathbf{89.0 \pm 11.94}$	83.6 ± 14.78	17
	100	WA	$\mathbf{92.0 \pm 7.51}$	89.0 ± 8.89	22	90.0 ± 8.22	88.3 ± 9.71	15
		TV	90.2 ± 8.18	88.1 ± 8.93	18	$\mathbf{91.4 \pm 8.03}$	86.8 ± 9.06	15
		SMax-W1	88.0 ± 9.52	87.9 ± 9.56	21	$\mathbf{88.8 \pm 9.31}$	85.4 ± 10.7	16
		SMax-W2	87.7 ± 9.17	86.1 ± 9.77	24	$\mathbf{90.1 \pm 9.73}$	85.1 ± 9.86	17
		SMin-W1	89.1 ± 9.15	88.6 ± 10.47	21	$\mathbf{90.2 \pm 9.3}$	85.6 ± 9.98	16
		SMin-W2	87.4 ± 9.7	83.2 ± 11.99	24	$\mathbf{89.2 \pm 9.27}$	84.8 ± 10.06	16
	200	WA	$\mathbf{94.7 \pm 5.11}$	90.2 ± 6.01	24	92.3 ± 5.2	90.0 ± 6.08	17
		TV	91.1 ± 6.33	89.0 ± 6.54	27	$\mathbf{91.9 \pm 6.06}$	89.65 ± 6.81	16
		SMax-W1	89.15 ± 6.76	88.65 ± 6.58	22	$\mathbf{89.7 \pm 6.84}$	86.5 ± 7.66	17
		SMax-W2	89.4 ± 6.45	88.15 ± 6.77	27	$\mathbf{91.7 \pm 5.65}$	86.65 ± 6.71	16
		SMin-W1	90.7 ± 5.79	89.25 ± 6.45	24	$\mathbf{91.3 \pm 5.1}$	88.8 ± 6.55	15
		SMin-W2	89.75 ± 5.64	88.0 ± 6.33	23	$\mathbf{91.55 \pm 5.68}$	87.8 ± 6.68	16

Table 3. Prediction accuracies for real datasets

Dataset	Size	APL3	APL3(NN)	k^*	APL1	FH18	N^*
Food	200	61.3 ± 8.32	$\mathbf{63.0 \pm 9.64}$	15	61.05 ± 9.34	57.55 ± 10.41	13
	1000	–	$\mathbf{73.16 \pm 3.99}$	20	63.11 ± 5.0	63.11 ± 5.54	20
Univ.	200	73.6 ± 9.67	$\mathbf{80.0 \pm 8.03}$	14	73.6 ± 8.47	75.7 ± 8.29	12
	1000	–	$\mathbf{87.9 \pm 3.04}$	17	76.76 ± 3.86	83.74 ± 3.26	12
Movie	200	51.9 ± 14.72	49.1 ± 15.2	19	$\mathbf{52.93 \pm 13.52}$	48.61 ± 14.26	15
	1000	–	$\mathbf{55.06 \pm 4.51}$	23	54.48 ± 4.7	53.38 ± 5.32	10

between $APL3$ and $APL1$ in case of synthetic datasets and $APL3(NN)$ and $APL1$ in case of real datasets. In this comparison, we want to check if the preference d is predicted in the same way by the two algorithms. For this purpose, we computed the frequency of the case where the two algorithms predict the correct preference for d (this case is noted TT), the frequency of the case where both algorithms predict an incorrect preference (noted FF), the frequency where $APL3$ or $APL3(NN)$ prediction is correct and $APL1$ prediction is wrong (TF) and the frequency where $APL3$ or $APL3(NN)$ prediction is wrong and $APL1$ prediction is correct (FT). We only consider the datasets generated from a weighted average in case of synthetic data in this new experimentation. Results are shown in Table 4. Results in this table show that the two compared algorithms predict preferences in the same way (the highest frequency is seen in column TT) for most cases. If we compare results in column TF and FT, it is clear that $APL3$ (or $APL3(NN)$) is significantly better than $APL1$ since the frequency of cases where $APL3$ (or $APL3(NN)$) yields the correct prediction while $APL1$ doesn't (column TF) is higher than the opposite case (column FT). These results confirm our previous study.

Table 4. Frequency of correctly/incorrectly predicted preferences d predicted same/not same by the two compared algorithms

Datasets	Size	TT	TF	FT	FF
D1	100	0.96	0.02	0.01	0.01
	200	0.93	0.02	0.02	0.03
D2	100	0.84	0.07	0.05	0.04
	200	0.9	0.04	0.01	0.05
Food	1000	0.528	0.18	0.106	0.186
Univ.	1000	0.735	0.158	0.033	0.074
Movie	1000	0.364	0.189	0.17	0.277

In Tables 5 and 6, we compare the best results obtained with our algorithms to the accuracies obtained by finding the weighted sum that best fits the data in the learning sets. The weights are found by using linear programming as explained in Appendix B.

Table 5 displays the results obtained using the synthetic datasets generated according to Tverski's model. The results for datasets generated by a weighted average or a Sugeno integral are not reproduced in this table because the weighted sum (WSUM) almost always reaches an accuracy of 100%. Only in three cases on thirty, its accuracy is slightly lower, with a worst performance of 97.5%. This is not unexpected for datasets generated by means of a weighted average, since WSUM is the right model in this case. It is more surprising for data generated by a Sugeno integral (even if we have only dealt here with particular cases), but we get here some empirical evidence that the Sugeno integral can be well-approximated by a weighted sum. The results are quite different for datasets generated by Tversky's model. WSUM shows the best accuracy in two cases; APL1 and APL3, also in two cases each. Tversky's model does not lead to transitive preference relations, in general, and this may be detrimental to WSUM that models transitive relations.

Table 5. Prediction accuracies for artificial datasets generated by the Tverski model

Dataset	Size	APL3	APL1	WSUM
TV (3 features)	50	**87.8 ± 12.86**	86.8 ± 14.28	82.00 ± 17.41
	100	89.6 ± 9.43	88.0 ± 9.6	**93.00 ± 8.23**
	200	81.25 ± 5.48	90.1 ± 5.18	**91.00 ± 6.43**
TV (5 features)	50	**89.4 ± 15.08**	89.2 ± 14.85	84.00 ± 15.78
	100	90.2 ± 8.18	**91.4 ± 8.03**	87.00 ± 9.49
	200	91.1 ± 6.33	**91.9 ± 6.06**	85.50 ± 6.53

Table 6 compares the accuracies obtained with the real datasets. WSUM yields the best results for all datasets except for the "Food" dataset, size 1000.

Table 6. Prediction accuracies for real datasets

Dataset	Size	APL3(NN)	APL1	WSUM
Food	200	63.0 ± 9.64	61.05 ± 9.34	**64.00 ± 20.11**
	1000	**73.16 ± 3.99**	63.11 ± 5.0	61.10 ± 10.19
Univ.	200	80.0 ± 8.03	73.6 ± 8.47	**99.50 ± 1.58**
	1000	87.9 ± 3.04	76.76 ± 3.86	**88.70 ± 21.43**
Movie	200	49.1 ± 15.2	52.93 ± 13.52	**69.50 ± 18.77**
	1000	55.06 ± 4.51	54.48 ± 4.7	**77.60 ± 16.93**

These examples suggest that analogy-based algorithms may surpass WSUM in some cases. However, the type of datasets for which it takes place is still to be determined.

5   Conclusion

The results presented in the previous section confirm the interest of considering analogical proportions for predicting preferences, which was the primary goal of this paper since such an approach has been proposed only recently. We observed that analogical proportions yield a better accuracy as compared to a weighted sum model for certain datasets (TV, among the synthetic datasets and Food, as a real dataset). Determining for which datasets this tends to be the case requires further investigation.

Analogical proportions may be a tool of interest for creating artificial examples that are useful for enlarging a training set, see, e.g., [2]. It would be worth investigating to see if such enlarged datasets could benefit to analogy-based preference learning algorithms as well as to the ones based on weighted sum.

Acknowledgements. This work was partially supported by ANR-11-LABX-0040-CIMI (Centre Inter. de Math. et d'Informatique) within the program ANR-11-IDEX-0002-02, project ISIPA.

A   Tversky's Additive Difference Model

Tversky's additive difference model functions used in the experiments are given below. Let d^1, d^2 be a pair of alternative that have to be compared. We denote by η_i the difference between d^1 and d^2 on the criterion i, i.e. $\eta_i = d_i^1 - d_i^2$. For the TV dataset in which 3 features are involved, we used the following piecewise linear functions:

$$
\Phi_1(\eta_1) = \begin{cases}
\operatorname{sgn}(\eta_1)\, 0.453 \cdot 0.143 \cdot \eta_1 & \text{if } |\eta_1| \in [0, 0.25], \\
\operatorname{sgn}(\eta_1)\, 0.453 \cdot [-0.168 + 0.815 \cdot \eta_1] & \text{if } |\eta_1| \in [0.25, 0.5], \\
\operatorname{sgn}(\eta_1)\, 0.453 \cdot [0.230 + 0.018 \cdot \eta_1] & \text{if } |\eta_1| \in [0.5, 0.75], \\
\operatorname{sgn}(\eta_1)\, 0.453 \cdot [-2.024 + 3.024 \cdot \eta_1] & \text{if } |\eta_1| \in [0.75, 1],
\end{cases}
$$

$$
\Phi_2(\eta_2) = \begin{cases}
\operatorname{sgn}(\eta_2)\, 0.053 \cdot 2.648 \cdot \eta_2 & \text{if } |\eta_2| \in [0, 0.25], \\
\operatorname{sgn}(\eta_2)\, 0.053 \cdot [0.371 + 1.163 \cdot \eta_2] & \text{if } |\eta_2| \in [0.25, 0.5], \\
\operatorname{sgn}(\eta_2)\, 0.053 \cdot [0.926 + 0.054 \cdot \eta_2] & \text{if } |\eta_2| \in [0.5, 0.75], \\
\operatorname{sgn}(\eta_2)\, 0.053 \cdot [0.866 + 0.134 \cdot \eta_2] & \text{if } |\eta_2| \in [0.75, 1],
\end{cases}
$$

$$
\Phi_3(\eta_3) = \begin{cases}
\operatorname{sgn}(\eta_3)\, 0.494 \cdot 0.289 \cdot \eta_3 & \text{if } |\eta_3| \in [0, 0.25], \\
\operatorname{sgn}(\eta_3)\, 0.494 \cdot [-0.197 + 1.076 \cdot \eta_3] & \text{if } |\eta_3| \in [0.25, 0.5], \\
\operatorname{sgn}(\eta_3)\, 0.494 \cdot [0.150 + 0.383 \cdot \eta_3] & \text{if } |\eta_3| \in [0.5, 0.75], \\
\operatorname{sgn}(\eta_3)\, 0.494 \cdot [-1.252 + 2.252 \cdot \eta_3] & \text{if } |\eta_3| \in [0.75, 1].
\end{cases}
$$

For the TV dataset in which 5 features are involved, we used the following piecewise linear functions:

$$\Phi_1(\eta_1) = \begin{cases} \text{sgn}(\eta_1) \cdot 0.294 \cdot 2.510 \cdot \eta_1 & \text{if } |\eta_1| \in [0, 0.25], \\ \text{sgn}(\eta_1) \cdot 0.294 \cdot [0.562 + 0.263 \cdot \eta_1] & \text{if } |\eta_1| \in [0.25, 0.5], \\ \text{sgn}(\eta_1) \cdot 0.294 \cdot [0.645 + 0.096 \cdot \eta_1] & \text{if } |\eta_1| \in [0.5, 0.75], \\ \text{sgn}(\eta_1) \cdot 0.294 \cdot [-0.130 + 1.130 \cdot \eta_1] & \text{if } |\eta_1| \in [0.75, 1], \end{cases}$$

$$\Phi_2(\eta_2) = \begin{cases} \text{sgn}(\eta_2) \cdot 0.151 \cdot 0.125 \cdot \eta_2 & \text{if } |\eta_2| \in [0, 0.25], \\ \text{sgn}(\eta_2) \cdot 0.151 \cdot [0.025 + 0.023 \cdot \eta_2] & \text{if } |\eta_2| \in [0.25, 0.5], \\ \text{sgn}(\eta_2) \cdot 0.151 \cdot [-0.545 + 1.164 \cdot \eta_2] & \text{if } |\eta_2| \in [0.5, 0.75], \\ \text{sgn}(\eta_2) \cdot 0.151 \cdot [-1.689 + 2.689 \cdot \eta_2] & \text{if } |\eta_2| \in [0.75, 1], \end{cases}$$

$$\Phi_3(\eta_3) = \begin{cases} \text{sgn}(\eta_3) \cdot 0.039 \cdot 2.388 \cdot \eta_3 & \text{if } |\eta_3| \in [0, 0.25], \\ \text{sgn}(\eta_3) \cdot 0.039 \cdot [0.582 + 0.057 \cdot \eta_3] & \text{if } |\eta_3| \in [0.25, 0.5], \\ \text{sgn}(\eta_3) \cdot 0.039 \cdot [-0.046 + 1.314 \cdot \eta_3] & \text{if } |\eta_3| \in [0.5, 0.75], \\ \text{sgn}(\eta_3) \cdot 0.039 \cdot [0.759 + 0.241 \cdot \eta_3] & \text{if } |\eta_3| \in [0.75, 1], \end{cases}$$

$$\Phi_1(\eta_4) = \begin{cases} \text{sgn}(\eta_4) \cdot 0.425 \cdot 0.014 \cdot \eta_4 & \text{if } |\eta_4| \in [0, 0.25], \\ \text{sgn}(\eta_4) \cdot 0.425 \cdot [-0.110 + 0.455 \cdot \eta_4] & \text{if } |\eta_4| \in [0.25, 0.5], \\ \text{sgn}(\eta_4) \cdot 0.425 \cdot [-0.341 + 0.917 \cdot \eta_4] & \text{if } |\eta_4| \in [0.5, 0.75], \\ \text{sgn}(\eta_4) \cdot 0.425 \cdot [-1.613 + 2.613 \cdot \eta_4] & \text{if } |\eta_4| \in [0.75, 1]. \end{cases}$$

$$\Phi_1(\eta_5) = \begin{cases} \text{sgn}(\eta_5) \cdot 0.091 \cdot 3.307 \cdot \eta_5 & \text{if } |\eta_5| \in [0, 0.25], \\ \text{sgn}(\eta_5) \cdot 0.091 \cdot [0.697 + 0.519 \cdot \eta_5] & \text{if } |\eta_5| \in [0.25, 0.5], \\ \text{sgn}(\eta_5) \cdot 0.091 \cdot [0.880 + 0.153 \cdot \eta_5] & \text{if } |\eta_5| \in [0.5, 0.75], \\ \text{sgn}(\eta_5) \cdot 0.091 \cdot [0.979 + 0.021 \cdot \eta_5] & \text{if } |\eta_5| \in [0.75, 1]. \end{cases}$$

B   Linear Program Used for Computing a Weighted Sum

We compared the performances of the algorithms presented in this paper to the results obtained with a linear program inferring the parameters of a weighted sum that fits as well as possible with the learning set. The linear program is given below:

$$\min \sum_{a \in E} \delta_a$$
$$\sum_{i=1}^{n} w_i \cdot (a_i^1 - a_i^2) + \delta_a \geq 0 \qquad \forall a \in E : a^1 \succ a^2$$
$$\sum_{i=1}^{n} w_i \cdot (a_i^1 - a_i^2) - \delta_a \leq \epsilon \qquad \forall a \in E : a^1 \prec a^2$$
$$w_i \in [0, 1] \qquad i = 1, ..., n$$
$$\delta_a \in [0, \infty[$$

with:

- n: number of features,
- E: learning set composed of pairs (a_1, a_2) evaluated on n features and a preference relation for each pair $(a_1 \succ a_2$ or $a_1 \prec a_2)$,
- w_i: weight associated to feature i,
- ϵ: a small positive value.

References

1. Bounhas, M., Pirlot, M., Prade, H.: Predicting preferences by means of analogical proportions. In: Cox, M.T., Funk, P., Begum, S. (eds.) ICCBR 2018. LNCS (LNAI), vol. 11156, pp. 515–531. Springer, Cham (2018). https://doi.org/10.1007/978-3-030-01081-2_34

2. Bounhas, M., Prade, H.: An analogical interpolation method for enlarging a training dataset. In: BenAmor, N., Theobald, M. (eds.) Proceedings of 13th International Conference on Scalable Uncertainty Management (SUM 2019), Compiègne, 16–18 December. LNCS, Springer (2019)

3. Bounhas, M., Prade, H., Richard, G.: Analogy-based classifiers for nominal or numerical data. Int. J. Approx. Reason. **91**, 36–55 (2017)

4. Chen, S., Joachims, T.: Predicting matchups and preferences in context. In: Proceedings of the 22nd ACM SIGKDD International Conference on Knowledge. Discovery and Data Mining (KDD 2016), pp. 775–784. ACM (2016)

5. Cohen, W.W., Schapire, R.E., Singer, Y.: Learning to order things. CoRR abs/1105.5464 (2011). http://arxiv.org/abs/1105.5464

6. Dubois, D., Prade, H., Richard, G.: Multiple-valued extensions of analogical proportions. Fuzzy Sets Syst. **292**, 193–202 (2016)

7. Fahandar, M.A., Hüllermeier, E.: Learning to rank based on analogical reasoning. In: Proceedings of 32nd National Conference on Artificial Intelligence (AAAI 2018), New Orleans, 2–7 February 2018 (2018)

8. Fürnkranz, J., Hüllermeier, E. (eds.): Preference Learning. Springer, Heidelberg (2010). https://doi.org/10.1007/978-3-642-14125-6

9. Hüllermeier, E., Fürnkranz, J.: Editorial: preference learning and ranking. Mach. Learn. **93**(2–3), 185–189 (2013)

10. Miclet, L., Bayoudh, S., Delhay, A.: Analogical dissimilarity: definition, algorithms and two experiments in machine learning. JAIR **32**, 793–824 (2008)

11. Pirlot, M., Prade, H., Richard, G.: Completing preferences by means of analogical proportions. In: Torra, V., Narukawa, Y., Navarro-Arribas, G., Yañez, C. (eds.) MDAI 2016. LNCS (LNAI), vol. 9880, pp. 135–147. Springer, Cham (2016). https://doi.org/10.1007/978-3-319-45656-0_12

12. Prade, H., Richard, G.: Homogeneous logical proportions: their uniqueness and their role in similarity-based prediction. In: Brewka, G., Eiter, T., McIlraith, S.A. (eds.) Proceedings of 13th International Conference on Principles of Knowledge Representation and Reasoning (KR 2012), Roma, 10–14 June, pp. 402–412. AAAI Press (2012)

13. Prade, H., Richard, G.: From analogical proportion to logical proportions. Log. Univers. **7**(4), 441–505 (2013)

14. Prade, H., Richard, G.: Analogical proportions: from equality to inequality. Int. J. Approx. Reason. **101**, 234–254 (2018)

15. Tversky, A.: Intransitivity of preferences. Psychol. Rev. **76**, 31–48 (1969)

Using Convolutional Neural Network in Cross-Domain Argumentation Mining Framework

Rihab Bouslama[1(✉)], Raouia Ayachi[2], and Nahla Ben Amor[1]

[1] LARODEC, ISG, Université de Tunis, Tunis, Tunisia
rihabbouslama@yahoo.fr, nahla.benamor@gmx.fr
[2] ESSECT, LARODEC, Université de Tunis, Tunis, Tunisia
raouia.ayachi@gmail.com

Abstract. Argument Mining has become a remarkable research area in computational argumentation and Natural Language Processing fields. Despite its importance, most of the current proposals are restricted to a text type (e.g., Essays, web comments) on a specific domain and fall behind expectations when applied to cross-domain data. This paper presents a new framework for Argumentation Mining to detect argumentative segments and their components automatically using Convolutional Neural Network (CNN). We focus on both (1) argumentative sentence detection and (2) argument components detection tasks. Based on different corpora, we investigate the performance of CNN on both in-domain level and cross-domain level. The investigation shows challenging results in comparison with classic machine learning models.

Keywords: Argumentation Mining · Arguments · Convolutional Neural Network · Deep learning

1 Introduction

An important sub-field of computational argumentation is Argumentation Mining (AM) which aims to detect argumentative sentences and argument components from different sources (e.g., online debates, social medias, persuasive essays, forums). In the last decade, AM field has gained the interest of many researchers due to its important impact in several domains [1,4,19]. AM process can be divided into sub-tasks taking the form of a pipeline as proposed in [3]. They presented three sub-tasks namely, *argumentative sentence detection, argument component boundary detection* and *argument structure prediction*. Argumentative sentence detection task is viewed as a classification problem where argumentative sentences are classified into two classes (i.e., argumentative, not argumentative). Argument component boundary detection is treated as a segmentation problem and may be presented either as a multi-class classification issue (i.e., classify each component) or as a binary classification issue (i.e., one classifier for each component) solved using machine learning classifiers.

© Springer Nature Switzerland AG 2019
N. Ben Amor et al. (Eds.): SUM 2019, LNAI 11940, pp. 355–367, 2019.
https://doi.org/10.1007/978-3-030-35514-2_26

Interesting applications in the AM area were proposed. An argument search engine was proposed in [19]. A web server for argumentation mining called MARGOT was proposed in [4] which is an online platform for non-expert of the argumentation domain. Considering the first two sub-tasks of AM (i.e., argument detection and argument component detection), an important study presented in [5] investigates different classifiers and different features in order to determine the best classifier and the most important features to take into consideration. Indeed, Classifiers' performances depend on the model, the data and the features. One of the most crucial steps, yet the most critical is the choice of the proper features to use. Recently, deep learning techniques overcome this restraint and present good results in text classification [8,9] which makes their exploration in AM field interesting [6,12,14,24]. In addition, features make the generalization of models over many corpora harder. Nonetheless, most of the existing argument mining approaches are limited to one specific domain (e.g., student essays [1], online debates [2]). Thus, generalizing AM approaches over heterogeneous corpora is still poorly explored.

On the other hand, Convolutional Neural Networks showed a considerable success in computer vision [7], speech recognition [17] and computational linguistics [8,9,11] where they have been proven to be competitive to traditional models without any need of knowledge on syntactic or semantic structure of a given language [8,10]. Considering the success of the CNN in text classification and more precisely its success in sentiment analysis field [8], its use in AM seems to give important results. Aker *et al.*, [5] tested Kim's model [8] to both *argumentative sentence detection* and *argument component boundary detection* tasks on two corpora namely, persuasive essays corpora and Wikipedia corpora. Without any changes in the model, the results were significantly important.

Deep learning algorithms interested many researchers in AM field. In [21], joint RNN was used to detect argument component boundary detection. In the latter, the argument component detection task was considered as a sequence labelling task while in this work we treat claim detection and premise detection separately. In [24], the authors used CNN and LSTM (Long short-Term Memory) to classify claims in online users comments. CNN was also used for bi-sequence classification [12]. Recently, Hua *et al.*, [6] proposed the use of CNN in arguments detection in peers review. Moreover, in [14] the authors proposed the use of deep learning networks, more specifically two variants of Kim's CNN and LSTM to identify claims in a cross-domain manner. Kim's model [8] was tested in many occasions in AM field. However, to the best of our knowledge, only one work studied the use of Zhang's character-based CNN model in AM [15]. This work presents models to classify argument components in classrooms discussions in order to automatically classify students' utterances into claims, evidences and warrants. Their results showed that convolutional networks (whether character or word level) are more robust than recurrent networks.

Moreover, most of the current proposed approaches for argument mining are designed to be used for specific text types and fall short when applied to heterogeneous texts. Only few proposals treated the cross-domain case such as the work of Ajjour *et al.* [22] where the authors studied the major parameters of unit segmentation systematically by exploring many features on a word-level setting on both in-domain level and cross-domain level. Recently, in [23] a new sentential annotation scheme that is reliably applicable by crowd workers to arbitrary Web texts was proposed.

Following recent advances in both AM and Deep Learning fields, we propose a cross-domain AM framework based on Convolutional Neural Networks, so-called *ArguWeb*, able to provide up to date arguments from the web in a cross-domain manner.

The rest of this paper is divided as follows: Sect. 2 presents the basic concepts of CNN in text classification. Section 3 describes *ArguWeb*: a new framework for argument mining from the web. Finally, Sect. 4 presents the conducted experimental study, discusses the results and presents an illustrative example on the use of *ArguWeb* framework.

2   Basics on Convolutional Neural Network

Convolutional neural networks (CNN) were originally developed for computer vision. CNNs utilize layers with convolving filters [8]. Later on, CNNs were adapted to Natural Language Processing (NLP) domain and showed remarkable results [8,9,11]. These algorithms have the advantage of no needing to syntactic and semantic knowledge. Kim *et al.,* proposed a new version of CNN which is word-based and uses *word2vec* technique.[1] Another revolution in text classification field is the work of Zhang et al., [9] where the authors investigate the use of character-level convolutional networks instead of word-based convolutional networks (e.g., ngrams) which presented interesting results. Zhang *et al.,* [9] succeeded to adapt the CNN from dealing with signals either in image or speech recognition to deal with characters and treat them as a kind of raw signals. Besides the advantage of no needing to syntactic and semantic knowledge, character-level CNN presents other advantages:

- Good managing of misspelled and out-of-vocabulary words as they can be easily learned.
- Ability to handle noisy data (especially texts extracted from forums and social media).
- No need to the text pre-processing phase (e.g., tokenization, stemming).
- No need to define features.

Figure 1 shows an overview of convolutional neural networks architecture. Although both word-based and char-level CNNs have similar global architectures, they differ in some details such as the input text representation and the number of convolutional and fully connected layers. Indeed, Character-level CNNs are based on temporal convolutional model that computes 1-D convolution and temporal max pooling which is a 1-D version of max pooling in computer vision [9]. It is based on a threshold function which is similar to Rectified Linear

---

[1] https://code.google.com/p/word2vec/.

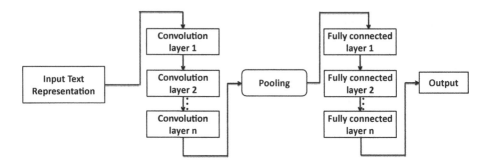

Fig. 1. An overview on the Convolutional Neural Network.

Units (ReLu) using stochastic gradient descent (SGD) with a minibatch of size 128. The input of the model is a sequence of encoded characters using one-hot encoding. These encoded characters present n vectors with fixed length l_0. The model proposed in [9] consists of 9 layers: 6 convolutional layers and 3 fully-connected layers. Two versions were presented: (i) a small version with an input length of 256 and (ii) a large version where the input length is 1024.

As in Kim's model [8], word-based CNNs, compute multi-dimentional convolution (n*k matrix) and max over time pooling. For the representation of a sentence, two input channels are proposed (i.e., static and non-static). Kim's model is composed of a layer that performs convolutions over the embedded word vectors predefined in Word2Vec, then max-pooling is applied to the result of the convolutional layer and similarly to char-level models, Rectified Linear Units (ReLu) is applied.

3   A New Framework for Argument Component Detection

In this section, we present *ArguWeb* which is a new framework that ensures: (1) data gathering from different online forums and (2) argumentative text and components detection.

In *ArguWeb*, arguments are extracted directly from the web. The framework is mainly composed of two phases, *pre-processing phase* where arguments are extracted from the web then segmented and *argument component detection phase* where arguments' components are detected (Fig. 2). The second phase is ensured by using trained character-level Convolutional Neural Network and word-based Convolutional Neural Network. An experimental study on the performance of character-level and word-based CNNs for Argumentation Mining was conducted for both sub-tasks *argumentative sentence detection* and *argument component detection*. In this work, we consider both situations: in-domain and cross-domain. We also, compare character-level CNNs performances to word-based CNNs, SVM and Naïve Bayes.

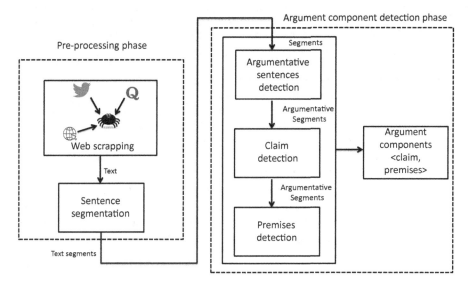

Fig. 2. An overview of *ArguWeb* architecture.

3.1 Pre-processing Phase

Using web scraping techniques, we extract users comments from several sources (e.g., social media, online forums). This step is ensured by a set of scrappers developed using Python. Then, extracted data enters the core of our system which is a set of trained models. More details about the argument component detection phase are presented in the next sub-section.

Existing AM contributions treat different input's granularity (i.e., paragraph, sentence, intra-sentence) and most of the existing work focuses on sentences and intra-sentences level [3]. In this work we are focusing on sentence level and we suppose that the whole sentence coincides with an argument component (i.e., claim or premise). Therefore, after scrapping users' comments from the web a text segmentation is required. Collected comments will be segmented into sentences based on the sentence tokenization using Natural Language Toolkit [20].

3.2 Argument Component Detection Phase

Argument component detection phase consists in: (1) detecting argumentative texts from non argumentative texts generated in the pre-processing phase and (2) detecting arguments components (i.e., claims and premises) from argumentative segments. We follow the premise/claim model where:

Definition 1. *An argument is a tuple:*

$$Argument = <Premises, claims>$$

Where the claims are the defended ideas and the premises present explanations, proofs, facts, etc. that backup the claims.

Figure 2 depicts the followed steps in the argument component detection process which consists of detecting argumentative sentences from non-argumentative sentences, detecting claims in argumentative sentences and then detecting premises presented to backup the claim. For the three tasks, word-based and character-based CNNs are trained on three different corpora.

Using char-level CNN, an alphabet containing all the letters, numbers and special characters is used to encode the input text. Each input character is quantified using *one-hot encoding*. A Stochastic Gradient Descent is used as an optimizer with mini-batches of size 32. As for word-based CNN, we do not use two input channels as proposed in the original paper and instead we use only one channel. We also ensure word embedding using an additional first layer that embeds words into low-dimensional vectors. We use the ADAM algorithm as an optimizer.

4 Experimental Study

In this section we detail the conducted experiments and we get deeper in *ArguWeb* components. We also describe the corpora used to train the framework models and we discuss the main results.

4.1 Experimental Protocol

We aim to evaluate the performance of *ArguWeb* in terms of arguments detection and argument component detection. For this end, we experiment two classic machine learning classifiers namely, SVM and Naïve Bayes as well as two deep learning classifiers namely, char-level CNN and word-level CNN.

Data. We perform our investigation on three different corpora:

– *Persuasive Essays* corpora [13]: which consists of over 400 persuasive essays written by students. All essays have been segmented by three expert annotators into three types of argument units (i.e., major claim, claim and premises). In this paper we follow a claim/premises argumentation model, so to ensure comparability between data sets, each major claim is considered as a claim. This corpora presents a domain specific data (i.e., argumentative essays). The first task that we evoke in this paper is argumentative sentences/texts detection. For this matter, we added non-argumentative data to argumentative essays corpora. Descriptive and narrative text are extracted from Academic Help[2] and descriptive short stories from The short story website[3].

---

[2] https://academichelp.net/.
[3] https://theshortstory.co.uk.

- *Web Discourse* corpora [16]: contains 990 comments and forum posts labeled as *persuasive* or *non-persuasive* and 340 documents annotated with the extended Toulming model to *claim, grounds, backing,* and *rebuttal and refutation*. For this data set we consider *grounds* and *backing* as premises since they backup the claim while *rebuttal and refutation* are ignored. This corpora presents domain-free data.
- *N-domain* corpora: we construct a third corpora by combining both persuasive essays and web-discourse corpora to make the heterogeneity of data even more intense with the goal to investigate the performance of the different models in a multi domain context.

Data description in term of classes distributions is depicted in Table 1, where the possible classes are: *Argumentative* (Arg), *Not-argumentative* (Not-arg), *Claim, Not-claim, Premise, Not-premise*. Each row depicts the number of instances in each class for each corpora.

Table 1. Classes distribution in each corpora

Corpora	Arg	Not-arg	Claim	Not-claim	Premise	Not-premise
Essays	402	228	2250	748	3706	2019
Web-discourse	526	461	275	526	575	543
N-domain	928	689	2525	1274	4281	2562

In order to train SVM and Naïve Bayes, a pre-processing data phase and a set of features (e.g., semantic, syntactic) are required. For this purpose, we apply word tokenization to tokenize all corpora into words and we also apply both word lemmatization and stemming. Thus, words such as "studies" will give us a stem "studi" and a lemma "study". This gives an idea on the meaning and the role of a given word. Indeed, before lemmatization and stemming, we use POS-tag (Part-Of-Speech tag) technique to indicate the grammatical category of each word in a sentence (i.e., noun, verb, adjective, adverb). We also consider the well known TF-IDF technique [25] that outperforms the bag-of-word techniques and stands for Term-Frequency of a given word in a sentence and for the Inverse-Document-Frequency (IDF) that measures a word's rarity in the vocabulary of each corpora.

As for word-based CNN, we only pad each sentence to the max sentence length in order to batch data in an efficient way and build a vocabulary index used to encode each sentence as a vector of integers. For character-level CNN, we only remove URL and hash tags from the original data.

Experiment Process Description. For each corpora (i.e., Essays, Web-Discourse, n-domain) three models of each classifier are constructed, one to detect argumentative segments, one to detect claims and the other one to detect premises. As for the cross-domain case, six models of each classifier are trained in each sub-task (i.e., arguments detection, claim detection and premise detection) where each model is trained on one corpora and tested on another one. This will guarantee the cross-domain context.

To train char-level CNN models, We start with a learning rate equal to 0.01 and halved each 3 epochs. For all corpora, the epoch size is fixed to 10 yet, the training process may be stopped if for 3 consecutive epochs the validation loss did not improve. The loss is calculated using *cross-entropy loss* function. Two versions of char-level CNN exist, a small version (i.e., the inputs length is 256) that is used for in-domain training and the large version (i.e., the inputs length is 1024) is used for cross-domain model training. As for word-based CNN models, we use the same loss function (i.e., cross-entropy loss) and we optimize the loss using the ADAM optimizer. We use a 128 dimension of characters for the embedding phase, filter sizes equal to 3,4,5, a mini-batch size equal to 128. In addition, a dropout regularization (L2) is applied to avoid overfitting set equal to 5 and a dropout rate equal to 0.5. The classification of the result is ensured using a softmax layer. In this paper, we do not use Word2Vec pre-trained word vector, instead we ensure word embedding from scratch.

In order to train char-based CNN, SVM and Naïve Bayes models, we split each data-set to 80% for training and 20% for validation while to train word-based CNN models we split data to 80% and 10% following the original paper of word-based CNN [8].

Implementation Environment. The whole *ArguWeb* framework is developed using Python. The web scrappers are developed to extract users' comments from forums websites such as Quora[4] using *BeautifulSoup* and *Requests* libraries on python. The extracted comments are segmented using Natural Language Toolkit [20] on python. In order to develop and train the SVM and Naïve Bayes models, we use the NLTK, Sklearn and collections predefined packages. Moreover, both char-level and word-based CNN are implemented using TensorFlow library.

Evaluation Metrics. To evaluate the models performances we use the most used and recommended metric in the state of the art (i.e., Macro F1-score) for both argument sentences detection and argument component detection since it treats all classes equally (e.g., argumentative and non-argumentative) and it is the most suitable in cases of imbalanced class distribution.

The macro F1-score is the harmonic mean of the macro average precision and the macro average recall:

$$F1 - Score = 2 * (Recall * Precision)/(Recall + Precision) \qquad (1)$$

where for each class i

$$Recall = \sum_{i=1}^{n}(TruePositive_i/TruePositive_i + FalseNegative_i)/n \qquad (2)$$

---

[4] https://www.quora.com/.

and

$$Precision = \sum_{i=1}^{n}(TruePositive_i/TruePositive_i + FalsePositive_i)/n \quad (3)$$

with n is the number of classes.

Precision refers to the percentage of the results which are relevant and *recall* refers to the percentage of total relevant results correctly classified which makes F1-score an efficient metric for models evaluation even in cases of uneven class distribution.

4.2 Experimental Results

We evaluate character-level CNN, word-based CNN, SVM and Naïve Bayes in two situations: in-domain and cross-domain based on macro F1-score. Table 2, depicts the macro F1-scores found using these models trained and tested on the same corpora if we consider in-domain setting or trained on one corpora and tested on another for cross-domain case. In Table 2, AD, CD, PD stands for Argument Detection, Claim Detection and Premise Detection, respectively. The in-domain results are presented in columns with a gray background. Other columns present the cross-domain macro F1-scores where models are trained on one of the training sets presented in the second header row and tested on corpora presented on the first header row. Then, each row presents the results of one of the proposed models in this paper. The highest value in each column is marked in bold.

In the first task (i.e., argument detection), word-based CNN outperforms all other models in the essays corpora while SVM and char-level CNN present close results in the web-discourse corpora with macro-F1 scores equal to 0.45 and 0.52, respectively. In n-domain corpora, SVM outperforms the other models. In case of cross-domain setting, word-based and char-level CNN present better results than SVM and Naïve Bayes except where models were tested on n-domain corpora. This may be explained by the fact that in this case, testing data is close to training data since the test set presents instances from both Essays corpora and Web-discourse corpora as explained before. Char-level and word-based CNNs present better results than SVM and Naïve Bayes in the second task (i.e., claim detection). As for the last task (i.e., premise detection), char-level CNN outperforms the rest of the models remarkably.

Web-discourse corpora contains domain-free data extracted from the web. Thus, this data contains many misspelled words, internet acronyms, out-of-vocabulary words, etc. This explains the fact that char-level CNN outperforms the rest of the models presented in this work in many cases and for both in-domain and cross-domain situations. This shows the importance of character level CNN and how it performs interesting results even if the model is trained on a noisy data.

Table 2. The in-domain and cross-domain macro F1-scores. Each row represents the results of one of the models (character-level CNN, word-level CNN, SVM and Naïve Bayes), the highest value is marked in bold.

Task	Models	Test on Essays			Test on Web-discourse			Test on n-domain		
		Essays	Web-discourse	n-domain	Essays	Web-discourse	n-domain	Essays	Web-discourse	n-domain
AD	Char-level CNN	0.72	**0.93**	0.56	**0.56**	0.52	0.81	0.51	0.46	0.62
	Word-based CNN	**0.74**	0.35	0.33	0.33	0.39	**0.83**	0.33	0.36	0.44
	SVM	0.66	0.30	**0.88**	0.37	**0.54**	0.61	**0.68**	0.29	**0.83**
	Naïve Bayes	0.11	0.63	0.57	0.25	0.34	0.37	0.36	**0.63**	0.75
CD	Char-level CNN	0.83	**0.59**	0.57	0.28	**0.54**	**0.88**	0.45	0.49	0.82
	Word-based CNN	**0.98**	0.34	**0.69**	**0.59**	0.45	0.51	**0.61**	0.34	0.44
	SVM	0.61	0.55	0.37	0.48	0.43	0.68	0.47	**0.55**	**0.89**
	Naïve Bayes	0.50	0.10	0.12	0.26	0.49	0.30	0.24	0.11	0.59
PD	Char-level CNN	**0.80**	**0.98**	**0.80**	**0.51**	**0.78**	0.75	**0.59**	**0.91**	**0.82**
	Word-based CNN	0.43	0.35	0.44	0.37	0.60	0.33	0.33	0.85	0.78
	SVM	0.44	0.06	0.12	0.43	0.71	**0.89**	0.35	0.84	0.69
	Naïve Bayes	0.39	0.10	0.16	0.34	0.57	0.87	0.35	0.80	0.62

ArguWeb presented coherent results in the argument sentence detection and the argument component detection tasks comparing to state of the art results [1,5,18]. The conducted experiments showed how character-level CNN outperforms word-based CNN, SVM and Naïve Bayes for noisy and web-extracted data. Both character-level and word-based CNNs presented interesting results compared to SVM and Naïve Bayes without any need of features selection.

4.3   Illustrative Example

In what follows we better explain the role of *ArguWeb*. In this illustrative example, we focus on character-level CNN since it showed interesting results when dealing with noisy, misspelled and out-of-vocabulary data and since we are dealing with data extracted from the web.

As mentioned before, the framework contains web scrappers responsible of extracting data from different websites. For instance, we focus on online forums such as Quora and Reddit and we extract users comments on different subjects.

Each extracted text from the web is classified based on the nine trained character-level CNNs (i.e., one character-level CNN is trained on each corpora: persuasive essays, web-discourse and n-domain and six others are trained on one corpora and tested on another). A text is considered as argumentative if at least six CNNs models classified it as argumentative. Similarly a segment is

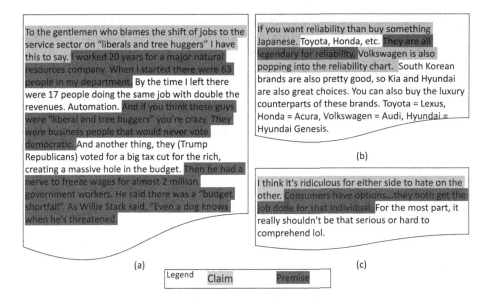

To the gentlemen who blames the shift of jobs to the service sector on "liberals and tree huggers" I have this to say. I worked 20 years for a major natural resources company. When I started there were 63 people in my department. By the time I left there were 17 people doing the same job with double the revenues. Automation. And if you think these guys were "liberal and tree huggers" you're crazy. They were business people that would never vote democratic. And another thing, they (Trump Republicans) voted for a big tax cut for the rich, creating a massive hole in the budget. Then he had a nerve to freeze wages for almost 2 million government workers. He said there was a "budget shortfall". As Willie Stark said, "Even a dog knows when he's threatened.

(a)

If you want reliability than buy something Japanese. Toyota, Honda, etc. They are all legendary for reliability. Volkswagen is also popping into the reliability chart. South Korean brands are also pretty good, so Kia and Hyundai are also great choices. You can also buy the luxury counterparts of these brands. Toyota = Lexus, Honda = Acura, Volkswagen = Audi, Hyundai = Hyundai Genesis.

(b)

I think it's ridiculous for either side to hate on the other. Consumers have options...they both get the job done for that individual. For the most part, it really shouldn't be that serious or hard to comprehend lol.

(c)

Legend   Claim   Premise

Fig. 3. Excerpt of arguments scrapped from online forum

classified as a claim (or premise) if at least six CNNs models labeled it as a claim (resp. premise).

Figure 3a and b contain examples of comments that were extracted from Quora. Figure 3c depicts a comment extracted from the Reddit forum platform. These comments were detected as arguments. Figure 3a, contains a comment extracted from an argument between many Quora's users, Fig. 3b contains a comment of a user convincing another one about Japenese cars and Fig. 3c contains a comment of a user arguing why iPhone and Samsung users hate on each other. Once arguments are detected, *ArguWeb* classify each comment's component to claim, premises or neither of them. Indeed, the Fig. 3 details the detected components (i.e., claims and premises) of these arguments. Uncoloured texts segments were not classified as claims neither as premises.

Comments like *"As announced by YouTube Music! Congrats, Taylor!!!"* were classified from the beginning as not-argumentative and were not processed by models responsible to detect the different components.

5   Conclusion

This paper proposes *ArguWeb* a cross-domain framework for arguments detection in the web. The framework is based on a set of web scrappers that extract users comments from the web (e.g., social media, online forums). Extracted data is classified as: (1) argumentative or not and (2) claims, premises or neither of them using character-level Convolutional Neural Networks and word-based Convolutional Neural Networks. An experimental study is conducted where both character-level and word-based CNN were compared to classic machine learning classifiers (i.e., SVM and Naïve Bayes). The study showed interesting results where both versions of CNN performed interesting and challenging results to classic machine learning techniques in both tasks. The framework is proposed to be used to extract arguments from different platforms on the web following the claim/premise model.

Future work will integrate *ArguWeb* framework in an automated Argumentation-Based Negotiation system. The integration of up to date arguments in such systems seems interesting. We will also handle arguments components detection in intra-sentence level rather than only in sentence level. Moreover, a semantic analysis of these components will be integrated in order to classify them to explanations, counter examples etc.

References

1. Stab, C., Gurevych, I.: Identifying argumentative discourse structures in persuasive essays. In: Proceedings of the 2014 Conference on Empirical Methods in Natural Language Processing (EMNLP), pp. 46–56 (2014)
2. Cabrio, E., Villata, S.: Combining textual entailment and argumentation theory for supporting online debates interactions. In: Proceedings of the 50th Annual Meeting of the Association for Computational Linguistics (vol. 2: Short Papers), pp. 208–212 (2012)
3. Lippi, M., Torroni, P.: Argumentation mining: state of the art and emerging trends. ACM Trans. Internet Technol. (TOIT) **16**(2), 10 (2016)
4. Lippi, M., Torroni, P.: MARGOT: a web server for argumentation mining. Expert Syst. Appl. **65**, 292–303 (2016)
5. Aker, A., et al.: What works and what does not: classifier and feature analysis for argument mining. In: Proceedings of the 4th Workshop on Argument Mining, pp. 91–96 (2017)
6. Hua, X., Nikolov, M., Badugu, N., Wang, L.: Argument mining for understanding peer reviews. arXiv preprint. arXiv:1903.10104 (2019)
7. Krizhevsky, A., Sutskever, I., Hinton, G.E.: Imagenet classification with deep convolutional neural networks. In: Proceedings of Advances in Neural Information Processing Systems, pp. 1097–1105 (2012)
8. Kim, Y.: Convolutional neural networks for sentence classification. arXiv preprint. arXiv:1408.5882 (2014)
9. Zhang, X., Zhao, J., LeCun, Y.: Character-level convolutional networks for text classification. In: Proceedings of Advances in Neural Information Processing Systems, pp. 649–657 (2015)
10. Dos Santos, C., Gatti, M.: Deep convolutional neural networks for sentiment analysis of short texts. In: Proceedings of COLING 2014, the 25th International Conference on Computational Linguistics: Technical Papers, pp. 69–78 (2014)
11. Kim, Y., Jernite, Y., Sontag, D., Rush, A.M.: Character-aware neural language models. In: Proceedings of 30th AAAI Conference on Artificial Intelligence (2016)
12. Laha, A., Raykar, V.: An empirical evaluation of various deep learning architectures for bi-sequence classification tasks. arXiv preprint. arXiv:1607.04853 (2016)
13. Stab, C., Gurevych, I.: Parsing argumentation structures in persuasive essays. Comput. Linguist. **43**(3), 619–659 (2017)

14. Daxenberger, J., Eger, S., Habernal, I., Stab, C., Gurevych, I.: What is the essence of a claim? Cross-domain claim identification. arXiv preprint. arXiv:1704.07203 (2017)
15. Lugini, L., Litman, D.: Argument component classification for classroom discussions. In: Proceedings of the 5th Workshop on Argument Mining, pp. 57–67 (2018)
16. Habernal, I., Eckle-Kohler, J., Gurevych, I.: Argumentation mining on the web from information seeking perspective. In: Proceedings of ArgNLP (2014)
17. Abdel-Hamid, O., Mohamed, A.R., Jiang, H., Penn, G.: Applying convolutional neural networks concepts to hybrid NN-HMM model for speech recognition. In: Proceedings of the 2012 IEEE International Conference on Acoustics, Speech and Signal Processing (ICASSP), pp. 4277–4280 (2012)
18. Al-Khatib, K., Wachsmuth, H., Hagen, M., Köhler, J., Stein, B.: Cross-domain mining of argumentative text through distant supervision. In: Proceedings of the 2016 Conference of the North American Chapter of the Association for Computational Linguistics: Human Language Technologies, pp. 1395–1404 (2016)
19. Wachsmuth, H., et al.: Building an argument search engine for the web. In: Proceedings of the 4th Workshop on Argument Mining, pp. 49–59 (2017)
20. Bird, S., Klein, E., Loper, E.: Natural Language Processing with Python: Analyzing Text with the Natural Language Toolkit. O'Reilly Media Inc., Newton (2009)
21. Li, M., Gao, Y., Wen, H., Du, Y., Liu, H., Wang, H.: Joint RNN model for argument component boundary detection. In: Proceeding of the 2017 IEEE International Conference on Systems, Man, and Cybernetics (SMC), pp. 57–62 (2017)
22. Ajjour, Y., Chen, W.F., Kiesel, J., Wachsmuth, H., Stein, B.: Unit segmentation of argumentative texts. In: Proceedings of the 4th Workshop on Argument Mining, pp. 118–128 (2017)
23. Stab, C., Miller, T., Gurevych, I.: Cross-topic argument mining from heterogeneous sources using attention-based neural networks. arXiv preprint. arXiv:1802.05758 (2018)
24. Guggilla, C., Miller, T., Gurevych, I.: CNN-and LSTM-based claim classification in online user comments. In: Proceedings of COLING 2016, the 26th International Conference on Computational Linguistics: Technical Papers, pp. 2740–2751 (2016)
25. Sparck Jones, K.: A statistical interpretation of term specificity and its application in retrieval. J. Doc. **28**(1), 11–21 (1972)

ConvNet and Dempster-Shafer Theory for Object Recognition

Zheng Tong[ID], Philippe Xu[ID], and Thierry Denœux[(✉)][ID]

Université de Technologie de Compiègne, CNRS, UMR 7253 Heudiasyc,
Compiègne, France
{zheng.tong,philippe.xu}@hds.utc.fr
thierry.denoeux@utc.fr

Abstract. We propose a novel classifier based on convolutional neural network (ConvNet) and Dempster-Shafer theory for object recognition allowing for ambiguous pattern rejection, called the ConvNet-BF classifier. In this classifier, a ConvNet with nonlinear convolutional layers and a global pooling layer extracts high-dimensional features from input data. The features are then imported into a belief function classifier, in which they are converted into mass functions and aggregated by Dempster's rule. Evidence-theoretic rules are finally used for pattern classification and rejection based on the aggregated mass functions. We propose an end-to-end learning strategy for adjusting the parameters in the ConvNet and the belief function classifier simultaneously and determining the rejection loss for evidence-theoretic rules. Experiments with the CIFAR-10, CIFAR-100, and MNIST datasets show that hybridizing belief function classifiers with ConvNets makes it possible to reduce error rates by rejecting patterns that would otherwise be misclassified.

Keywords: Pattern recognition · Belief function · Convolutional neural network · Supervised learning · Evidence theory

1 Introduction

Dempster-Shafer (DS) theory of belief functions [3,24] has been widely used for reasoning and making decisions with uncertainty [29]. DS theory is based on representing independent pieces of evidence by completely monotone capacities and aggregating them using Dempster's rule. In the past decades, DS theory has been applied to pattern recognition and supervised classification in three main directions. The first one is classifier fusion, in which classifier outputs are converted into mass functions and fused by Dempster's rule (e.g., [2,19]). Another direction is evidential calibration: the decisions of classifiers are transformed into

This research was carried out in the framework of the Labex MS2T, which was funded by the French Government, through the program "Investments for the future" managed by the National Agency for Research (Reference ANR- 11-IDEX-0004-02). It was also supported by a scholarship from the China Scholarship Council.

© Springer Nature Switzerland AG 2019
N. Ben Amor et al. (Eds.): SUM 2019, LNAI 11940, pp. 368–381, 2019.
https://doi.org/10.1007/978-3-030-35514-2_27

mass functions (e.g., [20,28]). The last approach is to design evidential classifiers (e.g., [6]), which represent the evidence of each feature as elementary mass functions and combine them by Dempster's rule. The combined mass functions are then used for decision making [5]. Compared with conventional classifiers, evidential classifiers can provide more informative outputs, which can be exploited for uncertainty quantification and novelty detection. Several principles have been proposed to design evidential classifiers, mainly including the evidential k-nearest neighbor rule [4,9], and evidential neural network classifiers [6]. In practice, the performance of evidential classifiers heavily depends on two factors: the training set size and the reliability of object representation. With the development of the "Big Data" age, the number of examples in benchmark datasets for supervised algorithms has increased from 10^2 to 10^5 [14] and even 10^9 [21]. However, little has been done to combine recent techniques for object representation with DS theory.

Thanks to the explosive development of deep learning [15] and its applications [14,25], several approaches for object representation have been developed, such as restricted Boltzmann machines [1], deep autoencoders [26,27], deep belief networks [22,23], and convolutional neural networks (ConvNets) [12,17]. ConvNet, which is maybe the most promising model and the main focus of this paper, mainly consists of convolutional layers, pooling layers, and fully connected layers. It has been proved that ConvNets have the ability to extract local features and compute global features, such as from edges to corners and contours to object parts. In general, robustness and automation are two desirable properties of ConvNets for object representation. Robustness means strong tolerance to translation and distortion in deep representation, while automation implies that object representation is data-driven with no human assistance.

Motivated by recent advances in DS theory and deep learning, we propose to combine ConvNet and DS theory for object recognition allowing for ambiguous pattern rejection. In this approach, a ConvNet with nonlinear convolutional layers and a global pooling layer is used to extract high-order features from input data. Then, the features are imported into a belief function classifier, in which they are converted into Dempster-Shafer mass functions and aggregated by Dempster's rule. Finally, evidence-theoretic rules are used for pattern recognition and rejection based on the aggregated mass functions. The performances of this classifier on the CIFAR-10, CIFAR-100, and MNIST datasets are demonstrated and discussed.

The organization of the rest of this paper is as follows. Background knowledge on DS theory and ConvNet is recalled in Sect. 2. The new combination between DS theory and ConvNet is then established in Sect. 3, and numerical experiments are reported in Sect. 4. Finally, we conclude the paper in Sect. 5.

2   Background

In this section, we first recall some necessary definitions regarding the DS theory and belief function classifier (Sect. 2.1). We then provide a description of the architecture of a ConvNet that will be combined with a belief function classifier later in the paper (Sect. 2.2).

2.1   Dempster-Shafer Theory

Evidence Theory. The main concepts regarding DS theory are briefly presented in this section, and some basic notations are introduced. Detailed information can be found in Shafer's original work [24] and some up-to-date studies [8].

Given a finite set $\Omega = \{\omega_1, \cdots, \omega_k\}$, called the *frame of discernment*, a *mass function* is a function m from 2^Ω to $[0,1]$ verifying $m(\emptyset) = 0$ and

$$\sum_{A \subseteq \Omega} m(A) = 1. \tag{1}$$

For any $A \subseteq \Omega$, given a certain piece of evidence, $m(A)$ can be regarded as the belief that one is willing to commit to A. Set A is called a *focal element* of m when $m(A) > 0$.

For all $A \subseteq \Omega$, a *credibility* function bel and a *plausibility* function pl, associated with m, are defined as

$$\text{bel}(A) = \sum_{B \subseteq A} m(B) \tag{2}$$

$$\text{pl}(A) = \sum_{A \cap B \neq \emptyset} m(B). \tag{3}$$

The quantity $\text{bel}(A)$ is interpreted as a global measure of one's belief that hypothesis A is true, while $\text{pl}(A)$ is the amount of belief that could potentially be placed in A.

Two mass functions m_1 and m_2 representing independent items of evidence can be combined by Dempster's rule $\oplus$ [3,24] as

$$(m_1 \oplus m_2)(A) = \frac{\displaystyle\sum_{B \cap C = A} m_1(B) m_2(C)}{\displaystyle\sum_{B \cap C \neq \emptyset} m_1(B) m_2(C)} \tag{4}$$

for all $A \neq \emptyset$ and $(m_1 \oplus m_2)(\emptyset) = 0$. Mass functions m_1 and m_2 can be combined if and only if the denominator on the right-hand side of (4) is strictly positive. The operator $\oplus$ is commutative and associative.

Belief Function Classifier. Based on DS theory, an adaptive pattern classifier, called belief function classifier, was proposed by Denœux [6]. The classifier uses reference patterns as items of evidence regarding the class membership. The evidence is represented by mass functions and combined using Dempster's rule.

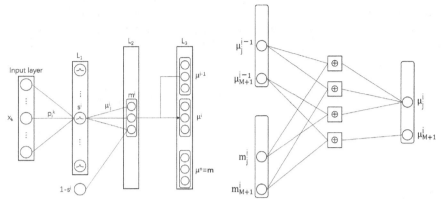

(a) Architecture of a belief function classifier (b) Connection between layers L_2 and L_3

Fig. 1. Belief function classifier

In this section, we describe the architecture of a belief function classifier. For a more complete introduction, readers are invited to refer to Denœux's original work [6].

We denote by $x \in \mathbb{R}^P$ a pattern to be classified into one of M classes $\omega_1, \cdots, \omega_M$, and by $\mathcal{X}$ a training set of N P-dimensional patterns. A belief function classifier quantifies the uncertainty about the class of x by a belief function on $\Omega = \{\omega_1, \cdots, \omega_M\}$, using a three-step procedure. This procedure can also be implemented in a multi-layer neural network illustrated in Fig. 1. It is based on n prototypes $p^1, \cdots, p^n$, which are the weight vectors of the units in the first hidden layer L_1. The three steps are defined as follows.

Step 1: The distance between x and each prototype p^i is computed as

$$d^i = \left\| x - p^i \right\| \quad i = 1, \cdots, n, \tag{5}$$

and the activation of the corresponding neuron is defined by introducing new parameters η^i ($\eta^i \in \mathbb{R}$) as $s^i = \alpha^i \exp(-(\eta^i d^i)^2)$, where $\alpha^i \in (0,1)$ is a parameter associated to the prototype p^i.

Step 2: The mass function m^i associated to prototype p^i is computed as

$$m^i = (m^i(\{\omega_1\}), \ldots, m^i(\{\omega_M\}), m^i(\Omega))^T \tag{6a}$$

$$= (u_1^i s^i, \ldots, u_M^i s^i, 1 - s^i)^T, \tag{6b}$$

where $u^i = (u_1^i, \ldots, u_M^i)$ is a vector of parameters associated to the prototype p^i verifying $\sum_{j=1}^{M} u_j^i = 1$.

As illustrated in Fig. 1a, Eq. (6) can be regarded as computing the activations of units in the second hidden layer L_2, composed of n modules of $M + 1$ units each. The units of module i are connected to neuron i of the previous layer. The output of module i in the hidden layer corresponds to the belief masses assigned by m^i.

Step 3: The n mass functions m^i, $i = 1, \cdots, n$, are combined in the final layer based on Dempster's rule as shown in Fig. 1b. The vectors of activations $\boldsymbol{\mu}^i = (\mu_1^i, \cdots, \mu_{M+1}^i)$, $i = 1, \ldots, n$ of the final layer L_3 is defined by the following equations:

$$\boldsymbol{\mu}^1 = \boldsymbol{m}^1, \tag{7a}$$

$$\mu_j^i = \mu_j^{i-1} m^i(\{\omega_j\}) + \mu_j^{i-1} m^i(\{\Omega\}) + \mu_{M+1}^{i-1} m^i(\{\omega_j\}) \tag{7b}$$

for $i = 2, \cdots, n$ and $j = 1, \cdots, M$, and

$$\mu_{M+1}^i = \mu_{M+1}^{i-1} m^i(\{\Omega\}) \quad i = 2, \cdots, n. \tag{7c}$$

The classifier outputs $\boldsymbol{m} = (m(\{\omega_1\}), \ldots, m(\{\omega_M\}), m(\Omega))^T$ is finally obtained as $\boldsymbol{m} = \boldsymbol{\mu}^n$.

Evidence-Theoretic Rejection Rules. Different strategies to make a decision (e.g., assignment to a class or rejection) based on the possible consequences of each action were proposed in [5]. For a complete training set $\mathcal{X}$, we consider actions α_i, $i \in \{1, \cdots, M\}$ assigning the pattern to each class and a rejection action α_0. Assuming the cost of correct classification to be 0, the cost of misclassification to be 1 and the cost of rejection to be λ_0, the three conditions for rejection reviewed in [5] can be expressed as

Maximum credibility: $\max_{j=1,\cdots,M} m(\{\omega_j\}) < 1 - \lambda_0$
Maximum plausibility: $\max_{j=1,\cdots,M} m(\{\omega_j\}) + m(\Omega) < 1 - \lambda_0$
Maximum pignistic probability: $\max_{j=1,\cdots,M} m(\{\omega_j\}) + \frac{m(\Omega)}{M} < 1 - \lambda_0$.

Otherwise, the pattern is assigned to class ω_j with $j = \arg\max_{k=1,\cdots,M} m(\{\omega_k\})$. For the maximum plausibility and maximum pignistic probability rules, rejection is possible if and only if $0 \leq \lambda_0 \leq 1 - 1/M$, whereas a rejection action for the maximum credibility rule only requires $0 \leq \lambda_0 \leq 1$.

2.2   Convolutional Neural Network

In this section, we provide a brief description of some state-of-the-art techniques for ConvNets including the nonlinear convolutional operation and global average pooling (GAP), which will be implemented in our new model in Sect. 3. Detailed information about the two structure layers can be found in [18].

Nonlinear Convolutional Operation. The convolutional layer [15] is highly efficient for feature extraction and representation. In order to approximate the representations of the latent concepts related to the class membership, a novel convolutional layer has been proposed [18], in which nonlinear multilayer perceptron (MLP) operations replace classic convolutional operations to convolve over the input. An MLP layer with nonlinear convolutional operations can be summarized as follows:

$$f_{i,j,k}^1 = ReLU\left(\left(\mathbf{w}_k^1\right)^T \cdot \mathbf{x} + b_k^1\right), k = 1, \cdots, C \tag{8a}$$

$$\vdots$$

$$f_{i,j,k}^m = ReLU\left(\left(\mathbf{w}_k^m\right)^T \cdot \mathbf{f}_{i,j}^{m-1} + b_k^m\right), k = 1, \cdots, C. \tag{8b}$$

Here, m is the number of layers in an MLP. Matrix $\mathbf{x}$, called receptive field of size $i \times j \times o$, is a patch of the input data with the size of $(rW - r - p + i) \times (rH - r - p + j) \times o$. An MLP layer with an r stride and a p padding can generate a $W \times H \times C$ tensor, called feature maps. The size of a feature map is $W \times H \times 1$, while the channel number of the feature maps is C. A rectified linear unit (ReLU) is used as an activation function as $ReLU(x) = \max(0, x)$. As shown in Eq. (8), element-by-element multiplications are first performed between $\mathbf{x}$ and the transpositions of the weight matrices $\mathbf{w}_k^1$ $(k = 1, \cdots, C)$ in the 1st layer of the MLP. Each weight matrix $\mathbf{w}_k^1$ has the same size as the receptive field. Then the multiplied values are summed, and the bias b_k^1 $(k = 1, \cdots, C)$ is added to the summed values. The results are transformed by a $ReLU$ function. The output vector is $\mathbf{f}_{i,j}^1 = (f_{i,j,1}^1, f_{i,j,2}^1, \cdots, f_{i,j,C}^1)$. The outputs then flow into the remaining layers in sequence, generating $\mathbf{f}_{i,j}^m$ of size $1 \times 1 \times C$. After processing all patches by the MLP, the input data is transformed into a $W \times H \times C$ tensor. As the channel number C of the last MLP in a ConvNet is the same as the input data dimension P in a belief function classifier, a $W \times H \times P$ tensor is finally generated by a ConvNet.

Global Average Pooling. In a traditional ConvNet, the tensor is vectorized and imported into fully connected layers and a softmax layer for a classification task. However, fully connected layers are prone to overfitting, though dropout [11] and its variation [10] have been proposed. A novel strategy, called global average pooling (GAP), has been proposed to remove traditional fully connected layers [18]. A GAP layer transforms the feature tensor $W \times H \times P$ into a feature vector $1 \times 1 \times P$ by taking the average of each feature map as follows:

$$x_k = \frac{\sum_{i=1}^{W}\sum_{j=1}^{H} f_{i,j,k}^m}{W \cdot H} \quad k = 1, \cdots, P. \tag{9}$$

The generated feature vector is used for classification. From the belief function perspective, the feature vector can be used for object representation and classified in one of M classes or rejected by a belief function classifier. Thus, a ConvNet can be regarded as a feature generator.

3   ConvNet-BF Classifier

In this section, we present a method to combine a belief function classifier and a ConvNet for objection recognition allowing for ambiguous pattern rejection. The architecture of the proposed method, called ConvNet-BF classifier, is illustrated in Fig. 2. A ConvNet-BF classifier can be divided into three parts: a ConvNet as a feature producer, a belief function classifier as a mass-function generator, and a decision rule. In this classifier, input data are first imported into a ConvNet with nonlinear convolutional layers and a global pooling layer to extract latent features related to the class membership. The features are then imported into a belief function classifier, in which they are converted into mass functions and aggregated by Dempster's rule. Finally, an evidence-theoretic rule is used for pattern classification and rejection based on the aggregated mass functions. As the background of the three parts has been introduced in Sect. 2, we only provide the details of the combination in this section, including the connectionist implementation and the learning strategy.

3.1   Connectionist Implementation

In a ConvNet-BF classifier, the Euclidean distance between a feature vector and each prototype is first computed and then used to generate a mass function. To reduce the classification error when P is large, we assign weights to each feature as

$$d^i = \sqrt{\sum_{k=1}^{P} w_k^i (x_k - p_k^i)^2}, \tag{10}$$

and the weights are normalized by introducing new parameters ζ_k^i ($\zeta_k^i \in \mathbb{R}$) as

$$w_k^i = \frac{(\zeta_k^i)^2}{\sum\limits_{l=1}^{P} (\zeta_l^i)^2}. \tag{11}$$

3.2   Learning

The proposed learning strategy to train a ConvNet-BF classifier consists in two parts: (a) an end-to-end training method to train ConvNet and belief function classifier simultaneously; (b) a data-driven method to select λ_0.

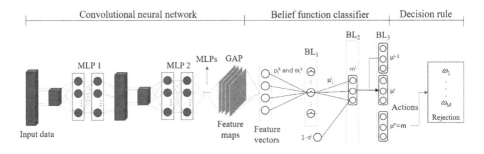

Fig. 2. Architecture of a ConvNet-BF classifier

End-to-End Training. Compared with the belief function classifier proposed in [6], we have different expressions for the derivatives w.r.t. w_k^i, ζ_k^i, and p_k^i in the new belief function classifier. A normalized error $E_\nu(x)$ is computed as:

$$E_\nu(x) = \frac{1}{2N} \sum_{i=1}^{I} \sum_{q=1}^{M} (Pre_{\nu,q,i} - Tar_{q,i})^2, \tag{12a}$$

$$Pre_{\nu,q,i} = m'_{q,i} + \nu m'_{M+1,i}, \tag{12b}$$

$$m'_i = \frac{m_i}{\sum_{k=1}^{M+1} m_i(\{\omega_k\})}. \tag{12c}$$

Here, $Tar_i = (Tar_{1,i}, \cdots, Tar_{M,i})$ and $m_i = (m_i(\{\omega_1\}), \ldots, m_i(\{\omega_M\}), m_i(\Omega))^T$ are the target output vector and the unnormalized network output vector for pattern x_i, respectively. We transform m_i to a vector $(Pre_{\nu,1,i}, \ldots, Pre_{\nu,M,i})$ by distributing a fraction ν of $m_i(\Omega)$ to each class under the constraint $0 \leq \nu \leq 1$. The numbers $Pre_{1,q,i}$, $Pre_{0,q,i}$ and $Pre_{1/M,q,i}$ represent, respectively, the credibility, the plausibility, and the pignistic probability of class ω_q. The derivatives of $E_\nu(x)$ w.r.t p_k^i, w_k^i, and ζ_k^i in a belief function classifier can be expressed as

$$\frac{\partial E_\nu(x)}{\partial p_k^i} = \frac{\partial E_\nu(x)}{\partial s^i} \frac{\partial s^i}{\partial p_k^i} = \frac{\partial E_\nu(x)}{\partial s^i} \cdot 2(\eta^i)^2 s^i \cdot \sum_{k=1}^{P} w_k^i(x_k - p_k^i), \tag{13}$$

$$\frac{\partial E_\nu(x)}{\partial w_k^i} = \frac{\partial E_\nu(x)}{\partial s^i} \frac{\partial s^i}{\partial w_k^i} = \frac{\partial E_\nu(x)}{\partial s^i} \cdot (\eta^i)^2 s^i \cdot (x_k - p_k^i)^2, \tag{14}$$

and

$$\frac{\partial E_\nu(x)}{\partial \zeta_k^i} = \frac{\partial E_\nu(x)}{\partial w_k^i} \frac{\partial w_k^i}{\partial \zeta_k^i} \tag{15a}$$

$$= \frac{2\zeta_k^i}{\left(\sum_{k=1}^{P} (\zeta_k^i)^2\right)^2} \left[\frac{\partial E_\nu(x)}{\partial w_k^i} \sum_{k=1}^{P} (\zeta_k^i)^2 - \sum_{k=1}^{P} (\zeta_k^i)^2 \frac{\partial E_\nu(x)}{\partial w_k^i} \right]. \tag{15b}$$

Finally, the derivatives of the error w.r.t. x_k, $w_{i,j,k}^m$ and b_k^m in the last MLP are given as

$$\frac{\partial E_\nu(x)}{\partial x_k} = \frac{\partial E_\nu(x)}{\partial s^i} \frac{\partial s^i}{\partial x_k} = -\frac{\partial E_\nu(x)}{\partial s^i} \cdot 2(\eta^i)^2 s^i \cdot \sum_{k=1}^{P} w_k^i(x_k - p_k^i), \tag{16}$$

$$\frac{\partial E_\nu(x)}{\partial w_{i,j,k}^m} = \frac{\partial E_\nu(x)}{\partial f_{i,j,k}^m} \cdot \frac{\partial f_{i,j,k}^m}{\partial w_{i,j,k}^m} = w_{i,j,k}^m \cdot \frac{\partial E_\nu(x)}{\partial f_{i,j,k}^m} \quad k = 1, \cdots, P, \tag{17}$$

and

$$\frac{\partial E_\nu(x)}{\partial b_k^m} = \frac{\partial E_\nu(x)}{\partial f_{i,j,k}^m} \cdot \frac{\partial f_{i,j,k}^m}{\partial b_k^m} = \frac{\partial E_\nu(x)}{\partial f_{i,j,k}^m} \quad k = 1, \cdots, P \tag{18}$$

with

$$\frac{\partial E_\nu(\boldsymbol{x})}{\partial f_{i,j,k}^m} = \frac{\partial E_\nu(\boldsymbol{x})}{\partial x_k} \cdot \frac{\partial x_k}{\partial f_{i,j,k}^m} = \frac{1}{W \cdot H} \frac{\partial E_\nu(\boldsymbol{x})}{\partial x_k} \quad k = 1, \cdots, P. \qquad (19)$$

Here, $w_{i,j,k}^m$ is the component of the weight matrix $\mathbf{w}_k^m$, while $f_{i,j,k}^m$ is the component of vector $\mathbf{f}_{i,j}^m$ in Eq. (8).

Determination of $\boldsymbol{\lambda_0}$. A data-driven method for determining λ_0 to guarantee a ConvNet-BF classifier with a certain rejection rate is shown in Fig. 3. We randomly select three-fifths of a training set χ to train a ConvNet-BF classifier, while random one-fifth of the set is used as a validation set. The remaining one-fifth of the set is used to draw a $\lambda_0^{(1)}$-rejection curve. We can determine the value of $\lambda_0^{(1)}$ for a certain rejection rate from the curve. We repeat the process and take the average of $\lambda_0^{(i)}$ as the final λ_0 for the desired rejection rate.

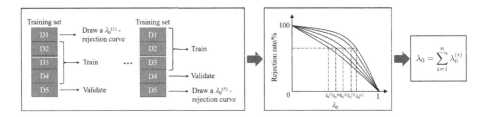

Fig. 3. Illustration of the procedure for determining λ_0

4 Numerical Experiments

In this section, we evaluate ConvNet-BF classifiers on three benchmark datasets: CIFAR-10 [13], CIFAR-100 [13], and MNIST [16]. To compare with traditional ConvNets, the architectures and training strategies of the ConvNet parts in ConvNet-BF classifiers are the same as those used in the study of Lin et al., called NIN [18]. Feature vectors from the ConvNet parts are imported into a belief function classifier in our method, while they are directly injected into softmax layers in NINs.

In order to make a fair comparison, a probability-based rejection rule is adopted for NINs as $\max_{j=1,\cdots,M} p_j < 1 - \lambda_0$, where p_j is the output probability of NINs.

4.1 CIFAR-10

The CIFAR-10 dataset [13] is made up of 60,000 RGB images of size 32×32 partitioned in 10 classes. There are 50,000 training images, and we randomly selected 10,000 images as validation data for the ConvNet-BF classifier. We then randomly used 10,000 images of the training set to determine λ_0.

The test set error rates without rejection of the ConvNet-BF and NIN classifiers are 9.46% and 9.21%, respectively. The difference is small but statistically significant according to McNemar's test (p-value: 0.012). Error rates without rejection mean that we only consider $\max_{j=1,\cdots,M} p_j$ and $\max_{j=1,\cdots,M} m(\{\omega_j\})$. If the selected class is not the correct one, we regard it as an error. It turns out in our experiment that using a belief function classifier instead of a softmax layer only slightly impacts the classifier performance.

The test set error rates with rejection of the two models are presented in Fig. 4a. A rejection decision is not regarded as an incorrect classification. When the rejection rate increases, the test set error decreases, which shows that the belief function classifier rejects a part of incorrect classification. However, the error decreases slightly when the rejection rate is higher than 7.5%. This demonstrates that the belief function classifier rejects more and more correctly classified patterns with the increase of rejection rates. Thus, a satisfactory $\lambda_0^{(i)}$ should be determined to guarantee that the ConvNet-BF classifier has a desirable accuracy rate and a low correct-rejection rate. Additionally, compared with the NIN, the ConvNet-BF classifier rejects significantly more incorrectly classified patterns. For example, the p-value of McNemar's test for the difference of error rates between the two classifiers with a 5.0% rejection rate is close to 0. We can conclude that a belief function classifier with an evidence-theoretic rejection rule is more suitable for making a decision allowing for pattern rejection than a softmax layer and the probability-based rejection rule.

Table 1 presents the confusion matrix of the ConvNet-BF classifier with the maximum credibility rule, whose rejection rate is 5.0%. The ConvNet-BF classifier tends to select rejection when there are two or more similar patterns, such as dog and cat, which can lead to incorrect classification. In the view of evidence theory, the ConvNet part provides conflicting evidence when two or more similar patterns exist. The maximally conflicting evidence corresponds to $m(\{\omega_i\}) = m(\{\omega_j\}) = 0.5$ [7]. Additionally, the additional mass function $m(\Omega)$ provides the possibility to verify whether the model is well trained because we have $m(\Omega) = 1$ when the ConvNet part cannot provide any useful evidence.

4.2   CIFAR-100

The CIFAR-100 dataset [13] has the same size and format at the CIFAR-10 dataset, but it contains 100 classes. Thus the number of images in each class is only 100. For CIFAR-100, we also randomly selected 10,000 images of the training set to determine λ_0. The ConvNet-BF and NIN classifiers achieved, respectively, 40.62% and 39.24% test set error rates without rejection, a small but statistically significant difference (p-value: 0.014). Similarly to CIFAR-10, it turns out that the belief function classifier has a similar error rate as a network with a softmax layer. Figure 4b shows the test set error rates with rejection for the two models. Compared with the rejection performance in CIFAR-10, the ConvNet-BF classifier rejects more incorrect classification results. We can conclude that the evidence-theoretic classifier still performs well when the classification task is difficult and the training set is not adequate. Similarly, Table 2 shows that the

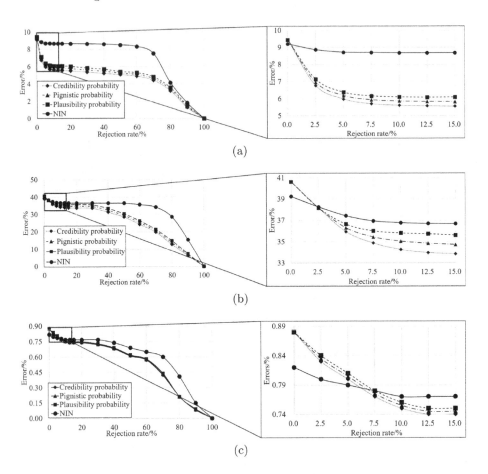

Fig. 4. Rejection-error curves: CIFAR-10 (a), CIFAR-100 (b), and MNIST (c)

Table 1. Confusion matrix for Cifar10.

	Airplane	Automobile	Bird	Cat	Deer	Dog	Frog	Horse	Ship	Truck
Airplane	-	0.03	0.03	0.01	0.02	0.05	0.04	0.01	0.04	0.05
Automobile	0	-	0.04	0.04	0.08	0.08	0.04	0.06	0.03	0.07
Bird	0.02	0.04	-	0.05	0.04	0.07	0.03	0.08	0	0.04
Cat	0.02	0.03	0.13	-	0.06	**0.44**	0.11	0.04	0.05	0.06
Deer	0.01	0.04	0.07	0.12	-	0.03	0.12	**0.34**	0.04	0.08
Dog	0.02	0.03	0.05	**0.49**	0.11	-	0.06	0.09	0.01	0.04
Frog	0.02	0.04	0.08	0.06	0.12	0.06	-	0.06	0.06	0.05
Horse	0.01	0.02	0.04	0.06	**0.31**	0.10	0.04	-	0.04	0.04
Ship	0.04	0.05	0.02	0.04	0.12	0.05	0.04	0.18	-	0.02
Truck	0.02	0	0.06	0.09	0.03	0.06	0.07	0.06	0.04	-
Rejection	0.20	0.13	0.14	**1.05**	**0.84**	**1.07**	0.14	**1.14**	0.18	0.11

ConvNet-BF classifier tends to select the rejection action when two classes are similar, in which case we have $m(\{\omega_i\}) \approx m(\{\omega_j\})$. In contrast, the classifier tends to produce $m(\Omega) \approx 1$ when the model is not trained well because of an inadequate training set.

4.3 MNIST

The MNIST database of handwritten digits consists of a training set of 60,000 examples and a test set of 10,000 examples. The training strategy for the ConvNet-BF classifier was the same as the strategy in CIFAR-10 and CIFAR-100. The test set error rates without rejection of the two models are close (0.88% and 0.82%) and weakly signifiant (p-value: 0.077). Again, using a belief function classifier instead of a softmax layer introduce no negative effect on the network in MNIST. The test set error rates with rejection of the two models are shown in Fig. 4c. The ConvNet-BF classifier rejects a small number of classification results because the feature vectors provided by the ConvNet part include little confusing information.

Table 2. Confusion matrix for the superclass *flowers*.

	Orchids	Poppies	Roses	Sunflowers	Tulips
Orchids	-	0.24	0.23	0.28	0.15
Poppies	0.14	-	**0.43**	0.10	**0.90**
Roses	0.27	0.12	-	0.16	0.13
Sunflowers	0.18	0.15	0.12	-	0.22
Tulips	0.08	**1.07**	**0.76**	0.17	-
Rejection	0.09	**0.37**	**0.63**	0.12	**0.34**

5   Conclusion

In this work, we proposed a novel classifier based on ConvNet and DS theory for object recognition allowing for ambiguous pattern rejection, called "ConvNet-BF classifier". This new structure consists of a ConvNet with nonlinear convolutional layers and a global pooling layer to extract high-dimensional features and a belief function classifier to convert the features into Dempster-Shafer mass functions. The mass functions can be used for classification or rejection based on evidence-theoretic rules. Additionally, the novel classifier can be trained in an end-to-end way.

The use of belief function classifiers in ConvNets had no negative effect on the classification performances on the CIFAR-10, CIFAR-100, and MNIST datasets. The combination of belief function classifiers and ConvNet can reduce the errors by rejecting a part of the incorrect classification. This provides a new direction to improve the performance of deep learning for object recognition. The classifier is prone to assign a rejection action when there are conflicting features, which easily yield incorrect classification in the traditional ConvNets. In addition, the proposed method opens a way to explain the relationship between the extracted features in convolutional layers and class membership of each pattern. The mass $m(\Omega)$ assigned to the set of classes provides the possibility to verify whether a ConvNet is well trained or not.

References

1. Bengio, Y.: Learning deep architectures for AI. Found. Trends® Mach. Learn. **2**(1), 1–127 (2009)
2. Bi, Y.: The impact of diversity on the accuracy of evidential classifier ensembles. Int. J. Approximate Reasoning **53**(4), 584–607 (2012)
3. Dempster, A.P.: Upper and lower probabilities induced by a multivalued mapping. In: Yager, R.R., Liu, L. (eds.) Classic Works of the Dempster-Shafer Theory of Belief Functions. STUDFUZZ, vol. 219, pp. 57–72. Springer, Heidelberg (2008). https://doi.org/10.1007/978-3-540-44792-4_3
4. Denœux, T.: A k-nearest neighbor classification rule based on Dempster-Shafer theory. IEEE Trans. Syst. Man Cybern. **25**(5), 804–813 (1995)
5. Denœux, T.: Analysis of evidence-theoretic decision rules for pattern classification. Pattern Recogn. **30**(7), 1095–1107 (1997)
6. Denœux, T.: A neural network classifier based on Dempster-Shafer theory. IEEE Trans. Syst. Man Cybern. Part A Syst. Hum. **30**(2), 131–150 (2000)
7. Denœux, T.: Logistic regression, neural networks and Dempster-Shafer theory: a new perspective. Knowl.-Based Syst. **176**, 54–67 (2019)
8. Denœux, T., Dubois, D., Prade, H.: Representations of uncertainty in artificial intelligence: beyond probability and possibility. In: Marquis, P., Papini, O., Prade, H. (eds.) A Guided Tour of Artificial Intelligence Research, Chap. 4. Springer (2019)
9. Denœux, T., Kanjanatarakul, O., Sriboonchitta, S.: A new evidential K-nearest neighbor rule based on contextual discounting with partially supervised learning. Int. J. Approximate Reasoning **113**, 287–302 (2019)
10. Gomez, A.N., Zhang, I., Swersky, K., Gal, Y., Hinton, G.E.: Targeted dropout. In: CDNNRIA Workshop at the 32nd Conference on Neural Information Processing Systems (NeurIPS 2018), Montréal (2018)
11. Hinton, G.E., Srivastava, N., Krizhevsky, A., Sutskever, I., Salakhutdinov, R.R.: Improving neural networks by preventing co-adaptation of feature detectors. arXiv preprint arXiv:1207.0580 (2012)
12. Kim, Y.: Convolutional neural networks for sentence classification. In: Proceedings of the 2014 Conference on Empirical Methods in Natural Language Processing, Doha, pp. 1746–1751 (2014)
13. Krizhevsky, A., Hinton, G.: Learning multiple layers of features from tiny images. University of Toronto, Technical report (2009)
14. Krizhevsky, A., Sutskever, I., Hinton, G.E.: ImageNet classification with deep convolutional neural networks. Commun. ACM **60**(6), 84–90 (2017)
15. LeCun, Y., Bengio, Y., Hinton, G.: Deep learning. Nature **521**(7553), 436 (2015)
16. LeCun, Y., Bottou, L., Bengio, Y., Haffner, P., et al.: Gradient-based learning applied to document recognition. Proc. IEEE **86**(11), 2278–2324 (1998)

17. Leng, B., Liu, Y., Yu, K., Zhang, X., Xiong, Z.: 3D object understanding with 3D convolutional neural networks. Inf. Sci. **366**, 188–201 (2016)
18. Lin, M., Chen, Q., Yan, S.: Network in network. In: International Conference on Learning Representations (ICLR 2014), Banff, pp. 1–10 (2014)
19. Liu, Z., Pan, Q., Dezert, J., Han, J.W., He, Y.: Classifier fusion with contextual reliability evaluation. IEEE Trans. Cybern. **48**(5), 1605–1618 (2018)
20. Minary, P., Pichon, F., Mercier, D., Lefevre, E., Droit, B.: Face pixel detection using evidential calibration and fusion. Int. J. Approximate Reasoning **91**, 202–215 (2017)
21. Sakaguchi, K., Post, M., Van Durme, B.: Efficient elicitation of annotations for human evaluation of machine translation. In: Proceedings of the Ninth Workshop on Statistical Machine Translation, Baltimore, pp. 1–11 (2014)
22. Salakhutdinov, R., Hinton, G.: Deep Boltzmann machines. In: Artificial Intelligence and Statistics, Florida, pp. 448–455 (2009)
23. Salakhutdinov, R., Tenenbaum, J.B., Torralba, A.: Learning with hierarchical-deep models. IEEE Trans. Pattern Anal. Mach. Intell. **35**(8), 1958–1971 (2012)
24. Shafer, G.: A Mathematical Theory of Evidence. Princeton University Press, Princeton (1976)
25. Tong, Z., Gao, J., Zhang, H.: Recognition, location, measurement, and 3D reconstruction of concealed cracks using convolutional neural networks. Constr. Build. Mater. **146**, 775–787 (2017)
26. Vincent, P., Larochelle, H., Bengio, Y., Manzagol, P.A.: Extracting and composing robust features with denoising autoencoders. In: Proceedings of the 25th International Conference on Machine Learning, pp. 1096–1103, New York (2008)
27. Vincent, P., Larochelle, H., Lajoie, I., Bengio, Y., Manzagol, P.A.: Stacked denoising autoencoders: learning useful representations in a deep network with a local denoising criterion. J. Mach. Learn. Res. **11**(Dec), 3371–3408 (2010)
28. Xu, P., Davoine, F., Zha, H., Denœux, T.: Evidential calibration of binary SVM classifiers. Int. J. Approximate Reasoning **72**, 55–70 (2016)
29. Yager, R.R., Liu, L.: Classic Works of the Dempster-Shafer Theory of Belief Functions, vol. 219. Springer, Heidelberg (2008). https://doi.org/10.1007/978-3-540-44792-4

On Learning Evidential Contextual Corrections from Soft Labels Using a Measure of Discrepancy Between Contour Functions

Siti Mutmainah[1,2(✉)], Samir Hachour[1], Frédéric Pichon[1], and David Mercier[1]

[1] EA 3926 LGI2A, Univ. Artois, 62400 Béthune, France
siti.mutmainah@univ-artois.fr, {samir.hachour,frederic.pichon,
david.mercier}@univ-artois.fr
[2] UIN Sunan Kalijaga, Yogyakarta, Indonesia
siti.mutmainah@uin-suka.ac.id

Abstract. In this paper, a proposition is made to learn the parameters of evidential contextual correction mechanisms from a learning set composed of soft labelled data, that is data where the true class of each object is only partially known. The method consists in optimizing a measure of discrepancy between the values of the corrected contour function and the ground truth also represented by a contour function. The advantages of this method are illustrated by tests on synthetic and real data.

Keywords: Belief functions · Contextual corrections · Learning · Soft labels

1 Introduction

In Dempster-Shafer theory [15,17], the correction of a source of information, a sensor for example, is classically done using the discounting operation introduced by Shafer [15], but also by so-called contextual correction mechanisms [10,13] taking into account more refined knowledge about the quality of a source.

These mechanisms, called contextual discounting, negating and reinforcement [13], can be derived from the notions of *reliability (or relevance)*, which concerns the competence of a source to answer the question of interest, and *truthfulness* [12,13] indicating the source's ability to say what it knows (it may also be linked with the notion of bias of a source). The contextual discounting is an extension of the discounting operation, which corresponds to a partially reliable and totally truthful source. The contextual negating is an extension of the negating operation [12,13], which corresponds to the case of a totally reliable but partially truthful source, the extreme case being the negation of a source [5]. At last, the contextual reinforcement is an extension of the reinforcement, a dual operation of the discounting [11,13].

In this paper, the problem of learning the parameters of these correction mechanisms from soft labels, meaning partially labelled data, is tackled. More specifically, in our case, soft labels indicate the true class of each object in an imprecise manner through a contour function.

© Springer Nature Switzerland AG 2019
N. Ben Amor et al. (Eds.): SUM 2019, LNAI 11940, pp. 382–389, 2019.
https://doi.org/10.1007/978-3-030-35514-2_28

A method for learning these corrections from labelled data (hard labels), where the truth is perfectly known for each element of the learning set, has already been introduced in [13]. It consists in minimizing a measure of discrepancy between the corrected contour functions and the ground truths over elements of a learning set. In this paper, it is shown that this same measure can be used to learn from soft labels, and tests on synthetic and real data illustrate its advantages to (1) improve a classifier even if the data is only partially labelled; and (2) obtain better performances than learning these corrections from approximate hard labels approaching the only available soft labels.

This paper is organized as follows. In Sect. 2, the basic concepts and notations used in this paper are presented. Then, in Sect. 3, the three applied contextual corrections as well as their learning from hard labels are exposed. The proposition to extend this method to soft labels is introduced. Tests of this method on synthetic and real data are presented in Sect. 4. At last, a discussion and future works are given in Sect. 5.

2   Belief Functions: Basic Concepts Used

Only the basic concepts used are presented in this section (See for example [3, 15, 17] for further details on the belief function framework).

From a frame of discernment $\Omega = \{\omega_1, ..., \omega_K\}$, a *mass function (MF)*, noted m^Ω or m if no ambiguity, is defined from 2^Ω to $[0, 1]$, and verify $\sum_{A \subseteq \Omega} m^\Omega(A) = 1$.

The focal elements of a MF m are the subsets A of Ω such that $m(A) > 0$.

A MF m is in one-to-one correspondence with a *plausibility function Pl* defined for all $A \subseteq \Omega$ by

$$Pl(A) = \sum_{B \cap A \neq \emptyset} m(B). \tag{1}$$

The *contour function pl* of a MF m is defined for all $\omega \in \Omega$ by

$$\begin{aligned} pl : \Omega &\to \quad [0, 1] \\ \omega &\mapsto pl(\omega) = Pl(\{\omega\}) . \end{aligned} \tag{2}$$

It is the restriction of the plausibility function to all the singletons of Ω.

The knowledge of the reliability of a source is classically taken into account by the operation called *discounting* [15, 16]. Let us suppose a source S provides a piece of information represented by a MF m_S. With $\beta \in [0, 1]$ the degree of belief of the reliability of the source, the discounting of m_S is defined by the MF m s.t.

$$m(A) = \beta \, m_S(A) + (1 - \beta) m_\Omega(A) , \tag{3}$$

for all $A \subseteq \Omega$, where m_Ω represents the total ignorance, i.e. the MF defined by $m_\Omega(\Omega) = 1$.

Several justifications for this mechanism can be found in [10, 13, 16].

The contour function of the MF m resulting from the discounting (3) is defined for all $\omega \in \Omega$ by (see for example [13, Prop. 11])

$$pl(\omega) = 1 - (1 - pl_S(\omega))\beta , \tag{4}$$

with pl_S the contour function of m_S.

3   Contextual Corrections and Learning from Labelled Data

In this Section, the contextual corrections we used are first exposed, then their learning from hard labels. The proposition to extend this method to soft labels is then introduced.

3.1   Contextual Corrections of a Mass Function

For the sake of simplicity, we only recall here the contour functions expressions resulting from the applications of contextual discounting, reinforcement and negating mechanisms in the case of K contexts where K is the number of elements in Ω.

It is shown in [13] that these expressions are rich enough to minimize the discrepancy measure used to learn the parameters of these corrections, this measure being presented in Sect. 3.2.

Let us suppose a source S providing a piece of information m_S.

The contour function resulting from the *contextual discounting (CD)* of m_S and a set of contexts composed of the singletons of Ω is given by

$$pl(\omega) = 1 - (1 - pl_S(\omega))\beta_{\{\omega\}} , \tag{5}$$

for all $\omega \in \Omega$, with the K parameters $\beta_{\{\omega\}}$ which may vary in $[0, 1]$.

For the *contextual reinforcement (CR)* and the *contextual negating (CN)*, the contour functions are respectively given, from a set of contexts composed of the complementary of each singleton of Ω, by

$$pl(\omega) = pl_S(\omega)\beta_{\overline{\{\omega\}}} , \tag{6}$$

and

$$pl(\omega) = 0.5 + (pl_S(\omega) - 0.5)(2\beta_{\overline{\{\omega\}}} - 1) , \tag{7}$$

for all $\omega \in \Omega$, with the K parameters $\beta_{\overline{\{\omega\}}}$ able to vary in $[0, 1]$.

3.2   Learning from Hard Labels

Let us suppose a source of information providing a MF m_S concerning the true class of an object among a set of possible classes Ω.

If we have a learning set composed of n instances (or objects) the true values of which are known, we can learn the parameters of a correction by minimizing a discrepancy measure between the output of the classifier which is corrected (a correction is applied to m_S) and the ground truth [7, 10, 13].

Introduced in [10], the following measure E_{pl} yields a simple optimization problem (a linear least-squares optimization problem, see [13, Prop. 12, 14 et 16]) to learn the vectors β_{CD}, β_{CR} and β_{CN} composed of the K parameters of corrections CD, CR and CN:

$$E_{pl}(\beta) = \sum_{i=1}^{n} \sum_{k=1}^{K} (pl_i(\omega_k) - \delta_{i,k})^2 , \tag{8}$$

where pl_i is the contour function regarding the class of the instance i resulting from a contextual correction (CD, CR or CN) of the MF provided by the source for this instance, and $\delta_{i,k}$ is the indicator function of the truth of all the instances $i \in \{1, \ldots, n\}$, i.e. $\delta_{i,k} = 1$ if the class of the instance i is ω_k, otherwise $\delta_{i,k} = 0$.

3.3 Learning from Soft Labels

In this paper, we consider the case where the truth is no longer given precisely by the values $\delta_{i,k}$, but only in an imprecise manner by a contour function $\tilde{\delta}_i$ s.t.

$$\tilde{\delta}_i : \Omega \rightarrow \quad [0,1] \atop \omega_k \mapsto \tilde{\delta}_i(\omega_k) = \tilde{\delta}_{i,k} \, . \tag{9}$$

The contour function $\tilde{\delta}_i$ gives information about the true class in Ω of the instance i.

Knowing then the truth only partially, we propose to learn the corrections parameters using the following discrepancy measure $\tilde{E}_{pl}$, extending directly (8):

$$\tilde{E}_{pl}(\boldsymbol{\beta}) = \sum_{i=1}^n \sum_{k=1}^K (pl_i(\omega_k) - \tilde{\delta}_{i,k})^2 \, . \tag{10}$$

The discrepancy measure $\tilde{E}_{pl}$ also yields, for each correction (CD, CR et CN), a linear least-squares optimization problem. For example, for CD, $\tilde{E}_{pl}$ can be written by

$$\tilde{E}_{pl}(\boldsymbol{\beta}) = \|Q\boldsymbol{\beta} - \tilde{d}\|^2 \tag{11}$$

with

$$Q = \begin{bmatrix} diag(\boldsymbol{pl}_1 - 1) \\ \vdots \\ diag(\boldsymbol{pl}_n - 1) \end{bmatrix} , \tilde{d} = \begin{bmatrix} \tilde{\delta}_1 - 1 \\ \vdots \\ \tilde{\delta}_n - 1 \end{bmatrix} \tag{12}$$

where $diag(\boldsymbol{v})$ is a square diagonal matrix whose diagonal is composed of the elements of the vector $\boldsymbol{v}$, and where for all $i \in \{1, \dots, n\}$, $\boldsymbol{\delta}_i$ is the column vector composed of the values of the contour function $\tilde{\delta}_i$, meaning $\boldsymbol{\delta}_i = (\tilde{\delta}_{i,1}, \dots, \tilde{\delta}_{i,K})^T$.

In the following, this learning proposition is tested with generated and real data.

4 Tests on Generated and Real Data

We first expose how soft labels can be generated from hard labels to make the tests exposed afterwards on synthetic and real data.

4.1 Generating Soft Labels from Hard Labels

It is not easy to find partially labelled data in the literature. Thus, as in [1,8,9,14], we have built our partially labelled data sets (soft labels) from perfect truths (hard labels) using the procedure described in Algorithm 1 (where $B\hat{e}ta, B,$ and $\mathcal{U}$ means respectively Bêta, Bernoulli and uniform distributions).

Algorithm 1. Soft labels generation

Input: hard labels δ_i with $i \in \{1,\dots,n\}$, where for each i, the integer $k \in \{1,\dots,K\}$ s.t. $\delta_{i,k} = 1$ is denoted by k_i.

Output: soft labels $\tilde{\delta}_i$ with $i \in \{1,\dots,n\}$.

1: **procedure** HARDTOSOFTLABELS
2:     **for** each instance i **do**
3:         Draw $p_i \sim B\hat{e}ta(\mu = .5, v = .04)$
4:         Draw $b_i \sim \mathcal{B}(p_i)$
5:         **if** $b_i = 1$ **then**
6:             Draw $k_i \sim \mathcal{U}_{\{1,\dots,K\}}$
7:         $\tilde{\delta}_{i,k_i} \leftarrow 1$
8:         $\tilde{\delta}_{i,k} \leftarrow p_i$ for all $k \neq k_i$

Algorithm 1 allows one to obtain soft labels that are all the more imprecise as the most plausible class is false.

4.2 Tests Performed

The chosen evidential classifier used as a source of information is the eviential k-nearest neighbor classifier (EkNN) introduced by Denœux in [2] with $k = 3$. We could have chosen another one with other settings, it can be seen as a black box.

The first test set we consider is composed of synthetic data composed of 3 classes built from 3 bivariate normal distributions with respective means $\mu_{\omega_1} = (1,2)$, $\mu_{\omega_2} = (2,1)$ and $\mu_{\omega_3} = (0,0)$, and a common covariance matrix Σ s.t.

$$\Sigma = \begin{bmatrix} 1 & 0.5 \\ 0.5 & 1 \end{bmatrix}. \tag{13}$$

For each class, 100 instances have been generated. They are illustrated in Fig. 1.

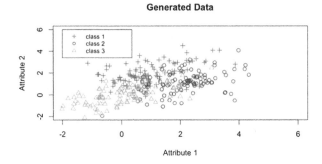

Fig. 1. Illustration of the generated dataset (3 classes, 2 attributes).

We have then considered several real data sets from the UCI database [6] composed of numerical attributes as the EkNN classifier is used. Theses data sets are described in Table 1.

Table 1. Characteristics of the UCI dataset used (number of instances without missing data, number of classes, number of numerical attributes used)

Data	#Instances	#Classes	#Attributes
Ionosphere	350	2	34
Iris	150	3	4
Sonar	208	2	60
Vowel	990	11	9
Wine	178	3	13

For each dataset, a 10-repeated 10-fold cross validation has been undertaken as follows:

- the group containing one tenth of the data is considered as the test set (the instances labels being made imprecise using Algorithm 1),
- the other 9 groups form the learning set, which is randomly divided into two groups of equal size:
  - one group to learn the EkNN classifier (learnt from hard truths),
  - one group to learn the parameters of the correction mechanisms from soft labels (the labels of the dataset are made imprecise using Algorithm 1).

For learning the parameters of contextual corrections, two strategies are compared.

1. In the first strategy, we use the optimization of Eq. (8) from the closest hard truths from the soft truths (the most plausible class is chosen). Corrections with this strategy are denoted by CD, CR and CN.
2. In the second strategy, Eq. (10) is directly optimized from soft labels (cf Sect. 3.3). The resulting corrections using this second strategy are denoted by CDsl, CRsl and CNsl.

The performances of the systems (the classifier alone and the corrections - CD, CR or CN - of this classifier according to the two strategies described above) are measured using $\tilde{E}_{pl}$ (10), where $\tilde{\delta}$ represents the partially known truth. This measure corresponds to the sum over the test instances of the differences, in the least squares sense, between the truths being sought and the system outputs.

The performances $\tilde{E}_{pl}$ (10) obtained from UCI and generated data for the classifier and its corrections are summed up in Table 2 for each type of correction. Standard deviations are indicated in brackets.

From the results presented in Table 2, we can remark that, for CD, the second strategy (CDsl) consisting in learning directly from the soft labels, allows one to obtain lower differences $\tilde{E}_{pl}$ from the truth on the test set than the first strategy (CD) where the correction parameters are learnt from approximate hard labels. We can also remark that this strategy yields lower differences $\tilde{E}_{pl}$ than the classifier alone, illustrating, in these experiments, the usefulness of soft labels even if hard labels are not available, which can be interesting in some applications.

The same conclusions can be drawn for CN.

Table 2. Performances $\tilde{E}_{pl}$ obtained for the classifier alone and the classifier corrected with CD, CR and CN using both strategies. Standard deviations are indicated in brackets.

Data	EkNN	CD	CDsl	CR	CRsl	CN	CNsl
Generated data	23.8 (3.8)	16.6 (2.8)	7.9 (1.5)	26.8 (3.0)	23.5 (3.7)	11.5 (1.6)	9.8 (0.6)
Ionosphere	16.2 (2.5)	9.6 (2.2)	5.3 (1.0)	17.2 (1.9)	15.9 (2.3)	9.3 (1.3)	8.4 (0.9)
Iris	12.5 (2.4)	8.4 (2.1)	3.3 (0.9)	13.1 (2.0)	12.3 (2.2)	6.7 (1.5)	4.8 (0.5)
Sonar	7.8 (2.0)	6.3 (1.9)	3.5 (0.9)	9.0 (1.6)	7.7 (1.9)	5.1 (0.8)	5.0 (0.9)
Vowel	279 (24)	278 (23)	62 (5)	310 (21)	279 (24)	240 (21)	65 (5)
Wine	13.3 (2.6)	10.4 (2.3)	4.3 (1.0)	15.0 (2.1)	13.3 (2.5)	7.2 (1.6)	5.7 (0.6)

For CR, the second strategy is also better than the first one but we can note that unlike the other corrections, there is no improvement for the first strategy in comparison to the classifier alone (the second strategy having also some close performances to the classifier alone).

5   Discussion and Future Works

We have shown that contextual corrections may lead to improved performances in the sense of measure $\tilde{E}_{pl}$, which relies on the plausibility values returned by the systems for each class for each instance. We also note that by using the same experiments as those in Sect. 4.2 but evaluating the performances using a simple 0–1 error criterion, where for each instance the most plausible class is compared to the true class, the performances remain globally identical for the classifier alone as well as all the corrections (the most plausible class being often the same for the classifier and each correction).

For future works, we are considering the use of other performance measures, which would also take fully into account the uncertainty and the imprecision of the outputs. For example, we would like to study those introduced by Zaffalon et al. [18].

It would also be possible to test other classifiers than the EkNN. We could also test the advantage of these correction mechanisms in classifiers fusion problems.

At last, we also intend to investigate the learning from soft labels using another measure than $\tilde{E}_{pl}$ and in particular the evidential likelihood introduced by Denœux [4] and already used to develop a CD-based EkNN [9].

Acknowledgement. The authors would like to thank the anonymous reviewers for their helpful and constructive comments, which have helped them to improve the quality of the paper and to consider new paths for future research.

Mrs. Mutmainah's research is supported by the overseas 5000 Doctors program of Indonesian Religious Affairs Ministry (MORA French Scholarship).

References

1. Côme, E., Oukhellou, L., Denœux, T., Aknin, P.: Learning from partially supervised data using mixture models and belief functions. Pattern Recogn. **42**(3), 334–348 (2009)

2. Denoeux, T.: A k-nearest neighbor classification rule based on Dempster-Shafer theory. IEEE Trans. Syst. Man Cybern. **25**(5), 804–813 (1995)
3. Denœux, T.: Conjunctive and disjunctive combination of belief functions induced by nondistinct bodies of evidence. Artif. Intell. **172**, 234–264 (2008)
4. Denœux, T.: Maximum likelihood estimation from uncertain data in the belief function framework. IEEE Trans. Knowl. Data Eng. **25**(1), 119–130 (2013)
5. Dubois, D., Prade, H.: A set-theoretic view of belief functions: logical operations and approximations by fuzzy sets. Int. J. Gen. Syst. **12**(3), 193–226 (1986)
6. Dua, D., Graff, C.: UCI Machine Learning Repository. School of Information and Computer Science, University of California, Irvine (2019). http://archive.ics.uci.edu/ml
7. Elouedi, Z., Mellouli, K., Smets, P.: The evaluation of sensors' reliability and their tuning for multisensor data fusion within the transferable belief model. In: Benferhat, S., Besnard, P. (eds.) ECSQARU 2001. LNCS (LNAI), vol. 2143, pp. 350–361. Springer, Heidelberg (2001). https://doi.org/10.1007/3-540-44652-4_31
8. Kanjanatarakul, O., Kuson, S., Denoeux, T.: An evidential K-nearest neighbor classifier based on contextual discounting and likelihood maximization. In: Destercke, S., Denoeux, T., Cuzzolin, F., Martin, A. (eds.) BELIEF 2018. LNCS (LNAI), vol. 11069, pp. 155–162. Springer, Cham (2018). https://doi.org/10.1007/978-3-319-99383-6_20
9. Kanjanatarakul, O., Kuson, S., Denœux, T.: A new evidential k-nearest neighbor rule based on contextual discounting with partially supervised learning. Int. J. Approx. Reason. **113**, 287–302 (2019)
10. Mercier, D., Quost, B., Denœux, T.: Refined modeling of sensor reliability in the belief function framework using contextual discounting. Inf. Fusion **9**(2), 246–258 (2008)
11. Mercier, D., Lefèvre, E., Delmotte, F.: Belief functions contextual discounting and canonical decompositions. Int. J. Approx. Reason. **53**(2), 146–158 (2012)
12. Pichon, F., Dubois, D., Denoeux, T.: Relevance and truthfulness in information correction and fusion. Int. J. Approx. Reason. **53**(2), 159–175 (2012)
13. Pichon, F., Mercier, D., Lefèvre, E., Delmotte, F.: Proposition and learning of some belief function contextual correction mechanisms. Int. J. Approx. Reason. **72**, 4–42 (2016)
14. Quost, B., Denoeux, T., Li, S.: Parametric classification with soft labels using the evidential EM algorithm: linear discriminant analysis versus logistic regression. Adv. Data Anal. Classif. **11**(4), 659–690 (2017)
15. Shafer, G.: A Mathematical Theory of Evidence. Princeton University Press, Princeton (1976)
16. Smets, P.: Belief functions: the disjunctive rule of combination and the generalized Bayesian theorem. Int. J. Approx. Reason. **9**(1), 1–35 (1993)
17. Smets, P., Kennes, R.: The transferable belief model. Artif. Intell. **66**(2), 191–234 (1994)
18. Zafallon, M., Corani, G., Mauá, D.-D.: Evaluating credal classifiers by utility-discounted predictive accuracy. Int. J. Approx. Reason. **53**(8), 1282–1301 (2012)

Efficient Möbius Transformations and Their Applications to D-S Theory

Maxime Chaveroche[(✉)][iD], Franck Davoine[iD], and Véronique Cherfaoui

Alliance Sorbonne Université, Université de Technologie de Compiègne, CNRS,
Laboratoire Heudiasyc, 57 Avenue de Landshut, 60200 Compiègne, France
{maxime.chaveroche,franck.davoine,veronique.cherfaoui}@hds.utc.fr

Abstract. Dempster-Shafer Theory (DST) generalizes Bayesian probability theory, offering useful additional information, but suffers from a high computational burden. A lot of work has been done to reduce the complexity of computations used in information fusion with Dempster's rule. The main approaches exploit either the structure of Boolean lattices or the information contained in belief sources. Each has its merits depending on the situation. In this paper, we propose sequences of graphs for the computation of the zeta and Möbius transformations that optimally exploit both the structure of distributive lattices and the information contained in belief sources. We call them the *Efficient Möbius Transformations* (EMT). We show that the complexity of the EMT is always inferior to the complexity of algorithms that consider the whole lattice, such as the *Fast Möbius Transform* (FMT) for all DST transformations. We then explain how to use them to fuse two belief sources. More generally, our EMTs apply to any function in any finite distributive lattice, focusing on a meet-closed or join-closed subset.

Keywords: Zeta transform · Möbius transform · Distributive lattice · Meet-closed subset · Join-closed subset · Fast Möbius Transform · FMT · Dempster-Shafer Theory · DST · Belief functions · Efficiency · Information-based · Complexity reduction

1 Introduction

Dempster-Shafer Theory (DST) [11] is an elegant formalism that generalizes Bayesian probability theory. It is more expressive by giving the possibility for a source to represent its belief in the state of a variable not only by assigning credit directly to a possible state (strong evidence) but also by assigning credit to any subset (weaker evidence) of the set Ω of all possible states. This assignment of credit is called a *mass function* and provides meta-information to quantify

This work was carried out and co-funded in the framework of the Labex MS2T and the Hauts-de-France region of France. It was supported by the French Government, through the program "Investments for the future" managed by the National Agency for Research (Reference ANR-11-IDEX-0004-02).

N. Ben Amor et al. (Eds.): SUM 2019, LNAI 11940, pp. 390–403, 2019.
https://doi.org/10.1007/978-3-030-35514-2_29

the level of uncertainty about one's believes considering the way one established them, which is critical for decision making.

Nevertheless, this information comes with a cost: considering $2^{|\Omega|}$ potential values instead of only $|\Omega|$ can lead to computationally and spatially expensive algorithms. They can become difficult to use for more than a dozen possible states (e.g. 20 states in Ω generate more than a million subsets), although we may need to consider large frames of discernment (e.g. for classification or identification). Moreover, these algorithms not being tractable anymore beyond a few dozen states means their performances greatly degrade before that, which further limits their application to real-time applications. To tackle this issue, a lot of work has been done to reduce the complexity of transformations used to combine belief sources with Dempster's rule [6]. We distinguish between two approaches that we call *powerset-based* and *evidence-based*.

The *powerset-based* approach concerns all algorithms based on the structure of the powerset 2^{Ω} of the frame of discernment Ω. They have a complexity dependent on $|\Omega|$. Early works [1,7,12,13] proposed optimizations by restricting the structure of evidence to only singletons and their negation, which greatly restrains the expressiveness of the DST. Later, a family of optimal algorithms working in the general case, i.e. the ones based on the *Fast Möbius Transform* (FMT) [9], was discovered. Their complexity is $O(|\Omega|.2^{|\Omega|})$ in time and $O(2^{|\Omega|})$ in space. It has become the de facto standard for the computation of every transformation in DST. Consequently, efforts were made to reduce the size of Ω to benefit from the optimal algorithms of the FMT. More specifically, [14] refers to the process of conditioning by the *combined core* (intersection of the unions of all *focal sets* of each belief source) and *lossless coarsening* (merging of elements of Ω which always appear together in focal sets). Also, Monte Carlo methods [14] have been proposed but depend on a number of trials that must be large and grows with $|\Omega|$, in addition to not being exact.

The *evidence-based* approach concerns all algorithms that aim to reduce the computations to the only subsets that contain information (*evidence*), called *focal sets* and usually far less numerous than $2^{|\Omega|}$. This approach, also refered as the *obvious* one, implicitly originates from the seminal work of Shafer [11] and is often more efficient than the powerset-based one since it only depends on information contained in sources in a quadratic way. Doing so, it allows for the exploitation of the full potential of DST by enabling us to choose any frame of discernment, without concern about its size. Moreover, the evidence-based approach benefits directly from the use of approximation methods, some of which are very efficient [10]. Therefore, this approach seems superior to the FMT in most use cases, above all when $|\Omega|$ is large, where an algorithm with exponential complexity is just intractable.

It is also possible to easily find evidence-based algorithms computing all DST transformation, except for the conjunctive and disjunctive decompositions for which we recently proposed a method [4].

However, since these algorithms rely only on the information contained in sources, they do not exploit the structure of the powerset to reduce the complexity, leading to situations in which the FMT can be more efficient if almost every subset contains information, i.e. if the number of focal sets tends towards $2^{|\Omega|}$ [14], all the most when no approximation method is employed.

In this paper, we fuse these two approaches into one, proposing new sequences of graphs, in the same fashion as the FMT, that are always more efficient than the FMT and can in addition benefit from evidence-based optimizations. We call them the *Efficient Möbius Transformations* (EMT). More generally, our approach applies to any function defined on a finite distributive lattice.

Outside the scope of DST, [2] is related to our approach in the sense that we both try to remove redundancy in the computation of the zeta and Möbius transforms on the subset lattice 2^{Ω}. However, they only consider the redundancy of computing the image of a subset that is known to be null beforehand. To do so, they only visit sets that are accessible from the focal sets of lowest rank by successive unions with each element of Ω. Here, we demonstrate that it is possible to avoid far more computations by reducing them to specific sets so that each image is only computed once. These sets are the focal points described in [4]. The study of their properties will be carried out in depth in an upcoming article [5]. Besides, our method is more general since it applies to any finite distributive lattice.

Furthermore, an important result of our work resides in the optimal computation of the zeta and Möbius transforms in any intersection-closed family F of sets from 2^{Ω}, i.e. with a complexity $O(|\Omega|.|F|)$. Indeed, in the work of [3] on the optimal computation of these transforms in any finite lattice L, they embedded L into the Boolean lattice 2^{Ω}, obtaining an intersection-closed family F as its equivalent, and found a meta-procedure building a circuit of size $O(|\Omega|.|F|)$ computing the zeta and Möbius transforms. However, they did not managed to build this circuit in less than $O(|\Omega|.2^{|\Omega|})$. Given F, our Theorem 2 in this paper directly computes this circuit in $O(|\Omega|.|F|)$, while being much simpler.

This paper is organized as follows: Sect. 2 will present the elements on which our method is built. Section 3 will present our EMT. Section 4 will discuss their complexity and their usage in DST. Finally, we will conclude this article with Sect. 5.

2   Background of Our Method

Let $(P, \leq)$ be a finite[1] set partially ordered by $\leq$.

Zeta Transform. The zeta transform $g : P \to \mathbb{R}$ of a function $f : P \to \mathbb{R}$ is defined as follows:

$$\forall y \in P, \quad g(y) = \sum_{x \leq y} f(x)$$

---

[1] The following definitions hold for lower semifinite partially ordered sets as well, i.e. partially ordered sets such that the number of elements of P lower in the sense of $\leq$ than another element of P is finite. But for the sake of simplicity, we will only talk of finite partially ordered sets.

For example, the commonality function q (resp. the implicability function b) in DST is the zeta transform of the mass function m for $(2^\Omega, \supseteq)$ (resp. $(2^\Omega, \subseteq)$).

Möbius Transform. The Möbius transform of g is f. It is defined as follows:

$$\forall y \in P, \quad f(y) = \sum_{x \leq y} g(x).\mu(x,y) \tag{1}$$

where μ is the Möbius function of P.

There is also a multiplicative version with the same properties in which the sum is replaced by a product. An example of this version would be the inverse of the conjunctive (resp. disjunctive) weight function in DST which is the multiplicative Möbius transform of the commonality (resp. implicability) function.

2.1 Sequence of Graphs and Computation of the Zeta Transform

Consider a procedure $\mathfrak{A} : (\mathbb{R}^P, \mathcal{G}_{P,\leq}, \{+, -, \cdot, /\}) \to \mathbb{R}^P$, where $\mathbb{R}^P$ is the set of functions of domain P and range $\mathbb{R}$, and $\mathcal{G}_{P,\leq}$ is the set of acyclic directed graphs in which every node is in P and every arrow is a pair $(x, y) \in P^2$ such that $x \leq y$. For any such function m and graph G, the procedure $\mathfrak{A}(m, G, +)$ outputs a function z such that, for every $y \in P$, $z(y)$ is the sum of every $m(x)$ where (x, y) is an arrow of G. We define its reverse procedure as $\mathfrak{A}(z, G, -)$, which outputs the function m' such that, for every $y \in P$, $m'(y)$ is the sum, for every arrow (x, y) of G, of $z(x)$ if $x = y$, and $-z(x)$ otherwise. If the arrows of G represent all pairs of P ordered by $\leq$, then $\mathfrak{A}(m, G, +)$ computes the zeta transform z of m. Note however that $\mathfrak{A}(z, G, -)$ does not output the Möbius transform m of z. For that, G has to be broken down into a sequence of subgraphs (e.g. one subgraph per rank of y, in order of increasing rank).

Moreover, the upper bound complexity of these procedures, if G represent all pairs of P ordered by $\leq$, is $O(|P|^2)$. Yet, it is known that the optimal upper bound complexity of the computation of the zeta and Möbius transforms if P is a finite lattice is $O(|^\vee \mathcal{I}(P)|.|P|)$ (see [3]). Thus, a decomposition of these procedures should lead to a lower complexity at least in this case.

For this, Theorem 3 of [9] defines a necessary and sufficient condition to verify that $\mathfrak{A}(\mathfrak{A}(\ldots(\mathfrak{A}(m, H_1, +), \ldots), H_{k-1}, +), H_k, +) = \mathfrak{A}(m, G_\leq, +)$, where H_i is the i-th directed acyclic graph of a sequence H of size k, and $G_\leq = \{(x, y) \in P^2 / x \leq y\}$. For short, it is said in [9] that H computes the Möbius transformation of $G_\leq$. Here, in order to dissipate any confusion, we will say instead that H computes the zeta transformation of $G_\leq$.

It is stated in our terms as follows: H computes the zeta transformation of $G_\leq$ if and only if every arrow from each H_i is in $G_\leq$ and every arrow g from $G_\leq$ can be decomposed as a unique path $(g_1, g_2, \ldots, g_{|H|}) \in H_1 \times H_2 \times \cdots \times H_{|H|}$, i.e. such that the tail of g is the one of g_1, the head of g is the one of $g_{|H|}$, and $\forall i \in \{1, \ldots, |H| - 1\}$, the head of g_i is the tail of g_{i+1}.

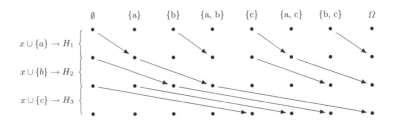

Fig. 1. Illustration representing the arrows contained in the sequence H computing the zeta transformation of $G_\subseteq = \{(X,Y) \in 2^\Omega \times 2^\Omega / X \subseteq Y\}$, where $\Omega = \{a,b,c\}$. For the sake of clarity, identity arrows are not displayed. This representation is derived from the one used in [9].

Application to the Boolean Lattice 2^Ω (FMT). Let $\Omega = \{\omega_1, \omega_2, \ldots, \omega_n\}$. The sequence H of graphs H_i computes the zeta transformation of $G_\subseteq = \{(X,Y) \in 2^\Omega \times 2^\Omega / X \subseteq Y\}$ if:

$$H_i = \{(X,Y) \in 2^\Omega \times 2^\Omega / Y = X \text{ or } Y = X \cup \{\omega_i\}\},$$

where $i \in \{1, \ldots, n\}$. Figure 1 illustrates the sequence H.

Dually, the sequence $\overline{H}$ of graphs $\overline{H}_i$ computes the zeta transformation of $G_\supseteq = \{(X,Y) \in 2^\Omega \times 2^\Omega / X \supseteq Y\}$ if:

$$\overline{H}_i = \{(X,Y) \in 2^\Omega \times 2^\Omega / X = Y \text{ or } X = Y \cup \{\omega_i\}\}.$$

The sequences of graphs H and $\overline{H}$ are the foundation of the FMT algorithms. Their execution is $O(n.2^n)$ in time and $O(2^n)$ in space.

2.2 Sequence of Graphs and Computation of the Möbius Transform

Now, consider that we have a sequence H computing the zeta transformation of $G_\leq$. It easy to see that the procedure $\mathfrak{A}(\ldots(\mathfrak{A}(\mathfrak{A}(z, H_k, -), H_{k-1}, -), \ldots), H_1, -)$ deconstructs $z = \mathfrak{A}(\mathfrak{A}(\ldots(\mathfrak{A}(m, H_1, +), \ldots), H_{k-1}, +), H_k, +)$, revisiting every arrow in H, as required to compute the Möbius transformation. But, to actually compute the Möbius transformation and get m back with H and $\mathfrak{A}$, we have to make sure that the images of z that we add through $\mathfrak{A}$ do not bear redundancies (e.g. if H is the sequence that only contains $G_\leq$, then H does compute the Möbius transformation of $G_\leq$ with Eq. 1, but not with $\mathfrak{A}$). For this, we only have to check that for each arrow (x,y) in $G_\leq$, there exists at most one path $(g_1, \ldots, g_p) \in H_{i_1} \times \cdots \times H_{i_p}$ where $p \in \mathbb{N}^*$ and $\forall j \in \{1, \ldots, p-1\}$, $1 \leq i_j \leq i_{j+1} \leq i_j + 1 \leq |H|$ and either $\text{tail}(g_j) \neq \text{head}(g_j)$ or $i_{j-1} < i_j < i_{j+1}$ (i.e. which moves right or down in Fig. 1). With this, we know that we do not subtract two images z_1 and z_2 to a same z_3 if one of z_1 and z_2 is supposed to be subtracted from the other beforehand. In the end, it is easy to see that, if for each graph H_i, all element $y \in P$ such that $(x,y) \in H_i$ and $(y, y') \in H_i$ where $x \neq y$ verifies $y' = y$ (i.e. no "horizontal" path of more than one arrow in each H_i), then the condition is already satisfied by the one of Sect. 2.1. So, if this condition is satisfied, we will say that *H computes the Möbius transformation of* $G_\leq$.

Application to the Boolean Lattice 2^Ω **(FMT).** Resuming the application of Sect. 2.1, for all $X \in 2^\Omega$, if $\omega_i \notin X$, then there is an arrow (X, Y) in H_i where $Y = X \cup \{\omega_i\}$ and $X \neq Y$, but then for any set Y' such that $(Y, Y') \in H_i$, we have $Y' = Y \cup \{\omega_i\} = Y$. Conversely, if $\omega_i \in X$, then the arrow $(X, X \cup \{\omega_i\})$ is in H_i, but its head and tail are equal. Thus, H also computes the Möbius transformation of $G_\subseteq$.

2.3   Order Theory

Irreducible Elements. We note $^\vee\mathcal{I}(P)$ the set of join-irreducible elements of P, i.e. the elements i such that $i \neq \bigwedge P$ for which it holds that $\forall x, y \in P$, if $x < i$ and $y < i$, then $x \vee y < i$. Dually, we note $^\wedge\mathcal{I}(P)$ the set of meet-irreducible elements of P, i.e. the elements i such that $i \neq \bigvee P$ for which it holds that $\forall x, y \in P$, if $x > i$ and $y > i$, then $x \wedge y > i$. For example, in the Boolean lattice 2^Ω, the join-irreducible elements are the singletons $\{\omega\}$, where $\omega \in \Omega$.

If P is a finite lattice, then every element of P is the join of join-irreducible elements and the meet of meet-irreducible elements.

Support of a Function in P. The support $\mathrm{supp}(f)$ of a function $f : P \to \mathbb{R}$ is defined as $\mathrm{supp}(f) = \{x \in P/f(x) \neq 0\}$.

For example, in DST, the set of focal elements of a mass function m is $\mathrm{supp}(m)$.

2.4   Focal Points

For any function $f : P \to \mathbb{R}$, we note $^\wedge\mathrm{supp}(f)$ (resp. $^\vee\mathrm{supp}(f)$) the smallest meet-closed (resp. join-closed) subset of P containing $\mathrm{supp}(f)$, i.e.:

$$^\wedge\mathrm{supp}(f) = \{x/\exists S \subseteq \mathrm{supp}(f), \ S \neq \emptyset, x = \bigwedge_{s \in S} s\}$$

$$^\vee\mathrm{supp}(f) = \{x/\exists S \subseteq \mathrm{supp}(f), \ S \neq \emptyset, x = \bigvee_{s \in S} s\}$$

The set of *focal points* $\mathring{\mathcal{F}}$ of a mass function m from [4] for the conjunctive weight function is $^\wedge\mathrm{supp}(m)$. For the disjunctive one, it is $^\vee\mathrm{supp}(m)$.

It has been proven in [4] that the image of 2^Ω through the conjunctive weight function can be computed without redundancies by only considering the focal points $^\wedge\mathrm{supp}(m)$ in the definition of the multiplicative Möbius transform of the commonality function. The image of all set in $2^\Omega \backslash ^\wedge\mathrm{supp}(m)$ through the conjunctive weight function is 1. The same can be stated for the disjunctive weight function regarding the implicability function and $^\vee\mathrm{supp}(m)$. In the same way, the image of any set in $2^\Omega \backslash ^\wedge\mathrm{supp}(m)$ through the commonality function is only a duplicate of the image of a set in $^\wedge\mathrm{supp}(m)$ and can be recovered by searching for its smallest superset in $^\wedge\mathrm{supp}(m)$. In fact, as generalized in an upcoming article [5], for any function $f : P \to \mathbb{R}$, $^\wedge\mathrm{supp}(f)$ is sufficient to define its zeta and Möbius transforms based on the partial order $\geq$, and $^\vee\mathrm{supp}(f)$ is sufficient to define its zeta and Möbius transforms based on the partial order $\leq$.

However, considering the case where P is a finite lattice, naive algorithms that only consider $^\wedge\mathrm{supp}(f)$ or $^\vee\mathrm{supp}(f)$ have upper bound complexities in $O(|^\wedge\mathrm{supp}(f)|^2)$ or $O(|^\vee\mathrm{supp}(f)|^2)$, which may be worse than the optimal complexity $O(|^\vee\mathcal{I}(P)|.|P|)$ for a procedure that considers the whole lattice P. In this paper, we propose algorithms with complexities always less than $O(|^\vee\mathcal{I}(P)|.|P|)$ computing the image of a meet-closed (e.g. $^\wedge\mathrm{supp}(f)$) or join-closed (e.g. $^\vee\mathrm{supp}(f)$) subset of P through the zeta or Möbius transform, provided that P is a finite distributive lattice.

3   Our Efficient Möbius Transformations

In this section, we consider a function $f : P \to \mathbb{R}$ where P is a finite distributive lattice (e.g. the Boolean lattice 2^Ω). We present here our Efficient Möbius Transformations as Theorems 1 and 2. The first one describes a way of computing the zeta and Möbius transforms of a function based on the smallest sublattice $^\mathcal{L}\mathrm{supp}(f)$ of P containing both $^\wedge\mathrm{supp}(f)$ and $^\vee\mathrm{supp}(f)$, which is defined in Proposition 2. The second one goes beyond this optimization by computing these transforms based only on $^\wedge\mathrm{supp}(f)$ or $^\vee\mathrm{supp}(f)$. Nevertheless, this second approach requires the direct computation of $^\wedge\mathrm{supp}(f)$ or $^\vee\mathrm{supp}(f)$, which has an upper bound complexity of $O(|\mathrm{supp}(f)|.|^\wedge\mathrm{supp}(f)|)$ or $O(|\mathrm{supp}(f)|.|^\vee\mathrm{supp}(f)|)$, which may be more than $O(|^\vee\mathcal{I}(P)|.|P|)$ if $|\mathrm{supp}(f)| \gg |^\vee\mathcal{I}(P)|$.

Lemma 1 (Safe join). *Let us consider a finite distributive lattice L. For all $i \in {}^\vee\mathcal{I}(L)$ and for all $x, y \in L$ such that $i \not\leq x$ and $i \not\leq y$, we have $i \not\leq x \vee y$.*

Proof. *By definition of a join-irreducible element, we know that $\forall i \in {}^\vee\mathcal{I}(L)$ and for all $a, b \in L$, if $a < i$ and $b < i$, then $a \vee b < i$. Moreover, for all $x, y \in L$ such that $i \not\leq x$ and $i \not\leq y$, we have equivalently $i \wedge x < i$ and $i \wedge y < i$. Thus, we get that $(i \wedge x) \vee (i \wedge y) < i$. Since L satisfies the distributive law, this implies that $(i \wedge x) \vee (i \wedge y) = i \wedge (x \vee y) < i$, which means that $i \not\leq x \vee y$.*

Proposition 1 (Iota elements of subsets of P). *For any $S \subseteq P$, the join-irreducible elements of the smallest sublattice L_S of P containing S are:*

$$\iota(S) = \left\{ \bigwedge \{s \in S / s \geq i\} / i \in {}^\vee\mathcal{I}(P) \text{ and } \exists s \in S, s \geq i \right\}.$$

Proof. *First, it can be easily shown that the meet of any two elements of $\iota(S)$ is either $\bigwedge S$ or in $\iota(S)$. Then, suppose that we generate L_S with the join of elements of $\iota(S)$, to which we add the element $\bigwedge S$. Then, since P is distributive, we have that for all $x, y \in L_S$, their meet $x \wedge y$ is either $\bigwedge S$ or equal to the join of every meet of pairs $(i_{S,x}, i_{S,y}) \in \iota(S)^2$, where $i_{S,x} \leq x$ and $i_{S,y} \leq y$. Thus, $x \wedge y \in L_S$, which implies that L_S is a sublattice of P. In addition, notice that for each nonzero element $s \in S$ and for all $i \in {}^\vee\mathcal{I}(P)$ such that $s \geq i$, we also have by construction $s \geq i_S \geq i$, where $i_S = \bigwedge \{s \in S / s \geq i\}$. Therefore, we* have $s = \bigvee\{i \in {}^{\vee}\mathcal{I}(P)/s \geq i\} = \bigvee\{i \in \iota(S)/s \geq i\}$, i.e. $s \in L_S$. Besides, if $\bigwedge P \in S$, then it is equal to $\bigwedge S$, which is also in L_S by construction. So, $S \subseteq L_S$. It follows that the meet or join of every nonempty subset of S is in L_S, i.e. $M_S \subseteq L_S$ and $J_S \subseteq L_S$, where M_S is the smallest meet-closed subset of P containing S and J_S is the smallest join-closed subset of P containing S. Furthermore, $\iota(S) \subseteq M_S$ which means that we cannot build a smaller sublattice of P containing S. Therefore, L_S is the smallest sublattice of P containing S.

Finally, for any $i \in {}^{\vee}\mathcal{I}(P)$ such that $\exists s \in S, s \geq i$, we note $i_S = \bigwedge\{s \in S/s \geq i\}$. For all $x, y \in L_S$, if $i_S > x$ and $i_S > y$, then by construction of $\iota(S)$, we have $i \not\leq x$ and $i \not\leq y$ (otherwise, i_S would be less than x or y), which implies by Lemma 1 that $i \not\leq x \vee y$. Since $i \leq i_S$, we have necessarily $i_S > x \vee y$. Therefore, i_S is a join-irreducible element of L_S.

Proposition 2 (Lattice support). The smallest sublattice of P containing both $^{\wedge}supp(f)$ and $^{\vee}supp(f)$, noted $^{\mathcal{L}}supp(f)$, can be defined as:

$$^{\mathcal{L}}supp(f) = \left\{\bigvee X/X \subseteq \iota(supp(f)), X \neq \emptyset\right\} \cup \left\{\bigwedge supp(f)\right\}.$$

More specifically, $^{\vee}supp(f)$ is contained in the upper closure $^{\mathcal{L},\uparrow}supp(f)$ of $supp(f)$ in $^{\mathcal{L}}supp(f)$:

$$^{\mathcal{L},\uparrow}supp(f) = \{x \in {}^{\mathcal{L}}supp(f)/\exists s \in supp(f), s \leq x\},$$

and $^{\wedge}supp(f)$ is contained in the lower closure $^{\mathcal{L},\downarrow}supp(f)$ of $supp(f)$ in $^{\mathcal{L}}supp(f)$:

$$^{\mathcal{L},\downarrow}supp(f) = \{x \in {}^{\mathcal{L}}supp(f)/\exists s \in supp(f), s \geq x\}.$$

These sets can be computed in less than respectively $O(|\iota(supp(f))|.|^{\mathcal{L},\uparrow}supp(f)|)$ and $O(|\iota(supp(f))|.|^{\mathcal{L},\downarrow}supp(f)|)$, which is at most $O(|^{\vee}\mathcal{I}(P)|.|P|)$.

Proof. The proof is immediate here, considering Proposition 1 and its proof. In addition, since $^{\wedge}supp(f)$ only contains the meet of elements of $supp(f)$, all element of $^{\wedge}supp(f)$ is less than at least one element of $supp(f)$. Similarly, since $^{\vee}supp(f)$ only contains the join of elements of $supp(f)$, all element of $^{\vee}supp(f)$ is greater than at least one element of $supp(f)$. Hence $^{\mathcal{L},\downarrow}supp(f)$ and $^{\mathcal{L},\uparrow}supp(f)$.

As pointed out in [8], a special ordering of the join-irreducible elements of a lattice when using the Fast Zeta Transform [3] leads to the optimal computation of its zeta and Möbius transforms. Here, we use this ordering to build our EMT for finite distributive lattices in a way similar to [8] but without the need to check the equality of the decompositions into the first j join-irreducible elements at each step.

Corollary 1 (Join-irreducible ordering). Let us consider a finite distributive lattice L and let its join-irreducible elements $^{\vee}\mathcal{I}(L)$ be ordered such that $\forall i_k, i_l \in {}^{\vee}\mathcal{I}(L), k < l \Rightarrow i_k \not\geq i_l$. We note $^{\vee}\mathcal{I}(L)_k = \{i_1, \ldots, i_{k-1}, i_k\}$.

For all element $i_k \in {}^{\vee}\mathcal{I}(L)$, we have $i_k \not\leq \bigvee {}^{\vee}\mathcal{I}(L)_{k-1}$.

If L is a graded lattice (i.e. a lattice equipped with a rank function $\rho : L \to \mathbb{N}$), then $\rho(i_1) \leq \rho(i_2) \leq \cdots \leq \rho(i_{|^{\vee}\mathcal{I}(L)|})$ implies this ordering. For example, in DST, $P = 2^{\Omega}$, so for all $A \in P$, $\rho(A) = |A|$.

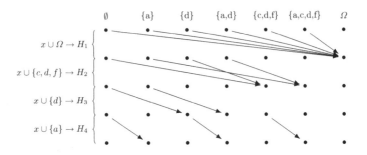

Fig. 2. Illustration representing the arrows contained in the sequence H when computing the zeta transformation of $G_\subseteq = \{(x,y) \in L^2/x \subseteq y\}$, where $L = \{\emptyset, \{a\}, \{d\}, \{a,d\}, \{c,d,f\}, \{a,c,d,f\}, \Omega\}$ with $\Omega = \{a,b,c,d,e,f\}$ and $^\vee\mathcal{I}(L) = \{\{a\}, \{d\}, \{c,d,f\}, \Omega\}$. For the sake of clarity, identity arrows are not displayed.

Proof. *Since the join-irreducible elements are ordered such that $\forall i_k, i_l \in {}^\vee\mathcal{I}(L)$, $k < l \Rightarrow i_k \not\geq i_l$, it is trivial to see that for any $i_l \in {}^\vee\mathcal{I}(L)$ and $i_k \in {}^\vee\mathcal{I}(L)_{l-1}$, we have $i_k \not\geq i_l$. Then, using Lemma 1 by recurrence, it is easy to get that $i_l \not\leq \bigvee {}^\vee\mathcal{I}(L)_{l-1}$.*

Theorem 1 (Efficient Möbius Transformation in a distributive lattice). *Let us consider a finite distributive lattice L (such as $^\mathcal{L}supp(f)$) and let its join-irreducible elements $^\vee\mathcal{I}(L)$ be ordered such that $\forall i_k, i_l \in {}^\vee\mathcal{I}(L)$, $k < l \Rightarrow i_k \not\geq i_l$. We note $n = |^\vee\mathcal{I}(L)|$.*

The sequence H of graphs H_k computes the zeta and Möbius transformations of $G_\leq = \{(x,y) \in L^2/x \leq y\}$ if:

$$H_k = \left\{(x,y) \in L^2/y = x \ \ or \ y = x \vee i_{\overline{k}}\right\},$$

where $\overline{k} = n + 1 - k$. This sequence is illustrated in Fig. 2. Its execution is $O(n.|L|)$.

Dually, the sequence $\overline{H}$ of graphs $\overline{H}_k$ computes the zeta and Möbius transformations of $G_\geq = \{(x,y) \in L^2/x \geq y\}$ if:

$$\overline{H}_k = \left\{(x,y) \in L^2/x = y \ \ or \ x = y \vee i_k\right\}.$$

Proof. *By definition, for all k and $\forall(x,y) \in H_k$, we have $x, y \in L$ and $x \leq y$, i.e. $(x,y) \in G_\leq$. Reciprocally, $\forall(x,y) \in G_\leq$, we have $x \leq y$, which can be decomposed as a unique path $(g_1, g_2, \ldots, g_n) \in H_1 \times H_2 \times \cdots \times H_n$:*

Similarly to the FMT, the sequence H builds unique paths simply by generating the whole lattice step by step with each join-irreducible element of L. However, unlike the FMT, the join-irreducible elements of L are not necessarily atoms. Doing so, pairs of join-irreducible elements may be ordered, causing the sequence H to skip or double some elements. And even if all the join-irreducible elements of L are atoms, since L is not necessarily a Boolean lattice, the join of two atoms may be greater than a third atom (e.g. if L is the diamond lattice), *leading to the same issue. Indeed, to build a unique path between two elements x, y of L such that $x \leq y$, we start from x. Then at step 1, we get to the join $x \vee i_n$ if $i_n \leq y$ (we stay at x otherwise, i.e. identity arrow), then we get to $x \vee i_n \vee i_{n-1}$ if $i_{n-1} \leq y$, and so on until we get to y. However, if we have $i_n \leq x \vee i_{n-1}$, with $i_n \nleq x$, then there are at least two paths from x to y: one passing by the join with i_n at step 1 and one passing by the identity arrow instead.*

More generally, this kind of issue may only appear if there is a k where $i_k \leq x \vee \bigvee {}^\vee\mathcal{I}(L)_{k-1}$ with $i_k \nleq x$, where ${}^\vee\mathcal{I}(L)_{k-1} = \{i_{k-1}, i_{k-2}, \ldots, i_1\}$. But, since L is a finite distributive lattice, and since its join-irreducible elements are ordered such that $\forall i_j, i_l \in {}^\vee\mathcal{I}(L)$, $j < l \Rightarrow i_j \ngeq i_l$, we have by Corollary 1 that $i_k \nleq \bigvee {}^\vee\mathcal{I}(L)_{k-1}$. So, if $i_k \nleq x$, then by Lemma 1, we also have $i_k \nleq x \vee \bigvee {}^\vee\mathcal{I}(L)_{k-1}$. Thereby, there is a unique path from x to y, meaning that the condition of Sect. 2.1 is satisfied. H computes the zeta transformation of $G_\leq$.

Also, $\forall x \in L$, if $i_k \nleq x$, then there is an arrow (x, y) in H_k where $y = x \vee i_k$ and $x \neq y$, but then for any element y' such that $(y, y') \in H_k$, we have $y' = y \vee i_k = y$. Conversely, if $i_k \leq x$, then the arrow $(x, x \vee i_k)$ is in H_k, but its head and tail are equal. Thus, the condition of Sect. 2.2 is satisfied. H also computes the Möbius transformation of $G_\leq$.

Finally, to obtain $\overline{H}$, we only need to reverse the paths of H, i.e. reverse the arrows in each H_k and reverse the sequence of join-irreducible elements.

The procedure described in Theorem 1 to compute the zeta and Möbius transforms of a function on P is always less than $O(|{}^\vee\mathcal{I}(P)|.|P|)$. Its upper bound complexity for the distributive lattice $L = {}^\mathcal{L}\text{supp}(f)$ is $O(|{}^\vee\mathcal{I}(L)|.|L|)$, which is actually the optimal one for a lattice.

Yet, we can reduce this complexity even further if we have ${}^\wedge\text{supp}(f)$ or ${}^\vee\text{supp}(f)$. This is the motivation behind the procedure decribed in the following Theorem 2. As a matter of fact, [3] proposed a meta-procedure producing an algorithm that computes the zeta and Möbius transforms in an arbitrary intersection-closed family F of sets of 2^Ω with a circuit of size $O(|\Omega|.|F|)$. However, this meta-procedure is $O(|\Omega|.2^{|\Omega|})$. Here, Theorem 2 provides a procedure that directly computes the zeta and Möbius transforms with the optimal complexity $O(|\Omega|.|F|)$, while being much simpler. Besides, our method is far more general since it has the potential (depending on data structure) to reach this complexity in any meet-closed subset of a finite distributive lattice.

Theorem 2 (Efficient Möbius Transformation in a join-closed or meet-closed subset of P). *Let us consider a meet-closed subset M of P (such as ${}^\wedge supp(f)$). Also, let the join-irreducible elements $\iota(M)$ be ordered such that $\forall i_k, i_l \in \iota(M)$, $k < l \Rightarrow i_k \ngeq i_l$.*

The sequence H^M of graphs H_k^M computes the zeta and Möbius transformations of $G_\geq^M = \{(x, y) \in M^2/x \geq y\}$ if:

$$H_k^M = \left\{ (x, y) \in M^2/x = y \right.$$

$$\left. or \left(x = \bigwedge \{s \in M/s \geq y \vee i_k\} \ and \ y \vee \bigvee \iota(M)_k \geq x \right) \right\},$$

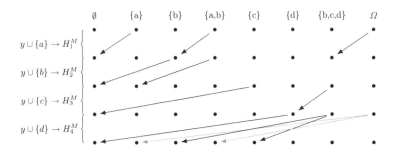

Fig. 3. Illustration representing the arrows contained in the sequence H^M when computing the zeta transformation of $G_{\supseteq}^M = \{(x,y) \in M^2/x \supseteq y\}$, where $M = \{\emptyset, \{a\}, \{b\}, \{a,b\}, \{c\}, \{d\}, \{b,c,d\}, \Omega\}$ with $\Omega = \{a,b,c,d\}$ and $\iota(M) = \{\{a\}, \{b\}, \{c\}, \{d\}\}$. For the sake of clarity, identity arrows are not displayed.

where $\iota(M)_k = \{i_1, i_2, \ldots, i_k\}$. This sequence is illustrated in Fig. 3. Its execution is $O(|\iota(M)|.|M|.\epsilon)$, where ϵ represents the number of operations required to obtain the proxy element $\bigwedge \{s \in M/s \geq y \vee i_k\}$ of x. It can be as low as 1 operation[2].

Dually, the expression of $\overline{H^M}$ follows the same pattern, simply reversing the paths of H^M by reversing the arrows in each H_k^M and reversing the sequence of join-irreducible elements.

Similarly, if P is a Boolean lattice, then the dual H^J of this sequence H^M of graphs computes the zeta and Möbius transformations of $G_{\leq}^J = \{(x,y) \in J^2/x \leq y\}$, where J is a join-closed subset of P (such as $\vee \operatorname{supp}(f)$). Let the meet-irreducible elements $\bar{\iota}(J)$ of the smallest sublattice of P containing J be ordered such that $\forall \bar{i}_k, \bar{i}_l \in \bar{\iota}(J)$, $k < l \Rightarrow \bar{i}_k \not\leq \bar{i}_l$. We have:

$$H_k^J = \Big\{ (x,y) \in J^2/x = y$$

$$or \ \Big(x = \bigvee \{s \in J/s \leq y \wedge \bar{i}_k\} \ \ and \ y \wedge \bigwedge \bar{\iota}(J)_k \leq x \Big) \Big\},$$

where $\bar{\iota}(J)_k = \{\bar{i}_1, \bar{i}_2, \ldots, \bar{i}_k\}$.

Dually, the expression of $\overline{H^J}$ follows the same pattern, simply reversing the paths of H^J by reversing the arrows in each H_k^J and reversing the sequence of meet-irreducible elements.

Proof. By definition, for all k and $\forall (x,y) \in H_k^M$, we have $x, y \in M$ and $x \geq y$, i.e. $(x,y) \in G_{\geq}^M$. Reciprocally, $\forall (x,y) \in G_{\geq}^M$, we have $x \geq y$, which can be decomposed as a unique path $(g_1, g_2, \ldots, g_{|\iota(M)|}) \in H_1^M \times H_2^M \times \cdots \times H_{|\iota(M)|}^M$:

---

[2] This unit cost can be obtained when $P = 2^\Omega$ using a dynamic binary tree as data structure for the representation of M. With it, finding the proxy element only takes the reading of a binary string, considered as one operation. Further details will soon be available in an extended version of this work [5].

The idea is that we use the same procedure as in Theorem 1 that builds unique paths simply by generating all elements of a finite distributive lattice L based on the join of its join-irreducible elements step by step, as if we had $M \subseteq L$, except that we remove all elements that are not in M. Doing so, the only difference is that the join $y \vee i_k$ of an element y of M with a join-irreducible $i_k \in \iota(M)$ of this hypothetical lattice L is not necessary in M. However, thanks to the meet-closure of M and to the synchronizing condition $y \vee \bigvee \iota(M)_k \geq p$, we can "jump the gap" between two elements y and p of M separated by elements of $L \backslash M$ and maintain the unicity of the path between any two elements x and y of M. Indeed, for all join-irreducible element $i_k \in \iota(M)$, if $x \geq y \vee i_k$, then since M is meet-closed, we have an element p of M that we call proxy such that $p = \bigwedge \{s \in M/s \geq y \vee i_k\}$. Yet, we have to make sure that (1) p can only be obtained from y with exactly one particular i_k if $p \neq y$, and (2) that the sequence of these particular join-irreducible elements forming the arrows of the path from x to y are in the correct order. This is the purpose of the synchronizing condition $y \vee \bigvee \iota(M)_k \geq p$.

For (1), we will show that for a same proxy p, it holds that $\exists!k \in [1, |\iota(M)|]$ such that $p \neq y$, $y \vee \bigvee \iota(M)_k \geq p$ and $y \not\geq i_k$. Recall that we ordered the elements $\iota(M)$ such that $\forall i_j, i_l \in \iota(M)$, $j < l \Rightarrow i_j \not\geq i_l$. Let us note k the greatest index among $[1, |\iota(M)|]$ such that $p \geq i_k$ and $y \not\geq i_k$. It is easy to see that the synchonizing condition is statisfied for i_k. Then, for all $j \in [1, k-1]$, Corollary 1 and Lemma 1 give us that $y \vee \bigvee \iota(M)_j \not\geq i_k$, meaning that $y \vee \bigvee \iota(M)_j \not\geq p$. For all $j \in [k+1, |\iota(M)|]$, either $y \geq i_j$ (i.e. $p = y \vee i_j = y$) or $p \not\geq i_j$. Either way, it is impossible to reach p from $y \vee i_j$. Therefore, there exists a unique path from y to p that takes the arrow (p, y) from H_k^M.

Concerning (2), for all $(x, y) \in G_{\geq}^M$, $x \neq y$, let us note the proxy element $p_1 = \bigwedge \{s \in M/s \geq y \vee i_{k_1}\}$ where k_1 is the greatest index among $[1, |\iota(M)|]$ such that $p_1 \geq i_{k_1}$ and $y \not\geq i_{k_1}$. We have $(p_1, y) \in H_{k_1}^M$. Let us suppose that there exists another proxy element p_2 such that $p_2 \neq p_1$, $x \geq p_2$ and $p_2 = \bigwedge \{s \in M/s \geq p_1 \vee i_{k_2}\}$ where k_2 is the greatest index among $[1, |\iota(M)|]$ such that $p_2 \geq i_{k_2}$ and $p_1 \not\geq i_{k_2}$. We have $(p_2, p_1) \in H_{k_2}^M$. Since $p_2 > p_1$ and $p_1 \geq i_{k_1}$, we have that $p_2 \geq i_{k_1}$, i.e. $k_2 \neq k_1$. So, two cases are possible: either $k_1 > k_2$ or $k_1 < k_2$. If $k_1 > k_2$, then there is a path $((p_2, p_1), (p_1, p_1), \ldots, (p_1, p_1), (p_1, y))$ from p_2 to y. Moreover, we know that at step k_1, we get p_1 from y and that we have $p_2 \geq i_{k_1}$ and $y \not\geq i_{k_1}$, meaning that there could only exist an arrow (p_2, y) in $H_{k_3}^M$ if $k_3 > k_1 > k_2$. Suppose this k_3 exists. Then, since $k_3 > k_1 > k_2$, we have that $p_2 \geq i_{k_3}$ and $y \not\geq i_{k_3}$, but also $p_1 \not\geq i_{k_3}$ since we would have $k_1 = k_3$ otherwise. This implies that $k_2 = k_3$, which is impossible. Therefore, there is no k_3 such that $(p_2, y) \in H_{k_3}^M$, i.e. there is a unique path from p_2 to y. Otherwise, if $k_1 < k_2$, then the latter path between p_2 and y does not exist. But, since $p_1 \not\geq i_{k_2}$ and $p_1 \geq y$, we have $y \not\geq i_{k_2}$, meaning that there exists an arrow $(p_2, y) \in H_{k_2}^M$, which forms a unique path from p_2 to y. The recurrence of this reasoning enables us to conclude that there is a unique path from x to y.

Thus, the condition of Sect. 2.1 is satisfied. H^M computes the zeta transformation of $G_{\geq}^M$. Also, for the same reasons as with Theorem 1, we have that H^M computes the Möbius transformation of $G_{\geq}^M$. The proof for H^J and $G_{\leq}^J$ is analog if P is a Boolean lattice.

4   Discussions for Dempster-Shafer Theory

In DST, we work with $P = 2^{\Omega}$, in which the singletons are its join-irreducible elements. If $|\text{supp}(f)|$ is of same order of magnitude as n or lower, where $n = |\Omega|$, then we can compute the focal points $^{\wedge}\text{supp}(f)$ or $^{\vee}\text{supp}(f)$ and use our Efficient Möbius Transformation of Theorem 2 to compute any DST transformation (e.g. the commonality/implicability function, the conjunctive/disjunctive weight function, etc, i.e. wherever the FMT applies) in at most $O(n.|\text{supp}(f)| + |\iota(\text{supp}(f))|.|^{R}\text{supp}(f)|)$ operations, where $R \in \{\wedge, \vee\}$, which is at most $O(n.2^n)$.

Otherwise, we can compute $^{\mathcal{L},\uparrow}\text{supp}(f)$ or $^{\mathcal{L},\downarrow}\text{supp}(f)$ of Proposition 2, and then use the Efficient Möbius Transformation of Theorem 1 to compute the same DST transformations in $O(n.|\text{supp}(f)| + |\iota(\text{supp}(f))|.|^{\mathcal{L},A}\text{supp}(f)|)$ operations, where $A \in \{\uparrow, \downarrow\}$, which is at most $O(n.2^n)$.

Therefore, we can always compute DST transformations more efficiently than the FMT with the EMT if $\text{supp}(f)$ is given.

Moreover, $^{\mathcal{L},\downarrow}\text{supp}(f)$ can be optimized if $\Omega \in \text{supp}(f)$ (which causes the equality $^{\mathcal{L},\downarrow}\text{supp}(f) = {}^{\mathcal{L}}\text{supp}(f)$). Indeed, one can equivalently compute the lattice $^{\mathcal{L},\downarrow}(\text{supp}(f)\backslash\{\Omega\})$, execute the EMT of Theorem 1, and then add the value on Ω to the value on all sets of $^{\mathcal{L},\downarrow}(\text{supp}(f)\backslash\{\Omega\})$. Dually, the same can be done with $^{\mathcal{L},\uparrow}(\text{supp}(f)\backslash\{\emptyset\})$. This trick can be particularly useful in the case of the conjunctive or disjunctive weight function, which requires that $\text{supp}(f)$ contains respectively Ω or $\emptyset$.

Also, optimizations built for the FMT, such as the reduction of Ω to the core $\mathcal{C}$ or its optimal coarsened version Ω', are already encoded in the use of the function ι (see Example 1), but optimizations built for the evidence-based approach, such as approximations by reduction of the number of focal sets, i.e. reducing the size of $\text{supp}(f)$, can still greatly enhance the EMT.

Finally, while it was proposed in [9] to fuse two mass functions m_1 and m_2 using Dempster's rule by computing the corresponding commonality functions q_1 and q_2 in $O(n.2^n)$, then $q_{12} = q_1.q_2$ in $O(2^n)$ and finally computing back the fused mass function m_{12} from q_{12} in $O(n.2^n)$, here we propose an even greater detour that has a lower complexity. Indeed, by computing q_1 and q_2 on $^{\wedge}\text{supp}(m_1)$ and $^{\wedge}\text{supp}(m_2)$, then the conjunctive weight functions w_1 and w_2 on these same elements, we get $w_{12} = w_1.w_2$ in $O(|^{\wedge}\text{supp}(m_1) \cup {}^{\wedge}\text{supp}(m_2)|)$ (all other set has a weight equal to 1). Consequently, we obtain the set $\text{supp}(1 - w_{12}) \subseteq {}^{\wedge}\text{supp}(m_1) \cup {}^{\wedge}\text{supp}(m_2)$ which can be used to compute $^{\wedge}\text{supp}(1 - w_{12})$ or $^{\mathcal{L},\downarrow}\text{supp}(1 - w_{12})$. From this, we simply compute q_{12} and then m_{12} in $O(n.|\text{supp}(1 - w_{12})| + |\iota(\text{supp}(1 - w_{12}))|.|^{\wedge}\text{supp}(1 - w_{12})|)$ or $O(n.|\text{supp}(1 - w_{12})| + |\iota(\text{supp}(1 - w_{12}))|.|^{\mathcal{L},\downarrow}\text{supp}(1 - w_{12})|)$.

Example 1 (Consonant case). If $\text{supp}(f) = \{F_1, F_2, \ldots, F_K\}$ such that $F_1 \subset F_2 \subset \cdots \subset F_K$, then the coarsening Ω' of Ω will have an element for each element of $\text{supp}(f)$, while $\iota(\text{supp}(f))$ will have a set of elements for each element of $\text{supp}(f)$. So, we get $|\Omega'| = |\iota(\text{supp}(f))| = K$. But, Ω' is then used to generate the Boolean lattice $2^{\Omega'}$, of size 2^K, where $\iota(\text{supp}(f))$ is used to generate an arbitrary lattice $^{\mathcal{L}}\text{supp}(f)$, of size K in this particular case ($K+1$ if $\emptyset \in \text{supp}(f)$).

5   Conclusion

In this paper, we proposed the *Efficient Möbius Transformations* (EMT), which are general procedures to compute the zeta and Möbius transforms of any function defined on any finite distributive lattice with optimal complexity. They are based on our reformulation of the Möbius inversion theorem with focal points only, featured in an upcoming detailed article [5] currently in preparation. The EMT optimally exploit the information contained in both the support of this function and the structure of distributive lattices. Doing so, the EMT always perform better than the optimal complexity for an algorithm considering the whole lattice, such as the FMT for all DST transformations, given the support of this function. In [5], we will see that our approach is still more efficient when this support is not given. This forthcoming article will also feature examples of application in DST, algorithms and implementation details.

References

1. Barnett, J.A.: Computational methods for a mathematical theory of evidence. In: Proceedings of IJCAI, vol. 81, pp. 868–875 (1981)
2. Björklund, A., Husfeldt, T., Kaski, P., Koivisto, M.: Trimmed moebius inversion and graphs of bounded degree. Theory Comput. Syst. **47**(3), 637–654 (2010)
3. Björklund, A., Husfeldt, T., Kaski, P., Koivisto, M., Nederlof, J., Parviainen, P.: Fast zeta transforms for lattices with few irreducibles. ACM TALG **12**(1), 4 (2016)
4. Chaveroche, M., Davoine, F., Cherfaoui, V.: Calcul exact de faible complexité des décompositions conjonctive et disjonctive pour la fusion d'information. In: Proceedings of GRETSI (2019)
5. Chaveroche, M., Davoine, F., Cherfaoui, V.: Efficient algorithms for Möbius transformations and their applications to Dempster-Shafer Theory. Manuscript available on request (2019)
6. Dempster, A.: A generalization of Bayesian inference. J. Roy. Stat. Soc. Ser. B (Methodol.) **30**, 205–232 (1968)
7. Gordon, J., Shortliffe, E.H.: A method for managing evidential reasoning in a hierarchical hypothesis space. Artif. Intell. **26**(3), 323–357 (1985)
8. Kaski, P., Kohonen, J., Westerbäck, T.: Fast Möbius inversion in semimodular lattices and U-labelable posets. arXiv preprint arXiv:1603.03889 (2016)
9. Kennes, R.: Computational aspects of the Mobius transformation of graphs. IEEE Trans. Syst. Man Cybern. **22**(2), 201–223 (1992)
10. Sarabi-Jamab, A., Araabi, B.N.: Information-based evaluation of approximation methods in Dempster-Shafer Theory. IJUFKS **24**(04), 503–535 (2016)
11. Shafer, G.: A Mathematical Theory of Evidence. Princeton University Press, Princeton (1976)
12. Shafer, G., Logan, R.: Implementing Dempster's rule for hierarchical evidence. Artif. Intell. **33**(3), 271–298 (1987)
13. Shenoy, P.P., Shafer, G.: Propagating belief functions with local computations. IEEE Expert **1**(3), 43–52 (1986)
14. Wilson, N.: Algorithms for Dempster-Shafer Theory. In: Kohlas, J., Moral, S. (eds.) Handbook of Defeasible Reasoning and Uncertainty Management Systems: Algorithms for Uncertainty and Defeasible Reasoning, vol. 5, pp. 421–475. Springer, Netherlands (2000). https://doi.org/10.1007/978-94-017-1737-3_10

Dealing with Continuous Variables in Graphical Models

Christophe Gonzales[✉]

Aix-Marseille Université, CNRS, LIS, Marseille, France
`christophe.gonzales@lis-lab.fr`

Abstract. Uncertain reasoning over both continuous and discrete random variables is important for many applications in artificial intelligence. Unfortunately, dealing with continuous variables is not an easy task. In this tutorial, we will study some of the methods and models developed in the literature for this purpose. We will start with the discretization of continuous random variables. A special focus will be made on the numerous issues they raise, ranging from which discretization criterion to use, to the appropriate way of using them during structure learning. These issues will justify the exploitation of hybrid models designed to encode mixed probability distributions. Several such models have been proposed in the literature. Among them, Conditional Linear Gaussian models are very popular. They can be used very efficiently for inference but they lack flexibility in the sense that they impose that the continuous random variables follow conditional Normal distributions and are related to other variables through linear relations. Other popular models are mixtures of truncated exponentials, mixtures of polynomials and mixtures of truncated basis functions. Through a clever use of mixtures of distributions, these models can approximate very well arbitrary mixed probability distributions. However, exact inference can be very time consuming in these models. Therefore, when choosing which model to exploit, one has to trade-off between the flexibility of the uncertainty model and the computational complexity of its learning and inference mechanisms.

Keywords: Continuous variable · Hybrid graphical model · Discretization

Since their introduction in the 80's, Bayesian networks (BN) have become one of the most popular model for handling "precise" uncertainties [24]. However, by their very definition, BNs are limited to cope only with discrete random variables. Unfortunately, in real-world applications, it is often the case that some variables are of a continuous nature. Dealing with such variables is challenging both for learning and inference tasks [9]. The goal of this tutorial is to investigate techniques used to cope with such variables and, more importantly, to highlight their pros and cons.

© Springer Nature Switzerland AG 2019
N. Ben Amor et al. (Eds.): SUM 2019, LNAI 11940, pp. 404–408, 2019.
https://doi.org/10.1007/978-3-030-35514-2_30

1   Mapping Continuous Variables into Discrete Ones

Probably, the simplest way to cope with continuous random variables in graphical models is to discretize them. Once the variables are discretized, these models can be learnt and exploited as usual [4,19,29]. However, appropriately discretizing variables raises many issues. First, when learning the graphical model, should all the variables be discretized independently or should the dependencies among variables learnt so far be taken into account to jointly discretize some sets of variables? The second alternative provides better results and is therefore advocated in the literature [7,18,21]. However, determining the best joint discretization is a complex task and only approximations are provided. In addition, when discretizing while learning the graphical model structure, it is tempting to define an overall function scoring both discretization and structure. Optimizing such a function therefore provides both an optimal structure and a discretization most suited for this structure. This is the approach followed in [7,21]. However, we shall see that this may prove to be a bad idea because many of the structure scoring functions represent posterior likelihoods and it is easy to construct discretizations resulting in infinite likelihoods whatever the structure. Choosing the criterion to optimize to determine the best discretization is also a challenge. Depending on the kind of observations that will be used subsequently in inferences, it may or may not be useful to consider uniform or non-uniform density functions within discretization intervals. As discretizing variables result in a loss of information, people often try to minimize this loss and therefore exploit entropy-based criteria to drive their search for the optimal discretization. While at first sight this seems a good idea, we will see that this may not be appropriate for structure learning and other criteria such as cluster-based optimization [18] or Kullback-Leibler divergence minimization [10] are probably much more appropriate. It should also be emphasized that inappropriate discretizations may have a significant impact on the learnt structure because, e.g., dependent continuous random variables may become independent when discretized. This is the very reason why it is proposed in [20] to compute independence tests at several different discretization resolutions.

2   Hybrid Graphical Models

As shown above, discretizations raise many issues. To avoid them, several models have been introduced to directly cope with continuous variables. Unfortunately, unlike in the discrete case, in the continuous case, there does not exist a universal representation for conditional probabilities [9, chap. 14]. In addition, determining conditional independencies among random variables is much more complicated in general in the continuous case than in the discrete one [1]. Therefore, one has to choose one such representation and one actually has to trade-off between the flexibility of the uncertainty model and the computational complexity of its learning and inference mechanisms.

Conditional Gaussian models and their mixing with discrete variables [14,16,17] lie on one side of the spectrum. They compactly represent multivariate Gaussian distributions (and their mixtures). In pure linear Gaussian models (i.e., when there are no discrete variables), junction-tree based exact inference mechanisms prove to be computationally very efficient (even more than in discrete Bayesian networks) [15,28]. However, their main drawback is their lack of flexibility: they can only model large multivariate Gaussian distributions. In addition, the relationships between variables can only be linear. To deal with more expressive mixed probability distributions, Conditional Linear Gaussian models (CLG) allow discrete variables to be part of the graphical model, with the constraint that the relations between the continuous random variables are still limited to linear ones. By introducing latent discrete variables, this limitation can be mitigated. This is a significant improvement, although CLGs are still not very well suited to represent models in which random variables are not distributed w.r.t. Normal distributions. Note that unlike inference in LG models, which contain no discrete variable, and can be performed exactly efficiently, in CLGs, for some part of the inference, one may have to resort to approximations (the so-called weak marginalization) [15]. Structure learning can also be performed efficiently in CLGs (at least when there are no latent variables) [6,8,11,16].

To overcome the lack of flexibility of CLGs, other models have been introduced that rely neither on Normal distributions nor on linear relations between the variables. Among the most popular, let us cite mixtures of exponentials (MTE) [3,22,27], mixtures of truncated basis functions (MoTBF) [12,13] and mixtures of polynomials (MoP) [30,31,33]. As their names suggest, these three models approximate mixed probability distributions by way of mixtures of specific types of probability density functions: in the case of MTEs and MoPs, those are exponentials and polynomials respectively. MoTBFs are more general and only require that the basis functions are closed under product and marginalization. Mixture distributions have been well studied in the literature, notably from the learning perspective [25]. However, unlike [25] in which the number of components of the mixture is implicitly assumed to be small, the design of MTEs, MoPs and MoTBFs allows them to compactly encode mixtures with exponential numbers of components. The rationale behind all these models is that, by cleverly exploiting mixtures, they can approximate very well (w.r.t. the Kullback-Leibler distance) arbitrary mixed probability distributions [2,3]. MoPs have several advantages over MTEs: their parameters for approximating density functions are easier to determine than those of MTEs. They are also applicable to a larger class of deterministic functions in hybrid Bayesian networks. These models are generally easy to learn from datasets [23,26]. In addition, they satisfy Shafer-Shenoy's propagation axioms [32] and inference can thus be performed using a junction tree-based algorithm [2,12,22]. However, in MTEs, MoPs and MoTBFs, combinations and projections are Algebraic operations over sums of functions. As such, as the inference progresses, the number of terms involved in these sums tends to grow exponentially, thereby limiting the use of this exact inference mechanism to problems with only a small number of cliques. To overcome this issue, approximate algorithms based on MCMC [22] or on the Penniless algorithm [27] have been provided in the literature.

3 Conclusion

Dealing with continuous random variables in probabilistic graphical models is challenging. Either one has to resort to discretization, but this raises many issues and the result may be far from the expected one, or to exploiting models specifically designed to cope with continuous variables. But choosing the best model is not easy in the sense that one has to trade-off between the flexibility of the model and the complexity of its learning and inference. Clearly, there is still room for improvements in such models, maybe by exploiting other features of probabilities, like, e.g., copula [5].

References

1. Bergsma, W.: Testing conditional independence for continuous random variables. Technical report, 2004-049, EURANDOM (2004)
2. Cobb, B., Shenoy, P.: Inference in hybrid Bayesian networks with mixtures of truncated exponentials. Int. J. Approximate Reasoning **41**(3), 257–286 (2006)
3. Cobb, B., Shenoy, P., Rumí, R.: Approximating probability density functions in hybrid Bayesian networks with mixtures of truncated exponentials. Stat. Comput. **16**(3), 293–308 (2006)
4. Dechter, R.: Bucket elimination: a unifying framework for reasoning. Artif. Intell. **113**, 41–85 (1999)
5. Elidan, G.: Copula Bayesian networks. In: Proceedings of NIPS 2010, pp. 559–567 (2010)
6. Elidan, G., Nachman, I., Friedman, N.: "ideal parent" structure learning for continuous variable Bayesian networks. J. Mach. Learn. Res. **8**, 1799–1833 (2007)
7. Friedman, N., Goldszmidt, M.: Discretizing continuous attributes while learning Bayesian networks. In: Proceedings of ICML 1996, pp. 157–165 (1996)
8. Geiger, D., Heckerman, D.: Learning Gaussian networks. In: Proceedings of UAI 1994, pp. 235–243 (1994)
9. Koller, D., Friedman, N.: Probabilistic Graphical Models: Principles and Techniques. MIT Press, Cambridge (2009)
10. Kozlov, A., Koller, D.: Nonuniform dynamic discretization in hybrid networks. In: Proceedings of UAI 1997, pp. 314–325 (1997)
11. Kuipers, J., Moffa, G., Heckerman, D.: Addendum on the scoring of Gaussian directed acyclic graphical models. Ann. Stat. **42**(4), 1689–1691 (2014)
12. Langseth, H., Nielsen, T., Rumí, R., Salmerón, A.: Inference in hybrid Bayesian networks with mixtures of truncated basis functions. In: Proceedings of PGM 2012, pp. 171–178 (2012)
13. Langseth, H., Nielsen, T., Rumí, R., Salmerón, A.: Mixtures of truncated basis functions. Int. J. Approximate Reasoning **53**(2), 212–227 (2012)
14. Lauritzen, S.: Propagation of probabilities, means and variances in mixed graphical association models. J. Am. Stat. Assoc. **87**, 1098–1108 (1992)

15. Lauritzen, S., Jensen, F.: Stable local computation with mixed Gaussian distributions. Stat. Comput. **11**(2), 191–203 (2001)
16. Lauritzen, S., Wermuth, N.: Graphical models for associations between variables, some of which are qualitative and some quantitative. Ann. Stat. **17**(1), 31–57 (1989)
17. Lerner, U., Segal, E., Koller, D.: Exact inference in networks with discrete children of continuous parents. In: Proceedings of UAI 2001, pp. 319–328 (2001)
18. Mabrouk, A., Gonzales, C., Jabet-Chevalier, K., Chojnaki, E.: Multivariate cluster-based discretization for Bayesian network structure learning. In: Beierle, C., Dekhtyar, A. (eds.) SUM 2015. LNCS (LNAI), vol. 9310, pp. 155–169. Springer, Cham (2015). https://doi.org/10.1007/978-3-319-23540-0_11
19. Madsen, A., Jensen, F.: LAZY propagation: a junction tree inference algorithm based on lazy inference. Artif. Intell. **113**(1–2), 203–245 (1999)
20. Margaritis, D., Thrun, S.: A Bayesian multiresolution independence test for continuous variables. In: Proceedings of UAI 2001, pp. 346–353 (2001)
21. Monti, S., Cooper, G.: A multivariate discretization method for learning Bayesian networks from mixed data. In: Proceedings of UAI 1998, pp. 404–413 (1998)
22. Moral, S., Rumi, R., Salmerón, A.: Mixtures of truncated exponentials in hybrid Bayesian networks. In: Benferhat, S., Besnard, P. (eds.) ECSQARU 2001. LNCS (LNAI), vol. 2143, pp. 156–167. Springer, Heidelberg (2001). https://doi.org/10.1007/3-540-44652-4_15
23. Moral, S., Rumí, R., Salmerón, A.: Estimating mixtures of truncated exponentials from data. In: Proceedings of PGM 2002, pp. 135–143 (2002)
24. Pearl, J.: Probabilistic Reasoning in Intelligent Systems: Networks of Plausible Inference. Morgan Kauffmann, Burlington (1988)
25. Poland, W., Shachter, R.: Three approaches to probability model selection. In: de Mantaras, R.L., Poole, D. (eds.) Proceedings of UAI 1994, pp. 478–483 (1994)
26. Romero, R., Rumí, R., Salmerón, A.: Structural learning of Bayesian networks with mixtures of truncated exponentials. In: Proceedings of PGM 2004, pp. 177–184 (2004)
27. Rumí, R., Salmerón, A.: Approximate probability propagation with mixtures of truncated exponentials. Int. J. Approximate Reasoning **45**(2), 191–210 (2007)
28. Salmerón, A., Rumí, R., Langseth, H., Madsen, A.L., Nielsen, T.D.: MPE inference in conditional linear Gaussian networks. In: Destercke, S., Denoeux, T. (eds.) ECSQARU 2015. LNCS (LNAI), vol. 9161, pp. 407–416. Springer, Cham (2015). https://doi.org/10.1007/978-3-319-20807-7_37
29. Shafer, G.: Probabilistic Expert Systems. Society for Industrial and Applied Mathematics (1996)
30. Shenoy, P.: A re-definition of mixtures of polynomials for inference in hybrid Bayesian networks. In: Liu, W. (ed.) ECSQARU 2011. LNCS (LNAI), vol. 6717, pp. 98–109. Springer, Heidelberg (2011). https://doi.org/10.1007/978-3-642-22152-1_9
31. Shenoy, P.: Two issues in using mixtures of polynomials for inference in hybrid Bayesian networks. Int. J. Approximate Reasoning **53**(5), 847–866 (2012)
32. Shenoy, P., Shafer, G.: Axioms for probability and belief function propagation. In: Proceedings of UAI 1990, pp. 169–198 (1990)
33. Shenoy, P., West, J.: Inference in hybrid Bayesian networks using mixtures of polynomials. Int. J. Approximate Reasoning **52**(5), 641–657 (2011)

Towards Scalable and Robust
Sum-Product Networks

Alvaro H. C. Correia and Cassio P. de Campos$^{(\boxtimes)}$

Eindhoven University of Technology, Eindhoven, The Netherlands
c.decampos@tue.nl

Abstract. Sum-Product Networks (SPNs) and their credal counterparts are machine learning models that combine good representational power with tractable inference. Yet they often have thousands of nodes which result in high processing times. We propose the addition of caches to the SPN nodes and show how this memoisation technique reduces inference times in a range of experiments. Moreover, we introduce class-selective SPNs, an architecture that is suited for classification tasks and enables efficient robustness computation in Credal SPNs. We also illustrate how robustness estimates relate to reliability through the accuracy of the model, and how one can explore robustness in ensemble modelling.

Keywords: Sum-Product Networks · Robustness

1 Introduction

Sum-Product Networks (SPNs) [15] (conceptually similar to Arithmetic Circuits [4]) are a class of deep probabilistic graphical models where exact marginal inference is always tractable. More precisely, any marginal query can be computed in time polynomial in the network size. Still, SPNs can capture high tree-width models [15] and are capable of representing complex and highly multidimensional distributions [5]. This promising combination of efficiency and representational power has motivated several applications of SPNs to a variety of machine learning tasks [1,3,11,16–18].

As any other standard probabilistic graphical model, SPNs learned from data are prone to overfitting when evaluated at poorly represented regions of the feature space, leading to overconfident and often unreliable conclusions. However, due to the probabilistic semantics of SPNs, we can mitigate that issue through a principled analyses of the reliability of each output. A notable example is Credal SPNs (CSPNs) [9], a extension of SPNs to imprecise probabilities where we can compute a measure of the robustness of each prediction. Such robustness values are useful tools for decision-making, as they are highly correlated with accuracy, and thus tell us when to trust the CSPN's prediction: if the robustness of a prediction is low, we can suspend judgement or even resort to another machine

© Springer Nature Switzerland AG 2019
N. Ben Amor et al. (Eds.): SUM 2019, LNAI 11940, pp. 409–422, 2019.
https://doi.org/10.1007/978-3-030-35514-2_31

learning model. Indeed, we show that robustness is also effective in ensemble modelling, opening up new avenues for reliable machine learning.

Unfortunately, computing robustness requires many passes through the network, which limits the scalability of CSPNs. We address that by introducing class-selective (C)SPNs, a type of architecture that enables efficient robustness computations due to their independent sub-networks: one for each label in the data. Class-selective (C)SPNs not only enable fast robustness estimation but also outperform general (C)SPNs in classification tasks. In our experiments, their accuracy was comparable to that of state-of-the-art methods, such as XGBoost [2].

We also study how to improve the scalability of (C)SPNs by reducing their inference time. Although (C)SPNs ensure tractable inference, they are often very large networks spanning thousands of nodes. In practice, that translates to computational costs that might be too high for some large-scale applications. A solution is to limit the network size, but that comes at the cost of the model's representational power. One way out of this trade-off is to notice that many operations reoccur often in SPNs. As we descend from the root, the number of variables in the scope of each node decreases. With a smaller feature space, these nodes are likely to be evaluated at identical instantiations of their respective variables, and we can avoid recomputing them by having a cache for previously seen instances. We investigated the benefit of such memoisation procedure across 25 UCI datasets [8] and observed that it reduces inference times considerably.

This paper is organised as follows. In Sect. 2, we give the necessary notation and definitions of (credal) sum-product networks. In Sect. 3, we introduce class-selective (C)SPNs and use them to derive a new algorithm for efficient robustness computation. We detail memoisation techniques for (C)SPNs in Sect. 4 and show their practical benefit in Sect. 5, where we report experiments on the effects of memoisation and the performance of class-selective (C)SPNs. We also discuss how robustness estimates translate into accuracy and how they can be exploited in ensemble models. Finally, we conclude and point promising directions for future work in Sect. 6.

2   (Credal) Sum-Product Networks

Before giving a formal definition of (C)SPNs, we introduce the necessary notation and background. We write integers in lowercase letters (e.g. i, j, k) and sets of integers in uppercase calligraphic letters (e.g. $\mathcal{E}, \mathcal{V}$). We denote by $X_{\mathcal{V}}$ the collection of random variables indexed by set $\mathcal{V}$, that is, $X_{\mathcal{V}} = \{X_i : i \in \mathcal{V}\}$. We reserve $\mathcal{V}$ to represent the indices of all the variables over which a model is defined, but when clear from the context, we omit the indexing set and use simply X and x. Note that there is no ambiguity, since individual variables and instantiations are always denoted with a subscript (e.g. X_i, x_i). The realisation of a set of random variables is denoted in lowercase letters (e.g. $X_{\mathcal{V}} = x_{\mathcal{V}}$). When only a subset of the variables is concerned, we use a different indexing set $\mathcal{E} \subseteq \mathcal{V}$ to identify the corresponding variables $X_{\mathcal{E}}$ and their realisations $x_{\mathcal{E}}$. Here $x_{\mathcal{E}}$ is what we call *partial evidence*, as not every variable is observed.

An SPN is a probabilistic graphical model defined over a set of random variables $X_\mathcal{V}$ by a weighted, rooted and acyclic directed graph where internal nodes perform either sum or product operations, and leaves are associated with indicator variables. Typically, an indicator variable is defined as the application of a function $\lambda_{i,j}$ such that

$$\lambda_{i,j}(x_\mathcal{E}) = \begin{cases} 0 & \text{if } i \in \mathcal{E} \text{ and } x_i \neq j \\ 1 & \text{otherwise,} \end{cases}$$

where $x_\mathcal{E}$ is any partial or complete evidence. The SPN and its root node are used interchangeably to mean the same object. We assume that every indicator variable appears in at most one leaf node. Every arc from a sum node i to a child j is associated with a non-negative weight $w_{i,j}$ such that $\sum_j w_{i,j} = 1$ (this constraint does not affect the generality of the model [14]).

Given an SPN S and a node i, we denote S^i the SPN obtained by rooting the network at i, that is, by discarding any non-descendant of i (other than i itself). We call S^i the sub-network rooted at i, which is an SPN by itself (albeit over a possibly different set of variables). If ω are the weights of an SPN S and i is a node, we denote by ω_i the weights in the sub-network S^i rooted at i, and by w_i the vector of weights $w_{i,j}$ associated with arcs from i to children j.

The *scope* of an SPN is the set of variables that appear in it. For an SPN which is a leaf associated with an indicator variable, the scope is the respective random variable. For an SPN which is not a leaf, the scope is the union of the scopes of its children. We assume that scopes of children of a sum node are identical (completeness) and scopes of children of a product node are disjoint (decomposability) [12].

The *value* of an SPN S at a given instantiation $x_\mathcal{E}$, written $S(x_\mathcal{E})$, is defined recursively in terms of its root node r. If r is a leaf node associated with indicator variable λ_{r,x_r} then $S(x_\mathcal{E}) = \lambda_{r,x_r}(x_\mathcal{E})$. Else, if r is a product node, then $S(x_\mathcal{E}) = \prod_j S^j(x_\mathcal{E})$, where j ranges over the children of r. Finally, if r is a sum node then $S(x_\mathcal{E}) = \sum_j w_{r,j} S^j(x_\mathcal{E})$, where again j ranges over the children of r. Given these properties, it is easy to check that S induces a probability distribution for $X_\mathcal{V}$ such that $S(x_\mathcal{E}) = \mathbb{P}(x_\mathcal{E})$ for any $x_\mathcal{E}$ and $\mathcal{E} \subseteq \mathcal{V}$. One can also compute expectations of functions over a variable X_i as $\mathbb{E}(f|X_\mathcal{E} = x_\mathcal{E}) = \sum_{x_i} f(x_i)\mathbb{P}(x_i|X_\mathcal{E} = x_\mathcal{E})$.

A Credal SPN (CSPN) is defined similarly, except for containing sets of weight vectors in each sum node instead of a single weight vector. More precisely, a CSPN C is defined by a set of SPNs $C = \{S_\omega : \omega \in \mathcal{C}\}$ over the same graph structure of S, where $\mathcal{C}$ is the Cartesian product of finitely-generated simplexes $\mathcal{C}_i$, one for each sum node i, such that the weights w_i of a sum node i are constrained by $\mathcal{C}_i$. While an SPN represents one joint distribution over its variables, a CSPN represents a set of joint distributions. Therefore, one can use CSPNs to obtain lower and upper bounds $\min_\omega \mathbb{E}_\omega(f|X_\mathcal{E} = x_\mathcal{E})$ and $\max_\omega \mathbb{E}_\omega(f|X_\mathcal{E} = x_\mathcal{E})$ on the expected value of some function f of a variable, conditional on evidence $X_\mathcal{E} = x_\mathcal{E}$. Recall that each choice of the weights ω of a CSPN $\{S_\omega : \omega \in \mathcal{C}\}$

defines an SPN and hence induces a probability measure $\mathbb{P}_\omega$. We can therefore compute bounds on the conditional expectations of a function over X_i:

$$\min_\omega \mathbb{E}_\omega(f|X_\mathcal{E} = x_\mathcal{E}) = \min_\omega \sum_{x_i} f(x_i)\mathbb{P}_\omega(X_i = x_i|X_\mathcal{E} = x_\mathcal{E}). \qquad (1)$$

The equations above are well-defined if $\min_\omega \mathbb{P}(X_\mathcal{E} = x_\mathcal{E}) > 0$, which we will assume to be true (statistical models often have some smoothing so that zero probability is not attributed to any assignment of variables – this is our assumption here, for simplicity, though this could be addressed in more sophisticated terms). Note also that we can focus on the computation of the lower expectation, as the upper expectation can be obtained from $\max_\omega \mathbb{E}_\omega(f|x_\mathcal{E}) = -\min_\omega \mathbb{E}_\omega(-f|x_\mathcal{E})$.

Computing the lower conditional expectation in Eq. (1) is equivalent to finding whether:

$$\min_\omega \mathbb{E}_\omega(f|X_\mathcal{E} = x_\mathcal{E}) > 0 \iff \min_\omega \sum_{x_i} f(x_i)\mathbb{P}_\omega(X_i = x_i, X_\mathcal{E} = x_\mathcal{E}) > 0, \quad (2)$$

as to obtain the exact value of the minimisation one can run a binary search for μ until $\min_\omega \mathbb{E}_\omega((f - \mu)|x_\mathcal{E}) = 0$ (to the desired precision). The following result will be used here. Corollary 1 is a small variation of the result in [9].

Theorem 1 (Theorem 1 in [9]). *Consider a CSPN $C = \{S_\omega : \omega \in \mathcal{C}\}$. Computing $\min_\omega S_\omega(x_\mathcal{E})$ and $\max_\omega S_\omega(x_\mathcal{E})$ takes $O(|C| \cdot K)$ time, where $|C|$ is the number of nodes and arcs in C, and K is an upper bound on the cost of solving a linear program of the form $\min_{w_i} \sum_j c_{i,j}w_{i,j}$ subject to $w_i \in \mathcal{C}_i$.*

Corollary 1. *Consider a CSPN $C = \{S_\omega : \omega \in \mathcal{C}\}$ with a bounded number of children per sum node specified by simplexes $\mathcal{C}_i$ of (finitely many) constraints of the form $l_{i,j} \leq w_{i,j} \leq u_{i,j}$ for given rationals $l_{i,j} \leq u_{i,j}$. Computing $\min_\omega S_\omega(x_\mathcal{E})$ and $\max_\omega S_\omega(x_\mathcal{E})$ can be solved in time $O(|S|)$.*

Proof. When local simplexes $\mathcal{C}_i$ have constraints $l_{i,j} \leq w_{i,j} \leq u_{i,j}$, then the local optimisations $S^i(x_\mathcal{E}) = \min_{w_i} \sum_j w_{i,j}S^j(x_\mathcal{E})$ are equivalent to fractional knapsack problems [7], which can be solved in constant time for nodes with bounded number of children. Thus, the overall running time is $O(|S|)$.

3   Efficient Robustness Measure Computation

We now define an architecture called class-selective (C)SPN that is provenly more efficient in computing robustness values. Figure 1 illustrates its structure.

Definition 1. *Consider a domain where variable X_c is called the class variable. A class-selective (C)SPN has a sum node as root node S with $|X_c|$ product nodes as its children, each of which has an indicator leaf node for a different value x_c of X_c (besides potentially other sibling (C)SPNs). These product nodes that are children of S have disjoint sets of internal descendant nodes.*

The name class-selective was inspired by selective SPNs [13], where only one child of each sum node is active at a time. In a class-selective SPN such property is restricted to the root node: for a given class value, only one of the sub-networks remains active. That is made clear in Fig. 1 as only one of the indicator nodes C_n is non-zero and all but one of children of the root node evaluate to zero.

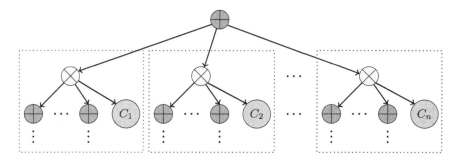

Fig. 1. Illustration of a class-selective SPN. In the graph, C_n is a leaf node applying the indicator function $\lambda_{c,n}(x_\mathcal{E})$.

In a class-selective CSPN, the computation of expectation of a function of the class variable can be achieved as efficiently as in standard SPNs:

$$
\min_\omega \sum_{x_c} f(x_c) \mathbb{P}_\omega(X_c = x_c, X_\mathcal{E} = x_\mathcal{E}) = \min_{w_r} \sum_{x_c : f(x_c) \geq 0} f(x_c) w_{r,x_c} \min_{\omega_{x_c}} S^{x_c}_{\omega_{x_c}}(x_c, x_\mathcal{E})
$$
$$
+ \sum_{x_c : f(x_c) < 0} f(x_c) w_{r,x_c} \max_{\omega_{x_c}} S^{x_c}_{\omega_{x_c}}(x_c, x_\mathcal{E}),
$$

where r is the root node index with children S^{x_c} for each value x_c. Note that each of these internal optimisations can be obtained by independent executions which take altogether time $O(|S|)$ by Corollary 1 (as each execution runs over non-overlapping sub-CSPNs corresponding to different class labels x_c). Moreover, note that in a non-credal class-selective SPN, finding the class label of maximum probability (and its probability) takes time $O(|S|)$ in the worst case. That is more efficient than general SPNs, where $|S| \cdot |X_c|$ nodes may need to be visited.

Let us turn our attention to the CSPN robustness estimation in a classification problem. Given input instance $X_\mathcal{E} = x_\mathcal{E}$ for which we want to predict the class variable value, we say that the classification using a CSPN C is robust if the class value $x_c = \arg\max_{x_c} \mathbb{P}(x_c|x_\mathcal{E})$ predicted by an SPN S_ω that belongs to $C = \{S_\omega : \omega \in \mathcal{C}\}$ is also the prediction of any other $S_{\omega'} \in C$ (hence it is unique for C), which happens if and only if

$$
\min_\omega \mathbb{E}_\omega(\mathcal{I}_{x_c} - \mathcal{I}_{x'_c}|X_\mathcal{E} = e) > 0 \text{ for every } x'_c \neq x_c. \tag{3}
$$

In the case of class-selective CSPNs, this task equates to checking whether

$$
\min_\omega \mathbb{P}_\omega(X_c = x_c, X_\mathcal{E} = x_\mathcal{E}) > \max_{x'_c \neq x_c} \max_\omega \mathbb{P}_\omega(X_c = x'_c, X_\mathcal{E} = x_\mathcal{E}),
$$

Algorithm 1. Efficient ϵ-robustness computation.

---

1 **Function** Robustness$(S,\ x_{\mathcal{E}},\ x_c,\ er)$:

    Data : Class-selective SPN S, Input $x_{\mathcal{E}}$, Prediction $x_c \mid X_{\mathcal{E}} = x_{\mathcal{E}}$,
           Precision $er < 1$

    Result: Robustness ϵ

2      $\epsilon_{\max} \leftarrow 1$;

3      $\epsilon_{\min} \leftarrow 0$;

4   **while** $\epsilon_{\min} < \epsilon_{\max} - er$ **do**

5          $\epsilon \leftarrow (\epsilon_{\min} + \epsilon_{\max})/2$;

6          $v \leftarrow \min_{\omega^{\epsilon}} S_{\omega^{\epsilon}}(x_c, x_{\mathcal{E}})$;

7   **for** $x'_c \neq x_c$ **do**

8              $v' \leftarrow \max_{\omega^{\epsilon}} S_{\omega^{\epsilon}}(x'_c, x_{\mathcal{E}})$;

9   **if** $v' \geq v$ **then**

10                 $\epsilon_{\max} \leftarrow \epsilon$;

11   **break**

12   **end**

13   **end**

14   **if** $\epsilon_{\max} > \epsilon$ **then**

15              $\epsilon_{\min} \leftarrow \epsilon$

16   **end**

17   **end**

18   **return** ϵ;

---

that is, regardless of the choice of weights $w \in \mathcal{C}$, we would have $\mathbb{P}_w(x_c|x_{\mathcal{E}}) > \mathbb{P}_w(x'_c|x_{\mathcal{E}})$ for all other labels x'_c.

General CSPNs may require $2 \cdot |S| \cdot (|X_c| - 1)$ node evaluations in the worst case to identify whether a predicted class label x_c is robust for instance $X_{\mathcal{E}} = x_{\mathcal{E}}$, while a class-selective CSPN obtains such result in $|S|$ node evaluations. This is because CSPNs will run over its nodes $(|X_c| - 1)$ times in order to reach a conclusion about Expression (3), while the class-selective CSPN can compute $\min_w S_w(x_c, x_{\mathcal{E}})$ and $\max_w S_w(x'_c, x_{\mathcal{E}})$ (the max is done for each x'_c) only once (taking overall $|S|$ node evaluations, since they run over non-overlapping sub-networks for different class values).

Finally, given an input instance $X_{\mathcal{E}} = x_{\mathcal{E}}$ and an SPN S learned from data, we can compute a robustness measure as follows. We define a collection of CSPNs $C_{S,\epsilon}$ parametrised by $0 \leq \epsilon < 1$ such that each w_i^{ϵ} of a node i in $C_{S,\epsilon}$ is allowed to vary within an ϵ-contaminated credal set of the original weight vector w_i of the same node i in S. A robustness measure for the issued prediction $C_{S,\epsilon}(x_{\mathcal{E}})$ is then defined by the largest ϵ such that $C_{S,\epsilon}$ is robust for $x_{\mathcal{E}}$. Finding such ϵ can be done using a simple binary search, as shown in Algorithm 1.

4 Memoisation

When evaluating an SPN on a number of different instances, some of the computations are likely to reoccur, especially in nodes with scopes of only a few variables. One simple and yet effective solution to reduce computing time is to cache the results at each node. Thus, when evaluating the network recursively, we do not have to visit any of the children of a node if a previously evaluated data point had the same instantiation over the scope of that node. To be more precise, consider a node S with scope $\mathcal{S}$, and two instances x, x' such that $x_\mathcal{S} = x'_\mathcal{S}$. It is clear that $S(x) = S(x')$, so once we have cached the value of one, there is no need to reevaluate node S, or any of its children, on the other. For a CSPN C, the same approach holds, but we need different caches for maximisation and minimisation, as well as for different SPNs S_ω that belong to C (after all, a change of ω may imply a change of the result). Notice that the computational overhead of memoisation is amortised constant, as it amounts to accessing a hash table.

Mei et al. proposed computing maximum a posteriori (MAP) by storing the values of nodes at a given evidence and searching for the most likely query in a reduced SPN, where nodes associated with the evidence are pruned [10]. Memoisation can be seen as eliminating such nodes implicitly (as their values are stored and they are not revisited) but goes further by using values calculated at other instances to save computational power. In fact, the application of memoisation to MAP inference is a promising avenue for future research, since many methods (e.g. hill climbing) evaluate the SPN at small variations of the same input that are likely to share partial instantiations in many of the nodes in the network.

5   Experiments

We investigated the performance of memoisation and class-selective (C)SPNs through a series of experiments over a range of 25 UCI datasets [8]. All experiments were run in a single modern core with our implementation of Credal SPNs, which runs LearnSPN [6] for structure learning. Source code is available on our pages and/or upon request.

Table 1 presents the UCI data sets on which we ran experiments described by their number of independent instances N, number of variables $|X|$ (including a class variable X_c) and number of class labels $|X_c|$. All data sets are categorical (or have been made categorical using discretisation by median value). We also show the 0–1 classification accuracy obtained by both General and Class-selective SPNs as well as the XGBoost library that provides a parallel tree gradient boosting method [2], considered a state-of-the-art technique for supervised classification tasks. Results are obtained by stratified 5-fold cross-validation, and as one can inspect, class-selective SPNs largely outperformed general SPNs while being comparable to XGBoost in terms of classification accuracy.

We leave further comparisons between general and class-selective networks to the appendix, where Table 3 depicts the two types of network in terms of their architecture and processing times on classification tasks. There one can see that class-selective SPNs have a higher number of parameters due to a larger number of sum nodes. However, in some cases general SPNs are deeper, which means class-selective SPNs tend to grow sideways, especially when the number of classes is high. Nonetheless, the larger number of parameters in class-selective networks

Table 1. Percent accuracy of XGBoost, General SPNs and Class-selective SPNs across several UCI datasets. All experiments consisted in stratified 5-fold cross validation.

Dataset	N	$|X|$	$|X_c|$	XGBoost	General SPN	Class-selective SPN
zoo	101	17	7	93.069	76.238	96.04
bridges	107	11	6	66.355	57.009	63.551
lymph	148	18	4	81.757	72.973	81.081
flags	194	29	8	66.495	43.299	57.732
autos	205	26	2	91.22	88.293	89.268
breast cancer	286	10	2	72.378	71.678	67.832
heart h	294	12	2	79.592	79.932	81.633
ecoli	336	6	8	72.321	63.393	74.405
liver disorders	345	7	2	65.797	57.391	64.638
dermatology	366	35	6	97.268	81.694	98.907
colic	368	23	2	83.696	77.717	78.533
balance scale	625	5	3	72	72.48	72.16
soybean	683	36	19	93.704	62.518	94.583
diabetes	768	9	2	70.313	70.703	69.922
vehicle	846	19	4	65.485	46.454	60.402
tic tac toe	958	10	2	84.969	69.937	73.382
vowel	990	14	11	64.747	33.737	59.394
solar flare 2	1,066	12	6	73.64	59.475	73.077
cmc	1,473	10	3	51.799	48.133	48.065
car	1,728	7	4	87.905	70.023	93.287
segment	2,310	17	7	82.771	67.662	80.823
sick	3,772	28	2	93.902	93.876	91.463
hypothyroid	3,772	28	4	92.285	92.285	91.569
spambase	4,601	8	2	78.918	78.505	76.766
nursery	12,960	9	5	94.961	81.505	92.299

does not translate into higher latency as both architectures have similar learning and inference times. We attribute that to the independence of the subnetwork of each class which facilitates inference. Notice that the two architectures are equally efficient only in the classification task (only aspect compared in Table 3) and not on robustness computations. We mathematically proved the latter to be more efficient in class-selective networks when using Algorithm 1.

In Table 2, we have the average inference time per instance for 25 UCI datasets. When using memoisation, the inference time dropped by at least 50% in all datasets we evaluated, proving that memoisation is a valuable tool to render (C)SPNs more efficient. We can also infer from the experiments, that the relative reduction in computing time tends to increase with the number of instances N. For datasets with more than 2000 instances, which are still relatively small, adding memoisation already cut the inference time by more than 90%. That is a promising result for large scale applications, as memoisation grows more effective with number of data points. We can better observe the effect of memoisation by plotting a histogram of the inference times as in Fig. 2, where we have 6 of the UCI datasets of Table 2. We can see that memoisation concentrates the distribution at lower time values, proving that most instances take advantage of the cached results in a number of nodes in the network.

Table 2. Average inference time and number of nodes visited per inference for a CSPN with (+M) and without (−M) memoisation across 25 UCI datasets. The respective ratios (%) are the saved time or node visits, that is, $1 - \frac{+M}{-M}$. Robustness was computed with precision of 0.004, which requires 8 passes through the network as per Algorithm 1.

Dataset	N	$\|X\|$	$\|X_c\|$	Inference Time (s) −M	+M	%	# Nodes Evaluated −M	+M	%
zoo	101	17	7	1.742	0.754	56.7	15,020	2,113	85.93
bridges	107	11	6	0.693	0.335	51.66	8,791	1,665	81.06
lymph	148	18	4	1.28	0.535	58.17	11,557	1,990	82.78
flags	194	29	8	5.44	1.641	69.84	55,670	5,973	89.27
autos	205	26	2	1.303	0.541	58.5	17,100	3,534	79.33
breast cancer	286	10	2	0.422	0.13	69.3	5,294	891	83.17
heart h	294	12	2	0.279	0.101	63.94	2,501	330	86.79
ecoli	336	6	8	0.891	0.164	81.62	8,703	663	92.38
liver disorders	345	7	2	0.172	4.936e−2	71.32	1,444	153	89.39
dermatology	366	35	6	5.747	1.276	77.8	57,239	3,623	93.67
colic	368	23	2	1.281	0.412	67.85	13,435	1,609	88.03
balance scale	625	5	3	0.15	2.264e−2	84.94	1,720	71	95.86
soybean	683	36	19	24.125	5.141	78.69	2.803e+5	5,831	97.92
diabetes	768	9	2	0.329	6.424e−2	80.49	3,028	201	93.37
vehicle	846	19	4	1.864	0.299	83.94	19,408	797	95.89
tic tac toe	958	10	2	0.679	0.132	80.56	7,693	519	93.25
vowel	990	14	11	12.126	1.791	85.23	1.252e+5	5,804	95.37
solar flare 2	1,066	12	6	3.467	0.364	89.51	22,320	707	96.83
cmc	1,473	10	3	1.43	0.209	85.36	17,688	775	95.62
car	1,728	7	4	0.769	0.124	83.92	9,720	412	95.76
segment	2,310	17	7	6.56	0.421	93.59	42,923	579	98.65
sick	3,772	28	2	1.844	0.15	91.89	11,412	146	98.72
hypothyroid	3,772	28	4	2.322	0.252	89.17	20,864	223	98.93
spambase	4,601	8	2	0.583	2.236e−2	96.16	6,430	75	98.83
nursery	12,960	9	5	8.27	0.661	92.01	74,418	1,088	98.54

If we consider the number of nodes visited instead of time, the results are even more telling. In Table 2 on average, less than 15% of node evaluations was necessary during inference with memoisation. That is a more drastic reduction than what we observed when comparing processing times, which means there still plenty of room for improvement in computational time with better data structures or faster programming languages.

It is worth noting that memoisation is only effective over discrete variables. When a subset of the variables is continuous, the memoisation will only be effective in nodes whose scope contains only discrete variables, which is likely to reduce the computational gains of memoisation. An alternative is to construct the cache with ranges of values for the continuous variables. To be sure, that is a form of discretisation that might worsen the performance of the model, but it occurs only at inference time and can be easily switched off when high precision is needed. A thorough study of the effect of memoisation on models with continuous variables is left for future work.

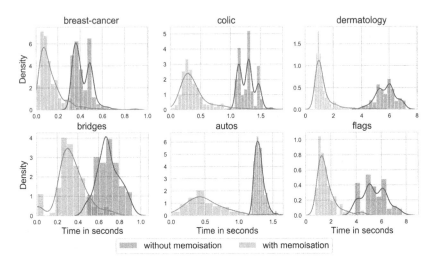

Fig. 2. Empirical distribution of CSPN inference times with and without memoisation.

5.1 Exploring Data on the Robustness Measure

We can interpret robustness as a measure of the model's confidence on its output. Roughly speaking, in a classification task, the robustness value ϵ of a prediction corresponds to how much we can tweak the networks parameters without changing the final result, that is, the class of maximum probability. Thus, a large ϵ means that many similar networks (in the parameter space) would give the same output for the instance in question. Similarly, we can think that small changes in the hyperparameters or the data sample would not produce a model whose prediction would be different for that given instance. Conversely, a small ϵ tell us that slightly different networks would already provide us with a distinct answer. In that case, the prediction is not reliable as it might fluctuate with any variation on the learning or data acquisition processes.

We can validate this interpretation by investigating how robustness relates to the accuracy of the model. In Fig. 3(a), we defined a number of robustness thresholds and, for each of them, we computed the accuracy of the model over instances for which ϵ was *above* the threshold. It is clear from the graph that the accuracy increases with the threshold, and we can infer that robustness does translate into reliability as the model is more accurate over instances with high ϵ values. We arrive at a similar conclusion in Fig. 3(b), where we consider instances for which ϵ was *below* a given threshold. In this case, the curves start at considerably lower accuracy values, where only examples with low robustness are considered, and then build up as instances with higher ϵ values are added.

Robustness values are not only a guide for reliable decision-making but are also applicable to ensemble modelling. The idea is to combine a CSPN $C = \{S_\omega : \omega \in \mathcal{C}\}$ with another model f by defining a robustness threshold t. We rely on the CSPN's prediction for all instances for which its robustness is above

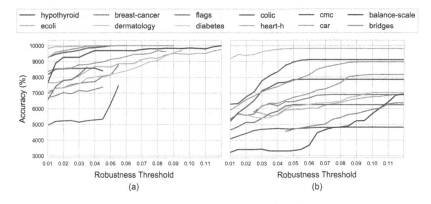

Fig. 3. Accuracy of predictions with robustness (a) *above* and (b) *below* different thresholds for 12 UCI datasets. Some curves end abruptly because we only computed the accuracy when 50 or more data points were available for a given threshold.

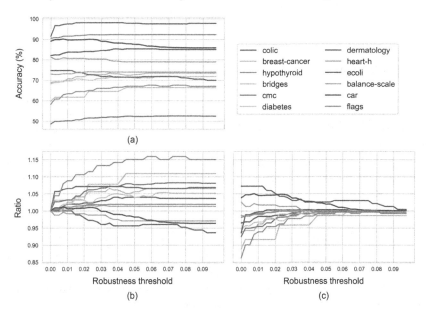

Fig. 4. Performance of the ensemble model against different robustness thresholds for 12 UCI datasets. (a) Accuracy of the ensemble model; (b) accuracy of the ensemble model over accuracy of the CSPN and (c) over accuracy of the XGBoost.

the threshold and, on f for the remaining ones. To be precise, we can define an ensemble $E_{C,f}$ as

$$
E_{C,f}(x_\varepsilon, t) = \begin{cases} x_c^* & \text{if } \epsilon \geq t \\ f(x_\varepsilon) & \text{otherwise,} \end{cases}
$$

where $x_c^* = \arg\max_{x_c} S(x_c, x_\mathcal{E})$ is the class predicted by a class-selective SPN S learned from data (from which the ϵ-contaminated CSPN C is built), and ϵ is the corresponding robustness value, $\epsilon = \texttt{Robustness}(S, x_\mathcal{E}, x_c^*)$—see Algorithm 1.

We implemented such an ensemble model by combining the ϵ-contaminated CSPN with an XGBoost model. We computed the accuracy of the ensemble for different thresholds t over a range of UCI data sets, as reported in Fig. 4. In plot (a), we see the accuracy varies considerably with the threshold t, which suggests there is an optimum value for t. In the other two plots, we compare the ensemble against the CSPN (b); and the XGBoost model (c). We computed the ratio of the accuracy of the ensemble over the accuracy of the competing model, so that any point above one indicates an improve in performance. For many datasets, the ensemble delivered better results and in some cases was superior to both original models for an appropriate robustness threshold. In spite of that, we have not investigated how to find good thresholds, which we leave for future work. Yet, we point out that the complexity of queries using the class-selective CSPN in the ensemble will be the same as that of class-selective SPNs (the robustness comes "for free"), since the binary search for the computation of the threshold will not be necessary (we can run it for the pre-decided t only).

6   Conclusion

SPNs and their credal version are highly expressive deep architectures with tractable inference, which makes them a promising option for large-scale applications. However, they still lag behind some other machine learning models in terms of architectural optimisations and fast implementations. We address that through the introduction of memoisation, a simple yet effective technique that caches previously computed results in a node. In our experiments, memoisation reduced the number of node evaluations by more than 85% and the inference time by at least 50% (often much more). We believe this is a valuable new tool to help bring (C)SPNs to large-scale applications where low latency is essential.

We also discussed a new architecture, class-selective (C)SPNs, that combine efficient robustness computation with high accuracy on classification tasks, outperforming general (C)SPNs. Even though they excel in discriminative tasks, class-selective SPNs are still generative models fully endowed with the semantics of graphical models. We demonstrated how their probabilistic semantics can be brought to bear through their extension to Credal SPNs. Namely, we explored how robustness values relate to the accuracy of the model and how one can use them to develop ensemble models guided through principled decision-making.

We finally point out some interesting directions for future work. As demonstrated here, class-selective (C)SPNs have proven to be powerful models in classification tasks, but they arbitrarily place the class variable in a privileged position in the network. Future research might investigate how well class-selective (C)SPNs fit the joint distribution and how they would fair in predicting other variables. Memoisation also opens up new promising research avenues, notably in how it performs on other inferences tasks such as MAP and how it can be extended to accommodate continuous variables.

A Appendix

In Table 3 we compare general and class-selective SPNs in terms of their architecture and processing times in classification tasks (no robustness computation).

Table 3. Comparison between General (Gen) and Class-Selective (CS) SPNs in learning and average inference times (s), number of nodes, height and number of parameters.

Dataset	N	$|X|$	$|X_c|$	Learning (s) Gen	Learning (s) CS	Inference (s) Gen	Inference (s) CS	# Nodes Gen	# Nodes CS	Height Gen	Height CS	# Parameters Gen	# Parameters CS
zoo	101	17	7	0.35	0.435	2.744e-2	4.522e-2	250	419	8	5	74	125
bridges	107	11	6	0.228	0.358	1.529e-2	2.74e-2	154	289	7	5	39	71
lymph	148	18	4	0.605	0.598	2.209e-2	2.686e-2	362	446	8	8	91	115
flags	194	29	8	1.582	2.311	0.115	0.197	1,013	1,744	11	7	188	328
autos	205	26	2	1.644	1.652	2.953e-2	3.02e-2	958	971	12	11	211	221
breast cancer	286	10	2	0.382	0.418	7.835e-3	9.793e-3	253	304	9	9	51	59
heart h	294	12	2	0.22	0.21	4.193e-3	4.307e-3	131	138	6	6	37	39
ecoli	336	6	8	0.121	0.242	1.223e-2	2.789e-2	101	233	5	5	25	59
liver disorders	345	7	2	0.107	0.108	2.467e-3	2.5e-3	77	78	6	6	24	25
dermatology	366	35	6	2.971	2.802	0.161	0.171	1,834	1,952	15	10	383	408
colic	368	23	2	1.084	1.326	1.885e-2	2.405e-2	625	791	11	12	149	183
balance scale	625	5	3	0.11	0.112	4.258e-3	4.068e-3	85	82	7	6	26	24
soybean	683	36	19	4.308	4.969	0.763	1.125	2,604	3,913	16	9	596	940
diabetes	768	9	2	0.263	0.252	5.672e-3	5.564e-3	176	172	9	8	58	56
vehicle	846	19	4	1.075	1.482	3.717e-2	5.226e-2	585	830	12	11	186	272
tic tac toe	958	10	2	0.725	0.66	1.623e-2	1.568e-2	496	470	11	11	128	116
vowel	990	14	11	3.498	3.879	0.417	0.49	2,444	2,800	17	13	502	663
solar flare 2	1,066	12	6	1.03	0.827	6.738e-2	6.088e-2	784	708	12	10	158	148
cmc	1,473	10	3	1.156	1.136	3.775e-2	3.685e-2	828	812	15	14	198	193
car	1,728	7	4	0.523	0.556	2.337e-2	2.794e-2	370	434	11	10	81	92
segment	2,310	17	7	1.558	2.028	9.55e-2	0.142	888	1,301	13	11	263	419
sick	3,772	28	2	2.665	2.229	2.493e-2	1.926e-2	802	616	13	11	249	196
hypothyroid	3,772	28	4	2.577	2.541	4.509e-2	5.393e-2	728	869	12	13	224	274
spambase	4,601	8	2	0.649	0.599	1.205e-2	1.247e-2	353	364	10	12	115	119
nursery	12,960	9	5	4.967	4.363	0.258	0.226	3,437	3,025	19	18	755	669

References

1. Amer, M.R., Todorovic, S.: Sum product networks for activity recognition. IEEE Trans. Pattern Anal. Mach. Intell. **38**(4), 800–813 (2016)
2. Chen, T., Guestrin, C.: XGBoost: A Scalable Tree Boosting System. In: Proceedings of the 22nd ACM SIGKDD International Conference on Knowledge Discovery and Data Mining, vol. 19, no. (6), pp. 785–794 (2016)
3. Cheng, W.C., Kok, S., Pham, H.V., Chieu, H.L., Chai, K.M.A.: Language modeling with sum-product networks. In: Proceedings of the Annual Conference of the International Speech Communication Association, INTERSPEECH, pp. 2098–2102 (2014)
4. Darwiche, A.: A differential approach to inference in bayesian networks. J. ACM **50**(3), 280–305 (2003)
5. Delalleau, O., Bengio, Y.: Shallow vs. deep sum-product networks. In: Advances in Neural Information Processing Systems, vol. 24. pp. 666–674. Curran Associates, Inc. (2011)
6. Gens, R., Domingos, P.: Learning the structure of sum-product networks. In: Proceedings of the 30th International Conference on Machine Learning, vol. 28, pp. 229–264 (2013)
7. Korte, B., Vygen, J.: Combinatorial Optimization: Theory and Algorithms, 5th edn. Springer Publishing Company, Incorporated, Heidelberg (2012)
8. Lichman, M.: UCI machine learning repository (2013). http://archive.ics.uci.edu/ml
9. Maua, D.D., Conaty, D., Cozman, F.G., Poppenhaeger, K., de Campos, P.C.: Robustifying sum-product networks. Int. J. Approximate Reasoning **101**, 163–180 (2018)
10. Mei, J., Jiang, Y., Tu, K.: Maximum a posteriori inference in sum-product networks. In: Thirty-Second AAAI Conference on Artificial Intelligence (2018)
11. Nath, A., Domingos, P.: Learning tractable probabilistic models for fault localization. In: 30th AAAI Conference on Artificial Intelligence, AAAI 2016, pp. 1294–1301 (2016)
12. Peharz, R.: Foundations of Sum-Product Networks for Probabilistic Modeling. Ph.D. thesis, Graz University of Technology (2015)
13. Peharz, R., Gens, R., Domingos, P.: Learning selective sum-product networks. In: Proceedings of the 31st International Conference on Machine Learning, vol. 32 (2014)
14. Peharz, R., Tschiatschek, S., Pernkopf, F., Domingos, P.: On Theoretical Properties of Sum-Product Networks. In: Proceedings of the 18th International Conference on Artificial Intelligence and Statistics (AISTATS), vol. 38, pp. 744–752 (2015)
15. Poon, H., Domingos, P.: Sum product networks: a new deep architecture. In: 2011 IEEE International Conference on Computer Vision Workshops (2011)
16. Pronobis, A., Rao, R.P.: Learning deep generative spatial models for mobile robots. In: IEEE International Conference on Intelligent Robots and Systems, pp. 755–762 (2017)
17. Sguerra, B.M., Cozman, F.G.: Image classification using sum-product networks for autonomous flight of micro aerial vehicles. In: 2016 5th Brazilian Conference on Intelligent Systems (BRACIS), pp. 139–144 (2016)
18. Wang, J., Wang, G.: Hierarchical spatial sum-product networks for action recognition in still images. IEEE Trans. Circuits Syst. Video Technol. **28**(1), 90–100 (2018)

Learning Models over Relational Data: A Brief Tutorial

Maximilian Schleich[1], Dan Olteanu[1(✉)], Mahmoud Abo-Khamis[2],
Hung Q. Ngo[2], and XuanLong Nguyen[3]

[1] University of Oxford, Oxford, UK
dan.olteanu@cs.ox.ac.uk
[2] RelationalAI, Inc., Berkeley, USA
[3] University of Michigan, Ann Arbor, USA
https://fdbresearch.github.io, https://www.relational.ai

Abstract. This tutorial overviews the state of the art in learning models over relational databases and makes the case for a first-principles approach that exploits recent developments in database research.

The input to learning classification and regression models is a training dataset defined by feature extraction queries over relational databases. The mainstream approach to learning over relational data is to materialize the training dataset, export it out of the database, and then learn over it using a statistical package. This approach can be expensive as it requires the materialization of the training dataset. An alternative approach is to cast the machine learning problem as a database problem by transforming the data-intensive component of the learning task into a batch of aggregates over the feature extraction query and by computing this batch directly over the input database.

The tutorial highlights a variety of techniques developed by the database theory and systems communities to improve the performance of the learning task. They rely on structural properties of the relational data and of the feature extraction query, including algebraic (semi-ring), combinatorial (hypertree width), statistical (sampling), or geometric (distance) structure. They also rely on factorized computation, code specialization, query compilation, and parallelization.

Keywords: Relational learning · Query processing

1 The Next Big Opportunity

Machine learning is emerging as general-purpose technology just as computing became general-purpose 70 years ago. A core ability of intelligence is the ability to predict, that is, to turn the information we have into the information we need. Over the last decade, significant progress has been made on improving

This project has received funding from the European Union's Horizon 2020 research and innovation programme under grant agreement No. 682588.

N. Ben Amor et al. (Eds.): SUM 2019, LNAI 11940, pp. 423–432, 2019.
https://doi.org/10.1007/978-3-030-35514-2_32

the quality of prediction by techniques that identify relevant features and by decreasing the cost of prediction using more performant hardware.

According to a 2017 Kaggle survey on the state of data science and machine learning among 16,000 machine learning practitioners [26], the majority of practical data science tasks involve relational data: in retail, 86% of used data is relational; in insurance, it is 83%; in marketing, it is 82%; while in finance it is 77%. This is not surprising. The relational model is the jewel in the data management crown. It is one of the most successful Computer Science stories. Since its inception in 1969, it has seen a massive adoption in practice. Relational data benefit from the investment of many human hours for curation and normalization and are rich with knowledge of the underlying domain modelled using database constraints.

Yet the current state of affairs in building predictive models over relational data largely ignores the structure and rich semantics readily available in relational databases. Current machine learning technology throws away this relational structure and works on one large training dataset that is constructed separately using queries over relational databases.

This tutorial overviews on-going efforts by the database theory and systems community to address the challenge of efficiently learning machine learning models over relational databases. It invariably only highlights some of the representative contributions towards this challenge, with an emphasis on recent contributions by the authors. The tutorial does not cover the wealth of approaches that use arrays of GPUs or compute farms for efficient machine learning. It instead puts forward the insight that an array of known and novel database optimization and processing techniques can make feasible a wide range of analytics workloads already on one commodity machine. There is still much to explore in the case of one machine before turning to compute farms. A key practical benefit of this line of work is energy-efficient, inexpensive analytics over large databases.

The organization of the tutorial follows the structure of the next sections.

2   Overview of Main Approaches to Machine Learning over Relational Databases

The approaches highlighted in this tutorial are classified depending on how tightly they integrate the data system, where the input data reside and the training dataset is constructed, and the machine learning library (statistical software package), which casts the model training problem as an optimization problem.

2.1   No Integration of Databases and Machine Learning

By far the most common approach to learning over relational data is to use two distinct systems, that is, the data system for managing the training dataset and the ML library for model training. These two systems are thus distinct tools on the technology stack with *no integration* between the two. The data system first computes the training dataset as the result of a *feature extraction query* and exports it as one table commonly in CSV or binary format. The ML library then imports the training dataset in its own format and learns the desired model.

For the first step, it is common to use open source database management systems, such as PostgreSQL or SparkSQL [57], or query processing libraries, such as Python Pandas [33] and R dplyr [56]. Common examples for ML libraries include scikit-learn [44], R [46], TensorFlow [1], and MLlib [34].

One advantage is the delegation of concerns: Database systems are used to deal with data, whereas statistical packages are for learning models. Using this approach, one can learn virtually any model over any database.

The key disadvantage is the non-trivial time spent on materializing, exporting, and importing the training dataset, which is commonly orders of magnitude larger than the input database. Even though the ML libraries are much less scalable than the data systems, in this approach they are thus expected to work on much larger inputs. Furthermore, these solutions inherit the limitations of both of their underlying systems, e.g., the maximum data frame size in R and the maximum number of columns in PostgreSQL are much less than typical database sizes and respectively number of model features.

2.2 Loose Integration of Databases and Machine Learning

A second approach is based on a *loose integration* of the two systems, with code of the statistical package migrated inside the database system space. In this approach, each machine learning task is implemented as a distinct user-defined aggregate function (UDAF) inside the database system. For instance, there are distinct UDAFs for learning: logistic regression models, linear regression models, k-means, Principal Component Analysis, and so on. Each of these UDAFs are registered in the underlying database system and there is a keyword in the query language supported by the database system to invoke them. The benefit is the direct interface between the two systems, with one single process running for both the construction of the training dataset and learning. The database system computes one table, which is the training dataset, and the learning task works directly on it. Prime example of this approach is MADLib [23] that extends PostgreSQL with a comprehensive library of machine learning UDAFs. The key advantage of this approach over the previous one is better runtime performance, since it does not need to export and import the (usually large) training dataset. Nevertheless, one has to explicitly write a UDAF for each new model and optimization method, essentially redoing the large implementation effort behind well-established statistical libraries. Approaches discussed in the next sections also suffer from this limitation, yet some contribute novel learning algorithms that can be asymptotically faster than existing off-the-shelf ones.

A variation of the second approach provides a *unified programming architecture*, one framework for many machine learning tasks instead of one distinct UDAF per task, with possible code reuse across UDAFs. Prime example of this approach is Bismark [16], a system that supports incremental (stochastic) gradient descent for convex programming. Its drawback is that its code may be less

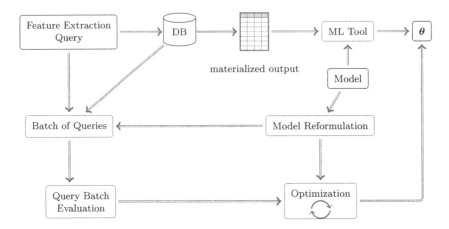

Fig. 1. Structure-aware versus structure-agnostic learning over relational databases.

efficient than the specialized UDAFs. Code reuse across various models and optimization problems may however speed up the development of new functionalities such as new models and optimization algorithms.

2.3 Tight Integration of Databases and Machine Learning

The aforementioned approaches do not exploit the structure of the data residing in the database. The next and final approach features a *tight integration* of the data and learning systems. The UDAF for the machine learning task is pushed into the feature extraction query and one single evaluation plan is created to compute both of them. This approach enables database optimizations such as pushing parts of the UDAFs past the joins of the feature extraction query. Prime examples are Orion [29],which supports generalized linear models, Hamlet [30], which supports logistic regression and naïve Bayes, Morpheus [11], which linear and logistic regression, k-means clustering, and Gaussian non-negative matrix factorization, F [40,41,51], which supports ridge linear regression, AC/DC [3], which supports polynomial regression and factorization machines [47–49], and LMFAO [50], which supports a larger class of models including the previously mentioned ones and decision trees [10], Chow-Liu trees [12], mutual information, and data cubes [19,22].

3 Structure-Aware Learning

The tightly-integrated systems F [51], AC/DC [3], and LMFAO [50] are *data structure-aware* in that they exploit the structure and sparsity of the database to lower the complexity and drastically improve the runtime performance of the learning process. In contrast, we call all the other systems *structure-agnostic*, since they do not exploit properties of the input database. Figure 1 depicts the difference between structure-aware (in green) and structure-agnostic (in red) approaches. The structure-aware systems compile the model specification into a set of aggregates, one per feature or feature interaction. This is called model reformulation in the figure. Data dependencies such as functional dependencies can be used to reparameterize the model, so a model over a smaller set of functionally determining features is learned instead and then mapped back to the original model. Join dependencies, such as those prevalent in feature extraction queries that put together several input tables, are exploited to avoid redundancy in the representation of join results and push the model aggregates past joins. The model aggregates over the feature extraction query define a batch of queries. In practice, for training datasets with tens of features, query batch sizes can be in the order of: hundreds to thousands for ridge linear regression; thousands for computing a decision tree node; and tens for an assignment step in k-means clustering [50]. The result of a query batch is then the input to an optimizer such as a gradient descent method that iterates until the model parameters converge.

Structure-aware methods have been developed (or are being developed) for a variety of models [4]. Besides those mentioned above, powerful models that can be supported are: Principal Component Analysis (PCA) [35], Support Vector Machines (SVM) [25], Sum Product Networks (SPN) [45], random forests, boosting regression trees, and AdaBoost. Newer methods also look at linear algebra programs where matrices admit a database interpretation such as the results of queries over relations. In particular, on-going work [17, 24] tackles various matrix decompositions, such as QR, Cholesky, SVD [18], and low-rank [54].

Structure-aware methods call for new data processing techniques to deal with large query batches. Recent work puts forward new optimization and evaluation strategies that go beyond the capabilities of existing database management systems. Recent experiments confirm this observation: Whereas existing query processing techniques are mature at executing one query, they miss opportunities for systematically sharing computation across several queries in a batch [50].

Tightly-integrated DB-ML systems commonly exploit four types of structure: algebraic, combinatorial, statistical, and geometric.

Algebraic Structure. The algebraic structure of semi-rings underlies the recent work on factorized databases [41, 42]. The distributivity law in particular allows to factor out data blocks common to several tuples, represent them once and compute over them once. Using factorization, relations can represented more succinctly as directed acyclic graphs. For instance, the natural join of two relations is a union of Cartesian products. Instead of representing such a Cartesian product of two relation parts explicitly as done by relational database systems, we can represent it symbolically as a tree whose root is the Cartesian product symbol and has as children the two relation parts. It has been shown that factorization can improve the performance of joins [42], aggregates [6,9], and more recently machine learning [2,4,41,51]. The additive inverse of rings allows to treat uniformly data updates (inserts and deletes) and enables incremental maintenance of models learned over relational data [27,28,39]. The sum-product abstraction in (semi) rings allows to use the same processing (computing and maintaining) mechanism for seemingly disparate tasks, such as database queries, covariance matrices, inference in probabilistic graphical models, and matrix chain multiplication [6,39]. The efficient maintenance of covariance matrices is a prerequisite for the availability of fresh models under data changes [39]. A recent tutorial overviews advances in incremental view maintenance [15].

Combinatorial Structure. The combinatorial structure prevalent in relational data has been formalized by notions such as width and data degree measures. If a feature extraction query has width w, then its data complexity is $\tilde{O}(N^w)$ for a database of size N, where $\tilde{O}$ hides logarithmic factors in N. Various width measures have been proposed recently, such as: the fractional edge cover number [8,20,37,38,55] to capture the asymptotic size of the results for join queries and the time to compute them; the fractional hypertree width [32] and the submodular width [7] to capture the time to compute Boolean conjunctive queries; the factorization width [42] to capture the size of the factorized results of conjunctive queries; the FAQ-width [6] that extends the factorization width from conjunctive queries to functional aggregate queries; and the sharp-submodular width [2] that improves on the previous widths for functional aggregate queries.

The degree information captures the number of occurrences of a data value in the input database [38]. Existing processing techniques adapt depending on the high or low degree of data values. A recent such technique has been shown to be worst-case optimal for incrementally maintaining the count of triangles in a graph [27]. Another such technique achieves a low complexity for computing queries with negated relations of bounded degree [5]. A special form of bounded degree is given by functional dependencies, which can be used to reparameterize (polynomial regression and factorization machine) models and learn simpler, equivalent models instead [4].

Statistical Structure. The statistical structure allows to sample through joins, such as the ripple joins [21] and the wander joins [31], and to sample for specific classes of machine learning models [43]. Sampling is employed whenever the input database is too large to be processed within a given time budget. It may nevertheless lead to approximation of both steps in the end-to-end learning task, from the computation of the feature extraction query to the subsequent optimization task that yields the desired model. Work in this space quantifies the loss in accuracy of the obtained model due to sampling.

Geometric Structure. Algorithms for clustering methods such as k-means [35] can exploit distance measures (such as the optimal transport distance between two probability measures) to obtain constant-factor approximations for the k-means objective by clustering over a small grid coreset instead of the full result of the feature extraction query [14].

4   Database Systems Considerations

Besides exploiting the structure of the input data and the learning task, the problem of learning models over databases can also benefit tremendously from database system techniques. Recent work [50] showed non-trivial speedups (several orders of magnitude) brought by code optimization for machine learning workloads over state-of-the-art systems such as TensorFlow [1], R [46], Scikit-learn [44], and mlpack [13]. Prime examples of code optimizations leading to such performance improvements include:

Code Specialization and Query Compilation. It involves generating code specific to the query and the schema of its input data, following prior work [36,52,53], and also specific to the model to be learned. This technique improves the runtime performance by inlining code and improving cache locality for the hot data path.

Sharing Computation. Sharing is best achieved by decomposing the aggregates in a query batch into simple views that are pushed down the join tree of the feature extraction query. Different aggregates may then need the same simple views at some nodes in the join tree. Sharing of scans of the input relations can also happen across views, even when they have different output schemas.

Parallelization. Parallelization can exploit multi-core CPU architectures but also large share-nothing distributed systems. It comprises both task parallelism, which identifies subqueries that are independent and can be computed in parallel, and domain parallelism, which partitions relations and computes the same subqueries over different parts in parallel.

This tutorial is a call to arms for more sustained and principled work on the theory and systems of structure-aware approaches to data analytics. What are the theoretical limits of structure-aware learning? What are the classes of machine learning models that can benefit from structure-aware learning over relational data? What other types of structure can benefit learning over relational data?

References

1. Abadi, M., et al.: Tensorflow: a system for large-scale machine learning. In: OSDI, pp. 265–283 (2016)
2. Abo Khamis, M., et al.: On functional aggregate queries with additive inequalities. In: PODS, pp. 414–431 (2019)
3. Abo Khamis, M., Ngo, H.Q., Nguyen, X., Olteanu, D., Schleich, M.: AC/DC: In-database learning thunderstruck. In: DEEM, pp. 8:1–8:10 (2018)
4. Abo Khamis, M., Ngo, H.Q., Nguyen, X., Olteanu, D., Schleich, M.: In-database learning with sparse tensors. In: PODS, pp. 325–340 (2018)
5. Abo Khamis, M., Ngo, H.Q., Olteanu, D., Suciu, D.: Boolean tensor decomposition for conjunctive queries with negation. In: ICDT, pp. 21:1–21:19 (2019)
6. Abo Khamis, M., Ngo, H.Q., Rudra, A.: FAQ: questions asked frequently. In: PODS, pp. 13–28 (2016)

7. Abo Khamis, M., Ngo, H.Q., Suciu, D.: What do shannon-type inequalities, submodular width, and disjunctive datalog have to do with one another? In: PODS, pp. 429–444 (2017)
8. Atserias, A., Grohe, M., Marx, D.: Size bounds and query plans for relational joins. In: FOCS, pp. 739–748 (2008)
9. Bakibayev, N., Kociský, T., Olteanu, D., Závodný, J.: Aggregation and ordering in factorised databases. PVLDB **6**(14), 1990–2001 (2013)
10. Breiman, L., Friedman, J., Olshen, R., Stone, C.: Classification and Regression Trees. Wadsworth and Brooks, Monterey (1984)
11. Chen, L., Kumar, A., Naughton, J.F., Patel, J.M.: Towards linear algebra over normalized data. PVLDB **10**(11), 1214–1225 (2017)
12. Chow, C., Liu, C.: Approximating discrete probability distributions with dependence trees. IEEE Trans. Inf. Theor. **14**(3), 462–467 (2006)
13. Curtin, R.R., Edel, M., Lozhnikov, M., Mentekidis, Y., Ghaisas, S., Zhang, S.: mlpack 3: a fast, flexible machine learning library. J. Open Source Soft. **3**, 726 (2018)
14. Curtin, R.R., Moseley, B., Ngo, H.Q., Nguyen, X., Olteanu, D., Schleich, M.: Rk-means: fast coreset construction for clustering relational data (2019)
15. Elghandour, I., Kara, A., Olteanu, D., Vansummeren, S.: Incremental techniques for large-scale dynamic query processing. In: CIKM, pp. 2297–2298 (2018). Tutorial
16. Feng, X., Kumar, A., Recht, B., Ré, C.: Towards a unified architecture for in-RDBMS analytics. In: SIGMOD, pp. 325–336 (2012)
17. van Geffen, B.: QR decomposition of normalised relational data (2018), MSc thesis, University of Oxford
18. Golub, G.H., Van Loan, C.F.: Matrix Computations, 4th edn. The Johns Hopkins University Press, Baltimore (2013)
19. Gray, J., Bosworth, A., Layman, A., Pirahesh, H.: Data cube: A relational aggregation operator generalizing group-by, cross-tab, and sub-total. In: ICDE, pp. 152–159 (1996)
20. Grohe, M., Marx, D.: Constraint solving via fractional edge covers. In: SODA, pp. 289–298 (2006)
21. Haas, P.J., Hellerstein, J.M.: Ripple joins for online aggregation. In: SIGMOD, pp. 287–298 (1999)
22. Harinarayan, V., Rajaraman, A., Ullman, J.D.: Implementing data cubes efficiently. In: SIGMOD, pp. 205–216 (1996)
23. Hellerstein, J.M., et al.: The madlib analytics library or MAD skills, the SQL. PVLDB **5**(12), 1700–1711 (2012)
24. Inelus, G.R.: Quadratically Regularised Principal Component Analysis over multi-relational databases, MSc thesis, University of Oxford (2019)
25. Joachims, T.: Training linear SVMS in linear time. In: SIGKDD, pp. 217–226 (2006)
26. Kaggle: The State of Data Science and Machine Learning (2017). https://www.kaggle.com/surveys/2017
27. Kara, A., Ngo, H.Q., Nikolic, M., Olteanu, D., Zhang, H.: Counting triangles under updates in worst-case optimal time. In: ICDT, pp. 4:1–4:18 (2019)
28. Koch, C., Ahmad, Y., Kennedy, O., Nikolic, M., Nötzli, A., Lupei, D., Shaikhha, A.: Dbtoaster: higher-order delta processing for dynamic, frequently fresh views. VLDB J. **23**(2), 253–278 (2014)
29. Kumar, A., Naughton, J.F., Patel, J.M.: Learning generalized linear models over normalized data. In: SIGMOD, pp. 1969–1984 (2015)

30. Kumar, A., Naughton, J.F., Patel, J.M., Zhu, X.: To join or not to join?: thinking twice about joins before feature selection. In: SIGMOD, pp. 19–34 (2016)
31. Li, F., Wu, B., Yi, K., Zhao, Z.: Wander join and XDB: online aggregation via random walks. ACM Trans. Database Syst. 44(1), 2:1–2:41 (2019)
32. Marx, D.: Approximating fractional hypertree width. ACM Trans. Algorithms 6(2), 29:1–29:17 (2010)
33. McKinney, W.: pandas: a foundational python library for data analysis and statistics. Python High Perform. Sci. Comput. 14 (2011)
34. Meng, X., et al.: Mllib: machine learning in apache spark. J. Mach. Learn. Res. 17(1), 1235–1241 (2016)
35. Murphy, K.P.: Machine Learning: A Probabilistic Perspective. MIT Press, Cambridge (2013)
36. Neumann, T.: Efficiently compiling efficient query plans for modern hardware. PVLDB 4(9), 539–550 (2011)
37. Ngo, H.Q., Porat, E., Ré, C., Rudra, A.: Worst-case optimal join algorithms. In: PODS, pp. 37–48 (2012)
38. Ngo, H.Q., Ré, C., Rudra, A.: Skew strikes back: New developments in the theory of join algorithms. In: SIGMOD Rec., pp. 5–16 (2013)
39. Nikolic, M., Olteanu, D.: Incremental view maintenance with triple lock factorization benefits. In: SIGMOD, pp. 365–380 (2018)
40. Olteanu, D., Schleich, M.: F: regression models over factorized views. PVLDB 9(10), 1573–1576 (2016)
41. Olteanu, D., Schleich, M.: Factorized databases. SIGMOD Rec. 45(2), 5–16 (2016)
42. Olteanu, D., Závodný, J.: Size bounds for factorised representations of query results. TODS 40(1), 2 (2015)
43. Park, Y., Qing, J., Shen, X., Mozafari, B.: Blinkml: efficient maximum likelihood estimation with probabilistic guarantees. In: SIGMOD, pp. 1135–1152 (2019)
44. Pedregosa, F., et al.: Scikit-learn: machine learning in python. J. Mach. Learn. Res. 12, 2825–2830 (2011)
45. Poon, H., Domingos, P.M.: Sum-product networks: a new deep architecture. In: UAI, pp. 337–346 (2011)
46. R Core Team: R: A Language and Environment for Statistical Computing. R Foundation for Stat. Comp. (2013). www.r-project.org
47. Rendle, S.: Factorization machines. In: Proceedings of the 2010 IEEE International Conference on Data Mining. ICDM 2010, pp. 995–1000. IEEE Computer Society, Washington, DC (2010)
48. Rendle, S.: Factorization machines with libFM. ACM Trans. Intell. Syst. Technol. 3(3), 57:1–57:22 (2012)
49. Rendle, S.: Scaling factorization machines to relational data. PVLDB 6(5), 337–348 (2013)
50. Schleich, M., Olteanu, D., Abo Khamis, M., Ngo, H.Q., Nguyen, X.: A layered aggregate engine for analytics workloads. In: SIGMOD, pp. 1642–1659 (2019)
51. Schleich, M., Olteanu, D., Ciucanu, R.: Learning linear regression models over factorized joins. In: SIGMOD, pp. 3–18 (2016)
52. Shaikhha, A., Klonatos, Y., Koch, C.: Building efficient query engines in a high-level language. TODS 43(1), 4:1–4:45 (2018)
53. Shaikhha, A., Klonatos, Y., Parreaux, L., Brown, L., Dashti, M., Koch, C.: How to architect a query compiler. In: SIGMOD, pp. 1907–1922 (2016)
54. Udell, M., Horn, C., Zadeh, R., Boyd, S.: Generalized low rank models. Found. Trends Mach. Learn. 9(1), 1–118 (2016)

55. Veldhuizen, T.L.: Triejoin: a simple, worst-case optimal join algorithm. In: ICDT, pp. 96–106 (2014)
56. Wickham, H., Francois, R., Henry, L., Müller, K., et al.: dplyr: a grammar of data manipulation. R package version 0.4 **3** (2015)
57. Zaharia, M., Chowdhury, M., et al.: Resilient distributed datasets: a fault-tolerant abstraction for in-memory cluster computing. In: NSDI, p. 2 (2012)

Subspace Clustering and Some Soft Variants

Marie-Jeanne Lesot$^{(\boxtimes)}$

Sorbonne Université, CNRS, LIP6, 75005 Paris, France
`Marie-Jeanne.Lesot@lip6.fr`

Abstract. Subspace clustering is an unsupervised machine learning task that, as clustering, decomposes a data set into subgroups that are both distinct and compact, and that, in addition, explicitly takes into account the fact that the data subgroups live in different subspaces of the feature space. This paper provides a brief survey of the main approaches that have been proposed to address this task, distinguishing between the two paradigms used in the literature: the first one builds a local similarity matrix to extract more appropriate data subgroups, whereas the second one explicitly identifies the subspaces, so as to dispose of more complete information about the clusters. It then focuses on soft computing approaches, that in particular exploit the framework of the fuzzy set theory to identify both the data subgroups and their associated subspaces.

Keywords: Machine learning · Unsupervised learning · Subspace clustering · Soft computing · Fuzzy logic

1 Introduction

In the unsupervised learning framework, the only available input is a set of data, here considered to be numerically described by feature vectors. The aim is then to extract information from the data, e.g. in the form of linguistic summaries (see e.g. [17]), frequent value co-occurrences, as expressed by association rules, or as clusters. The latter are subgroups of data that are both compact and distinct, which means that any data point is more similar to points assigned to the same group than to points assigned to other groups. These clusters provide insight to the data structure and a summary of the data set.

Subspace clustering [3,28] is a refined form of the clustering task, where the clusters are assumed to live in different subspaces of the feature space: on the one hand, this assumption can help identifying more relevant data subgroups, relaxing the need to use a single, global, similarity relation; on the other hand, it leads to refine the identified data summaries, so as to characterise each cluster through its associated subspace. These two points of view have led to the two main families of subspace clustering approaches, that have slightly different aims and definitions.

© Springer Nature Switzerland AG 2019
N. Ben Amor et al. (Eds.): SUM 2019, LNAI 11940, pp. 433–443, 2019.
https://doi.org/10.1007/978-3-030-35514-2_33

This paper first discusses in more details the definition of the subspace clustering task in Sect. 2 and presents in turn the two main paradigms, in Sects. 3 and 4 respectively. It then focuses on soft computing approaches that have been proposed to perform subspace clustering, in particular fuzzy ones: fuzzy logic tools have proved to be useful to all types of machine learning tasks, such as classification, extraction of association rules or clustering. Section 5 describes their applications to the case of subspace clustering. Section 6 concludes the paper.

2   Subspace Clustering Task Definition

Clustering. Clustering aims at decomposing a data set into subgroups that are both compact and separable: compactness imposes a high internal similarity for points assigned to the same cluster; separability imposes a high dissimilarity for points assigned to different clusters, so that the clusters are distinct one from another. These two properties thus jointly justify the individual existence of each of the extracted clusters.

There exist many approaches to address this task that can broadly be structured into five main families: hierarchical, partitioning, density-based, spectral and, more recently, deep approaches. In a nutshell, hierarchical clustering identifies multiple data partitions, represented in a tree structure called dendrogram, that allows to vary the desired granularity level of the data decomposition into subgroups. Partitioning approaches, that provide numerous variants to the seminal k-means method, optimise a cost function that can be interpreted as a quantisation error, i.e. assessing the approximation error when a data point is represented by the centre of the cluster it is assigned to. Density-based approaches, exemplified by DBSCAN [9], group points according to a transitive neighbour relation and define cluster boundaries as low density regions. Spectral methods [23] rely on diagonalising the pairwise similarity matrix, considering that two points should be assigned to the same cluster if they have the same similarity profile to the other points. They can also be interpreted as identifying connex components in the similarity graph, whose nodes correspond to the data points and edges are weighted by the pairwise similarity values. Deep clustering approaches [24] are often based on an encoder-decoder architecture, where the encoder provides a low dimension representation of the data, corresponding to the cluster representation, that must allow to reconstruct the data in the decoding phase.

Subspace Clustering. Subspace clustering refines the clustering task by considering that each cluster lives in its own subspace of the whole feature space. Among others, this assumption implies that there is not a single, global, distance (or similarity) measure to compare the data points, defined in the whole feature space: each cluster can be associated to its own, local, comparison measure, defined in its corresponding subspace.

Subspace clustering cannot be addressed by performing local feature selection for each cluster: such an approach would first identify the clusters in the global feature space before characterising them. Now subspace clustering aims at extracting more appropriate clusters that can be identified in lower dimensional spaces only. Reciprocally, first performing feature selection and then clustering the data would impose a subspace common to all clusters. Subspace clustering addresses both subgroup and feature identification simultaneously, so as to obtain better subgroups, defined locally.

Subspace clustering is especially useful for high dimensional data, due to the curse of dimensionality that makes all distances between pairs of points have very close values: it can be the case that there exists no dense data subgroup in the whole feature space and that clusters can only be identified when considering subspaces with lower dimensionality.

Two Main Paradigms. Numerous approaches for subspace clustering have been proposed in the literature, offering a diversity similar to that of the general clustering task. Two main paradigms can be distinguished, that focus on slightly different objectives and lead to independent method developments, as sketched below and described in more details in the next two sections.

The first category, presented in Sect. 3, exploits the hypothesis that the available data has been drawn from distinct subspaces so as to improve the clustering results: the subspace existence is viewed as a useful intermediary tool for the clustering aim, but their identification is not a goal in itself and they are not further used. Methods in this category rely on deriving an affinity matrix from the data, that captures local similarity between data points and can be used to cluster the data, instead of a predefined global similarity (or distance) measure.

The second category, described in Sect. 4, considers that the subspaces in which the clusters live provide useful information in themselves: methods in this category aim at explicitly identifying these subspaces, so as to characterise the identified clusters and extract more knowledge from the data. Methods in this category rely on predefined forms of the subspaces and extract from the data their optimal instanciation.

3   Learning an Affinity Matrix

As mentioned in the previous section, subspace clustering can be defined as a clustering task in the case where the data have been drawn from a union of low dimensional subspaces [15,28]. Such a priori knowledge is for instance available in many computer vision applications, such as image segmentation, motion segmentation or image clustering [7,15,18,22] to name a few.

Methods in this category rely on a two-step procedure that consists in first learning a local affinity matrix and then performing spectral clustering on this matrix: the affinity matrix learns local similarity (or distance) values for each couple of data, instead of applying a global, predefined, measure.

Among others [15,28], self-expressive approaches first represent the data points as linear combination of other data points, that must thus be in the same subspace. More formally, they learn a self-representation matrix C, minimising the reconstruction cost $\|X - XC\|_F$ where $\| \|_F$ is the Frobenius norm. An affinity matrix can then be defined as $W = \frac{1}{2}(|C| + |C^T|)$ and used for spectral clustering. Various constraints on the self-representation matrix C can be considered, adding penalisation terms to the reconstruction cost with various norms. Some examples include Sparse Subspace Clustering, SSC [7,8], or Low-Rank Representation, LRR [22]. In order to extend to the case of non-linear subspaces, kernel approaches have been proposed [26,30,31], as well as, more recently, deep learning methods that do not require to set a priori the considered non-linear data transformation. The latter can consist in applying SSC to the latent features extracted by an auto-encoder architecture [27] or, more intrinsically, to integrate a self-expressive layer between the encoder and decoder steps [15,32].

4   Identifying the Subspaces

A second approach to subspace clustering considers that the subspaces in which the clusters live are also interesting as such and provide useful insight to the data structure. It thus provides an explicit representation of these subspaces, whereas the methods described in the previous section only exploit the matrix of the local distances between all data pairs.

Many approaches have been proposed to identify the subspaces associated to each cluster, they can be organised into three categories discussed in turn below. They also differ by the form of the subspaces they consider that can be hyper-rectangles [3], vector subspaces [1,28] or hyperplanes of low dimension [29], to name a few.

Bottom-Up Strategy. A first category of approaches starts from atomic clusters with high density and very low dimensionality that are then iteratively fused to build more complex clusters and subspaces.

This is for instance the case of the CLIQUE algorithm [3] that starts from dense unit cubes, progressively combined to define clusters and subspaces as maximal sets of adjacent cells, parallel to the axes. A final step provides a textual description of each cluster that aims at being both concise and informative for the user: it contains information about the involved dimensions and their value boundaries, offering an enriched result as compared to the list of the points assigned to the considered cluster.

ENCLUS [6] and MAFIA [5] follow the same bottom-up principle. The differences come from the fact that ENCLUS minimises the cell entropy instead of maximising their densities and that MAFIA allows to consider an adaptive grid to define the initial cube units according to the data distribution for each attribute.

Top-Down Strategy: Projected Clustering. A second category of methods applies a top-down exploration method, that progressively refines subspaces initially defined as the whole feature space. The refinement step consists in projecting the data to subspaces, making it possible to identify the cluster structure of the data even when there is no dense clusters in the full feature space.

PROCLUS [1] belongs to this framework of projected clustering, it identifies three components: (i) clusters, (ii) associated dimensions that define axes-parallel subspaces, as well as (iii) outliers, i.e. points that are assigned to none of the clusters. The candidate projection subspaces are defined by the dimensions along which the cluster members have the lowest dispersion. ORCLUS [2] is a variant of PROCLUS that allows to identify subspaces that are not parallel to the initial axes.

Partitioning Strategy: Optimising a Cost Function. A third category of methods relies on the definition of a cost function that extends the classical and seminal k-means cost function so as to integrate the desired subspaces associated with the clusters. They are based on replacing the Euclidean distance used to compare the data by weighted variants thereof, where the weights are attached to each cluster so as to dispose of the local definition of the distance function: a dimension associated with a large weight is interpreted as playing a major role in the cluster definition, as small variations in this dimension lead large increases in the distance value. The subspaces can thus be defined indirectly by the weights attached to the dimensions. Some approaches impose the subspaces to be axes-parallel, others allow for rotations.

The subspace clustering methods usually require as hyperparameter the desired number of clusters, set in advance. Most of them apply an alternated optimisation scheme, as performed by the k-means algorithm: given candidate cluster definitions, they optimise the data assignment to the clusters and, given a candidate assignment, they optimise the cluster description. The latter adds to the traditional cluster centres the associated distance weights. The approaches belonging to this category vary by the constraints imposed to these weights, as illustrated by some examples below.

It can first be observed that, although proposed in another framework, the Gaussian Mixture Model (GMM) clustering approach can be interpreted as addressing the subspace clustering task: GMM associate each cluster with its covariance matrix and computes the distance between a data point and the centre of the cluster it is assigned to as a local Malahanobis distance. As a consequence, for example, a dimension associated with a low variance can be interpreted as highly characterising the cluster, and the cluster subspace can be defined as the one spanned by the dimensions with minimal variances. Full covariance matrices allow for general cluster subspaces, diagonal matrices impose the subspaces to be parallel to the initial axes.

Another example is provided by the Fuzzy Subspace Clustering algorithm, FSC [10], that, despite his name, provides crisp assignment of the data to the clusters: it is named fuzzy because of the weights in $[0, 1]$ attached to the dimensions. Denoting n the number of data points, d the number of features, $x_i = (x_{i1}, \cdots, x_{id})$ for $i = 1..n$ the data, c the desired number of clusters, $c_r = (c_{r1}, \cdots, c_{rd})$ for $r = 1..c$ the cluster centres, u_{ri} the assignment of data x_i

to cluster r, $w_r = (w_{r1}, \cdots, w_{rd})$ for $r = 1..c$ the dimension weights for cluster r and η and q two hyperparameters, FSC considers the cost function

$$J_{FSC} = \sum_{i=1}^{n} \sum_{r=1}^{c} u_{ri} \sum_{p=1}^{d} w_{rp}^q (x_{ip} - c_{rp})^2 + \eta \sum_{r=1}^{c} \sum_{p=1}^{d} w_{rp}^q \qquad (1)$$

under the constraints $u_{ri} \in \{0,1\}$, $\sum_{r=1}^{c} u_{ri} = 1$ for all i and $\sum_{p=1}^{d} w_{rp} = 1$ for all r. In this cost, the first term is identical to the k-means cost function when replacing the Euclidean distance by a weighted one, the second term is required so that the update equations are well defined [10]. The two terms are balanced by the η hyperparameter. The first two constraints are identical to the k-means ones, the third one forbids the trivial solution where all weights $w_{rp} = 0$. The q hyperparameter defining the exponent of the weights w_{rp}^q is similar to the fuzzifier m used in the fuzzy c-means algorithm to avoid converging to binary weights $w_{rp} \in \{0,1\}$ [20].

The Entropy Weighted k-means algorithm, EWKM [16], is an extension that aims at controlling the sparsity of the dimension weights w_{rp}, so that they tend to equal 0, instead of being small but non-zero: to that aim, it replaces the second term in J_{FSC} with an entropy regularisation term, balanced with a γ parameter

$$J_{EWKM} = \sum_{i=1}^{n} \sum_{r=1}^{c} u_{ri} \sum_{p=1}^{d} w_{rp} (x_{ip} - c_{rp})^2 + \gamma \sum_{r=1}^{c} \sum_{p=1}^{d} w_{rp} \log w_{rp} \qquad (2)$$

under the same constraints. When γ tends to 0, it allows to control the sparsity level of the w_{rp} weights.

5   Soft Variants

This section aims at detailing partitioning approaches, introduced in the previous section, in the case where the point assignment to the cluster is not binary but soft, i.e., using the above notations, $u_{ri} \in [0,1]$ instead of $u_{ri} \in \{0,1\}$: they constitute subspace extensions of the fuzzy c-means algorithm, called *fcm*, and its variants (see e.g. [14,21] for overviews).

First the Gustafson-Kessel algorithm [13] can be viewed as the fuzzy correspondent of GMM discussed in the previous section: both use the Mahalanobis distance and consider weighted assignments to the clusters. They differ by the interpretation of these weights and by the cost function they consider: GMM optimises the log-likelihood of the data, in a probabilistic modelling framework, whereas Gustafson-Kessel considers a quantisation error, in a *fcm* manner.

Using the same notations as in the previous section, with the additional hyperparameters m, called fuzzifier, and $(\alpha_r)_{r=1..c} \in \mathbb{R}$, the Attribute Weighted Fuzzy c-means algorithm, AWFCM [19], is based on the cost function

$$J_{AWFCM} = \sum_{i=1}^{n} \sum_{r=1}^{c} u_{ri}^m \sum_{p=1}^{d} w_{rp}^q (x_{ip} - c_{rp})^2 \qquad (3)$$

with $u_{ri} \in [0, 1]$ and under the constraints $\sum_{r=1}^{c} u_{ri} = 1$ for all i, $\sum_{i=1}^{n} u_{ri} > 0$ for all r and $\sum_{p=1}^{d} w_{rp} = \alpha_r$ for all r. The cost function is thus identical to the *fcm* one, replacing the Euclidean distance with its weighted variant. The first two constraints also are identical to the *fcm* ones, the third one forbids the trivial solution $w_{rp} = 0$. The (α_r) hyperparameters can also allow to weight the relative importance of the c clusters in the final partition, but they are usually set to be all equal to 1 [19].

Many variants of AWFCM have been proposed, for instance to introduce sparsity in the subspace description: AWFCM indeed produces solutions where none of the w_{rp} parameters equals zero, even if they are very small. This is similar to a well-known effect of the *fcm*, where the optimisation actually leads to $u_{ri} \in \;]0, 1[$, except for data points that are equal to a cluster centre: the membership degrees can be very small, but they cannot equal zero [20]. Borgelt [4] thus proposes to apply the sparsity inducing constraints introduced for the membership degrees u_{ri} [20], considering

$$J_{BOR} = \sum_{i=1}^{n} \sum_{r=1}^{c} g(u_{ri}) \sum_{p=1}^{d} g(w_{rp})(x_{ip} - c_{rp})^2 \qquad (4)$$

under the same constraints as AWFCM, where

$$g(x) = \frac{1 - \beta}{1 + \beta} x^2 + \frac{2\beta}{1 + \beta} x \text{ with } \beta \in [0, 1[$$

Two different β values can be respectively considered for the membership degrees u_{ri} and the dimension weights w_{rp}: setting $\beta = 0$ leads to the same function as AWFCM with $m = 2$ and $q = 2$, which are the traditional choices [19]. Considering a non-zero β value allows to get $u_{ri} = 0$ or $w_{rp} = 0$ [20], providing a sparsity property, both for the membership degrees and the dimension weights.

The Weighted Laplacian Fuzzy Clustering algorithm, WLFC [11], is another variant of AWFCM that aims at solving an observed greediness of this algorithm: AWFCM sometimes appears to be over-efficient and to fail to respect the global geometry of the data, because of its adaptation to local structure [11]. To address this issue, WLFC proposes to add a regularisation term to the cost function, so as to counterbalance the local effect of cluster subspaces:

$$J_{WLFC} = \sum_{i=1}^{n} \sum_{r=1}^{c} u_{ri}^2 \sum_{p=1}^{d} w_{rp}^q (x_{ip} - c_{rp})^2 + \gamma \sum_{i,j=1}^{n} \sum_{r=1}^{c} (u_{ri} - u_{si})^2 s_{ij} \qquad (5)$$

under the same constraints as AWFCM. In this cost, s_{ij} is a well-chosen global similarity measure [11] that imposes that neighbouring points in the whole feature space still have somewhat similar membership degrees. The γ hyperparameter allows to balance the two effects and to prevent some discontinuity in the solution among point neighbourhood.

The Proximal Fuzzy Subspace C-Means, PFSCM [12] considers the cost function defined as

$$J_{PFSCM} = \sum_{i=1}^{n} \sum_{r=1}^{c} u_{ri}^{m} \sum_{p=1}^{d} w_{rp}^{2}(x_{ip} - c_{rp})^{2} + \gamma \sum_{r=1}^{c} |\sum_{p=1}^{d} (w_{rp}) - 1| \qquad (6)$$

under the first two constraints of AWFCM: the second term can be interpreted as an inline version of the third constraint that is thus moved within the cost function. As it is not differentiable, PFSCM proposes an original optimisation scheme that does not rely on standard alternate optimisation but on proximal descent (see e.g. [25]). This algorithm appears to identify better the number of relevant dimensions for each cluster, where AWFCM tends to underestimate it. Moreover, the proposition to apply proximal optimisation techniques to the clustering task opens the way for defining a wide range of regularisation terms: it allows for more advanced penalty terms that are not required to be differentiable.

6   Conclusion

This paper proposed a brief overview of the subspace clustering task and the main categories of methods proposed to address it. They differ in the understanding of the general aim and offer a large variety of approaches that provide different types of outputs and knowledge extracted from the data.

Still they have several properties in common. First most of them rely on a non-constant distance measure: the comparison of two data points does not rely on a global measure, but on a local one, that somehow takes the assignment to the same cluster as a parameter to define this measure. As such, subspace clustering constitutes a task that must extract from the data, in an unsupervised way, both compact and distinct data subgroups, as well as the reasons why these subgroups can be considered as compact. This makes it clear that subspace clustering is a highly demanding and difficult task, that aims at exploiting inputs with little information (indeed, inputs reduce to the data position in the feature space only) to extract very rich knowledge.

Moreover, it can be observed that many subspace clustering methods share a constraint of sparsity: it imposes subspaces to be as small as possible so as to contain the clusters, while avoiding to oversimplify their complexity. A large variety of criteria to define sparsity and integrate it into the task objective is exploited across the existing approaches.

Among the directions for ongoing works in the subspace clustering domain, a major one deals with the question of evaluation: as is especially the case for any unsupervised learning task, there is no consensus about the quality criteria to be used to assess the obtained results. The first category of methods, that exploit the subspace existence to learn an affinity matrix, usually focuses on evaluating the cluster quality: they resort to general clustering criteria, such as the clustering error, measured as accuracy, the cluster purity or Normalised Mutual Information. Thus they are often evaluated in a supervised manner, considering a reference of expected data subgroups. When subspace clustering is understood as also characterising the clusters using the subspaces in which they live, the evaluation must also assess these extracted subspaces, e.g. taking into account both their adequacy and sparsity. The definition of corresponding quality criteria still constitutes an open question in the subspace clustering domain.

Acknowledgements. I wish to thank Arthur Guillon and Christophe Marsala with whom I started exploring the domain of subspace clustering.

References

1. Aggarwal, C.C., Wolf, J.L., Yu, P.S., Procopiuc, C., Park, J.S.: Fast algorithms for projected clustering. In: Proceedings of the International Conference on Management of Data, SIGMOD, pp. 61–72. ACM (1999)
2. Aggarwal, C.C., Yu, P.S.: Finding generalized projected clusters in high dimensional spaces. In: Proceedings of the International Conference on Management of Data, SIGMOD, pp. 70–81. ACM (2000)
3. Agrawal, R., Gehrke, J., Gunopulos, D., Raghavan, P.: Automatic subspace clustering of high dimensional data for data mining applications. In: Proceedings of the ACM SIGMOD International Conference on Management of Data, SIGMOD, pp. 94–105. ACM (1998)
4. Borgelt, C.: Fuzzy subspace clustering. In: Fink, A., Lausen, B., Seidel, W., Ultsch, A. (eds.) Advances in Data Analysis, Data Handling and Business Intelligence. Studies in Classification, Data Analysis, and Knowledge Organization, pp. 93–103. Springer, Heidelberg (2010). https://doi.org/10.1007/978-3-642-01044-6_8
5. Burdick, D., Calimlim, M., Gehrke, J.: MAFIA: a maximal frequent itemset algorithm for transactional databases. In: Proceedings of the 17th International Conference on Data Engineering, pp. 443–452 (2001)
6. Cheng, C.H., Fu, A.W., Zhang, Y.: Entropy-based subspace clustering for mining numerical data. In: Proceedings of the 5th ACM International Conference on Knowledge Discovery and Data Mining, pp. 84–93 (1999)
7. Elhamifar, E., Vidal, R.: Sparse subspace clustering. In: Proceedings of the IEEE International Conference on Computer Vision and Pattern Recognition, CVPR, pp. 2790–2797 (2009)
8. Elhamifar, E., Vidal, R.: Sparse subspace clustering: algorithm, theory, and applications. IEEE Trans. Pattern Anal. Mach. Intell. **35**(11), 2765–2781 (2013)
9. Ester, M., Kriegel, H.P., Sander, J., Xu, X.: A density-based algorithm for discovering clusters in large spatial databases with noise. In: Proceedings of the 2nd ACM International Conference on Knowledge Discovery and Data Mining, KDD, pp. 226–231 (1996)
10. Gan, G., Wu, J.: A convergence theorem for the fuzzy subspace clustering algorithm. Pattern Recogn. **41**(6), 1939–1947 (2008)
11. Guillon, A., Lesot, M.J., Marsala, C.: Laplacian regularization for fuzzy subspace clustering. In: Proceedings of the IEEE International Conference on Fuzzy Systems, FUZZ-IEEE 2017 (2017)
12. Guillon, A., Lesot, M.J., Marsala, C.: A proximal framework for fuzzy subspace clustering. Fuzzy Sets Syst. **366**, 24–45 (2019)

13. Gustafson, D., Kessel, W.: Fuzzy clustering with a fuzzy covariance matrix. In: Proceedings of the IEEE Conference on Decision and Control, vol. 17, pp. 761–766. IEEE (1978)
14. Höppner, F., Klawonn, F., Kruse, R., Runkler, T.: Fuzzy Cluster Analysis: Methods for Classification, Data Analysis and Image Recognition. Wiley, New York (1999)
15. Ji, P., Zhang, T., Li, H., Salzmann, M., Reid, I.: Deep subspace clustering networks. In: Proceedings of the 31st International Conference on Neural Information Processing Systems, NIPS (2017)
16. Jing, L., Ng, M.K., Huang, J.Z.: An entropy weighting k-means algorithm for subspace clustering of high-dimensional sparse data. IEEE Trans. Knowl. Data Eng. **19**(8), 1026–1041 (2007)
17. Kacprzyk, J., Zadrozny, S.: Linguistic database summaries and their protoforms: towards natural language based knowledge discovery tools. Inf. Sci. **173**(4), 281–304 (2005)
18. Kanatani, K.: Motion segmentation by subspace separation and model selection. In: Proceedings of the 8th International Conference on Computer Vision, ICCV, vol. 2, pp. 586–591 (2001)
19. Keller, A., Klawonn, F.: Fuzzy clustering with weighting of data variables. Int. J. Uncertain. Fuzziness Knowl. Based Syst. **8**(6), 735–746 (2000)
20. Klawonn, F., Höppner, F.: What is fuzzy about fuzzy clustering? Understanding and improving the concept of the fuzzifier. In: Proceedings of the 5th International Symposium on Intelligent Data Analysis, pp. 254–264 (2003)
21. Kruse, R., Döring, C., Lesot, M.J.: Fundamentals of fuzzy clustering. In: de Oliveira, J., Pedrycz, W. (eds.) Advances in Fuzzy Clustering and its Applications. Wiley, New York (2007)
22. Liu, G., Lin, Z., Yan, S., Sun, J., Yu, Y., Ma, Y.: Robust recovery of subspace structures by low-rank representation. IEEE Trans. Pattern Anal. Mach. Intell. **35**(1), 171–184 (2013)
23. von Luxburg, U.: A tutorial on spectral clustering. Stat. Comput. **17**(4), 395–416 (2007)
24. Min, E., Guo, X., Liu, Q., Zhang, G., Cui, J., Lun, J.: A survey of clustering with deep learning: from the perspective of network architecture. IEEE Access **6**, 39501–39514 (2018)
25. Parikh, N., Boyd, S.: Proximal algorithms. Found. Trends Optim. **1**(3), 123–231 (2014)
26. Patel, V.M., Vidal, R.: Kernel sparse subspace clustering. In: Proceedings of ICIP, pp. 2849–2853 (2014)
27. Peng, X., Xiao, S., Feng, J., Yau, W.Y., Yi, Z.: Deep subspace clustering with sparsity prior. In: Proceedings of the 25th International Joint Conference on Artificial Intelligence, IJCAI, pp. 1925–1931 (2016)
28. Vidal, R.: A tutorial on subspace clustering. IEEE Sig. Process. Mag. **28**(2), 52–68 (2010)
29. Wang, D., Ding, C., Li, T.: K-subspace clustering. In: Buntine, W., Grobelnik, M., Mladenić, D., Shawe-Taylor, J. (eds.) ECML PKDD 2009. LNCS (LNAI), vol. 5782, pp. 506–521. Springer, Heidelberg (2009). https://doi.org/10.1007/978-3-642-04174-7_33
30. Xiao, S., Tan, M., Xu, D., Dong, Z.Y.: Robust kernel low-rank representation. IEEE Trans. Neural Netw. Learn. Syst. **27**(11), 2268–2281 (2016)

31. Yin, M., Guo, Y., Gao, J., He, Z., Xie, S.: Kernel sparse subspace clustering on symmetric positive definite manifolds. In: Proceedings of the IEEE International Conference on Computer Vision and Pattern Recognition, CVPR, pp. 5157–5164 (2016)
32. Zhou, L., Bai, X., Wang, D., Liu, X., Zhou, J., Hancock, E.: Latent distribution preserving deep subspace clustering. In: Proceedings of the 28th International Joint Conference on Artificial Intelligence, IJCAI, pp. 4440–4446 (2019)

Invited Keynotes

From Shallow to Deep Interactions Between Knowledge Representation, Reasoning and Machine Learning

Kay R. Amel[✉]

GDR "Aspects Formels et Algorithmiques de l'Intelligence Artificielle",
CNRS, Gif-sur-Yvette, France

Abstract. Reasoning and learning are two basic concerns at the core of Artificial Intelligence (AI). In the last three decades, Knowledge Representation and Reasoning (KRR) on the one hand and Machine Learning (ML) on the other hand, have been considerably developed and have specialised in a large number of dedicated sub-fields. These technical developments and specialisations, while they were strengthening the respective corpora of methods in KRR and in ML, also contributed to an almost complete separation of the lines of research in these two areas, making many researchers on one side largely ignorant of what was going on the other side.

This state of affairs is also somewhat relying on general, overly simplistic, dichotomies that suggest great differences between KRR and ML: KRR deals with knowledge, ML handles data; KRR privileges symbolic, discrete approaches, while numerical methods dominate ML. Even if such a rough picture points out things that cannot be fully denied, it is also misleading, as for instance KRR can deal with data as well (e.g., formal concept analysis) and ML approaches may rely on symbolic knowledge (e.g., inductive logic programming). Indeed, the frontier between the two fields is actually much blurrier than it appears, as both share approaches such as Bayesian networks, or case-based reasoning and analogical reasoning, as well as important concerns such as uncertainty representation. In fact, one may well argue that similarities between the two fields are more numerous than one may think.

This talk proposes a tentative and original survey of meeting points between KRR and ML. Some common concerns are first identified and discussed such as

---

Kay R. Amel is the pen name of the working group "Apprentissage et Raisonnement" of the GDR ("Groupement De Recherche") "Aspects Formels et Algorithmiques de l'Intelligence Artificielle", CNRS, France (https://www.gdria.fr/presentation/). The contributors to this paper include: Zied Bouraoui (CRIL, Lens, Fr, zied.bouraoui@cril.fr), Antoine Cornuéjols (AgroParisTech, Paris, Fr, antoine.cornuejols@agroparistech.fr), Thierry Denoeux (Heudiasyc, Compiègne, Fr, thierry.denoeux@utc.fr), Sébastien Destercke (Heudiasyc, Compiègne, Fr, sebastien.destercke@hds.utc.fr), Didier Dubois (IRIT, Toulouse, Fr, dubois@irit.fr), Romain Guillaume (IRIT, Toulouse, Fr, Romain.Guillaume@irit.fr), Jérôme Mengin (IRIT, Toulouse, Fr, Jerome.Mengin@irit.fr), Henri Prade (IRIT, Toulouse, Fr, prade@irit.fr), Steven Schockaert (School of Computer Science and Informatics, Cardiff, UK, SchockaertS1@cardiff.ac.uk), Mathieu Serrurier (IRIT, Toulouse, Fr, mathieu.serrurier@gmail.com), Christel Vrain (LIFO, Orléans, Fr, Christel.Vrain@univ-orleans.fr).

N. Ben Amor et al. (Eds.): SUM 2019, LNAI 11940, pp. 447–448, 2019.
https://doi.org/10.1007/978-3-030-35514-2

the types of representation used, the roles of knowledge and data, the lack or the excess of information, the need for explanations and causal understanding.

Then some methodologies combining reasoning and learning are reviewed (such as inductive logic programming, neuro-symbolic reasoning, formal concept analysis, rule-based representations and machine learning, uncertainty assessment in prediction, or case-based reasoning and analogical reasoning), before discussing examples of synergies between KRR and ML (including topics such as belief functions on regression, EM algorithm versus revision, the semantic description of vector representations, the combination of deep learning with high level inference, knowledge graph completion, declarative frameworks for data mining, or preferences and recommendation).

The full paper will be the first step of a work in progress aiming at a better mutual understanding of researches in KRR and ML, and how they could cooperate.

Algebraic Approximations for Weighted Model Counting

Wolfgang Gatterbauer[✉]

Khoury College of Computer Sciences, Northeastern University, Boston, USA
`wgatterbauer@northeastern.edu`

Abstract. It is a common approach in computer science to approximate a function that is hard to evaluate by a simpler function. Finding such fast approximations is especially important for probabilistic inference, which is widely used, yet notoriously hard. We discuss a recent algebraic approach for approximating the *probability of Boolean functions* with upper and lower bounds. We give the intuition for these bounds and illustrate their use with three applications: (1) *anytime approximations* of monotone Boolean formulas, (2) *approximate lifted inference* with relational databases, and (3) *approximate weighted model counting*.

1 Probabilistic Inference and Weighted Model Counting

Probabilistic inference over large data sets has become a central data management problem. It is at the core of a wide range of approaches, such as graphical models, statistical relational learning or probabilistic databases. Yet a major drawback of exact probabilistic inference is that it is computationally intractable for most real-world problems. Thus developing general and scalable approximate schemes is a subject of fundamental interest. We focus on *weighted model counting*, which is a generic inference problem to which all above approaches can be reduced. It is essentially the same problem as computing the *probability of a Boolean formula*. Each truth assignment of the Boolean variables corresponds to one model whose weight is the probability of this truth assignment. Weighted model counting then asks for the sum of the weights of all satisfying assignments.

2 Optimal Oblivious Dissociation Bounds

We discuss recently developed deterministic upper and lower bounds for the probability of Boolean functions. The bounds result from treating multiple occurrences of variables as independent and assigning them new individual probabilities, an approach called *dissociation*. By performing several dissociations, one can transform a Boolean formula whose probability is difficult to compute, into one whose probability is easy to compute. Appropriately executed, these steps can give rise to a novel class of inequalities from which upper and lower bounds can be derived efficiently. In addition, the resulting bounds are *oblivious*, i.e. they

© Springer Nature Switzerland AG 2019
N. Ben Amor et al. (Eds.): SUM 2019, LNAI 11940, pp. 449–450, 2019.
https://doi.org/10.1007/978-3-030-35514-2

require only limited observations of the structure and parameters of the problem. This technique can yield fast approximate schemes that generate upper and lower bounds for various inference tasks.

3   Talk Outline

We discuss Boolean formulas and their connection to weighted model counting. We introduce dissociation-based bounds and draw the connection to *approximate knowledge compilation*. We then illustrate the use of dissociation-based bounds with three applications: (1) *anytime approximations* of monotone Boolean formulas [7]. (2) *approximate lifted inference* with relational databases [4, 5], and (3) *approximate weighted model counting* [3]. If time remains, we will discuss the similarities and differences to four other techniques that similarly fall into Pearl's classification of *extensional* approaches to uncertainty [9, Ch 1.1.4]: (*i*) relaxation-based methods in logical optimization [8, Ch 13], (*ii*) relaxation & compensation for approximate probabilistic inference in graphical models [2], (*iii*) probabilistic soft logic that uses continuous relaxations in a smart way [1], and (*iv*) quantization on algebraic decision diagrams [6]. The slides will be made available at https://northeastern-datalab.github.io/afresearch/.

Acknowledgements. This work is supported in part by National Science Foundation grant IIS-1762268. I would also like to thank my various collaborators on the topic: Li Chou, Floris Geerts, Vibhav Gogate, Peter Ivanov, Dan Suciu, Martin Theobald, and Maarten Van den Heuvel.

References

1. Bach, S.H., Broecheler, M., Huang, B., Getoor, L.: Hinge-loss markov random fields and probabilistic soft logic. J. Mach. Learn. Res. **18**, 109:1–109:67 (2017)
2. Choi, A., Darwiche, A.: Relax then compensate: on max-product belief propagation and more. In: NIPS, pp. 351–359 (2009)
3. Chou, L., Gatterbauer, W., Gogate, V.: Dissociation-based oblivious bounds for weighted model counting. In: UAI (2018)
4. Gatterbauer, W., Suciu, D.: Approximate lifted inference with probabilistic databases. PVLDB **8**(5), 629–640 (2015)
5. Gatterbauer, W., Suciu, D.: Dissociation and propagation for approximate lifted inference with standard relational database management systems. VLDB J. **26**(1), 5–30 (2017)
6. Gogate, V., Domingos, P.: Approximation by quantization. In: UAI, pp. 247–255 (2011)
7. den Heuvel, M.V., Ivanov, P., Gatterbauer, W., Geerts, F., Theobald, M.: Anytime approximation in probabilistic databases via scaled dissociations. In: SIGMOD, pp. 1295–1312 (2019)
8. Hooker, J.: Logic-Based Methods for Optimization: Combining Optimization and Constraint Satisfaction. John Wiley & sons (2000)
9. Pearl, J.: Probabilistic Reasoning in Intelligent Systems: Networks of Plausible Inference. Morgan Kaufmann (1988)

Author Index

Abo-Khamis, Mahmoud 423
Amel, Kay R. 447
Amor, Nahla Ben 355
Antoine, Violaine 66
Ayachi, Raouia 355

Bartashevich, Palina 310
Benabbou, Nawal 221
Benferhat, Salem 207
Bonifati, Angela 250
Bounhas, Myriam 136, 339
Bouraoui, Zied 207
Bourdache, Nadjet 93
Bouslama, Rihab 355

Chaveroche, Maxime 390
Cherfaoui, Véronique 390
Correia, Alvaro H. C. 409
Couso, Ines 266
Crosscombe, Michael 310

Davoine, Franck 390
de Campos, Cassio P. 409
Denœux, Thierry 368
Destercke, Sébastien 266, 280, 289
Diaz, Amaia Nazabal Ruiz 250
Dubois, Didier 153, 169
Dumbrava, Stefania 250
Dusserre, Gilles 107

Gatterbauer, Wolfgang 449
Gonzales, Christophe 404
Guillot, Pierre-Louis 289

Hachour, Samir 382
Harispe, Sébastien 107
Hüllermeier, Eyke 266

Imoussaten, Abdelhak 107, 122

Jacquin, Lucie 122
Jamshidi, Pooyan 324
Javidian, Mohammad Ali 324

Kawasaki, Tatsuki 79
Kuhlmann, Isabelle 24

L'Héritier, Cécile 107
Labreuche, Christophe 192
Lagrue, Sylvain 280
Lawry, Jonathan 310
Lesot, Marie-Jeanne 433
Lust, Thibaut 221

Martin, Hugo 52
Mercier, David 382
Montmain, Jacky 122
Moriguchi, Sosuke 79
Mutmainah, Siti 382

Ngo, Hung Q. 423
Nguyen, XuanLong 423

Olteanu, Dan 423

Papini, Odile 207
Perny, Patrice 52, 93
Perrin, Didier 122
Pichon, Frédéric 382
Pirlot, Marc 339
Potyka, Nico 236
Prade, Henri 136, 153, 169, 339

Renooij, Silja 38
Roig, Benoît 107

Salhi, Yakoub   184
Schleich, Maximilian   423
Sobrie, Olivier   339
Spanjaard, Olivier   93

Takahashi, Kazuko   79
Thimm, Matthias   1, 9, 24
Tong, Zheng   368
Trousset, François   122

Valtorta, Marco   324
van der Gaag, Linda C.   38
Vuillemot, Romain   250

Wilson, Nic   169
Würbel, Eric   207

Xie, Jiarui   66
Xu, Philippe   368

Printed in the United States
By Bookmasters